EXTINCT
AND VANISHING BIRDS
OF THE WORLD

By
JAMES C. GREENWAY, JR.
American Museum of Natural History

SECOND REVISED EDITION

Dover Publications, Inc., New York

Published in Canada by General Publishing Company, Ltd., 30 Lesmill Road, Don Mills, Toronto, Ontario.
Published in the United Kingdom by Constable and Company, Ltd., 10 Orange Street, London WC 2.

This Dover edition, first published in 1967, is a revised republication of the work originally published in 1958 by the American Committee for International Wild Life Protection as Special Publication No. 13. It is reprinted by special arrangement with the American Committee for International Wild Life Protection.

Cover Illustrations, by D. M. Reid-Henry.
Upper left: *Cyanoramphus zealandicus* (Latham).
Upper right: *Cyanoramphus ulietanus* (Gmelin).
Lower left: *Raphus cucullatus* (Gmelin). Lower right: *Aplonis mavornata* Buller. The two Cyanoramphus parrots once lived on Riaitea and Tahiti. The dodo lived on Mauritius. The starling probably occurred on one of the South Sea islands and was collected by a naturalist of one of Captain Cook's voyages.

International Standard Book Number: 0-486-21869-4
Library of Congress Catalog Card Number: 67-29307

Manufactured in the United States of America
Dover Publications, Inc.
180 Varick Street
New York, N. Y. 10014

As if God had made the innocency of so poor a creature to become such an admirable instrument for the sustention of man.

—CAPITAINE RICHARD WHITBOURNE

Preface to the Dover Edition

DURING THE thirteen years that have gone since the manuscript of the first printing was finished, six forms of birds that had been thought to be extinct in all probability have been found in reality to have been alive all the time. They are the Eskimo curlew of North America, the scrub-bird of far western Australia, and the Eyrean grass-wren of Central Australia. All three are full species. Add to these the plain or blue pigeon of Puerto Rico, the Puerto Rican nighthawk and Darwin's ground finch of Chatham Island in the Galápagos, all of which are subspecies with more or less closely related forms flourishing elsewhere. Other cheerful events and discoveries can be recorded, such as the finding of breeding birds, and in numbers, of the black-capped petrel in Haiti and Newell's shearwater on Kauai, Hawaii.

On the other side of the ledger we list sadly one species and four subspecies as almost certainly extinct. These are the seedsnipe of Barrier Island, off New Zealand, the Canada goose of the Commander Islands in the Northwestern Pacific, the Scops owl of Anjouan Island in the Comoros, between Madagascar and Africa, and a thrush from Maré Island in the Loyaltiesoff New Caledonia. To these four subspecies the huia or wattlebird of New Zealand, a full species, must be added.

All this seems to leave us the better by two species, which is good enough for these days or any others perhaps.

In this revised edition, the status of the species and subspecies has been brought up-to-date wherever new information has become available. A separate bibliography of works cited in this revised edition is found on page 484. Wherever these works are mentioned in the text, they are marked with an asterisk (*).

For a modern and comprehensive list of rare and endangered birds see the Red Data Book, published by the International Union for Conservation of Nature and Natural Resources, vol. 2, 1966, Morges, Switzerland.

JAMES C. GREENWAY, JR.

New York, N. Y.
April, 1967

Foreword

WE ARE now witnessing the most tremendous changes in the world, and one of the saddest consequences is the awful threat to the existence of many forms of wildlife. Human populations increase; weapons are improved; new poisons are found and used; and remote areas, so far inaccessible, are penetrated more and more easily. As a result, plants and animals are fast decreasing and some may eventually disappear altogether.

Birds, conspicuous and easily killed as many of them are, become particularly affected. Numerous species have been dwindling in numbers in the course of the last twenty years and have reached a dangerously low level. Furthermore, a large proportion of them are narrowly adapted to certain types of habitats, the destruction of which they cannot survive.

In order to try to avoid the total disappearance of the threatened species, a thorough study of the extinct and nearly extinct ones is a first necessity. No modern and comprehensive work of this kind was so far available, and Mr. Greenway's book fills a regrettable gap.

One of the difficulties of such a work was to make a sensible choice among the rarer species of those which are considered as definitely and immediately threatened. So many are in a serious position that the list could actually include hundreds. I believe that Mr. Greenway has succeeded in making an excellent selection. My association with the author has been a long and close one. It goes back to the very beginning of his ornithological career during our expeditions to Madagascar and to Indo-China in 1929 and 1930. I am, therefore, particularly happy to congratulate him on his excellent work. It supplies us with the best possible information on the extinct and threatened birds, including the reasons of the disappearance as much as they could be ascertained.

As far as extinct birds are concerned, we can only express deep regret that measures had not been taken in time to save their existence, thus avoiding an appalling impoverishment of life on earth. But when it comes to threatened species, we face a different problem. Can they be saved, and how? The answer varies according to species and circumstances. In many cases it will be possible to save them through appropriate protective measures the success of which entirely depends upon the wisdom and the good will of men. It is only a matter of stopping destruction. Such is the case with *Diomedea albatrus, Phalacrocorax harrisi, Megapodius laperouse, Grus japonensis, Tricholimnas silvestris, Rhynochetos jubatus,*

Columba jouyi, Psittacula krameri echo, Campephilus principalis bairdi, Dryocopus javensis richardi, Saphaeopipo noguchii, Acrocephalus luscinia rehsi, Loxigilla portoricensis grandis. These are only some of the species that have greatly decreased in number mainly owing to human interference. They could become numerous again if proper protection was afforded them.

But to be able to take efficient steps, investigations should be carefully and thoroughly conducted. We still do not know the present nesting localities of a number of very rare birds: *Struthio camelus syriacus, Pterodroma hasitata, Puffinus newelli, Ophrysia superciliosa, Rhinoptilus bitorquatus, Psephotus pulcherrimus,* and *Geopsittacus occidentalis,* in particular; and we still know nothing of the condition and status of the rarer birds of the Island of Kaui, in the Hawaiian group, such as *Phaeornis palmeri, P. obscurus myadestinus, Hemignathus obscurus procerus, H. lucidus hanapepe, Moho braccatus,* and *M. bishopi.* Scientific data are the only sound bases for efficient conservation measures.

If some birds can be maintained when prohibition of their killing is properly enforced, others will be safe if adequate refuges are set apart, and if their indispensable habitat is preserved. Still others can be propagated in captivity or at controlled liberty according to circumstances; such is the case of many species of game birds, waterfowl, cranes, pigeons and doves, and parrots.* Each case has to be carefully studied before adequate measures can be proposed. The basis for such studies is an accurate picture of the present situation.

It is sad to realize that so many beautiful creatures have vanished forever or may very soon die out. It should be a lesson for mankind and remind us of our duty toward Nature. We play on earth a leading role, which should not be that of a villain.

JEAN DELACOUR,
President, International Committee for Bird Preservation
November 1954

---

* The Carolina parakeet and the passenger pigeon, to cite only two, could easily have been maintained in captivity.

Acknowledgments

The background for this book is intimately connected with the American Committee for International Wild Life Protection, for the bases of it are the card catalogues kept by the late John C. Phillips, a founder of the Committee and its president for seven years, and the late Thomas Barbour, for long an interested member of the advisory board. Dr. Barbour encouraged me to keep the list up to date, and no doubt it was knowledge of this fact that prompted the Committee to ask me to prepare an account of extinct and vanishing birds. Without the interest and generosity of Childs Frick, for many years chairman of the Committee's executive board, the venture would have been all but impossible. Richard H. Pough of the board of directors served admirably as chairman of a subcommittee to see to it that the book appeared.

Illustrations were made possible by a grant from the Walcott Fund. D. M. Reid-Henry made all the illustrations except one, that of the St. Kitts bullfinch, which was drawn by Henry Seidel.

The manuscript was completed in October 1954.

I am grateful indeed to Dr. Ernst Mayr and Dr. Erwin Stresemann, who read much of the manuscript and helped me generously; to Dr. Robert Cushman Murphy and Dr. Dean Amadon, who read parts and made most helpful suggestions; and to Mr. Paul H. Oehser who edited the book and saw it through the press.

A list of persons who have helped with special knowledge is much too long to reproduce here. Curators of museums all over the world have taken trouble to advise me in many matters, but principally in listing specimens in their care. I thank them from the heart.

Contents

ACCOUNTS OF EXTINCT AND VANISHING FORMS—*Continued*

EXTINCT AND VANISHING BIRDS OF THE WORLD

Explanations and Definitions

Species. It is with the category called a "species" that we must be most concerned in this account of extinct birds. What were these entities that once existed but exist no longer? Obviously they were groups of individuals which, although differing slightly from one another, resembled one another in certain characteristics that set them apart from all other groups. Furthermore, these individuals were able to reproduce their kind so that the unit was continuous for perhaps a million years or much more. Lastly, because of innate, sometimes intangible, but nevertheless inherited characteristics, these individuals did not breed with members of other groups, or, if they did, no population resulting from such unions ever established itself. In short and in essence, it appears that they were genetically complex units, which did not breed with others, and did not resemble them, that exist no longer. Characteristics special and particular to the group, and which may be seen, are the images of these unseen attributes. In other words, among other characters a long, mobile thumb, as well as the intelligence to use it, distinguishes men from other anthropoids, but should men extirpate themselves by some means, not only would these and of course the other characters disappear, but also the continuing mechanism by which the characters are transferred from generation to generation. Evolution is a continuous process. When a species is extirpated it is as though a single tree were cut from a continuously flourishing forest.

For the purpose of this book the foregoing paragraph is a statement of the essential nature of a species. More concisely, a species can be defined as a group of animals or plants with one or more characters distinguishing them from other similar groups. Members of such groups interbreed and reproduce their characters in the offspring but do not breed successfully with other groups.[1]

Subspecies are also interbreeding groups of similar creatures possessing attributes differentiating them from other similar groups. Basically it is in the matter of breeding with other similar groups that the difference between species and subspecies lies. Individuals of a subspecies do breed

---

[1] This definition is in close accord with that of Mayr, 1942.

with neighboring subspecies, and the resulting progeny usually occupy more or less territory between such forms on large land masses. The nature of their characters is intermediate. Where, however, subspecies are separated by seas, deserts, or other geographical barriers, no such populations of intermediate form exist, and it is in these cases that much difficulty arises in distinguishing species from subspecies. It must be said in truth that in many cases the decision as to whether a well-differentiated isolated population is a species or a subspecies is an arbitrary one when no numbers of related individuals with clearly intermediate characters can be found. In fact such decisions have often been called guesses, and perhaps they gave rise to a nonsense definition formerly much in vogue: "A species is what a good zoologist says it is."

Study of the honeyeaters of the Hawaiian Islands (*Moho*) brings the nature of these decisions strongly into focus. Differing from one another sharply in size, shape of tail, and distribution of yellow patches in the black plumage, they would probably not breed with one another if brought together and are therefore probably specifically distinct, yet they are at first glance very similar.

A species is sometimes, but not always, composed of closely related and quite similar groups called subspecies. Such are called polytypic species or "form circles." These are the groups listed with similar second names; thus one of two rare and extinct West Indian green parrots is named *Amazona vittata vittata,* and a similar subspecies of another island *Amazona vittata graciliceps.* They differ only in their relative size. Both are subspecies and members of a single species called *vittata.* Species embracing as many as 70 subspecies are known. In the case of the California condor, so few in numbers and with such a small range, no subspecies exist, and it is called *Gymnogyps californianus.*

Higher Classifications. According to convention all animals, including birds, are divided in categories which might be represented as squares or circles each within another. The American heath-hen (*Tympanuchus*) is classified thus:

CLASS—Aves or birds
ORDER—Galliformes, including pheasants, chickens, quail, grouse, ptarmigan.
FAMILY—Tetraonidae, including grouse and ptarmigan.
GENUS—Tympanuchus, including only the North American prairie chickens and the extinct heath hen.
SPECIES—cupido, including three forms or subspecies.
SUBSPECIES—cupido, including only the heath hen of eastern North America.

Genus. The concept "genus" as the next higher category of classification than a species probably should not concern us. It is subject to extraordinary variations of human opinion and so may well be defined as in the "nonsense definition" of species above. Moreover, many of the extinct forms listed in this book, although they may be named as ducks or hawks (that is, classified in a family), stand alone as a single species in one genus. Therefore in these cases the genus partakes of the essential nature of a species rather than the larger category that it should be. In too many cases this is due to lack of knowledge of relationships. So many of these birds, long isolated on islands, stand apart as might men from Mars.

Outstanding examples of this dilemma when dealing with genera of extinct forms are the Labrador duck (*Camptorhynchus*); the rails *Cabalus, Porzanula, Pennula, Aphanolimnas,* and *Pareudiastes;* the shorebird *Prosobonia;* and the Hawaiian honeycreeper *Ciridops.* Furthermore, many learned ornithologists will disagree violently with the disuse of the single-species genera *Pseudotadorna, Nesolimnas, Notornis,* and many genera of Hawaiian honey-creepers.

Family. Certainly the concept of "family" is also subject to much individual variation of judgment, especially in treatment of small perching birds (Passeres). Only one "family," the dodos (Raphidae), has been extirpated in modern times, and since it contained but a single genus of three species it may, in this philosophical sense, be considered to be comparable rather to a superspecies than to a "family."

Population. Subspecies contain many populations although these are never recognized formally. So many have disappeared that it would be quite impossible to mention them all.

Synonyms. "Discrete population" has been used for many classifications in this book, as has "form." These are useful neutral covering terms, having no exact meaning.

Introduction and Summation

To be complete this book should contain accounts of all the populations of birds that have disappeared from the earth since time began. Actually it gives some account of a few discrete populations that have been lost to us since the beginning of recorded history, but the vast majority have disappeared during the 280 years since the last dodo is thought to have died.

With the exception of that poor, flightless creature, the form and color of all the 42 species and 44 subspecies in the list that follows are known by their representative types in museums, and their histories are well documented. Experts have sought them for years without success. We can be sure that they are extinct.

EXTINCT

EMUS (Dromaeidae):
    Dromaeius novaehollandiae diemenensis — Tasmania
    Dromaeius novaehollandiae diemenianus — Kangaroo Island, Australia
PETRELS (Hydrobatidae):
    Oceanodroma macrodactyla — Guadalupe Island, eastern Pacific
CORMORANTS (Phalacrocoracidae):
    Phalacrocorax perspicillatus — Bering Island, North Pacific
HERONS (Ardeidae):
    Nycticorax caledonicus crassirostris — Bonin Islands, western Pacific
DUCKS (Anatidae):
    Anas strepera couesi — Washington Island, central Pacific
    Branta canadensis asiatica — Kurile Islands, Northeastern Pacific
    Camptorhynchus labradorium — Eastern North America
    Mergus australis — Auckland Island, New Zealand
    Tadorna cristata — Korea
HAWKS (Falconidae):
    Polyborus lutosus — Guadalupe Island, eastern Pacific
QUAILS and PARTRIDGES (Phasianidae):
    Coturnix novae-zelandiae novae-zelandiae — New Zealand
    Tympanuchus cupido cupido — North America
RAILS (Rallidae):
    Rallus pacificus — Tahiti, South Pacific
    Amaurolimnas concolor concolor — Jamaica, West Indies
    Rallus dieffenbachii — Chatham Islands, New Zealand
    Rallus wakensis — Wake Island, Pacific
    Rallus muelleri — Auckland Islands, New Zealand
    Cabalus modestus — Chatham Islands, New Zealand

Porzanula palmeri — Laysan Island, northwest Hawaiian Islands

Pennula sandwichensis — Island of Hawaii
Aphanolimnas monasa — Kusaie Island, Caroline Islands, Pacific
Poliolimnas cinereus brevipes — Iwo Jima, Vulcan Islands, western Pacific

Gallinula nesiotis nesiotis — Tristan da Cunha Island, south Atlantic
Porphyrio porphyrio albus — Lord Howe Island, western Pacific

SHOREBIRDS (Charadriidae):
Coenocorypha aucklandica barrierensis — New Zealand
Prosobonia leucoptera — Tahiti, South Pacific

AUKS (Alcidae):
Alca (Pinguinus) impennis — North Atlantic

PIGEONS (Columbidae):
Ectopistes migratorius — North America
Alectroenas nitidissima — Mauritius, Indian Ocean
Hemiphaga novaeseelandiae spadicea — Norfolk Island, western Pacific
Columba versicolor — Bonin Islands, northwestern Pacific

DODO (Raphidae):
Raphus cucullatus — Mauritius Island, Indian Ocean

PARROTS (Psittacidae):
Nestor meridionalis productus — Norfolk Island, western Pacific
Amazona vittata graciliceps — Culebra Island, West Indies
Ara tricolor — Cuba, West Indies
Aratinga chloroptera maugei — Puerto Rico and Mona Island, West Indies

Conuropsis carolinensis carolinensis — East coast of North America
Conuropsis carolinensis ludoviciana — Central North America
Mascarinus mascarinus — Réunion Island, Indian Ocean
Psittacula eupatria wardi — Seychelles Island, Indian Ocean
Cyanoramphus novaezelandiae subflavescens — Lord Howe Island, western Pacific
Cyanoramphus novaezelandiae erythrotis — Macquarie Island, south of New Zealand
Cyanoramphus zealandicus — Tahiti, South Pacific
Cyanoramphus ulietanus — Raiatea Island, Society Islands, South Pacific

OWLS (Strigidae):
Speotyto cunicularia guadeloupensis — Marie Galante Island, West Indies
Speotyto cunicularia amaura — Antigua and Nevis Islands, West Indies
Sceloglaux albifacies rufifaces — North Island, New Zealand

NIGHTHAWKS (Caprimulgidae):
Siphonorhis americanus — Jamaica, West Indies

KINGFISHERS (Alcedinidae):
Halcyon miyakoensis — Riu Kiu Islands, western Pacific

WOODPECKERS (Picidae):
Colaptes cafer rufipileus — Guadalupe Island, eastern Pacific

NEW ZEALAND WRENS (Xenicidae):
Xenicus [Traversia] lyalli — Stephens Island, New Zealand

WRENS (Troglodytidae):
Troglodytes musculus guadeloupensis — Guadeloupe Island, West Indies

Troglodytes musculus martinicensis	Martinique Island, West Indies
Thryomanes bewickii brevicauda	Guadalupe Island, eastern Pacific

THRUSHES (Turdinae):
Turdus poliocephalus vinitinctus	Lord Howe Island, western Pacific
Turdus ulietensis	Raiatea, Society Islands, Pacific
Zoothera terrestris	Bonin Islands, northwestern Pacific
Phaeornis obscurus oahensis	Oahu, Hawaiian Islands

OLD WORLD WARBLERS (Sylviinae):
Acrocephalus familiaris familiaris	Laysan, northwestern Hawaiian Islands
Bowdleria rufescens	Pitt or Mangare, Chatham Islands

FLYCATCHERS (Muscicapinae):
Gerygone igata insularis	Lord Howe Island, western Pacific
Rhipidura flabellifera cervina	Lord Howe Island, western Pacific

HONEYEATERS (Meliphagidae):
Anthornis melanura melanocephala	Chatham Islands, off New Zealand
Chaetoptila angustipluma	Island of Hawaii
Moho apicalis	Oahu, Hawaiian Islands

WHITE-EYES (Zosteropidae):
Zosterops strenua	Lord Howe Island, western Pacific
Zosterops semiflava	Seychelles Islands, Indian Ocean

HAWAIIAN HONEYCREEPERS (Drepanididae):
Himatione sanguinea freethii	Laysan Island, northwestern Hawaiian
Drepanis pacifica	Island of Hawaii
Drepanis funerea	Molokai Island, Hawaiian Islands
Hemignathus obscurus wolstenholmei	Oahu, Hawaiian Islands
Hemignathus obscurus lanaiensis	Lanai Island, Hawaiian Islands
Hemignathus lucidus lucidus	Oahu, Hawaiian Islands
Loxops coccinea rufa	Oahu, Hawaiian Islands
Ciridops anna	Island of Hawaii
Psittirostra [Rhodacanthis] palmeri	Island of Hawaii
Psittirostra [Rhodacanthis] flaviceps	Island of Hawaii

SPARROWS AND FINCHES (Fringillidae):
Pipilo erythrophthalmus consobrinus	Guadalupe Island, eastern Pacific
Chaunoproctus ferreirostris	Bonin Islands, northwestern Pacific
Loxigilla portoricensis grandis	St. Christopher Island, West Indies

STARLINGS (Sturnidae):
Fregilupus varius	Réunion Island, Mascarene Islands, Indian Ocean
Aplonis fuscus hullianus	Lord Howe Island, western Pacific
Aplonis corvina	Kusaie Island, Caroline Islands, southwestern Pacific
Aplonis mavornata	Society Islands, South Pacific

PROBABLY EXTINCT

The second list contains forms that are not certainly extinct, although probably they are. Searching for rare birds in rugged mountains and dripping forests does not appeal to many men. Perhaps a few individuals of the following forms may still be found.

Rails (Rallidae):
    Nesoclopeus poecilopterus   Viti Levu, Fiji Islands
    Pareudiastes pacificus   Savaii, Samoa
Pigeons (Columbidae):
    Microgoura meeki   Choiseul, Solomon Islands
Parrots (Psittacidae):
    Psittacula krameri exsul   Rodriguez Island, Indian Ocean
Owls (Strigidae):
    Otus rutilus capnodes   Anjouan Island, Comoros
Cuckoos (Cuculidae):
    Coua delalandei   Madagascar
Thrushes (Turdidae):
    Turdus poliocephalus mareensis   Maré, Loyalty Islands
    Phaeornis obscurus rutha   Molokai Island, Hawaiian Islands
    Phaeornis obscurus lanaiensis   Lanai Island, Hawaiian Islands
Hawaiian Honeycreepers
    (Drepanididae):
    Loxops (Viridonia) sagittirostris   Island of Hawaii
    Loxops (Paroreomyza) maculata   Molokai Island, Hawaiian Islands
      flammea
    Loxops (Paroreomyza) maculata   Lanai Island, Hawaiian Islands
      montana
    Hemignathus obscurus obscurus   Island of Hawaii
    Hemignathus lucidus affinis   Maui Island, Hawaiian Islands
    Psittirostra [Chloridops] kona   Island of Hawaii
Honeyeaters (Meliphagidae):
    Moho bishopi   Molokai Island, Hawaiian Islands
    Moho nobilis   Island of Hawaii
Finches (Fringillidae):
    Neospiza concolor   São Tomé Island, west coast of Africa
Wattlebirds (Calleidae):
    Heterolocha acutirostris   North Island, New Zealand

TOTALS

	Extinct		Probably Extinct	
	Species	Subsp.	Species	Subsp.
North Atlantic Islands	1	—	—	—
North America	2	3	—	—
Guadalupe Island	2	3	—	—
West Indies	2	8	—	—
South Atlantic Islands	—	1	2	—
Hawaiian Islands	9	7	4	6
Central and Western Pacific	12	12	2	1
Australia	—	2	—	—
New Zealand	1	3	1	—
New Zealand (outlying islands)	5	2	—	—
Asiatic Islands	2	1	1	—
Asia	1	—	—	—
Islands of Indian Ocean	5	2	1	1
	42	44	11	8

KNOWN ONLY FROM RECENT OSSEOUS REMAINS

All the forms listed next below are known from bones and/or descriptions of travelers who were not necessarily experts in natural history. There is little doubt that the birds bred near the caves and marshes where the remains were found and little doubt that those not preceded by question marks differed from all other populations. Their exact form and color are quite unknown, so that their precise relationships are not clear. It has been necessary to be somewhat arbitrary in listing certain species. In the case of the Mascarene herons, for example, there must still be considerable doubt about relationships, but they are nevertheless listed as valid, for we can be certain that some such forms were once to be found on the islands, whereas, in the case of the grebe *Podiceps "gadowi,"* information is too scanty to record the name as anything but hypothetical. Likewise the dodo is listed as a species about which we can be certain, even though tangible evidence is fragmentary. However, contemporary description, unsatisfactory as it may be, allows this. On the other hand, no such description is available for the moa (*Megalapteryx*), which is listed as being known only by its bones, although bits of skin and feathers have actually been found. All are thought to have lived within the period of recorded history.

MOAS (Dinornithidae):
    Megalapteryx didinus   South Island, New Zealand
HERONS (Ardeidae):
    (?)*Ardea mauritiana*   Mauritius Island, Indian Ocean
    (?)*Ardea megacephala*   Rodriguez Island, Indian Ocean
RAILS (Rallidae):
    Aphanapteryx bonasia   Mauritius Island, Indian Ocean
    Aphanapteryx leguati   Rodriguez Island, Indian Ocean
    Notornis mantelli mantelli   North Island, New Zealand
    Fulica newtoni   Rodriguez Island, Indian Ocean
DARTER (Anhingidae):
    Anhinga nanus   Mauritius Island, Indian Ocean
DUCKS (Anatidae)
    Anas theodori   Mauritius Island, Indian Ocean
    Sarcidiornis mauritianus   Mauritius Island, Indian Ocean
PIGEONS (Columbidae):
    Columba (?)*rodericana*   Rodriguez Island, Indian Ocean
DODOS (Raphidae):
    Raphus solitarius   Reunion Island, Indian Ocean
    Pezophaps solitaria   Rodriguez Island, Indian Ocean
PARROTS (Psittacidae):
    Ara autochthones   St. Croix, West Indies
    Lophopsittacus mauritianus   Mauritius Island, Indian Ocean
    Necropsittacus rodericanus   Rodriguez Island, Indian Ocean
OWLS (Strigidae):
    Tyto sauzieri   Mauritius Island, Indian Ocean
    Athene murivora   Rodriguez Island, Indian Ocean

Appended to this list are the names given to two specimens. Localities at which they were collected are doubtful. Neither can be assigned to any other known species.

Thinornis rossi:
    Unique specimen (type) in the
    British Museum   ?Auckland Island, New Zealand

Necropsar leguati:
    Unique specimen (type) in   ?Île au Mât, near Rodriguez Island
    Liverpool

HYPOTHETICAL

In the fourth category are the birds known only by pictures and the descriptions of travelers of long ago, the majority of whom had no training in natural history and who could not leave accurate descriptions, although sometimes they tried. In some instances it is possible to correlate such descriptions with finds of bones, but only in such instances can theory be regarded as anything more than highly imaginative. For example we cannot question the fact that populations of rails and parrots lived on the Mascarene Islands in the Indian Ocean and in the West Indian Islands 300 years ago, but fanciful reconstruction of their exact appearance cannot be taken seriously. The following names are listed as hypothetical:

Podiceps gadowi	Mauritius Island, Indian Ocean
Pterodroma hindwoodi	Norfolk Island, western Pacific
Ardea duboisi	Réunion Island, Indian Ocean
Astur alphonsi	Mauritius Island, Indian Ocean
Leguatia gigantea	Mauritius Island, Indian Ocean
Kuina mundyi	Mauritius Island, Indian Ocean
Zenaida plumbea	Jamaica Island, West Indies
Nesoenas duboisi	Réunion Island, Indian Ocean
Victoriornis imperialis	Réunion Island, Indian Ocean
Ara guadeloupensis	Guadeloupe Island, West Indies
Ara gossei	Jamaica, West Indies
Ara erythrocephala	Jamaica, West Indies
Ara martinica	Martinique, West Indies
Ara erythrura	? West Indies
Ara atwoodi	Dominica, West Indies
Anodorhynchus purpurascens	Guadeloupe Island, West Indies
Aratinga labati	Guadeloupe Island, West Indies
Amazona violacea	Guadeloupe Island, West Indies
Amazona martinicana	Martinique Island, West Indies
Necropsittacus borbonicus	Réunion Island, Indian Ocean
Necropsittacus francicus	Mauritius Island, Indian Ocean
Bubo leguati	Rodriguez Island, Indian Ocean
Scops commersoni	Mauritius Island, Indian Ocean
Strix newtoni	Mauritius Island, Indian Ocean
Dysmorodrepanis munroi	Lanai Island, Hawaiian Islands
Testudophaga bicolor	Rodriguez Island, Indian Ocean
Foudia bruante	Réunion Island, Indian Ocean

SMALL POPULATIONS

The fifth and longest list contains the names of small populations that still exist but which, because of special circumstances, usually intimately connected with interference by man, are thought to be in danger of extinction.

Struthio camelus syriacus	Saudi Arabia
Diomedea albatrus	Seven Islands of Izu and Bonin Islands, western Pacific
Pterodroma cahow	Bermuda
Pterodroma hasitata	Hispaniola, West Indies
Puffinus puffinus newelli	Kauai, Hawaiian Islands
Phalacrocorax [Nannopterum] harrisi	Narborough Island, Galápagos
Branta (Nesochen) sandvicensis	Island of Hawaii
Anas platyrhynchos wyvilliana	Hawaiian Islands
Anas platyrhynchos laysanensis	Laysan Island, northwestern Hawaiian Islands
Anas castanea nesiotis	Campbell Island, New Zealand
Anas platyrhynchos oustaleti	Marianas Islands, western Pacific
Rhodonessa caryophyllacea	India
Gymnogyps californianus	West coast North America
Megapodius lapérouse lapérouse	Marianas Islands, western Pacific
Megapodius lapérouse senex	Palau Islands, western Pacific
Tympanuchus cupido attwateri	Southwestern North America
Tympanuchus pallidicinctus	Southwestern North America
Colinus virginianus ridgwayi	Northern Mexico
Ophrysia superciliosa	Northwestern India
Grus americana	North-central North America
Grus japonensis	Eastern Siberia, Japan
Porphyrio (Notornis) mantelli hochstetteri	South Island, New Zealand
Tricholimnas sylvestris	Lord Howe Island, western Pacific'
Tricholimnas lafresnayanus	New Caledonia Island, western Pacific
Rhynochetos jubatus	New Caledonia Island, western Pacific
Numenis borealis	North America
Thinornis novae-seelandiae	Chatham Islands, New Zealand
Rhinoptilus bitorquatus	Eastern India
Laterallus jamaicensis jamaicensis	Cuba
Columba inornata wetmorei	Puerto Rico
Columba jouyi	Okinawa Archipelago, western Pacific
Vini diadema	New Caledonia Island, western Pacific
Amazona imperialis	Dominica Island, West Indies
Amazona versicolor versicolor	St. Lucia Island, West Indies
Amazona versicolor arausiaca	Dominica Island, West Indies
Amazona guildingi	St. Vincent Island, West Indies
Psittacula krameri echo	Mauritius Island, Indian Ocean
Neophema chrysogaster	Southern Australia and Tasmania
Neophema splendida	Southern Australia
Neophema pulchella	Eastern Australia
Psephotus pulcherrimus	Northeastern Australia
Geopsittacus occidentalis	Western Australia
Otus insularis	Sechelle Islands
Asio flammeus portoricensis	Puerto Rico, West Indies

Campephilus principalis principalis	Southeastern North America
Sceloglaux albifacies albifacies	South Island, New Zealand
Caprimulgus vociferus noctitherus	Puerto Rico, West Indies
Campephilus principalis bairdi	Cuba
Dryocopus javensis richardsi	Korea
Saphaeopipo noguchii	Okinawa Island, western Pacific
Xenicus longipes stokesii	North Island, New Zealand
Atrichornis rufescens	Eastern Australia
Atrichornis clamosus	Western Australia
Troglodytes musculus musicus	St. Vincent Island, West Indies
Cinclocerthia ruficauda gutturalis	Martinique Island, West Indies
Ramphocinclus brachyurus brachyurus	Martinique Island, West Indies
Ramphocinclus brachyurus sanctae-luciae	St. Lucia Island, West Indies
Phaeornis palmeri	Kauai Island, Hawaiian Islands
Phaeornis obscurus myadestina	Kauai Island, Hawaiian Islands
Turdus poliocephalus pritzbueri	Lifu Island, Loyalty Islands, western Pacific
Nesocichla eremita eremita	Tristan da Cunha Island, South Atlantic
Turnagra capensis tanagra	North Island, New Zealand
Turnagra capensis capensis	South Island, New Zealand
Acrocephalus luscinia rehsei	Nauru Island (between Carolines and Marshalls, southwestern Pacific)
Dasyornis brachypterus longirostris	West Australia
Dasyornis broadbenti littoralis	Southwestern Australia
Amytornis goydera	South Australia
Leucopeza semperi	St. Lucia, West Indies
Loxops coccinea ochracea	Maui Island, Hawaiian Islands
Hemignathus obscurus procerus	Kauai Island, Hawaiian Islands
Hemignathus lucidus hanapepe	Kauai Island, Hawaiian Islands
Hemignathus wilsoni	Island of Hawaii
Pseudonestor xanthophrys	Maui Island, Hawaiian Islands
Palmeria dolei	Maui Island, Molokai Island, Hawaiian Islands
Psittirostra psittacea	Hawaiian Islands
Psittirostra [Loxioides] bailleui	Island of Hawaii
Notiomystis cincta	North Island, New Zealand
Moho braccatus	Kauai Island, Hawaiian Islands
Loxigilla portoricensis grandis	St. Christopher Island, West Indies
Heteralocha acutirostris	North Island, New Zealand
Creadion carunculatus rufusates	North Island, New Zealand
Creadion carunculatus carunculatus	South Island, New Zealand
Geospyza magnirostris	Charles Island, Galápagos

SOME RARE BIRDS PROBABLY NOT IN IMMEDIATE DANGER

The birds treated in this section fall into two categories: (1) forms that are rare in museum collections and about which nothing, except the bare identity, is known; and (2) some small populations that do not appear to be in immediate danger of extinction but because they are so few in numbers may be endangered should they be much disturbed in the future.

PETRELS (Procellariidae)

Pterodroma aterrima (Bonaparte)

Procellaria aterrima BONAPARTE, Consp. Av., **2**, 1857, p. 191 (Réunion).

Status. Uncertain. According to Alexander (1954), three specimens of this petrel have been captured in the Indian Ocean in recent years. One was taken off the coast of Arabia in February 1953, a second in the Gulf of Aden in July 1946, and a third at the approximate position lat. 12° N., long. 60° E. (300 miles east of Sokotra) in August 1953. Only one of these was preserved. The species was formerly known from only three specimens, one of which came from Mauritius, and two (the types) were thought to have come from Réunion (see Berlioz, 1946). Alexander (*l.c.*) casts doubt on the accuracy of the latter locality on the grounds that the collector may have been deceived by a local name and because the birds never have been found there since 1834. However, because little or no effort has been made to find the birds and because the name may have applied to this species, it is possible that it does breed on Réunion. Alexander suggests that Sokotra will be found to be a breeding place.

Specimens are in Cambridge, Leiden, Oxford, and Paris.

Range. Breeding place unknown. To be seen in the western Indian Ocean.

Pterodroma leucoptera longirostris (Stejneger)

Aestrelata longirostris STEJNEGER, Proc. U.S. Nat. Mus., **16**, 1893, p. 618 (Mutsu Bay, Japan).

Common name. Gadfly petrel.

Status and range. Unknown. According to Austin and Kuroda (1953) only 10 specimens exist. All were taken at sea off the coasts of Japan. Austin and Kuroda hazard the guess that the breeding place is an island in the South Pacific or in Antarctic seas.

Specimens are in the Yamashina collection in Tokyo (cotypes), in Chicago, and in New York.

Pterodroma rostrata becki Murphy

Pterodroma rostrata becki MURPHY, Amer. Mus. Nov., no. 322, 1928, p. 1 (3° S., 155° E., east of New Ireland and north of the Solomons).

Common name. Beck's petrel.

Status and range. Unknown. According to Murphy (1952) the form is known only by two specimens, a male and female, both taken at sea near the Solomons and both in New York. Murphy (*l.c.*) suggests that they lived on one of the Solomon Islands.

They closely resemble the petrel of Tahiti (*rostrata*) but are smaller, according to the same author.

IBISES (Threskiornithidae)

Bostrychia olivacea rothschildi (Bannerman)

Lampribis rothschildi BANNERMAN, Bull. Brit. Orn. Club, **40**, 1919, p. 6 (Infant d'Henrique, Principe Island).

Status. Whether or not this subspecies can be maintained as a taxonomic reality (see Amadon, 1953, p. 407), it is probable that the population of Principe Island, Gulf of Guinea, has been extirpated. Closely related if not identical birds occur in Guinea. According to Bannerman (1930, p. 121) the birds were last seen in 1901, when they were said to be rare. Boyd Alexander (in Bannerman, 1915) could not find them in 1909 nor could Correia in 1928 (Amadon, *l.c.*) nor the Oxford Expedition of 1948 (Snow, 1950).

Although these birds were found last in almost inaccessible places, they had a loud distinctive call like a crow and therefore are not so likely to have eluded the naturalists who have searched for them.

KITES (Milvinae)

Chondrohierax wilsonii (Cassin)

Cymindis Wilsonii CASSIN, Journ. Acad. Nat. Sci. Philadelphia, **1**, 1847, p. 21, pl. 7.

Common names. Hook-billed kite; caguarero.

Status. Apparently a rare bird, said by Ramsden (in Bond, 1950) to occur in small numbers in woods south of Guantánamo. Further search should be made. It is quite possible that the species is in danger of extinction.

Range. Cuba: formerly south coasts west to Cochinos Bay, now very localized near Guantánamo. The species *uncinatus* is also rare in Grenada and Trinidad, according to Bond. It is still to be found in Central America and South America, however.

CURASSOWS (Cracidae)

Crax blumenbachii Spix

Crax blumenbachii SPIX, Aves Brazil, **2**, 1825, p. 50 (State of Rio de Janeiro).

Common name. Mutu.

Status. Insufficiently known; mentioned by Stresemann (1954) as being on the road to extinction. However, he quotes Pinto to the effect that birds were collected on the Rio São José, Espiritu Santo, in 1940 and at Ilheos, Bahia, in 1944. It may be questioned that with such a wide range, in many places little disturbed, the species is in great danger.

Range. Forests of eastern Brazil. Recorded from southern Bahia, Espiritu Santo, Rio de Janeiro, and eastern Minas Gerais (Mayrink, Minas Novas, Rio da Pomba).

Crax rubra griscomi Nelson

Crax rubra griscomi NELSON, Proc. Biol. Soc. Washington, **39**, 1926, p. 106 (Cozumel Island).

Status. Rare and perhaps in danger. However, according to Paynter (1955), the few birds that remain are not much molested by shooters. He tells me that it is difficult to reach much of the birds' still undisturbed habitat.

Range. Cozumel Island off the coast of Yucatán, Mexico.

PHEASANTS (Phasianidae)

Lophura inornata hoogerwerfi (Chasen)

Houppifer hoogerwerfi CHASEN, Treubia, **17**, 1939, p. 184 (Meloewak, Atjeh, North Sumatra).

Common name. Atjeh pheasant.
Status. Unknown. Only two specimens exist. Both are females. The type is in Buitenzorg, Java, and a second skin in Philadelphia.
Range. Known only from Meloewak, Atjeh, Java.

Pucrasia macrolopha castanea Gould

Pucrasia macrolopha castanea GOULD, Proc. Zool. Soc. London, 1854 [1855], p. 99 (Kafirstan).

Common name. Western koklass pheasant.
Status. Unknown. It is a rare bird in collections. Delacour (1951, p. 84) says that there is a good deal of individual variation and gradation toward the race of Kashmir.
Range. Mountains of Chitral, Kafirstan, and Afghanistan, adjacent to the northwestern frontier of India (Peters).

BUTTON QUAIL (Turnicidae)

Turnix varia novaecaledonia Ogilvie-Grant

> *Turnix varia novaecaledoniae* OGILVIE-GRANT, Cat. Birds British Mus., **22**, 1893, p. 552, footnote (New Caledonia).

Common name. Painted button quail.

Status. Quite rare in collections, this representative of the Australian painted quail is probably not rare in the grasslands of the lee side of the island.

RAILS (Rallidae)

Pardirallus maculatus inoptatus (Bangs)

> *Limnopardalis maculatus inoptatus* BANGS, Proc. New England Zool. Club, **4**, 1913, p. 90 (near Jaruco, Cuba).

Common names. Cuban spotted rail; gallinuela escribando.

Status. Barbour (1923) reported that this bird was so excessively rare that he thought it must have been extirpated, and for long it was known by only two specimens. However, Bond (1950) has found it to be common near Santo Tomás.

Range. Cuba, in the Provinces of Habana, Matanzas, and Las Villas.

Ortygonax (Cyanolimnas) cerverai (Barbour and Peters)

> *Cyanolimnas cerverai* BARBOUR AND PETERS, Proc. New England Zool. Club, **9**, 1927, p. 95 (Santo Tomás, Ciénaga de Zapata, Cuba).

Common name. Zapata rail.

Status and range. Bond (1950) describes the restricted range of this isolated and very distinct population. "Known only from that part of the Zapata Swamp directly north of the heavily wooded territory known as Santo Tomás where apparently confined to within a mile of the high ground."

Aramidopsis plateni (Blasius)

> *Rallus plateni* Blasius, Braunschweig Anz., **3**, no. 52, 1886 (Rurukan, Celebes).

Common name. Platen's Celebes rail.

Status. Rare and localized in extremely dense lliana and bamboo second-growth bush on the borderline of lowland and highland forests of Celebes. Only 10 specimens are known, all found in the north, central, and southeast after long painful searches. Of three obtained near Rurukan; two are in Braunschweig and one in Halberstadt; two from Tomohon are in

Dresden and Basle; one from Kantewoe in Leiden; two from Wamo in Berlin and New York; two from Minahassa (see Stresemann, 1940, pp. 12, 13, for map and 1941, p. 31, for an account by the collector Heinrich). *Range.* Celebes.

SHOREBIRDS (Scolopacidae)

Capella undulata gigantea (Temminck)

> *Scolopax gigantea* TEMMINCK, Nouv. Rec. Pl. Col., livr. 68, pl. 403 (Brazil = São Paulo, Pinto, 1944).

Common names. Narcejao, batuirão, gallinhola, giant snipe.

Status. Uncertain. Pinto (in Stresemann, 1954) reports that he himself saw the birds about 20 years ago. The species is well represented in all the larger museums.

Range. Southern Brazil from Minas Gerais, Goyaz, and Mato Grosso south to Rio Grande do Sul, Paraguay, and perhaps Argentina.

The subspecies *undulata* occurs on Mount Roraima in Venezuela, French and British Guiana, and northern Brazil.

AUKS (Alcidae)

Synthliboramphus wumisuzumi (Temminck)

> *Uria wumisuzumi* TEMMINCK, Pl. Col., livr. 98, 1835, pl. 579 (shores of Japan and Korea).

Common names. Kanmuri umisuzumi (crested sea sparrow), Japanese murrelet.

Status. Localized but not rare on its range. Recently it was discovered that the United States Air Force had been using Sanbondake, Izu Islands, one of the principal nesting places, as a bombing target. As soon as the matter was brought to the attention of authorities, however, this practice was stopped (Austin and Kuroda, 1953).

Range. Nests on rocks and headlands on the Seven Islands of Izu and, at least formerly, on Mikomato Island near Shimoda and Ipponmatsu in Kanagawa Prefecture.

PIGEONS (Columbidae)

Leptotila wellsi (Lawrence)

> *Engyptila wellsi* LAWRENCE, Auk, 1, 1884, p. 180 (Grenada).

Common name. Grenada dove.

Status. Unknown. According to *Bond (1963), there is a small population "behind Grande Anse," Grenada. Recent searches elsewhere have not been successful.

Between 1920 and 1925 several were brought to London alive, and they were bred in captivity (Seth-Smith, Avic. Mag., 1926, p. 221).

Range. Grenada and possibly St. Vincent and Tobago (Bond, 1950). Specimens are in London, New York, and Washington.

Ptilinopus porphyraceus pelewensis Hartlaub and Finsch

Ptilinopus pelewensis HARTLAUB AND FINSCH, Proc. Zool. Soc. London, 1868, p. 7 (Pelew Islands).

Common name. Crimson-crowned fruit dove.

Status. According to Baker (1951, p. 186), this race was found in small numbers and restricted to undisturbed forests in 1945. Coultas, collector for the Whitney Expedition of the American Museum, reported in 1931 that they were persistently shot by Japanese, who sold the birds for 25 sen apiece. Certainly no such population can survive commercial shooting and trapping, and this prompted Mayr (1945b) to remark that they were in need of protection.

Range. Palau Islands: Babelthuap, Koror, Garakayo, Peleliu, Ngabad, Angaur.

It may well be that other subspecies, *ponapensis* of Truk and Ponape and *hernsheimi* of Kusaie in the Carolines, will also one day be in need of protection.

Ducula goliath (G. R. Gray)

Carpophaga goliath Gray, Proc. Zool. Soc. London, 1859, p. 165, pl. 155 (Isle of Pines).

Status. According to Mayr, 1945b (who had his information from Mac-Millan, a collector for the American Museum), this species because of steady persecution becomes more and more restricted to mountain forests and other remote locations.

Range. Forests of New Caledonia.

Ducula aurorae (Peale)

Carpophaga aurorae PEALE, U.S. Expl. Exped., **8**, 1848, p. 201 (Makatea Island).

Status. Rare and localized. Beck and Quayle, collectors for the Whitney Expedition of the American Museum, found birds on Makatea in the Tuamotu Archipelago. Neither they nor other naturalists since then were able to find them on Tahiti, where they were reported last in 1904 by Wilson (1907). He found them rare and subject to considerable shooting and trapping pressure.

Range. Makatea, Tuamotus, and formerly Tahiti.

Others of this large Micronesian pigeon are localized and have been thought to be in need of protection:

Ducula oceanica ratakensis (Takatsukasa and Yamashina): Wotje and Arno, Ratak Chain, Marshall Islands.

Ducula oceanica oceanica (Lesson and Garnot): Kusaie, eastern Carolines; Jaluit and Elmore, Marshall Islands.

Ducula oceanica monacha (Momiyama): Yap, Babelthuap, Koror, Angaur, and Current Islands.

Ducula oceanica teraokai (Momiyama): Truk, Caroline Islands.

Ducula oceanica townsendi (Wetmore): Ponape, Caroline Islands.

Gallicolumba canifrons (Hartlaub and Finsch)

> *Phlegoenas canifrons* HARTLAUB AND FINSCH, Proc. Zool. Soc. London, 1872, p. 101 (Palau Islands).

Common name. Ground dove.

Status. Not uncommon but wary and hard to find, according to Baker (1951). In 1931 collectors for the American Museum had thought the species to be nearing extinction.

Range. Palau Islands: Babelthuap, Koror, Garakayo, Peleliu, Ngabad, Angaur.

Gallicolumba erythroptera (Gmelin)

> *Columba erythroptera* GMELIN, Syst. Nat., **1**, pt. 2, 1789, p. 775 (Eimeo = Moorea).

Common name. Ground dove.

Status. Large populations were found by Beck and Quayle, collectors for the American Museum, on most of the Tuamotus. The type series is said to have been collected on Tahiti and Moorea, from which the species has been extirpated. A single specimen from Moorea is in Leyden.

Range. Tuamotus and, formerly, Tahiti and Moorea.

PARROTS (Psittacidae)

Strigops habroptilus G. R. Gray

> *Strigops habroptilus* GRAY, Genera birds, **2**, 1845, p. [427], pl. LV ("one of the islands of the South Pacific = Dusky Sound, South Island, New Zealand, Mathews and Iredale, 1913).

Common name. Kakapo.

Status. Probably extinct on the North Island of New Zealand; rare and localized on the South Island, according to *Williams (1962, p. 21). Recorded recently from Nelson, Westland, and Fiordland.

This species appears to be particularly sensitive to interference. Whether or not it was ever to be found in numbers on the North Island, it was apparently restricted to a few isolated places such as the headwaters of the Wanganui River and the vicinity of Lake Taupo by 1850 while much forest still remained.

Range. New Zealand: North and South Islands.

Prosopeia personata (G. R. Gray)

> *Coracopsis ? personata* GRAY, Proc. Zool. Soc. London, 1848, p. 21 ("?New Guinea" = Fiji).

Common name. Yellow-breasted musk parrot.

Status and range. Perhaps extirpated from the smaller islands of Ovalau and Mbau, Fiji Islands, but probably still exists in considerable numbers on Viti Levu where the Whitney Expedition of the American Museum obtained large series. Bahr (1912) and Wood and Wetmore (1926) predicted extinction for the species because the birds are partial to fruit and are consequently shot whenever they appear.

Prosopeia tabuensis, with subspecies confined to small islands in the Fijis and Tongas (where introduced), should perhaps be mentioned although they are apparently not rare in Fiji.

Eunymphicus cornutus cornutus (Gmelin)

> *Psittacus cornutus* GMELIN, Syst. Nat., **1**, pt. 1, 1788, p. 327 (Nova Caledonia).

Common name. Crested parakeet.

Status. Not uncommon but difficult to find in the tall dense forests except in seasons when they resort to flowering trees. MacMillan collected a number for the American Museum in 1938. Sarasin, who had made a previous effort in 1912, secured only three captive birds and reported the species to be almost extinct.

Range. New Caledonia.

MacMillan (1939, p. 31) reported that he secured the subspecies *uveansis* on Uvéa, Loyalty Islands, only, not on Lifu where it had been reported.

OWLS (Strigidae)

Otus nudipes newtoni (Lawrence)

> *Gymnoglaux newtoni* LAWRENCE, Ann. Lyc. Nat. Hist. New York, **7**, 1860, p. 259 (St. Croix).

Common name. Newton's owl.

Status. According to Bond (1950) this is a rare bird wherever it occurs.

Range. Vieques Island, Savana Island (?), St. Thomas, St. John, and St. Croix (Bond, 1950).

HUMMINGBIRDS (Trochilidae)

Augastes lumachellus (Lesson)

> *Ornisyma lumachella* LESSON, Rev. Zool., 1883, p. 315 ("Bahia" = unknown).

Status. Unknown. Because this species is known only from specimens shipped to Europe long ago for the millinery trade (of course without any information attached) it is impossible to know anything about it. Berlioz (1944, p. 152) said "perhaps one may fear for its survival," but because few have searched the hinterland of Bahia and Mato Grosso one may hope that the species will be found again.

WOODPECKERS (Picidae)

Melanerpes superciliaris nyeanus (Ridgway)

> *Centurus nyeanus* RIDGWAY, Auk, **3**, 1886, p. 336 (Watling's Island = San Salvador).

Status. A small population confined to about 2 square miles of the northern end of San Salvador. The area is remote and undisturbed at present.

Range. Northern San Salvador (Watling Island, Bahamas).

LYREBIRDS (Menuridae)

Menura alberti Bonaparte

> *Menura alberti* BONAPARTE, Consp. Gen. Av., **1**, 1850, p. 215 ("ex. Austr." = Richmond River, N.S.W.).

Common name. Lyrebird.

Status. Not uncommon in undisturbed forests but localized and restricted in range. The Australian Government protects the species (J. R. Kinghorn, *in litt.*).

Range. Rain forests of southeastern Australia in northern New South Wales and southern Queensland, Australia.

BULBULS (Pycnonotidae)

Microscelis borbonica borbonica (Gmelin)

> *Turdus borbonicus* GMELIN, Syst. Nat., **1**, pt. 2, 1789, p. 821 ("in insula Bourbon" = Réunion).

Common name. Merle de Réunion.

Status. Rare and local (Berlioz, 1946). It may be assumed that this sub-

INTRODUCTION AND SUMMATION

species is in no immediate danger of extinction. Even so small a population should survive unless some great change should be brought about in its habitat which now seems unlikely.

Range. Réunion, Indian Ocean.

Microscelis borbonica olivacea (Jardine and Selby)

Hypsipetes olivacea JARDINE AND SELBY, Illust. Orn., **4**, 1837, pl. 2, text (no locality = Mauritius).

Common name. Merle.

Status. Localized and rather rare (Georges Antelme, *in litt.*). Meinertzhagen (1912) was of the opinion that this population was dangerously small.

Range. Mauritius, Indian Ocean.

ANTBIRDS (Formicariidae)

Myrmotherula erythronotos (Hartlaub)

Formicivora erythronotos HARTLAUB, Rev. Mag. Zool. (ser. 2), **4**, 1852, p. 4 ("Brazil").

Status. Unknown. The only locality where birds are certainly known to have been collected is Nova Friburgo, Prov. Rio de Janeiro, a junction point on the railway between Rio de Janeiro and Campos. No collectors, either for the Museu Paulista or the American Museum, have been able to find the species. If, as is possible, this was a very small population confined to the Province of Rio de Janiero it may be extinct. However, there is still hope that it may be found elsewhere.

WOODHEWERS (Dendrocolaptidae)

Asthenes sclateri (Cabanis)

Synallaxis Sclateri CABANIS, Journ. für Orn., **26**, 1878, p. 196 (Sierra de Córdova).

Status. Unknown. Stresemann (1954, p. 53) surmises that the species may be extinct. However, it has been thought to be a synonym of *Asthenes hudsoni* Sclater, 1874, until recently (Stresemann, 1948, p. 445), and no special search has been made for it.

The collector Dr. Adolf Döring found it in rocky country, whereas *A. hudsoni* is a bird of the grasslands. It differs from *hudsoni* in having a darker back. The paler markings of the feathers are less pronounced than in *anthoides* but more so than *humilis,* and in the more rounded tail feathers which have less pronounced and darker markings (see

Stresemann, *l.c.*). It should be said that *hudsoni* has also a yellow throat and streaked underparts in certain plumages.

Range. Known only from Sierra Córdova, Argentina.

THRUSHES (Turdidae)

Myadestes elizabethae retrusus Bangs and Zappey

Myadestes elizabethae retrusus BANGS AND ZAPPEY, Amer. Nat., **39**, 1905, p. 208 Pasidita, Isle of Pines).

Status. Very rare and localized according to Bond (1950).

Range. Isle of Pines (Cuba): Known only from the vicinity of Paso Piedros (at sea level), which connects the southern portion of the Isle of Pines with the northern; found here on both the north and south borders of the Ciénaga; said to occur at Hato and Punto del Este in the south (Bond, *l.c.*).

Mimocichla ravida Cory

Mimocichla ravida CORY, Auk, **3**, 1886, p. 499 (Grand Cayman).

Common name. Cayman thrush.

Status. Rare and apparently restricted to the east and northeast parts of the island. Sighted in 1938 (Bond, 1950).

Range. Grand Cayman, West Indies.

Copsychus seychellarum Newton

Copsychus seychellarum NEWTON, Ibis, 1865, p. 332, pl. 8 (in insularis Seychellarum).

Status. This population has increased in recent years. For a time it was thought to be extinct (Vesey-Fitzgerald).

Range. Seychelles Islands, Indian Ocean.

Timaliidae

Psophodes nigrogularis nigrogularis Gould

Psophodes nigrogularis GOULD, Proc. Zool. Soc. London, 1844, p. 5; and Birds Australia, pt. 15, 1844, pl. 11, p. 11 (Wongan Hills, southwestern Australia).

Status. Localized and rather rare. During the years 1901–1939 there had been reports of the probable extinction of this subspecies (Leach, 1929, p. 23). Whittell (1939) discovered them "in some strength" near Gnowangerup, West Australia. Formerly they occurred near Perth, Busselton, Margaret River, and King Georges Sound.

Range. Southwestern Australia: vicinity of Gnowangerup and in the Border Country (Serventy and Whittell, 1951).

Psophodes nigrogularis leucogaster Howe and Ross

> *Psophodes nigrogularis leucogaster* HOWE AND ROSS, Emu, **32**, 1933, p. 147 Manya, northwestern Victoria).

Common name. Mallee whipbird.

Status. Rare and local according to Howe and Ross (*l.c.*) and McGilp and Parsons (1937, p. 3) who secured a female in 1936.

This is a shy bird, difficult to find even when in favored habitat, and there have been long periods during which no bird was recorded as having been seen.

Range. Australia on the border of Victoria and South Australia perhaps west to the Murray River.

Specimens of this species (including *leucogaster*) are in Adelaide, Cambridge (Massachusetts), London, Melbourne, New York, Perth (West Australia), Philadelphia, and Sydney.

FLOWERPECKERS (Dicaeidae)

Melanocharis arfakiana (Finsch)

> *Dicaeum arfakiana* FINSCH, Notes Leyden Mus., **22**, 1900, p. 70 (Arfak Mts., New Guinea).

Status. Rare in collections but because of its undisturbed habitat and probably wide range in no danger of extinction. The only two specimens known are in Leyden and New York (Mayr).

KINGLETS (Regulidae)

Regulus calendula obscurus Ridgway

> *Regulus calendula obscurus* RIDGWAY, Bull. U.S. Geol. Geogr. Surv. Terr., **2**, no. 2, 1876, p. 184 (Guadalupe Island, Pacific).

Common name. Guadalupe Island kinglet.

Status. Localized and rare. It is quite possible that this small population is in danger of extinction because of the slow destruction of its habitat, a cypress grove (*Cupressus guadalupensis*), but probably it is in no immediate danger.

In spite of the fact that several parties of naturalists visited the island, and even the cypress grove at the top of the mountain, the kinglets were not reported between 1906 and 1953 and were said to be extinct (Slevin, 1936). In 1953 Howell and Cade (1954, p. 289) found five singing males but were unable to give an estimate of the actual size of the population. They reported that the cypresses were all old trees; seedlings are eaten by the introduced goats. A smaller grove, which grew near a spring on a plateau below the summit, reported by Bryant (1887) and others, has been destroyed.

Range. Guadalupe Island, off Baja California, where confined to a cypress grove at the northern summit of the 4,000-foot ridge.

WRENS (Troglodytidae)

Thryomanes (*Ferminia*) *cerverai* (Barbour)

Ferminia cerverai BARBOUR, Proc. New England Zool. Club, 9, 1926, p. 74 (Santo Tomás, Ciénaga de Zapata, Cuba).

Common names. Zapata wren; fermina.

Status. Localized but common in its range (Bond, 1950).

Range. Zapata Swamp, Cuba. Known only from the vicinity of Santo Tomás in an area of no more than 5 square miles of the Zapata Swamp.

WARBLERS (Parulidae)

Dendroica (*Catharopeza*) *bishopi* (Lawrence)

Leucopeza bishopi LAWRENCE, Ann. New York Acad. Sci., 1, 1878, p. 151 (St. Vincent).

Common name. Whistling warbler.

Status. Not uncommon and even common in certain localities. Bond (1950) reports that "the eruption of the Soufrière in 1902 apparently wiped out the species in the northern mountains, where in recent years it has become reestablished." Father Devas (in Bond, 1950) says he has found birds in all mountain forests.

Range. St. Vincent, West Indies.

Vermivora crissalis (Salvin and Godman)

Helminthophila crissalis SALVIN AND GODMAN, Ibis, 1889, p. 380 (Sierra Nevada, Colima, Mexico).

Figure. Sutton, in Van Tyne, 1936.

Common name. Colima warbler.

Status. This bird is often mentioned by ornithologists as the rarest of the wood warblers, for it was known by few specimens in collections until recently and is localized in mountain ranges above 6,000 feet. Apparently it is common in the Chisos Mountains of Texas (Van Tyne, 1936) and elsewhere in its range.

Range. Texas, in the Chisos Mountains; Mexico, in Sierra Madre Occidental of Coahuila, Miguihuana in Tamaulipas; in winter Sierra Nevada of Colima, Michoacán, and Guerrero.

The Colima warbler is a mountain representative of a superspecies which includes Virginia's warbler (*Vermivora virginiae*) and Lucy's warbler (*V. luciae*).

Vermivora bachmanii (Audubon)

Fringilla bachmanii AUDUBON, Birds America, **2**, 1833, pl. 165 (near Charleston, S. C.).

Common name. Bachman's warbler.

Status. This species is characterized by the peculiar, unpredictable desertions of known breeding places and apparently whole regions. For example, the birds disappeared from their habitat in South Carolina, reappeared in 1938, and are still to be found breeding there (Sprunt, 1954). None have been seen in Florida since 1909.

Range. Breeds in certain unpredictable years in the southeastern United States, namely in Missouri, western Kentucky, Alabama, and coast of South Carolina; south to Cuba in winter.

Calospiza cabanisi Sclater

Calospiza cabanisi SCLATER, Ibis, 1868, p. 71 ("Costa Cuca, Guatemala").

Common name. Cabanis's tanager.

Status. A small population known only by two immature specimens. The type was perhaps taken in western Guatemala. It is in Berlin. Stresemann (1954) reported that this specimen was purchased by Dr. Barth about 1866 and that therefore it is not certain that it was in the collection of Dr. Bernoulli (see Salvin and Godman, 1883). It follows that the type locality, which is in the dry lowlands, may not be correct.

A second specimen was taken by Brodkorb in cloud forests on Mount Ovando near Escuintla, Chiapas, Mexico, on August 29, 1937, thus fulfilling a prediction of Salvin and Godman.

Tangavius armenti (Cabanis)

Molothrus armenti CABANIS, Mus. Hein., **1**, 1851, p. 192 (note) ("Cartagena," Colombia).

Common name. Arment's cowbird.

Status. Unknown. Schauensee (1951, p. 992) quotes Dugand (1947) to the effect that he did not meet with this species in its supposed range.

According to Stresemann (1954) there are only two specimens known; they are in Berlin (type) and New York. See also Friedmann, 1929, p. 318, and 1927, p. 506.

Range. Only known specimens from Savanilla and Cartagena on the Caribbean coast of Colombia. This may actually not represent the range.[1]

Nemosia rourei Cabanis

> *Nemosia rourei* CABANIS, Journ. für Orn., **18**, 1870, p. 459 (Muriahié, north bank of Rio Parahyba do Sul).

Status. Unknown. The type in the Berlin Museum is the only known specimen.

Range. Known only from Brazil in Rio de Janeiro, Muriahié, north bank of Rio Parahyba do Sul.

Pseudodacnis hartlaubi (Sclater)

> *Dacnis hartlaubi* SCLATER, Proc. Zool. Soc. London, 1854 (= 1855), p. 251 ("Nova Grenada" = Colombia).

Status and range. Unknown. The only locality from which this species is known is the valley of the Dagua on the Pacific slope of the western Andes above Juntas, now called Cisneros according to de Schauensee (1951, p. 1024). Apparently this locality is on the southern slopes of the Chancos range. The same author (*l.c.,* p. 294) says that the excellent collector von Sneidern searched this region without finding this species. It was last collected by Raap at a date unknown. Other than this specimen there are eight "Bogota trade skins" of unknown origin in museums. There is much ornithologically unexplored country in Colombia, and we may well hope that this species still exists.

FINCHES (Fringillidae)

Sporophila falcirostris (Temminck)

> *Pyrrhula falcirostris* TEMMINCK, Nouv. Rec. Pl. Col., livr. **2**, 1820, pl. 11 ("Bresil" = Bahia, as proposed by Pinto, 1944).

Common name. Papa-capim.

Status. Not well known, but apparently populations are to be found in eastern São Paulo and Espiritu Santo (Pinto, 1944, p. 611). See also Stresemann (1954, p. 51).

Range. Forests of eastern Brazil to São Paulo.

---

[1] In December 1956 Dr. Herbert Friedmann identified this bird in a collection of live birds in the National Zoological Park. It presumably came from the interior of Colombia.

Neospiza concolor (Bocage)

Amblyospiza concolor BOCAGE, Jorn. Sci. Lisboa, **12**, 1888, p. 229.

Common names. Enjolo; São Thomé grosbeak.

Status. Not certainly known, but according to Snow (Ibis, 1950, p. 583) and Amadon (1953, p. 431), no recent collectors have found the species. Boyd Alexander searched in 1909; Correia, for the American Museum, in 1928; and the Oxford University Expedition in 1949. Both Snow (*l.c.*) and Amadon (*l.c.*) suggest that it may be extinct. According to Alexander (in Bannerman, 1915) and Snow (1950) forests of the lowland have been destroyed and much cut up by coffee plantations up to 5,000 feet on the mountain sides. It may be that the bird's habitat has been destroyed, but so little is known of this subject that there may still be hope of survival especially because there is much almost impenetrable bush remaining.

Range. São Thomé Island, Gulf of Guinea, off west coast of Africa.

Francisco Newton collected the types on the east coast, on the Rio Quija, in forests near Angolares and sighted a bird near San Miguel on the west coast.

Position Doubtful (? Nectariniidae)

Neodrepanis hypoxantha Salomonsen

Neodrepanis hypoxantha SALOMONSEN, Bull. British Orn. Club, **53**, 1933, p. 182.

Status and range. No accurate information is now available. This is probably a rather rare species confined to forests of northeastern Madagascar. Seven specimens are known. Two in the British Museum (cotypes) are from an unspecified locality "east of Antananarivo," now Tananarive, collected in July 1881 (Salomonsen, *l.c.*, and 1934). Three in Germany were collected at Andrangoloaka, a locality on the edge of the eastern forests east of Antsirabé. This locality, although it does not appear on modern maps, is listed in Grandidier's Histoire de Madagascar, Geographie (texte p. 159), and described "village (1400 m.) sur la limite de partage des eaux de l'Ikopa et du Mangoro," which accords with the location given by Delacour to Stresemann, 1937). It was probably not far from the graphite mining village of Tsinjoarivo where there was a beautiful stand of original forest in 1929. The Expedition Franco-Anglo-American spent only a part of one day in that region. Two of these specimens are in Berlin (Stresemann, 1954).

A sixth specimen is in Washington. According to Wetmore (1953) it was secured by the Rev. James Wills in October or November 1895 and the label is marked "E. Imerina." Imerina is usually considered to be the grassy plateau country surrounding the capital city of Tananarive (see Grandidier, *l.c.*, and Atlas des Colonies Françaises, Soc. Editions Geographiques Maritimes et Coloniales, Paris, 1934). Probably the specimen came from the forests which still remain on the mountain slopes east of Imerina proper.

A male in the Museum of Comparative Zoology (no. 233140) marked "février 25, foret Sianaka," was purchased from a dealer, Karl Fritsche of Bremerhaven, together with 74 birds from that general region. There is no reason to doubt that the locality is correct. It is possible that the specimen may have been collected a few years earlier than 1925, for it is also probable that Fritsche bought them from a dealer in Madagascar, but some are marked 1925 in the same handwriting, and it may be assumed that the date is approximately correct also. Because the Forest of Sianaka, lying to the northeast of the Plateau of Imerina, is very extensive and has not been disturbed greatly, *Neodrepanis hypoxantha* will be found again. Austin Rand (*in litt.*) agrees to this. He writes, "It's a small inconspicuous bird and years might pass without one being collected. . . . I think what evidence we have is much too scanty to conclude that the bird is extinct."

In the past this species has been placed with the sunbirds (Nectariniidae), but this allocation is questionable.

Geography of Extinction

Because this work is limited to discussion of the discrete populations that have disappeared during the past 280 years, the geographical perspective is distorted. It will be seen that there is no account of extinction of birds on the continents of Europe save in discussion of the ostriches (pp. 138–141) and the great auk (pp. 271–291); actually not one form known by more than its bones has been extirpated.

Perhaps it may fancifully be said that in prehistoric times men and birds arranged a means of living together to the end that no birds were extirpated. Possibly a few vulnerable species disappeared from Europe before the sixteenth century, as did the great auk from that part of its range. Records of the large duck traps of eighteenth century England show that great numbers were captured. Their use had been long continued. Their use was abandoned, but perhaps some now unknown species had disappeared two centuries before. It is conceivable, in the light of the history of the Labrador duck, that a species so small in numbers might have been extirpated. In more recent times the range of the great bustard (*Otis tarda*) has become much restricted, and other birds less numerous, but it is unlikely that any have become extinct.

The disappearance of the crested shelldrake from the continent of Asia (p. 175) is a striking example, but we have no other from that continent unless perhaps the pink-headed duck, the mountain quail, or Jourdain's courser, all Indian birds, are extinct. Destruction of habitat on the Pacific coasts of China has been great, and it may well be that some distinct form unknown to us has gone.

As far as is known neither Africa nor South America has suffered losses in recent times. It should be said that birds are often to be found where ornithologists are. Perhaps here, as elsewhere, there were once species to be found that exist no more.

Island faunas have obviously been most vulnerable. It is these, together with North American species, that bore the brunt of sudden and devastating incursion of modern Europeans, to which we are limited.

The West Indies

The West Indian Archipelago extends in a 1,500-mile arc from latitude 25° N., only 45 miles east of the Florida Peninsula, to 12° N., close to the coast of Venezuela. Its islands are usually divided into three groups. Most northerly are the Bahamas, all low, dry islands of coral limestone. The central group, or Greater Antilles (Cuba, Jamaica, Hispaniola, and Puerto Rico), are large, mountainous, and more heavily forested than the Bahamas, although forests have been much reduced. Largest of these is Cuba, 750 miles long and averaging about 50 miles in width. Jamaican mountains rise to 7,360 feet. The third group, the Lesser Antilles, form the barrier between the Atlantic Ocean and the Caribbean Sea. All of them are smaller, but many are mountainous and well wooded at higher altitudes. For example, the total area of Guadeloupe is 688 and that of Grenada 138 square miles; mountains are 2,500 to 3,000 feet in height and precipitous. A tropical climate is tempered by trade winds during most of the year, but disastrous storms, with winds of over 150 miles per hour (hurricanes), occur regularly in the autumn months. These storms have been important factors in the distribution of animals but not in their extinction. No islands (except Trinidad) were ever connected with the mainland. Their faunas are those of oceanic islands.

The following list illustrates the geographical distribution of the extinct birds.

EXTINCT

Bahamas—none
Greater Antilles
　　Cuba—*Ara tricolor*
　　Jamaica—*Amaurolimnas concolor concolor; Siphonorhis americanus*
　　Puerto Rico—*Aratinga chloroptera maugei; Amazona vittata graciliceps*

Lesser Antilles
　　St. Christopher—*Loxigilla portoricensis grandis*
　　Antigua ⎫
　　Nevis　 ⎭—*Speotyto cunicularia amaura*
　　Martinique—*Troglodytes musculus martinicensis*
　　Guadeloupe—*Speotyto cunicularia guadeloupensis; Troglodytes musculus guadeloupensis*
Trinidad—none

No doubt the concomitants of civilization are responsible for the disappearance of all, but it must be admitted that in no case are all the true reasons for the extinction of any of these forms known without question. Only hypotheses are possible, those based on what is known of the habits and habitats of the birds, their enemies, and certain comparisons with other extinct forms in other parts of the world.

The first obvious fact is that all are lowland forms living relatively close to civilization and therefore particularly subject to predation and loss of habitat. Only two (excluding fossils) belong to endemic, long-established, and relatively specialized species, of which there are 66 in the West Indies.

Naturalists who have lived in the islands and seen the birds [1] have outlined the probable reasons for the fate of these forms in somewhat the following manner. A rail, two owls, and two nighthawks, five of the twelve, nested on the ground; quite probably they were victims of the rats and mongooses introduced by man. A sixth, the small finch of St. Christopher (St. Kitts), is thought to have been too much disturbed at nesting time by a large band of monkeys (presumably *Cercopethicus aethiops sabaeus*), also imported by man. However, since another species of bullfinch (*Loxigilla noctis*) has been able to survive such disturbance on the island of Barbados, other unknown factors may have been involved.

A single dove and two parrots probably were shot and trapped too long and incessantly to allow them to survive on islands so heavily populated by human beings as Puerto Rico. The Cuban macaw, never recorded as numerous, probably suffered the same fate in spite of the fact that the human population of Cuba is relatively not as large as Puerto Rico.

Two wrens, formerly inhabiting Guadeloupe and Martinique in the Lesser Antilles, were also extirpated by some agencies for which civilization is responsible. No doubt rats and mongooses were at least partly responsible, for these birds also nested on or near the ground. However, the naturalist Frederick Ober noted (in Lawrence, 1879) that all birds were rare on Martinique and that they were much persecuted by a very large population of humans. He did not find the wrens near houses and towns as on other islands, the implication being that they had been extirpated from such habitat on Martinique.

This observation is direct evidence for a general rule for the West Indies that, with the exception of the Windward Islands, birds have been extirpated where human beings are many and forests relatively small. The chart on page 32 gives a rough indication of a ratio between available habitat and predation. It shows that on Hispaniola where there are 5.8 acres of forest per human being not one species has disappeared, although from those islands where less than 1 acre per human remains, two to four bird forms are now extinct. This is true in spite of the fact that all these islands have large populations of rats and mongooses, even including the Windward Islands of St. Lucia, St. Vincent, and Grenada, all three being exceptions to the general rule in that none of their birds have been extirpated.

---

[1] Gosse, Gundlach, Danforth, and especially Bond.

Witness of naturalists as well as common sense indicates that rats and mongooses were probably primarily responsible for the extinction of five ground-nesting forms. It appears also that both destruction of lowland forests and predation by human beings were probably not only contribut-

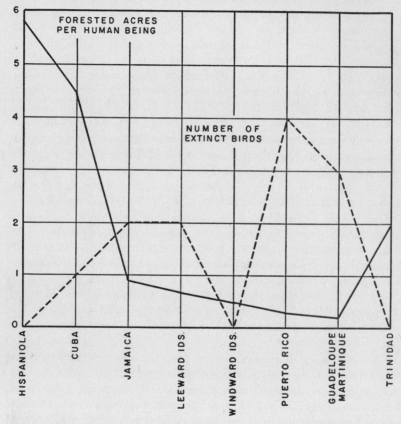

CHART 1.—An illustration of the relationship between the numbers of human beings, the extent of forest at about the time of the disappearance of certain birds, and the numbers of birds extirpated in historical times from the West Indies. Survival on the Windward Islands (St. Lucia, St. Vincent, Grenada) is thought to be a result of the disposition of men toward birds.

ing factors to the extinction of these but also the most important reasons for the extirpation of the wrens from Guadeloupe and Martinique. Why subspecies of those wrens still exist on St. Lucia, St. Vincent, and Grenada may be explained by the precipitous mountains and the attitude of the people.

First, the gradient of the mountains of the Windward Islands is much steeper at lower altitudes than others, making the land impossible to cultivate and leaving habitat for the birds which exists no longer on Guadeloupe and Martinique. Second, there can be no doubt that Ober was correct in his observations made just before the final disappearance of birds. They were much harassed on Martinique, but less so on the Windward Islands.

The number of human beings, regardless of forest, appears to be related to the number of extinct birds. The extinct pigeon and conure (*Aratinga*) of Puerto Rico were representative subspecies of forms common on neighboring Hispaniola. At about the time of their extinction the human population of Puerto Rico was 322 per square mile, while that of Hispaniola was only 4. Of the two Puerto Rican Amazonas, one is extinct (that of tiny Culebra Island) while the other is extremely rare. Species of that genus are common on less heavily populated islands.

It is also a fact that the wrens have disappeared from Guadeloupe and Martinique where there were, respectively, 536 and 478 human beings per square mile, rare on St. Lucia and St. Vincent (214 and 314), but quite common on Dominica, where there were only 99 persons per square mile in 1900.

As is most usual when glib rules are formulated for nature, another rude fact intrudes. Wrens were (and still are) not uncommon on Grenada, where the human population was 459 per square mile, and at the same time there is no evidence that the people treated the birds with any more harmless indifference than did those of St. Lucia and St. Vincent, where the birds are rare. Furthermore, the mongoose is quite as common on Grenada as it is on other islands, and there is no reason to think there are fewer rats. The only obvious difference between Grenada and its neighbors is that for many years cacao has been a favorite crop rather than sugarcane and cotton, but there is no good reason to believe that this would be an important factor in survival.

A second striking fact of the geographical distribution of extinct forms is that they are confined to the Greater and Lesser Antilles; none have been extirpated from the Bahamas or Trinidad. Survival of Bahaman species is more probably due to sociological factors and the absence of the mongoose than to zoogeographical position of the islands. These birds are derived partly form North America, and this applies particularly to those of the northern group, which is only 45 miles from Florida, and partly from Central America and the West Indies, some of which species arrived from Cuba and some were probably distributed by hurricanes. North American populations may well receive new recruits from the mainland sporadically, but it is much less likely that species such as the swallows (*Callichelidon cyaneoviridis*) and the hummingbird (*Calliphlox evelynae* and subsp.), derived from Central America, do so. It may be assumed that small numbers of human beings and their attitude toward birds have been the important factors that preserved those small populations.

Both Trinidad and Tobago are exceptions to the rule that the West Indies are oceanic islands. Both were evidently connected within rather recent geologic time to South America; their faunas are Venezuelan rather than West Indian. Even more important for survival, they are so close to the mainland that their bird populations may be replenished from time to time.

Except for these two most southern islands all the West Indies were populated by birds that flew, and mammals, reptiles, and insects arrived by various means over water. There are 66 endemic, long-isolated, and somewhat specialized species. No such extreme specialization is to be seen in recent birds as on Pacific islands; there are no flightless birds, for example. Nevertheless, it is remarkable that only 2 of the 66 endemics are extinct, when the West Indian fauna is compared with that of other oceanic islands. However, if the time limit is extended to include fossil forms, it becomes obvious that continuing processes of evolution and extinction were in force long before the advent of man in the West Indies. These islands are therefore no exception to the rule that specialized forms are vulnerable.

A list of 11 apparently endemic fossil birds includes three very large and undoubtedly specialized forms, a hawk and two barn owls:

KNOWN ONLY FROM BONES

Titanohierax gloveralleni Wetmore, 1937: Known only from Great Exuma Island, Bahamas.
Tyto ostologa Wetmore, 1922: Known only from St. Michel, Haiti, Hispaniola.
Tyto (?ostologa) pollens Wetmore, 1937: Known only from Great Exuma Island, Bahamas.

Of these forms Wetmore remarks (1937) that their span of life as species was probably coincident with that of their prey, a once ubiquitous rodent (*Geocapromys*) now confined to French Cay in the Bahamas, Jamaica, and Swan Island.

A list of 10 forms, still to be found in very small numbers, also contains three of the endemic subspecies:

SMALL POPULATIONS

Bermuda—Pterodroma cahow
Greater Antilles
    ?Hispaniola, ?Dominica; formerly Guadeloupe and Jamaica—*Pterodroma hasitata*
    Cuba—*Campephilus principalis bairdi*
Lesser Antilles
    St. Lucia—*Amazona versicolor versicolor; Ramphocinclus brachyurus sanctae-*
    luciae; Leucopeza semperi
    St. Vincent—*Amazona guildingi; Troglodytes musculus musicus*
    Martinique—*Cinclocerthia ruficauda gutturalis; Ramphocinclus brachyurus*
    brachyurus

The most specialized species are those of the genera *Ramphocinclus* and *Cinclocerthia,* but both of these have apparently suffered such marked diminution in numbers for the same reasons as have the wrens. However, they inhabit dense forests and have escaped extinction. Whether the parrots were ever common is not known; they also inhabit the higher mountain forests where they are protected. In neither case does it appear that their endemicity and relative specialization have been contributing causes for their rarity.

It has been shown that the rarity of the petrel of Bermuda is due directly to interference by men and by rats. Certainly also there has been very much predation of the high mountain petrel (*Pterodroma hasitata*) in the past. It may be hoped that both species may recover.

North America

Within the past 200 years the continent of North America has lost more bird species than any other land area of comparable size. This dreary fact is due directly to a relatively sudden onset by large human populations that made a practice of using wild birds as a staple and, at the last, commercializing them without restriction. Exactly the same result had been brought about on small islands of the Indian Ocean during the eighteenth century, and the difference in the two cases lies only in the size of the habitats and the populations of birds. Relatively greater numbers of enemies, unknown to and unexpected by them, equalized that difference.

Underlying the immediate cause for extinction was the essentially specialized nature of the species, and this phenomenon had its exact homologue in the Pacific islands. That the degree of morphological specialization, culminating in a loss of flight in some species, was less marked in most of the extinct American forms matters little, for both large continental populations and small island ones were handicapped in their efforts to maintain the birth rate above the death rate by quite peculiar traits. The endemic American passenger pigeons, because of their peculiarly gregarious habits, were no more able to protect themselves from the depredations of men were the confiding, flightless rails of Laysan Island able to withstand the rats.

In fact, all the species of birds that have been extirpated from the continent of North America were specialized endemics, just as were most of those of the Pacific Islands, the differences being only of degree. Indeed the great auk, flightless and nesting only on small islands, was in precisely the same evolutionary cul-de-sac as the rails of the Pacific islands.

The fossil record, unsatisfactory as it is, shows that species handicapped in like manner disappeared thousands of years before the arrival of Europeans in America. Such forms as the presumably flightless seabird *Mancalla* of Pleistocene, California, and the giant raptor *Teratornis* are part of the proof that extinction is, in some species, the result of a restricted channel of evolution.

Ranges of the rare and extinct species represent all the major geographical features of the continent: (1) small islands and coast of the Atlantic Ocean, (2) hardwood forests of the East, (3) the great plains of the midsections, (4) the jungles of Florida and Louisiana, (5) the mountains of the Pacific coast, and (6) Guadalupe Island, Pacific.

Atlantic Islands
Extinct:
> Great auk (*Alca impennis*)
> Labrador duck (*Camptorhynchus labradorium*)
> Heath-hen (*Tympanuchus cupido cupido*)
Small Populations:
> Ipswich sparrow (*Passerculus savannah princeps*)

Eastern Hardwood Forests (Atlantic coasts west to the Mississippi River)
Extinct:
> Passenger pigeon (*Ectopistes migratorius*)
> Carolina parakeet (*Conuropsis carolinensis carolinensis* and *C. c. ludovicianus*)
Small Populations:
> Wild turkey (*Meleagris gallopavo silvestris*)
> Ivory-billed woodpecker (*Campephilus principalis principalis*)
> Kirtland's warbler (*Dendroica kirtlandii*)

The Great Plains
Probably Extinct:
> Giant Canada goose (*Branta canadensis maxima*)
> Eskimo curlew (*Numenius borealis*)
Small Populations:
> Whooping crane (*Grus americana*)
> Lesser prairie chicken (*Tympanuchus pallidicinctus*)
> Attwater's prairie chicken (*Tympanuchus cupido attwateri*)

Northeastern coasts of North America, from Newfoundland and Labrador as far south as New England, are characterized by steep headlands and thousands of small granite islands. It is possible that great auks lived on many of these as far south as Maine, but the only known breeding place within historical times was tiny Funk Island, off the coast of Newfoundland, and the skerries off Iceland. Hungry fishermen killed them all.

So little is known of the Labrador duck that observations on its disappearance must be quite speculative, although it may be said that the lamellae of the bill may indicate specialized food habits, as Outram Bangs remarked. The reported rarity of the birds and their rather rapid extinction suggest a small island habitat and possible raiding by humans during the breeding season. It may well be that they nested on islets off the coasts as do eider ducks (*Somateria mollissima dresseri*) today.

What happened to the population of eider ducks in Maine is a good example of how uncontrolled shooting, as well as miscellaneous forms of molestation, can exterminate birds, although these were fortunately not all killed. According to Gross (1944), breeding eiders on the coast of Maine had been reduced to but a single pair by 1907. In that year the Audubon Society put a guard on the nest and began a propaganda campaign to protect the birds; shortly afterward a law was passed to prevent shooting this species. At present eiders may be found nesting on more than 50 barren islets off the Maine coast in the month of June.

In 1600, the approximate date of the arrival of Europeans in North America, the eastern half of the continent from the Mississippi River to the Atlantic Ocean, and from about the 47° parallel in southern Canada southward to the Coastal Plains of the Carolinas, was almost entirely covered with forest. It has been estimated that 431,000,000 acres were forested but that now only 19,000,000 acres remain, and it is not probable that more than a few quite small stands of timber are on land that has never been cut. Of course this was no impenetrable jungle—there are few places in the world where such a thing exists.

In 1635 William Wood said of the forest near Cape Cod Bay (Massachusetts): "Whereas it is generally conceived that the woods grow so thicke, that there is no more cleare ground than is hewed out by the labour of man; it is nothing so; in many places, divers Acres being cleare, so that one may ride a hunting in most places of the land, if he will venture himselfe for being lost: there is no underwood saving in swamps. . . . Of these swamps, some be ten some twenty some thirty miles long." Wood continues to say that the Indians burned the undergrowth in the month of November, and that where there were no Indians there was undergrowth, by which it appears that the changes wrought by man came long before Europeans.

In general the highlands were covered with oaks, hickories, and chest-

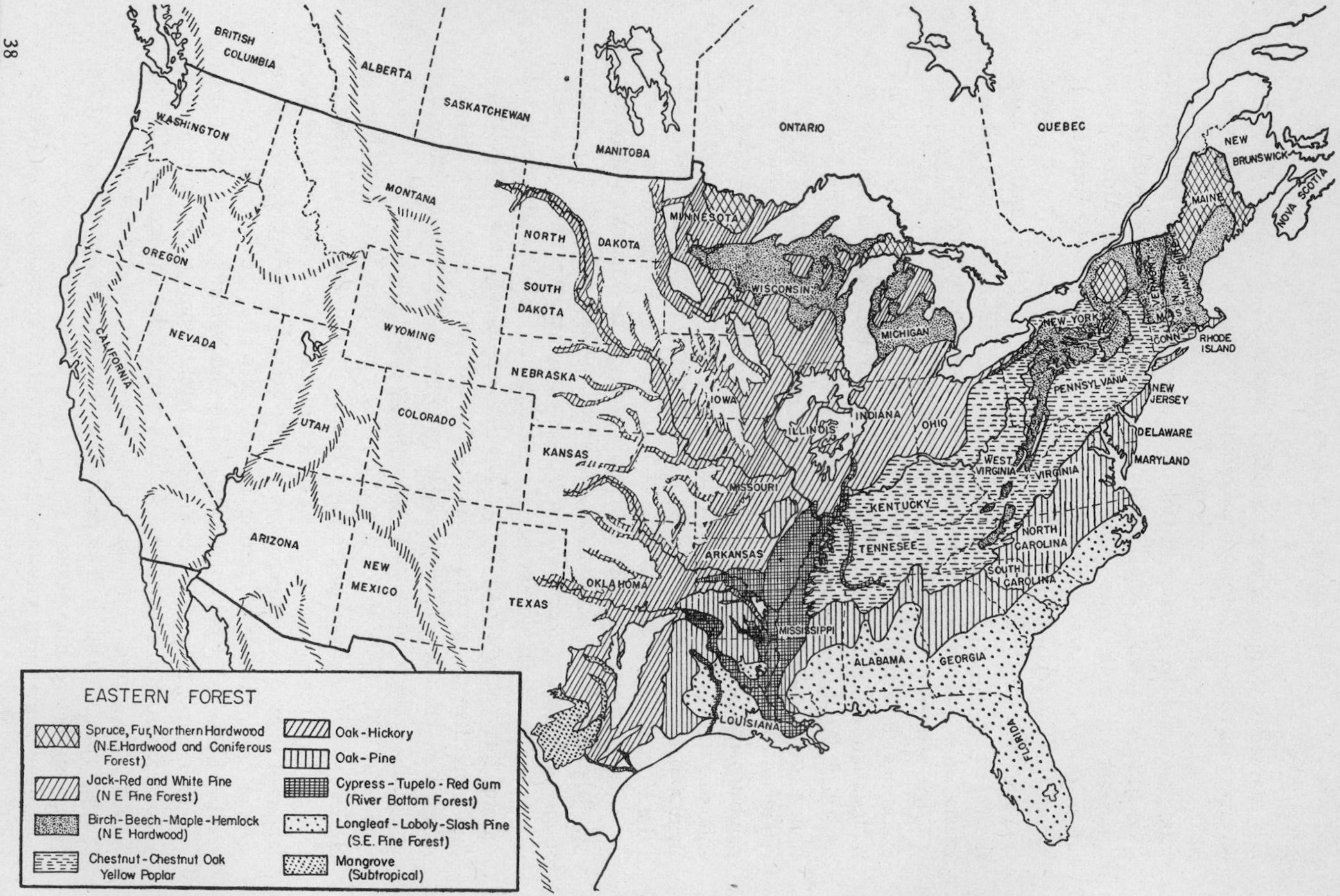

MAP 1.—Forest types of eastern United States

EASTERN FOREST

	Spruce, Fir, Northern Hardwood (N.E. Hardwood and Coniferous Forest)		Oak-Hickory
	Jack-Red and White Pine (N E Pine Forest)		Oak-Pine
	Birch-Beech-Maple-Hemlock (N E Hardwood)		Cypress-Tupelo-Red Gum (River Bottom Forest)
	Chestnut-Chestnut Oak Yellow Poplar		Longleaf-Loblolly-Slash Pine (S.E. Pine Forest)
			Mangrove (Subtropical)

38

nuts. The most common oak (*Quercus alba*) of those days is now rather rare, and the chestnuts (*Castanea dentata*), having suffered blight, never reach maturity. The sugar maple (*Acer saccharum*) and the red maple (*A. rubrum*) remain. Lower hills and valleys were characterized chiefly by maples and beeches (*Fagus grandifolia*). None of this original forest remains.

The history of the recent extinction of birds on the North American Continent is so closely related in time to the penetration of the continent by Europeans and their civilization that it is impossible not to believe that the one is the result of the other.

Destruction of habitat, loss of food supply, and finally persecution by men are undoubtedly the causes, and this last is the most important.

Because passenger pigeons wandered far in search of food and their appearances at any given place were sporadic, the dates of the last large colonial nestings best illustrate the timing of their disappearance. Breeding birds disappeared from the Atlantic coasts and from the more heavily populated midwestern regions between 1830 and 1860, from the more isolated and undisturbed places between 1885 and 1894.

On the northeastern seaboard the first permanent settlement of Europeans was established in 1620. Early records (Wood, Josselyn, Winthrop, et al.) agree that in certain seasons and certain years passenger pigeons were extremely abundant. The resident nonmigratory birds, the turkey and heath hen, were so common that, in the case of the latter, servants complained of being fed them too often. Even though they were staples of the colonists these species were not entirely extirpated from the region for 220 years. The last large colonial nesting of passenger pigeons in Massachusetts took place about 1851. The last heath hen was shot on the mainland of New England in 1830 (they lingered, by dint of introductions, on Marthas Vineyard Island for a hundred years longer), and the last wild turkey about 1850. It should be said, however, that these remaining stray birds were shot in the rocky hills remote from the coastal towns and cultivated river valleys. Forest habitat for birds had been destroyed in places more habitable for men years before; in fact premiums had been offered for growing trees as early as 1804.

Forests of the hills had been almost free from molestation until the decade 1790–1800. As settlers moved back into small, rocky hill farms, they often set fire to the woods, and much of them was destroyed as the human populations increased.

In 1754 there were about 24 acres per man in Massachusetts; in 1776 there were 17 acres, and in 1800 there were 11 acres. By 1830, when the last heath hen was shot on the mainland, there were only 8 acres, and by

1850, when the last wild turkey was gone, only 4 acres per man remained. Probably less than an acre of forest per man still stood at that time.

During the years 1830–1850 men moved from hill forests of New England (and the cities as well) in great numbers and traveled westward across the Appalachian Mountains and through the great hardwood forests of Ohio, by wagon, on horseback, and on foot. The first settlements in that State had been Indian and French, but those were few and little more than trading posts. In 1788 the first few English-Americans came from the east and south, but in increasing numbers. Their farms were small and domestic animals few; little of the birds' habitat was destroyed at first, but after 1820 a great freshet of migration arrived and increased in 1825, when a canal was dug, and 1834, when a highway was cut through from the east. The flood increased here and flowed on to new frontiers for 60 years.

After 1830 the Carolina parakeet was seen no longer in Ohio. The last great colonial nesting of passenger pigeons took place in 1855, and, although perhaps a few pairs nested until about 1893, the end came for the species when original forests were destroyed. It may be that a few wild turkeys could be found until after 1860, but it is also probable that such populations were reinforced by escapes from captivity. Human populations of Licking County, Ohio, where the final passenger pigeon nesting took place, increased at the rate of 60 percent during the 20 years 1820–1840. Estimations of destruction of bird habitat agree fairly well in that more than 60 percent of original forests had been removed before 1870, leaving but 4 acres of woodland per man in that year. Actually the situation of the birds was even more dangerous than such figures indicate, for the large trees such as beeches, chestnuts, and maples were quickly cut or burned away, and others, of no consequence to the birds, replaced them.

It is remarkable that these species were extirpated so quickly on the frontiers of the years 1830–1860, although they had survived in New England for more than 200 years.

Perhaps even more surprising it is that human populations were apparently less dense at the time. In Ohio there were about 13 acres per man in the year the pigeons ceased to nest, whereas there were only 4 in Massachusetts in the corresponding year. An answer to this lies in the fact that the populations of New England did not occupy the highlands to any great extent until after the Revolution, and by no means were they heavily populated even later as were the coasts and the river valleys. Thus at least a part of the birds' habitat remained relatively undisturbed. In 1850 there were about 18 acres per man in New Hampshire and 20 in Vermont.

Even though the large colonial nestings, which seem to have been necessary for continued successive breeding, appear to have ended in 1850,
the pigeons apparently nested in pairs and small colonies for 30 to 40
years longer, for the last recorded nest was found in 1890 in New England
and 1893 in Ohio. Here specialization as a predisposing factor is illustrated. The pigeons were unable to perpetuate themselves in pairs and
small colonies, after breeding the large colonies became impossible for
them.

On their westward migrations but few settlers wandered into the hills
and remained to farm the land. Such rough, unproductive regions as was
Potter County, in northern Pennsylvania, retained their forests and birds
long after they had been cut, burned, and extirpated from the fertile plains
of Ohio. There was a large roost and apparently many birds lived there in
the spring of 1886, and single pairs nested in Luzerne County in 1889.
At that time there mere as many as 29 acres per man, and it is said that
81 percent of Potter County remained forested (about 20 acres per man),
but it is doubtful that more than a few trees were of the original stand,
for timber had long been an industry in Potter County. By that time many
changes had taken place in the country to the northwest; the last remaining habitat was under attack and constant predation had taken its toll of
the birds.

Just as the westward human migrations passed south of the mountainous counties of Pennsylvania, so they passed by the northern States
bordering the Great Lakes. Open country and treeless prairies lying to
the southward and westward in Indiana and Illinois were more desirable
places for farmers than were the heavily wooded regions of northern
Michigan and Wisconsin. This tendency was more marked in the northernmost parts of both States. These sections, characterized as well by
evergreens as by hardwoods, were the last breeding places of passenger
pigeons. Large colonial nestings were seen in Michigan in an isolated
region of mixed hardwoods near the eastern shore of the lake in 1881,
and for the last time in 1886. A few pairs nested not far to the northward
until 1896, but the last bird was killed in 1898. Wild turkeys, which apparently never lived in the northern half of the state, were last seen in 1886.

During this time human populations of Michigan increased rapidly in
relation to available land. In 1850 there were 92 acres per man, 48 in 1860,
38 in 1870, 22 in 1880, 16 in 1890, and 15 in 1900. At that time there were
about 8 acres of forest per man.

Probably the jack pine, or Kirtland, warbler (*Dendroica kirtlandii*) is
in no great danger of immediate extinction, but because its breeding range
is confined to scattered pineland habitats within an area of only about

160 square miles in northeastern Michigan it should be mentioned. Its total population has been estimated at less than 1,000 by painstaking counts of singing males on the breeding ground (Mayfield, 1953, pp. 17–20). The species has attracted further attention because only one single bird has been seen on the wintering grounds in the Bahama Islands during the past 40 years, in spite of careful search by several professional ornithologists over the period. Nor have any birds ever been seen elsewhere than the Bahamas during the winter (Van Tyne, 1951, pp. 537–544). Naturalists found them "not uncommon" in 1880.

The probability is that this population is the remnant of a much larger one and that, for unknown reasons, it has undergone drastic reduction recently. Perhaps birds will be found on islands off the north coast of Cuba or the Turks Islands in winter. Certainly they were not molested to any extent either in the Bahamas or in Michigan in the past.

The State of Wisconsin, about 60 miles to the westward, across Lake Michigan, is physically similar to Michigan. The last great passenger-pigeon nestings took place in central and northern sections, characterized by small oaks and pines, at almost the same time. There were two quite close together in 1882, and a final rather smaller one in 1885 to the northward in Langlade County. The last wild bird was seen in September 1899. When it had been noticed first that the birds were becoming less numerous (about the year 1870), there were 30 acres per man in Wisconsin. There were 24 acres in 1880 and 18 in 1890. There had been not quite 12 acres per man in Ohio when passenger pigeons bred there for the last time. In Langlade County, Wis., there were as many as 59 acres per man at the time of the last colonial nesting, although there had been only 11 in Licking County, Ohio, and about 7 in Worcester County, Mass., when the birds ceased to breed normally in those places.

The difference no doubt lies in two facts. First, by this time most of the birds' natural habitat had been destroyed. By 1886 all the large hardwood stands of the southern counties had been cut except for small "wood lots" on the farms, and about one-fourth of the predominantly pine forests to the north had either been cut or burned. Obviously there was insufficient natural food for the birds. Colonial nestings, apparently so necessary for the continued existence of the species, were no longer possible, for not enough food could be found. This was their last stand. There was nowhere for them to go. Unfortunately there are no good records of the destruction of forests. However, an indication of the time element can be deduced from some details of the history of the lumbering industry that followed the movement of pioneers from east to west. On the Atlantic coast, 144 million board feet of pine, spruce, and hemlock were cut in the State of Maine in 1851. During the years 1870–1890 (apparently the crucial ones for the extinct species) the industry moved westward. Not quite 4 billion board feet were cut in the States bordering the Great Lakes in 1875. That figure was doubled in 1890, after which less and less was cut as the supply dwindled, and only four years later the passenger pigeon ceased to nest in the area, and nine years later the species was extinct.

The second and more important reason was the unbridled destruction of birds, not only on their feeding grounds but also in their roosts and nesting colonies. Having cleared away the trees and planted wheat or Indian corn, farmers of the decade 1860–1870 in these places were plagued by great flocks of pigeons. They shot them, netted them, ate some gratefully, and threw the rest to the pigs, if they could not sell them. Birds were so many that there was no thought of their extinction. But during those years the new railroads and the telegraph made it possible to commercialize the last vast flocks.

Schorger (1937) has estimated that at least 1,200,000 were killed by 600 professional pigeon netters known to have registered in hotels of the region. There may have been as many as 5,000 in the area in 1885, according to Brewster. These professional netters and dynamiters, aided by much improved transportation and communication, attacking the last possible breeding areas, probably account for the relatively rapid and total disappearance of the birds from a region much less heavily populated by humans than New England had been when pigeons had ceased to breed in numbers there.

So little evidence of large nesting colonies of pigeons in the Mississippi Valley east of the river remains to us that it may be said only that they probably ceased to do so before 1850. It was a few years later that requirements of wood-burning steamers began to cause much forest to be cut along the river banks. Commercial lumbering was apparently not undertaken in Kentucky and Tennessee to any great extent until 1865. Perhaps predation was more severe than now seems probable, but the fact is that their disappearance from this region so early is mysterious. However, it is most probable that the same sort of human interference drove them from that region, but records of the events are not to be found.

The great alluvial swamps of the Mississippi Valley, as well as the river valleys of the Coastal Plains from the Mississippi River eastward along the coasts of the Gulf of Mexico, on the peninsula of Florida, and northward through the Carolinas, were characterized by "river bottom forests." Tall cypresses (*Taxodium*), tupelos (*Nyassa*), and red gums rose from stagnant waters. Higher land surrounding them was covered, as most of it is still, by pines, but it was the great swamp trees that were the homes of the extinct Carolina parakeet and of the ivory-billed woodpecker, now reduced to a few pairs in northwestern Florida.

The Carolina parakeets, like the passenger pigeons, because of their habit of feeding in great numbers on cultivated plants, were a great nuisance to an increasing number of farmers and fruitgrowers, and they were shot whenever they appeared. Their disappearance was probably due to overshooting and to other interference, rather than to the destruction of their habitat. However, like the passenger pigeons, their complete extinction from Louisiana when there were still about 30 acres per human being cannot readily be explained. Possibly a great many more than has been supposed were captured for a lively cage-bird market. Certainly the birds roosted in hollow trees in large "colonies" and were on occasion taken by placing a bag over the hole, but there is little evidence of the extent of such activity. Although Florida was even more sparsely settled than Louisiana had been at the time of the bird's disappearance, considerable predation by man is assumed to be the direct cause of the extirpation of the species.

According to Tanner (1942) extinction of populations of the ivory-billed woodpeckers and reduction of the whole subspecies to a dangerously low level of numbers came about when forests were destroyed by lumbering speculations. He writes: "In many cases the disappearance of the birds almost coincided with the logging operations. In others there is no close correlation, but there are no records of ivory-bills inhabiting areas for any length of time after those have been cut over." Very little of such habitat remains. Original forests, almost continuous in the East, gave way to open parklands and small scattered prairies along the upper Mississippi River north of the 35° parallel in the States of Illinois, Kentucky, and Iowa. Large trees persist to the westward only in the river valleys. Clumps of oaks grew on the uplands, which for the most part were covered with coarse grass. Here began the country of the buffalo and the prairie chicken, and this extended westward over almost treeless plains to the Rocky Mountains.

The Eskimo curlew and the whooping crane appear to have suffered to the point of extinction by the increase of human habitation in spite of the fact that they were to be found in those regions only as migrants in spring and fall. Although it must be said that nothing is known of what may have befallen the curlews on their breeding range in the tundra bordering the Arctic Sea, it is probable that their extirpation was due largely to overshooting on their migration route, but perhaps especially at its narrowest point, where the North Platte River crossed it.

About the year 1870 all eye-witness accounts [1] of the northward migrating flocks of Eskimo curlews on the Great Plains of North America agree to their often enormous size. During April and May farmers and shooters, from Louisiana and Texas on the coast and northward through Kansas and Nebraska, gathered about newly planted fields, watched the birds circle, plane, and land. Close-packed feeding birds were shot by the thousands, and more could be bagged when the survivors returned as they always did. Even though we accept accounts of those gunners with mental reservation, there is no doubt that thousands of birds were shipped to market in the towns. These wonderful shoots lasted for about 30 years, but by 1900 the flocks were very much smaller; to see 75 birds together was remarkable.

During those 30 years, 1870–1900, the railroads were completed and a rapid increase in the human population of the Prairie States just west of the Missouri River took place. Their numbers in central Kansas and Nebraska, where the migrating birds were most concentrated, had increased from 2 persons or less per square acre to 16 or even 45 in the year 1900. There were more than three times as many people in those States. In Buffalo County, Nebr., about the center of this area, there were 2,850 acres per man in 1870. During the next 20 years the human population increased at the rate of 94 percent, and there remained only 22 acres in 1890.

The concentration of birds west of the Missouri River between the 97th and 98th parallels has been noticed by farmers, shooters, and ornithologists alike. Indeed all authorities have recorded them to have been very rare west of the lower Mississippi and the Missouri Rivers. Perhaps then it may be assumed that the vast majority of the curlews passed that way, for although a few may have taken a more westerly route, it seems unlikely because of the relative scarcity of food. No birds have ever been recorded from Colorado, and the only indication that any passed to the westward of the usual roads of migration is a single specimen collected at Lake Palomas on the borders of New and Old Mexico on April 8, 1892. [2]

It appears that when the eager gunners of those years saw a field of 50 square acres covered with Eskimo curlews, as has been recorded, they had a considerable portion of the whole species before their eyes. If this were so then it would not be impossible for the population to have been so reduced that a recovery could not be made. Add to the numbers killed in North America those that were shot in Argentina, about which noth-

[1] The excellent account of Swenk (1916) should be consulted for exact references.
[2] Bailey, F. M., 1928.

ing is known, and a fair explanation for their disappearance emerges. They were curious and unwary. They could not be protected. Too many were shot.

Through these same regions the whooping cranes migrated from winter quarters on the Gulf coast north to the marshy plains of Canada. No doubt these populations, so much fewer in numbers, suffered considerably on migration. Indeed the record of birds actually reported as killed, which is quite certainly incomplete, shows that 66 percent of the birds that died were killed on migration, and the majority of these in Nebraska, according to R. P. Allen (1952). Furthermore, the greatest numbers were reported killed during the years 1880–1900, during the rapid expansion of population there. It was in these years that they ceased to breed in North America.

This is not to say that the cranes that migrated along the Missouri River to the eastward escaped persecution, or that they were not shot during the winter in Louisiana and Texas. Allen estimates that even as late as 1939, 33 percent of the total population of 39 was shot in Louisiana.

The astonishing fact is that cranes survived such heavy shooting pressure, and that it was not until the great Canadian marshes were drained that the population reached the point of extinction. However, the reduction of nesting areas and consequent diminution in numbers had begun 40 years before in Illinois presumably by the draining of marshes as well as by overshooting, egg collecting, and disturbance in the nesting area. Here at about latitude 41° N., on both sides of the Mississippi, were hundreds of square miles of shallow lakes and reedy marshes surrounded by straggling aspens. The grasslands extended to level horizons. These were presumably the center of the breeding populations. With the encroachment of men the birds disappeared. They ceased to nest in North Dakota in 1884 and in Minnesota in 1889, and the last nest in this region was found in Iowa in 1894. However, a very small number are known to have bred in similar country far to the north in Alberta until 1922. Where the three or four remaining pairs nest now is not known.

According to Delacour (1951), a population of very large Canada geese (*Branta canadensis maxima*) once lived on these plains, from North Dakota eastward to Michigan and south to Tennessee and Missouri. Males of this race weighed as much as 18 pounds. Occasionally smaller birds breed in this region still, but it is probable that the original population is extinct.

There is a curious parallel course in the history of the cranes and the passenger pigeons: the last nests of both to be found in the United States were found in the same year, and in both events the human population had reached almost exactly the same density. There were 18 acres per man both in Iowa and Wisconsin in 1890. The identical numbers are, of course, coincidence.

Beyond the prairies lie the Rocky Mountains, and on their western slopes, not far from the Pacific Ocean, the last few California condors nest. Only about 60 of them are left to breed on the sheer cliffs of the canyons among the San Rafael Mountain range, and occasionally in the forests of giant redwoods to the north. Like the whooping cranes they are remnants of species that were much greater in numbers 10,000 years ago, when there were more mammals and presumably more carcasses for them to feed upon. Certainly they were more numerous a hundred years ago when the Spanish cattle ranged California; and perhaps after 1846, when Americans took possession of the country, and the land was farmed, or groves of fruit trees supplanted cattle ranges, the birds were deprived of food again. It may be that their number was much increased during the seventeenth and first half of the eighteenth century by the introduced cattle, which died in open country, whereas the corpses of deer were hidden in the forests.

With the increase in human population, construction of roads, and the great improvement of the rifle, the birds have been sadly depleted in numbers. Shooting indeed has probably been the most important factor, but disturbance of all kinds, such as the building of roads, and nearness of vehicles and even the cautious approach of parties of bird-watchers, appears to agitate them and probably impedes breeding activity. The closing of critical areas may have the population but, as in the case of the whooping crane, the birds cannot be guarded at all times. We may hope.

Guadalupe Island (lat. 20° N., long. 118° 22′ W.), when compared to North America, illustrates the extreme vulnerability of small island faunas. It lies 140 miles off the coast of Lower (Baja) California and 180 miles southwest of San Diego. The island is about 20 miles long by 6 in width, and the central ridge is 4,500 feet above the sea. At the northern end of this ridge there was a large grove of endemic cypress trees (*Cupressus guadalupensis*), and on the old lava flows of its steep sides the white sage (*Senecio*) and pines (*Pinus radiata*) flourished. The bush has almost disappeared and the groves of trees are thinned and dying because of an enormous herd of hungry goats, estimated between 40,000 to 50,000 in number. It is not known exactly when these were introduced, but it is thought to have been long ago. These groves were the habitats of three subspecies of small birds of Rocky Mountain origin—they are now extinct. Indeed, 39 percent of the breeding bird fauna of the island has been extirpated by the destruction of habitat and by introduced house cats which infest the island.

EXTINCT

PETREL (*Oceanodroma macrodactyla*)
CARACARA (hawk) (*Polyborus lutosus*)
FLICKER (woodpecker) (*Colaptes cafer rufipileus*)

WREN (*Thryomanes bewickii brevicauda*)
TOWHEE (*Pipilo erythrophthalmus consobrinus*)

The Hawaiian Islands

The Hawaiian Islands stretch in a long chain across the middle of the North Pacific Ocean for 1,578 miles in the direction west-northwest to east-southeast. The most westerly island is Ocean (Kure), about 3,000 miles from Tokyo; and the most easterly is Hawaii, about 2,000 miles from San Francisco. Ocean Island, and also Midway, Pearl and Hermes Reef, Lisiansky, Laysan Island and other islands, are low, coral-limestone atolls, none more than 40 feet high. Next to the eastward are the tops of four extinct volcanoes rising 120 to 895 feet above the sea; these are Gardner, French Frigate, Necker, and Nihoa. No men or any imported animals are to be found on them. Finally there are the eight large, mountainous, and forested islands, the Hawaiian group proper. These are Niihau, Kauai, Oahu, Molokai, Lanai, Maui, Kahoolawe, and Hawaii. Lowlands of all are heavily populated by men, although less so than in the West Indian islands, only Oahu and Lanai having less than 1 acre of forest per human being.[1] The forested mountains, scored by deep ravines, rise to 5,170 feet on Kauai, 4,025 on Oahu, 4,970 on Molokai. These peaks are the eroded remains of extinct volcanoes, and the three western islands are thought to be older than the two most easterly. Maui possesses a great quiescent volcano 10,032 feet high. The largest island, Hawaii, has three great mountains, a range 5,505 feet in the north, a recent volcano of 8,269 feet in the south, and between them two enormous mountains that rise gradually from the sea to 13,784 feet in about 50 miles. The highest is "Mauna Kea," White Mountain, so called because of its winter snows.

The forests that grow upon their sides are as extraordinary as the birds that live in them; here, there are more peculiar, special native species than anywhere on earth. Because the climate is controlled by regular northeast trade winds as well as the height of the mountains, tree species replace one another not only with altitude but also according to rainfall,

---

[1] This is only a rough estimate. I have not been able to find a good estimate of forested acres of any of the Hawaiian Islands.

which is extremely heavy on the windward sides. On the lee of the mountains the character of the country varies considerably. In one or two places on Hawaii the trees descend from 6,000 feet almost to sea level— on Molokai and Oahu not so; indeed the lowlands are extremely dry. Native trees grow sparsely if at all there, but there are still great numbers of species though not of individuals, and in places land shells, now extinct, have been found. For these reasons it appears that forest growth in the lowlands was much heavier and more extensive at some remote period, but whether trees were destroyed by the Hawaiians since the twelfth century or by natural causes before their arrival has not been determined surely. Many naturalists [1] have stated that they were all destroyed by Hawaiians between the twelfth century and the arrival of Captain Cook. No doubt Hawaiians did burn much forest; it is the usual practice of colonists. However, recent investigation of tree spores in mountain bogs indicate that there have been changes of climate and rainfall during several thousand years past. Such theory might account for the great rarity and localization of certain trees such as *Sophora unifoliata* and *Acacia koa,* and the extinction of about two dozen forms of very distinct land shells.

Three-fourths of the forests of the islands are now to be found between 2,000 and 6,500 feet above sea level, and it is in this region that the native birds are now. On the windward sides of the mountains the rainfall is the highest in the world. Even on the lee sides the weather is often foggy. Dense wet undergrowth of ferns makes walking difficult, and the trees are wound about with air plants and vines. Forests of fern trees are fairly common on Hawaii, but the characteristic tree of the islands is the ohia-lehua (*Metrosideros*), whose thin, black, and twisted trunks are topped with umbrellas of silky scarlet flowers. Since the coming of Europeans much of such forest has died and many of the native birds have gone with it.

In fact, an astonishing total of 60 percent of the 68 known forms of native land birds are either extinct or listed here as "probably extinct" (because of lack of exact knowledge of their status), or exist still in such small populations that they are thought to be in danger of extinction. They are listed below in these three categories.

EXTINCT

Western atolls (Laysan chain)
    Laysan Island—*Porzanula palmeri* (also Midway, see text); *Acrocephalus familiaris familiaris; Himatione sanguinea freethii*

---

[1] Schauinsland (1899), Hall (1904), Zimmerman (1951); for contrary evidence see Selling (1948).

Middle chain of small volcanic islands—none

Large mountainous islands (*Hawaiian Islands proper*)

Oahu—*Phaeornis obscurus oahensis; Loxops coccinea rufa; Hemignathus lucidus; Hemignathus obscurus lichtensteinii; Moho apicalis*

Molokai—*Drepanis funerea*

Lanai—*Hemignathus obscurus lanaiensis*

Hawaii—*Pennula sandwichensis; Drepanis pacifica; Ciridops anna; Psittirostra [Rhodacanthis] palmeri; Psittirostra [Rhodacanthis] flaviceps; Chaetoptila angustipluma*

In addition to these, three species and seven subspecies are quite probably extinct, although it is still just possible that very small populations may still be found.

PROBABLY EXTINCT

Molokai—*Phaeornis obscurus rutha; Loxops (Paroreomyza) maculata flammea; Moho bishopi*

Lanai—*Phaeornis obscurus lanaiensis; Loxops (Paroreomyza) maculata montana*

Maui—*Hemignathus lucidus affinis*

Hawaii—*Loxops (Viridonia) sagittirostris; Hemignathus obscurus obscurus; Psittirostra (Chloridops) kona; Moho nobilis*

It is quite possible that certain forms of Kauai are now extinct, but so little is known of the island and the possibility of survival is relatively so much better that they are included in the following list.

SMALL POPULATIONS (PROBABLY IN DANGER OF EXTINCTION)

Laysan Island—*Anas platyrhynchos laysanensis*

Kauai—*Phaeornis palmeri; Phaeornis obscurus myadestina; Hemignathus obscurus procerus; Hemignathus lucidus hanapepe; Moho braccatus*

Molokai—*Palmeria dolei*

Maui—*Loxops coccinea ochracea; Pseudonestor xanthophrys; Palmeria dolei*

Hawaii—*Branta [Nesochen] sandvicensis; Psittirostra [Loxioides] bailleui; Hemignathus wilsoni*

In addition *Puffinus newelli,* which breeds on Kauai, and *Anas platyrhynchos wyvilliana,* which breeds on Niihau, Kauai, islets off Oahu, ? Molokai, Maui, and Hawaii, should be mentioned. *Psittirostra psittacea* has been extirpated from Oahu and Lanai and is becoming increasingly rare elsewhere.

It is apparent that birds have been extirpated from the islands that are heavily populated by human beings and upon which animals dangerous to native birds have been introduced. No birds have disappeared from uninhabited islands—as far as is known. That the impact of European civilization has been sudden and terrible cannot be doubted. For example, the islets of Laysan and Midway, farthest of the long chain northwest of Honolulu, are the only ones of that group from which birds have been extirpated; also they are the only ones that have been inhabited by men for long periods. Once there were five native birds—a duck, a rail, a warbler with Asiatic relations, and two Hawaiian creepers; only the duck and one of the creepers remain. The facts are clear. These were destroyed by rabbits that ate all vegetation on Laysan, so depriving the birds of habitat, and by rats, which ate eggs and young birds on Midway.[1]

Forty years after Laysan was permanently settled and about 25 after the introduction of rabbits, three-fifths of its birds had disappeared. Norway rats, accidentally coming ashore from vessels during the war of 1941–1945 and finding the place to their liking, killed all the flightless rails on Midway in two years.

Of the larger, more mountainous, and more heavily populated islands, Lanai illustrates the point. It contains only 141 square miles. In 1890 there were 348 humans and six subspecies of native land birds there. By 1940 there were 3,720 people, but more than half the birds had disappeared, for not only had three subspecies been extirpated but also the island population of a species still to be found on most of the other islands was extinct. In the meantime pineapple farmers had built a model city, cultivated the lowlands, and contrived to destroy all but a few acres of the original forests.

Quite probably events on other islands of the group were similar to these. No doubt if enough information were available to devise a time chart of the destruction of forest habitat by cattle, swine, sheep, and goats, showing exactly where and when woods have given way to grasslands, it could be shown clearly that certain birds have disappeared concurrently with forests. Likewise, if the distribution of introduced rats and cats were known in time, as well as space, their effect of bird populations would be demonstrable.

We do know that the hoofed animals have been the means of forest destruction at all altitudes in the mountains, but particularly at lower altitudes. We know that cats and rats are destructive, but they do not frequent the rain forests above 3,000 feet. We know that native birds have disappeared from the forests of lower altitudes (where such are still to be found) or are extremely rare there. Naturalists who have worked on the Hawaiian Islands have made it plain that these ugly handmaidens of civilization should not be underrated as destructive forces. All combined, or even one alone, may have been the cause of the loss of species small in numbers.

Without doubt Eurasian tree rats got ashore from ships early in the

---

[1] See Wetmore, 1925; Fisher and Baldwin, 1946.

nineteenth century. They have a theoretical breeding potential of 800 per pair per year, and they avidly devour young birds and eggs. They were at first a danger to lowland birds and later to those of the mountain forests. Brown rats were reported by Peale, naturalist of the United States Exploring Expedition in 1840. Since the roof or tree rats almost always precede the brown, they were both probably quite common then. They occur now on all larger islands. Perkins (1903) said of them in the decade 1890–1900: "The imported rats now abound in many parts of the forest and lead largely an arboreal life, feeding on such as are to be had, especially that of the Ieie (*Freycinetia*) and the mountain apple and the brightly colored arboreal mollusks of the genus *Achatinella* and the duller ground-frequenting *Amastra*." And again, in a discussion of the rarity of birds, he says: "Now over extensive areas it is often difficult to find a single red Ieie fruit, which the foreign rats have not more or less eaten and befouled." Today dead *Achatinella* shells can be found in great numbers 2,500 feet above the sea—killed and eaten by rats. *Achatinella* is becoming somewhat scarce, and *Amastra* can be found only in crevices on the cliffs. Flowers of the *Freycinetia* vine were favorites of the extinct nectar-feeding birds. If the rats nested within the birds' range they might well have extirpated small populations of very restricted range and low altitude. However, it is impossible to account for the survival of the remaining species of birds now, so long after the extinction of others. The black mamo, so prized by Hawaiians for its yellow feathers, was not uncommon in the higher mountain forests of Molokai 50 years ago. The status of the apapane (*Himatione*) was the same. Today the former is extinct, the latter still to be found, from which fact it is apparent that there are other factors to be reckoned with in the process of extinction.

The effect of the introduction of "domestic" cats by Europeans on bird populations is difficult to estimate, for even a guess at the numbers that have taken to the mountain forests has never been possible. That they have done so is a fact, but because they are nocturnal and take to the trees when alarmed, they are almost never seen. Neither is there any record of the years in which they were introduced on the islands, but no doubt they came with the first permanent white settlers. Following the pattern of successful acclimatization of foreign animals, they would have bred to numbers above the point of subsistence after a time, and during this period would naturally be a great danger to birds. This period would have been succeeded by a great drop in numbers after which populations would not be so large that they would be likely to extirpate an entire species unless it were very small in numbers.

Perkins, who spent ten years in the forests, has this to say: "Cats were introduced into the Hawaiian Islands at a very early time, and, no doubt, increased excessively, while, as their owners moved from place to place, many strayed into the woods and began to feed on mice, rats, and birds. They are now found wild on all the islands, apparently only the wettest portions of the forest being free from them. On Lanai, in walking up a single ravine, I counted the remains of no less than twenty-two native birds killed by cats, and these must all have been destroyed within two days, as previously the whole gulch had been washed out by a heavy flood. Two cats were actually shot on this occasion as they were devouring their prey, and several others seen, but, owing to the fact that they are extremely shy and most nocturnal in habits, few people who have not lived much in the woods have any idea of their numbers."

Because the mongoose (*Mungos,* presumably *burmanicus*) was not introduced into Hawaii from Jamaica until September 1883, it is not probable that it was responsible for the extirpation of any bird species or subspecies. The Hawaiian rail or moho (*Pennula*), a ground-nesting bird vulnerable to mongooses, probably disappeared from Hawaii long before, since none of the collectors of the years 1880–1900 were able to obtain one, and in fact no specimen was taken later than 1864. If, as Munro (1944) states, there was a form on Molokai and perhaps other islands, it is much more likely that rats extirpated it.

A few Hawaiians trapped native birds for the manufacture of cloaks and helmets covered with feathers. It was often stated by writers and lecturers of 50 years ago that species were extirpated in this way. Actually Hawaiians are said to have used six species for this purpose. They were the mamo (*Drepanis*), the o-o (*Moho*), the iwa (*Hemignathus*), the iiwi (*Vestiaria*), the ou (*Psittirostra*), and the apapane (*Himatione*). To be sure, the first two are extinct. The third, although no longer to be found on Oahu and other islands, still exists on Hawaii. The fourth is rare now but was quite common during several decades after the last feather cloak was made; and the last two are still quite common in spite of the fact that their red feathers were much prized. It is not probable that this trade was any more than a minor limiting factor upon populations. A second theory is that the royal "tabu" put upon birds by the kings was removed during the last years of the nineteenth century and that many birds were killed by the people in consequence. So difficult is the terrain and so disagreeable the climate at the altitudes where the birds were to be found, and so small are they, that it does not appear to be probable that the people destroyed great numbers. By that time it would have been far more profitable to trap the cattle, goats, and pigs that roamed the mountains, as indeed they did. W. A. Bryan (1908), although he gives some credence to these hypotheses, points out that on Kauai and Molokai they (*Moho*) were dying out "apparently of their own accord, or at least from other unassignable causes."

Theoretically certain species may well have been exterminated by diseases transmitted through imported birds, of which there are many. Attractive as this hypothesis is—Munro (1944) and Amadon (1950) consider it of great importance—there is unfortunately little evidence to show that such has been the case, for few have worked on the problem. Amadon has summarized what is known. Both Perkins and Henshaw record having found birds with swollen, distorted feet, legs, and heads and in one case the bird was quite unable to fly. Considering how few birds are picked up dead in their native habitat, this evidence is important, especially since the cause was thought to be "bird pox," a disease known to be epizootic and dangerous to populations. However, there is no real evidence that bird pox was actually the cause of death. The second disease which has been considered as a possible agent of extinction is avian malaria. Recently the parasite *Plasmodium vaughani* has been found in the blood of the Pekin robin (*Leiothrix lutea*), an importation from Asia, now common in Hawaiian forests. However, the parasite has not yet been found in the blood of native species. Bird diseases should be a promising field for research.

According to Perkins (1903) the loss of lowland forests was "a most efficient cause of the destruction of native birds." All the true reasons for the almost total loss of these woods of sea level are not known. However, trees of the middle or rain forests, where three species of native birds are yet to be found, have long been dying and are dying still because of the ravages of pigs, probably imported by the Hawaiians, as well as goats, sheep, and cattle put ashore by Captain Cook and Captain Vancouver between 1778 and 1794. This they did by damaging the roots of the native air plants, by hardening the earth about them, thus depriving them of air, by destroying all undergrowth, and, worst of all, by killing the young trees. Much forest at lower levels has been cut down by Europeans in order to plant sugar, pineapples, and coffee. The distinguished botanist J. F. Rock (1913) described such a place. "To windward the mountain (Mauna Kea) slopes rather gently, forming the Hamaku coast, which at the lowlands has been planted with sugar cane exclusively up to an elevation of 2,000 ft. . . .

"Between 2,000 and 3,000 ft. elevation the forest has disappeared and only stragglers of tree ferns can be found standing, though ten times as many are lying dead on the ground and overgrown with all possible weeds, which the ranchmen have imported with their grass seeds. . . . At 3,000 feet a few Koa Trees can be found, together with Naio, and here also was found a single native palm, *Pritchardia sp.,* windswept and half dead. If one considers the natural condition in which this palm flourishes, as for example in the dense tropical rain forests in Kohala . . . it stands a witness to the fact that there surrounding it was once a beautiful tropical jungle." None of the forest birds remained in such a place.

All through the foregoing account of depredation by accompaniments of European civilization appears the probability that small populations are especially vulnerable to imported predators. The subtler, underlying (and perhaps most important) fact is that many of these extinct species were rare when civilized men discovered them. Such were *Chaetoptila angusti-pluma,* the large green honeyeater, and *Loxops (Viridonia) sagittirostris* of Hawaii. Certainly also *Ciridops anna,* the small gray and red bird, was rare in 1890, but according to Perkins its range had formerly been much greater. We have no information to the effect that these were any more vulnerable to rats than species that survived longer. It appears, therefore, that these very distinct and presumably specialized species had been disappearing from regions quite untouched by civilization and before the arrival of foreign predators. This is probably also true of *Psittirostra* [*Rhodacanthis*] *palmeri* and *flaviceps* as well as the famous mamo (*Drepanis pacifica*).

In addition to these very specialized and long-established species of Hawaii, two subspecies of more widely distributed species had disappeared from Oahu before European civilization is likely to have affected the lives of birds to any extent. These were *Phaeornis obscurus oahensis* and *Hemignathus lucidus lucidus.* In 1850 there were less than 90,000 human inhabitants on all the islands. The few hundred Europeans were concentrated in the towns of Honolulu on Oahu and Hilo on Hawaii, and it was not until 1852 that Asiatic laborers were imported. Sugar plantations were still insignificant and the pineapple industry did not exist.

Following is a timetable of the birds' disappearances. Because of the probability of a lapse of time between the last report and the death of the last bird, the decade following the final sighting of a given species is shown.

1830–1840	Oahu (1)		1910–1920	Oahu (1)
1840–1850	Oahu (3)			Molokai (4)
1860–1870	Hawaii (1)			Maui (1)
1890–1900	Hawaii (1)		1930–1940	Lanai (2)
1900–1910	Lanai (1)			Hawaii (1)
	Hawaii (7)			

MAP 2

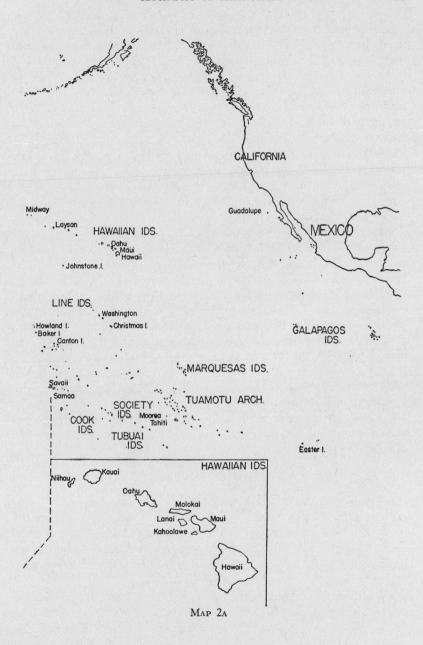

Midway
• •

• Laysan

HAWAIIAN IDS.

• Oahu
Maui
Hawaii

• Johnstone I.

LINE IDS.

• Washington

Howland I.
• Baker I
• Canton I.

• Christmas I.

CALIFORNIA

Guadalupe •

MEXICO

GALAPAGOS
IDS.

• MARQUESAS IDS.

Savaii
Samoa

SOCIETY
IDS. Moorea
Tahiti

TUAMOTU ARCH.

COOK
IDS.

TUBUAI
IDS.

Easter I.

HAWAIIAN IDS.

Niihau   Kauai

Oahu

Molokai

Lanai   Maui
Kahoolawe

Hawaii

MAP 2A

In addition to the time factor in some of the birds' disappearances and the apparently inexplicable rarity of others, a third fact must be reckoned with. It is that all 26 extinct forms were highly specialized in form and in habit. Fifteen belonged to the family Drepanididae, confined to the Hawaiian Islands. Ten were very distinct species. Theoretically they had been predisposed to extinction. This phenomenon is closely bound to the evolution of the entire fauna, an excellent discussion of which is by Amadon (1950).

The ancestors of these species, immigrants of long ago, were forced into very specialized food-hunting habits, often under severe competition from other species, and since the habitats offered by the small islands were not large or varied, the numbers of birds were severely limited. Biological theory holds that the evolution of small populations, accidentally isolated as these were, is characterized by rapidly inherited changes often giving rise to quite new forms. Inheritance of certain characteristics is often of a random and accidental nature, and the creatures thus born may have been cast in a mold of form not at all well suited to their surroundings, and this is accentuated when the birds are few in number. Where populations are larger—on continents where the great rivers of evolution flow strongly—new forms of this sort are not possible, because individuals with peculiar new characteristics seldom live to breed, or, if the new genetical strain continues at all, it disappears after a few generations. But on islands small parties of new arrivals have formed new species, and these in certain cases have through these special processes of evolution become extinct. Evolution and extinction are here one and inseparable. When more competitors and more predators are suddenly added to the powerful agencies already controlling the numbers of individuals, such misfit populations disappear rapidly.

Pacific Islands

All who have seen Pacific islands have divided them into two classes: the low coral-limestone atolls, homes for sea birds, as against high volcanic peaks, covered with forest, affording habitat for small numbers of land and fresh-water birds. Sea birds have generally survived while land birds have not, and so we are concerned principally with the high islands and their birds.

All such emigrated originally from Asia, Australia, and New Zealand and by flights of about 600 miles or often less, and by varying routes from island to island, reached archipelagoes that suited them and remained to breed. These mass movements appear to have taken place from time to time, often separated by long periods. Why such displacements have taken place is not known, but that they do sometimes cannot be questioned. Appearance of small British perching birds, introduced into New Zealand, on the satellite islands, and extension of range of a pigeon (*Streptopelia decaocto*) in Europe and the cattle egret (*Ardeola ibis*) in America within the past 20 years are examples.

Whether by the peculiar genetical constitution of the first immigrants, by the press of evolution in new surroundings, or, as apparently often happens, by removal of pressures of continental existence, the progeny are differentiated physically and often in other attributes. As this process continues in geographical isolation, the populations often reach a point of sexual isolation from their one-time relatives of the continent, and at that point they become species. Obviously those that reach this point can no longer benefit genetically by immigration from large land masses should such occur.

In consequence of such isolation, many characteristics acquired during generations on islands free from predators, and even competitors, have rendered them helpless to protect themselves when at last confronted by unexpected enemies. Especially the island rails have so suffered, for many of them have lost their powers of flight.

Sudden and disastrous as the onset of civilization has been everywhere, it has been the more distant most isolated populations that have suffered most.

Geographers, ethnologists, and zoogeographers [1] have subdivided the islands further according to their special interests. Such classifications are always somewhat arbitrary, and usually according to the distribution of the races of men in the islands: Melanesia, Micronesia, and Polynesia.

In order to illustrate the geography of the extinction of these birds, Pacific islands may be likened to arcs [2] roughly parallel to the coasts of Asia, distributed in enormous festoons across the Pacific from west to east for 5,000 miles from Asia to the Gambier Islands, beyond Tahiti.

Arc I: Asiatic Continental Shelf

Closest to the continent of Asia, this arc begins at the tip of the Kamchatkan Peninsula of Siberia, at latitude 50° N., with the relatively small

---

[1] See Mayr (1933, 1941b) who shows that the western islands are richer in species than are the eastern.

[2] Modification of "Pacific Islands," Encyclopedia Britannica, ed. 1948, vol. 17. The Hawaiian Islands and Aleutian Islands are treated separately because their avifaunas are of North American origin for the most part.

Kurile Islands, and continues southward through the large land masses of Japan, Formosa, the Philippines, and New Guinea, New Caledonia, and New Zealand. Except for New Zealand, none of these "almost continents" have had any form of bird extirpated from them within the past 200 years, as far as we know. They are connected by groups of smaller islands: the Riu Kius between Japan and Formosa, and the Solomons, Santa Cruz, New Hebrides, and Loyalties between New Guinea and New Caledonia. Norfolk and Lord Howe Islands are isolated about 600 miles to the south. It is from these smaller islands that birds have recently been extirpated.

ARC I

	Total species	Endemic species	Extinct species	Extinct forms	Area (sq. mi.)
Bering Island ⎱ Kurile Islands ⎰	71	1	1	2(?)	6,165
Riu Kui Islands	84	4	1	1	935
Solomon Islands	126	40	1(?)	1(?)	13,000
New Caledonia ⎱ Santa Cruz ⎬ New Hebrides ⎰	—	—	0	0	—
Norfolk Island	15	0	0	2	13
Lord Howe Island	16	0	0	7	5
TOTALS	—	74 12.7%	2(3?) .33%	10(12?) 1.9%	—

"Total of species" is obtained by adding all breeding land and fresh water species and representatives of superspecies, i.e.: only one subspecies of a single species is counted. "Endemics" are those confined to the group in question. "Forms" include species and subspecies. Areas are approximate.

ARC II: WESTERN PACIFIC

(Islands off Japan south through the Palaus and Carolines)

Varying from 400 to 500 miles to the eastward is a shorter festoon of much smaller islands beginning with the Seven Islands of Izu, the most

ARC II

	Total species	Endemic species	Extinct species	Extinct forms	Area (sq. mi.)
Seven Islands	11	0	0	0	—
Bonin and Sulphur Islands	14	4	3	5	29
Marianas	18	6	0	0	400
Palaus	31	5	0	0	307
Carolines	28	9	2	2	306
TOTALS	102	24 (23.5%)	5 (4.9%)	7 (6.7%)	—

For explanation, see table under Arc I.

northerly of which are just outside Tokio Bay. They continue southward through the Bonins, the Volcanoes, the Marianas, Carolines, and Palaus, 5° north of the Equator.

Arc III: Central Pacific

(Wake and the Marshalls south through Fiji, Samoa, and the islands on the continental shelf of New Zealand)

This is a very much longer arc beginning with Wake Island at lat. 19° N., and the low, dry Marshall Islands near the Equator, through the almost desert Gilbert and Ellice Atolls, and including the large, heavily forested and fertile Fiji and Samoan Islands. The satellite islands of New Zealand, *i.e.,* the Kermadecs and Chatham, much farther south, have fewer species; and the Aucklands, Campbell, Bounty, Antipodes, and Macquarie, lying in a region of cold, howling gales, are most of them almost without vegetation or birds.

ARC III

	Total species	Endemic species	Extinct species	Extinct forms	Area (sq. mi.)
Wake Island 1	1	1	1	9
Fiji 54	22	1(?)	1(?)	7,083
Samoa 33	10	1(?)	1(?)	1,173
Chatham Islands ⎫ Auckland Islands ⎬ .... 35 Macquarie Island ⎭	6	5	7	610
Totals .... —	39 (31.7%)	6(8?) (4.9%)	7(9?) (5.6%)	—

For explanation, see table under Arc I.

Arc IV: Eastern Pacific

(The Line Islands south to the Cook group)

This category includes the long string of widely separated atolls extending from Palmyra, about 360 miles north of the Equator, southward

ARC IV

	Total species	Endemic species	Extinct species	Extinct forms	Area (sq. mi.)
Line Islands 2	2	0	1	190
Marquesas 11	9	0	0	480
Society Islands 17	16	6	6	780
Tuamotus 8	2	0	0	330
Henderson Island 4	4	0	0	50
Totals 42	33 (78.6%)	6 (14.5%)	7 (16.6%)	—

For explanation, see table under Arc I. Question marks indicate that species have probably been extirpated but that questions of provenance have been raised.

for a thousand miles to Starbuck—the so-called Line Islands. It includes also the beautiful Marquesas, the lonely Tuamotus, Tahiti and the other Society Islands, and finally the Austral and Cook groups, the most isolated of all. It is 3,124 miles from there to the coast of South America.

Such an arrangement shows that within the past 180 years the islands have suffered loss of bird populations in an almost regular progression from west to east and roughly in proportion to their distance from Asia. The following table summarizes this:

EXTINCTION AS RELATED TO RELATIVE DISTANCE FROM ASIA

	Total species	Percentage extinct forms	Land area (sq. mi.) approx.
Arc I: Nearest Asiatic coasts	503	1.9	32,033
Arc II: Western Pacific	102	6.7	1,042
Arc III: Central Pacific	123	5.6	8,662
Arc IV: Eastern Pacific	42	16.6	1,830

The number of extinct species and subspecies increases with the distance from the continent of origin.

The obvious exception to the regular progression is the difference between Arc II and III. An explanation lies in the fact that the islands of Arc II are smaller than those of Arc III. As Mayr (1938b) and Amadon (1944) have strongly suggested, the very size of such islands as Fiji and Samoa (included in Arc III) affords at least relatively good assurance of survival. Actually only 2.4 percent of their birds have been extirpated, and indeed it is possible that the two rails that represent this figure may be found again. Compare this with the situation of the Bonins in Arc II, which have only an area of 29 square miles, as against 7,083 of Fiji, and where 30 percent of the birds have been long gone.

Certainly small size of islands restricts the available habitat and consequently the numbers of individuals. Kittlitz (1832–33), writing of the Bonin Islands before the onset of a new and devastating regime, remarked the rarity of the small birds now extinct. Those were all distant species to be found only on that small island. For thousands of generations they had bred without disturbance, but also without benefit of gene exchange with more wary relatives from a more dangerous land.

EXTINCTION AS RELATED TO ENDEMICITY AND DISTANCE FROM ASIA

	Number species	Percentage endemic species	Percentage extinct species	Land mass (sq. mi.)
Nearest Asiatic coasts	503	12.7	0.33	32,033
Western and central Pacific	225	27.0	5.7	9,704
Eastern Pacific	42	78.6	14.5	2,200

The relation of the number of species to land mass remains about equal, but the number of endemic species (those confined to the island groups involved) varies with the distance from the continent of origin as do the percentages of extinct endemics.

The table shows that isolation, evolution, and extinction are related phenomena.

No geographical arrangement can illustrate the great pressure of introduced predators under which these island species have worked in their efforts to reproduce themselves during the 180 years since Captain Cook landed in Tahiti.

Whaling vessels were notoriously infested by rats. Melville said of them in such a ship: "They stood in their holes peering at you like old grandfathers in a doorway. Often they darted in upon us at meal-times and nibbled our food . . . every chink and cranny swarmed with them; they did not live among you but you among them." When such a vessel was grounded on a beach for cleaning, the crew went ashore and lived in tents, and the rats, following the food supply, established themselves by the destructive thousands on the islands.

And there can be no doubt that sailing vessels, sheathed though they may have been against the ravages of shipworms, were beached and cleaned on out-of-the-way islands. As early as 1687 the pirate Dampier wrote in his journal that the "Cygnet" was hove down on an islet in the Batan group, north of Luzon.

Wrecks also played a part, and sometimes a disastrous one, as will be shown in the account of Lord Howe Island, and apparently the Bonins suffered in the same way. Melville, again, speaks of a small whaling ship all among the trees on the shore at Papeete: "Having sprung a leak at sea," he says, "she had made all sail for the island to heave down for repairs. Found utterly unseaworthy, however, her oil was taken out and sent home in another vessel; the hull was then stripped and sold for a trifle." Her rats, of course, went ashore.

Sea birds have been comparatively fortunate. Their powers of flight and the necessity for wandering far in search of food have protected species from total extinction. These attributes have allowed occasional birds to join colonies far from islands where they were hatched: so, as a rule, there has been more genetical sharing among them, and there are consequently few isolated and specialized species. They are better able to meet predators on their own ground as well.

How one family of sea birds protected themselves on tiny Howland Island of the Line group is told in the report of an agent of the Howland family who visited the island in 1854. "Next to the birds and living in close and constant proximity to them, the only fauna we observed were armies of rats," he wrote. "They appeared to be of the gray Scandinavian variety and to be subsisting and thriving on the eggs and fledglings it was only too easy for them to obtain. Occasionally the birds would make a raid on a band of rats engaged in these depredations and succeed in seizing some of the younger robbers. They would, we observed, kill the rats by tearing them to pieces with their beaks and talons, or more often where the rat was of a small size, fly off to a point at a considerable height above the reefs and drop their prey into the pounding surf where it would seem the rat would perish."

Steller's albatross is the one exception to this rule. Other than this, no species has been reduced to the point of extinction. On such devasted islands as Tahiti, Norfolk, and Lord Howe, where men have actually tried to make a staple of the birds during the breeding seasons, certain populations have been exterminated. From Norfolk Island *Pterodroma solandri,* which still breeds on Lord Howe Island, has disappeared; and from Tahiti, *Pterodroma rostrata rostrata,* which has much wider range.

Evidence afforded by guano deposits on the Line Islands (*q.v.*) indicates that populations of terns and boobies have been extirpated by changes of climate since Pleistocene time. Perhaps many species of land birds suffered the same fate through this climatic change, but no evidence of them remains to us.

Arc I: Asiatic Continental Shelf (Siberia south to Norfolk and Lord Howe Islands).

The *Kurile Islands* (lat. 50° 56′ to 44° 45′ N.) are links between the Peninsula of Kamchatka, in Siberia, and Hokkaido, northernmost of the Japanese islands. No more than 8 miles of water separates the most northern and the most southern islands from larger land masses, and no strait between them is more than 30 miles wide. There are 32 islands with 6,159 square miles of land area. The scenery is wild and awesome, with heavy forests of pine and birch, and on some islands there are active volcanoes 3,000 to 7,382 feet in height. Heavy gales and much fog make for difficult navigation among them. The majority of birds are Kamchatkan, but there are some, particularly in the southern islands, that are Japanese.

In September 1879, Henry James Snow, a sea-otter hunter, landed on Ushishiru, a small island just south of Rashowa (Raschau). He says, "We took two young wild geese from Ushishir where a few resort to breed." Neither these nor any other specimens are preserved, but he did collect eggs in another year. These are in the British Museum, registered as "*Branta hutchinsi.*" [1] The subspecies was a small one. Since no specimens have been preserved, the exact identity of the population cannot be determined. It is now extinct.

The *Riu-Kiu* (Ryukiu, Liu-Kiu, Loo-choo, Luchu) *Archipelago* connects much larger land masses, the Japanese islands to the north with Formosa to the south. There are 55 islands, with a total area of about 900 square miles. The archipelago may be divided into four groups: (1) Satsunan Shoto, containing the rather large islands of Yaku Shima and Tanega Shima, only 20 miles south of Kiushu; (2) the Linshoten Islands, the nearest of which is only about 30 miles south of Yaku Shima; (3) a middle group, containing the largest islands, Amami and Okinawa; (4) Sakishima Gunto, separated from Okinawa by a strait 25 miles wide and from northeastern Formosa by about 75 miles.

The mean temperatures vary between 60° and 80° F., and the vegetation is by no means tropical, being characterized by rather sparse pine woods.

The distribution of birds reflects the distances and the climate, for about 23 percent of the total number of the islands' forms are closely related to Japanese birds, but 8 percent are Formosan. Only a single species is known to have been extirpated from this group. It was a kingfisher known only from Miyako Shima, a somewhat isolated island in the southern group.

EXTINCT

Halcyon miyakoensis Kuroda

A thrush confined to the Yakushima in the northern group has not been collected since 1904.

POSSIBLY EXTINCT

Turdus celaenops yakushimensis (Ogawa)

A large wood pigeon (*Columba jouyi*) has apparently been extirpated from the island of Okinawa. It is perhaps still to be found on small islands near the coasts and on the Daito Islands about 200 miles to the eastward.

Two rather specialized species occur only on the Riu Kiu Islands; a jay, *Garrulus* [*Lalocitta*] *lidthi,* is said to be quite common still on Amami Shima and Tokuno Shima in the north-central group. A peculiar woodpecker, *Saphaeopipo noguchii,* is still to be found on Okinawa, but it is by no means common.

---

[1] The exact identity of the population is in doubt. See note on Canada geese of the Aleutian Islands below.

The *Solomon Islands* (between lat. 6° 45′ and 11° S., and long. 158° 15′ to 163° E.) are separated from the Bismarcks (New Britain and New Ireland) by a strait about 100 miles in width. Two small islands, Emirau and Nissan, are links between them. The Bismarcks, in turn, are only about 30 miles from the coast of New Guinea. There are seven principal islands: Bougainville, Choiseul, New Georgia, Guadalcanal, Santa Isabel, Malaita, and San Cristobal, as well as hundreds of smaller islands, islets, and rocks, a total of about 13,000 square miles. The largest of them, Bougainville-Buka, is 150 miles long and 20 to 40 miles wide. Ridges of volcanic mountains 6,000 to 9,000 feet in height form the "backbone" of the larger islands. Dense tropical vegetation covers their slopes; bird habitats are like those of the northeastern coasts of New Guinea. Of the smaller islands, Rennell, isolated a hundred miles west of San Cristobal, is the most important ornithologically. The birds of that island are so special that is must be treated separately.

As might be expected, Solomon Island birds are related to those of neighboring coasts and island groups, but there are an astonishing number of species that are to be found nowhere else, considering the proximity to New Guinea and other islands. Based upon a conservative estimate of the limits of species as against subspecies, there are 40 endemic species, or about 32 percent of the total of 126 land and fresh-water birds now known to breed on the islands.

PROBABLY EXTINCT

Microgoura meeki Rothschild

A ground-living pigeon, quite unlike any other in all the world, once was to be found on the island of Choiseul but has not been found since 1904 and is now most probably extinct. It was the most specialized in form of any of the islands' birds.

Of specialized species, few in numbers and confined to single islands, and for those reasons perhaps in danger, a rail and a thrush and a warbler should be mentioned. They are terrestrial, or live close to the forest floor, as did *Microgoura* and so many other extinct birds of Pacific islands.

SMALL POPULATIONS

Mountains of San Cristobal—*Edithornis silvestris; Zoothera margaretae; Vitia parens*

In less danger because they live in the trees and are better fliers, are a warbler, a flycatcher, two white-eyes, three honey-eaters, a flower-pecker, a species of crow, and no fewer than 86 subspecies that are confined to single islands.

Bougainville—*Meliphaga bougainvillei; Corvus meeki*
Kolombangara—*Phylloscopus amoenus; Zosterops murphyi*
Malaita—*Rhipidura malaitae; Zosterops stresemanni*
Guadalcanal—*Guadalcanaria inexpectata*
San Cristobal—*Rhipidura tenebrosa; Meliarchus sclateri; Dicaeum tristrami*

It is remarkable that in this small group a general rule for the Pacific is repeated: The most isolated islands farthest from large land masses have the largest number of distinct species to be found nowhere else. This is particularly noticeable on the small island Rennell.

Enjoying, as they did until no more than 25 years ago, a terrifying reputation for ferocity, the Solomon islanders have probably preserved their native birds and animals better than many islanders.

Rennell Island, about 90 miles west of the southern tip of San Cristobal, is a small plateau 42 miles long by 15 wide, 300 to 400 feet above the sea. It is a raised atoll, according to Hamlin (1931), and it possesses a fresh-water lake in lieu of a lagoon, as does Washington Island of the Line Islands far to the eastward. The basic rock being coral limestone, with sufficient beds of soil deposited by rotting vegetation to be found only here and there, the great trees of the Solomons are scattered, but the island is covered elsewhere with low bush.

It is an extraordinary, an even unique, island. The human population is Polynesian, isolated in the midst of Melanesia. Theoretically they are a remnant left behind in the long migrations of their relatives who colonized New Zealand, Tahiti, and the Hawaiian Islands. If this is true, however, it is strange that, except for a bat, there are no mammals,[1] not even dogs or Polynesian rats, as there are on even the most remote Polynesian islands. No cattle or hogs were landed by the Spaniards. No traders of other nations brought ship rats or house cats. The result has been that the bird fauna has not been disturbed appreciably. Of the 33 land and fresh-water forms known to nest on the island, 19 (57 percent) are to be found nowhere else. The Australian Government has made great efforts to keep the island and its fauna inviolate. As a result of this action, as well as its good luck and freedom from predators, the birds are as nearly in their natural state as any in the world.

The *Santa Cruz, New Hebrides,* and *Loyalty Islands* (between lat. 10° to 20° 40′ S., and long. 162° to 169° 46′ E.) form an 850-mile chain of volcanic islands. The most northern of the Santa Cruz group is 260 miles southeast

---

[1] According to Hamlin, 1931.

of the most southern of the Solomons. Continuing southward, the New Hebrides group is separated from the Santa Cruz by a shallow channel about 100 miles wide on the north, and by 150 miles of water from the Loyalty Islands on the south, which in turn are only 150 miles from the east coast of New Caledonia.

The northern islands (Wilson and Swallow groups) are very small. Five larger islands to the south of them are Ndeni (Santa Cruz or Egmont), Utupua (Edgecumbe), and Vanikoro (La Perouse). Three smaller islands, Tucupia, Cherry, and Mitre, lie 180 miles to the eastward. The total land area of the whole group is no more than 400 square miles. Theirs are extremely steep and rugged mountains, rising to 3,000 feet on Vanikoro and 1,800 on Ndeni, covered for the most part with heavy rain forest.

There are 40 islands in the New Hebrides group, including the Banks and Torres Islands, with a total area of 5,700 square miles. Largest is Espiritu Santo with 870 square miles. Like the Santa Cruz Islands, they are volcanic, heavily forested, hot, and damp, but endemic malaria is said to be of a less lethal variety. Northernmost are two groups of coral atolls, Banks and Torres. Espiritu Santo, Malekula, Efate, Erromango, Tanna, and Aneitum are the largest and most important islands.

Birds of the Santa Cruz, New Hebrides, and Loyalties are to a great extent shared in common, and are preponderantly Papuan in origin, according to Mayr (1939a, 1941b). These relationships are especially interesting because of the proximity of the much larger land mass of New Caledonia. They illustrate importance of prevailing winds as well as the special habitats offered by the Loyalties, which have atolls.

Of the 118 species of land birds recorded from the three groups, only 13, or about 11 percent, are confined to the three groups (endemic). It may be that another should be added, for J. R. Forster records a pigeon (*Columba ferruginea* Wagler) as having been collected on Tanna in August 1744, during Captain Cook's second voyage. Nothing resembling this has been seen since. That the species was confined to Tanna, a small island in the southern group with relatively few birds, is unlikely. So little is known of the bird that no explanation for its disappearance is possible. Certainly its habitat has born an evil reputation for treacherous natives and deadly disease. It has been little disturbed.

Mayr (1939a) has suggested that in the more distant past eruptions of volcanoes as well as cyclonic storms may have extirpated species on the three southern islands, of which Tanna is one. Tanna itself has an active volcano, and the formation of both Aneitum and Erromanga, as well as recent seismic disturbances, indicates the possibility of destructive cataclysms. Because these three islands are farthest east and so more nearly in the direct tract of typhoons, it is possible also that such storms have been more destructive here than elsewhere, but that species have thus been extirpated is less probable. Relatively little ornithological work has been done on these unhealthy and inhospitable islands, and possibly the mysterious dove of Tanna may be found again, but this hope is a faint one indeed.

Seven species are confined to single islands: two white-eyes on Santa Cruz, a flycatcher on Vanikoro, a warbler and a starling on Espiritu Santo, and two white-eyes on Lifu in the Loyalties. They are mentioned here because of their restricted habitat and presumably small numbers, although they do not appear to be in danger.

SMALL POPULATIONS

Santa Cruz—*Gallicolumba sanctaecrucis; Zosterops sanctaecrucis; Woodfordia lacertosa*
Vanikoro—*Mayrornis schistaceus*
Espiritu Santo—*Cichlornis whitneyi; Aplonis santovestris*
Loyalty Islands: Lifu—*Zosterops inornata; Zosterops minuta*

A beautiful parrot (*Eunymphicus cornutus uveaensis*), with representative subspecies on New Caledonia, also is said to be rather rare.

New Caledonia, largest single island, barring New Guinea, of either Melanesia or Polynesia, has a total area of 6,221 square miles. Peaks of its "backbone," the mountain massif running the whole 250 miles of its length, rise to 5,954 feet at the northwestern end. Mountains of the southwest are no more than 1,500 feet, but here as elsewhere they are a barrier to the prevailing easterly winds and the lee slopes are comparatively arid. Except for river valleys, these are covered almost exclusively by a single species of eucalypt of an Australian family. Forests of the wetter windward slopes are luxuriant.

Given such a large land mass with a much greater variety of bird habitats, it follows that there are a great number of species. The total is 68, of which 16 (25 percent) occur nowhere else. This includes the kagu, a family represented by a single form existing only on this island.

An endemic parrot (*Vini diadema*) and a subspecies of wood rail (*Tricholimnas l. lafresnayanus*), of which a representative subspecies is to be found also on Lord Howe Island, are extremely rare if not quite extinct.

SMALL POPULATIONS

RAIL: *Tricholimnas lafresnayanus lafresnayanus*
KAGU: *Rhynochetos jubatus*

PIGEONS: *Drepanoptila holosericea; Ducula goliath*
PARROTS: *Vini diadema; Eunymphicus cornutus cornutus*
OWLET-NIGHTJAR: *Aegotheles savesi*
CUCKOO-SHRIKE: *Coracina analis*
WARBLER: *Megalurulus mariei*
FLYCATCHERS: *Eopsaltria flaviventris; Clytorhynchus pachycephaloides*
WHISTLER: *Pachycephala rufiventris*
CROW: *Corvus moneduloides*
HONEYEATERS: *Gymnomyza aubryana; Guadalcanaria undulata*
WEAVER-FINCH: *Erythrura psittacea*

Norfolk Island (lat. 29° 02' S., long. 167° 56' E.) lies about 400 miles equidistant between New Caledonia and New Zealand, and about 600 miles from the east coast of Australia. Even including its satellites, Nepean and Philip, its total land area is only about 13 square miles, and its hills are no more than 400 feet high. These were formerly covered with a fine stand of Norfolk pines (*Araucaria*), a hundred feet in height. The climate is cool and pleasant for Europeans.

No evidence was found of any human habitation in 1777 when Captain Cook discovered the island, but only 11 years later a penal colony was established, and for short periods attempts were made to live off the land. It is quite likely that species that have never been recorded were killed off during those years.

Fifteen species of land birds are known to have inhabited the island. Parenthetically it should be said that 37.5 percent are New Zealand birds, 18.5 percent are Papuan, and the remaining are both Australian and New Zealand or widespread species of obscure origins. This preponderance of New Zealand species is no doubt due to prevailing south winds and the direction of storm winds which are almost all from that direction.

Two of the total of 15 are extinct, and this, considering the history of other small and isolated islands, especially the neighboring Lord Howe Island, is a surprisingly low figure. These two are representative of New Zealand species—a pigeon and a parrot. Representatives of both are still found in numbers on New Zealand, although in somewhat restricted ranges.

EXTINCT

Hemiphaga novaeseelandiae spadicea
Nestor meridionalis productus

In addition to these, a population of large petrels (*Pterodroma solandri* [1] =*P. phillipii*=*P. melanopus*) was extirpated from Norfolk by the convicts and their guards between 1788 and 1790.[2] This species still nests in burrows on the higher mountain slopes of Lord Howe Island. The history

---
[1] See Murphy, 1952.
[2] See Whitley, 1934, pp. 42–49, for a complete account from contemporary diaries.

of this event is exactly that of the cahow of Bermuda on the opposite side of the earth. When men make a staple of wild birds on islands the results are disastrous.

It is probable that a second species occurred on the island at the end of the eighteenth century and that it was extirpated in the same way. A letter from Lt. Ralph Clark written at Norfolk Island on February 11, 1791 (see Whitley, 1934, p. 46), says: "Everybody here owes their existence to the Mount Pit Birds [presumably *Pterodroma solandri*], before they entirely left us, another bird came in and supplied their place, but was not attached to Mount Pit alone but was found in holes of the ground in the day time, all over the island. They resemble the Mount Pit Bird in plumage, make, etc. all but the feet, which are white, which the Mount Pit birds are not . . ." Iredale remarks (1930, p. 115) that the color of the feet suggests the shearwater, *Puffinus carneipes* Gould, which is still to be found nesting on Lord Howe Island but not on Norfolk. He finds that differences in contemporary drawings by Watling support this theory.

Lord Howe Island (lat. 31° 32′ S., long. 159° 04′ E.) is only about 7 miles long by a mile wide, with a total area of 5 square miles. Two great well-wooded peaks, Mount Lidgbird and Mount Gower (2,800 feet), make it visible for many miles from seaward. The nearest land is the east coast of Australia, 300 miles distant. Norfolk Island is 550 miles northeastward.

The history of this bird fauna is as intimately known as any in the Pacific. It illustrates perfectly what has happened on all small islands everywhere; so many island species elsewhere have disappeared without ever having been known to man.

Excluding those possibly introduced by man, 16 representatives of Australian, New Zealand, and more widely distributed species of land birds breed now (or are known to have bred) on the island. If we add with a question a pigeon, known only by a contemporary drawing (*Columba vitiensis godmanae*), one half of these—a rail, a pigeon, a parrot, two flycatchers, a thrush, a white-eye, and a starling—are extinct.

EXTINCT

Porphyrio porphyrio albus
Cyanoramphus novaezelandiae subflavescens
Gergyone igata insularis
Rhipidura fuliginosa cervina
Turdus poliocephalus vinitinctus
Zosterops strenua
Aplonis fuscus hullianus

The three largest and most striking birds—the large white rail, perhaps a fat pigeon, and the parrot—were all killed between 1834 and 1874, during the 40 years following permanent occupation of the islands by human beings. The rail and the pigeon are said to have been killed for food by the first settlers and the parrot soon afterward because of its depredations upon gardens. The four small perching birds were all extirpated between 1918 and 1925 by large populations of rats which came ashore from the wreck of a ship, beached to save her from sinking, according to Hindwood (1940), who has well summarized these pathetic events. The rats were no doubt assisted by unknown numbers of owls imported to prey upon them.

It will be interesting to see what species will occupy island habitats now presumably left empty. In the past Australia and New Zealand have each contributed 18.7 percent of the birds, while the remaining have been widely distributed species with obscure origins. Because winds blow from the easterly quadrant during eight months of the year with considerable force, it is to be expected that the island was populated by birds from that direction. Australia being the nearest land and not only the prevailing light winds but also numbers of storms being from the westward, it is perhaps surprising that only 18 percent came from there.

Arc II: Western Pacific (islands off Japan south through the Palaus and Carolines).

The *Seven Islands of Izu:* An 800-mile chain of volcano peaks, some but not all quiescent, extends south from Fujiyama and Tokyo Bay to the Marianas Islands. The most northerly group is called the "Seven Islands," although there are actually 5 sizable islands and about 16 islets. O-Shima is in the entrance of the Sagami Sea, only about 15 miles from the coast. South of it are several small islands, and Miyake Jima 20 miles beyond them. The six islands south of Miyake are separated from it and from each other by straits 50 to 60 miles wide—considerable barriers for birds. Because European explorers and American whalers gave them names before the Japanese ones were known to mapmakers, these southern islands have several synonyms. The islands are called Hachijo Shima, Aoga (Awoga) Shima, Bayonnaise Rocks, Simisu Shima (Smith Island), Tori Shima (Mitsugo Shima or St. Peters, Ponafidian), and Sofu Gan (Lot's Wife).

On Tori Shima (Bird Island) a great colony of Steller's albatrosses (*Diomedea albatrus*) bred until about 1932, by which time most, if not all, had been killed by employees of feather merchants. The island is a volcano cone which erupted in 1933 and 1941, destroying the birds' habitat. It was feared that the species might be on the verge of extinction, but birds were again found breeding on Tori Shima in 1954.

A small warbler of Tori Shima (*Cettia cantans panafidinicus*)[1] is extremely rare, if not indeed extinct. A Japanese naval detachment, quartered on Tori Shima during the war 1942–1945, left a population of house cats which have done much damage to small bird populations.

Bonin and Sulphur Islands: Between lat. 24° and 27° N., about 300 miles south of the southernmost of the "Seven Islands" '(an isolated rock called Lot's Wife because it is white with guano), lie 27 islands and rocks called the Bonins (Ogasawara Shoto). Three groups of islands, separated from each other by about 50 miles, have only 29 square miles of land sur-

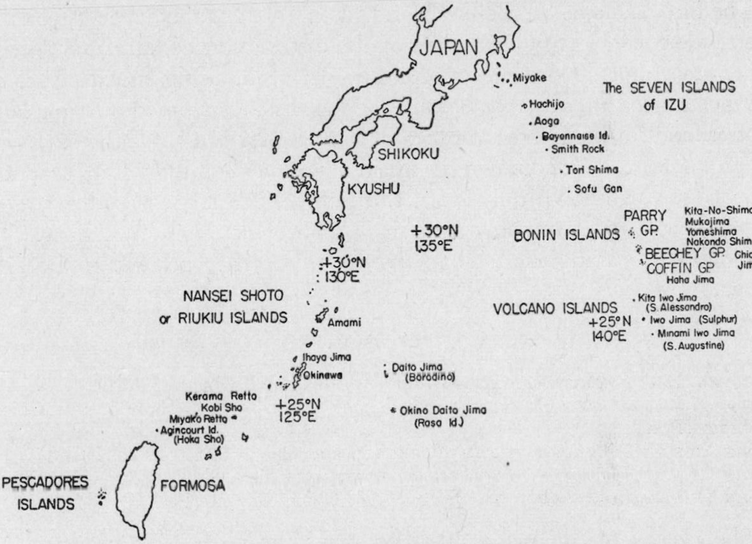

JAPAN

SHIKOKU

KYUSHU

Miyake

Hachijo

Aoga

Bayonnaise Id.

Smith Rock

Tori Shima

Sofu Gan

The SEVEN ISLANDS of IZU

+30°N 135°E

+30°N 130°E

BONIN ISLANDS

PARRY GP.

Kita-No-Shima
Mukojima
Yomeshima
Nakondo Shima

BEECHEY GP Chichi
COFFIN GP Jima
Haha Jima

NANSEI SHOTO or RIUKIU ISLANDS

Amami

Ihaya Jima

Okinawa

Kerama Retto
Kobi Sho
Miyako Retto
Agincourt Id.
(Hoka Sho)

+25°N 125°E

Daito Jima
(Borodino)

Okino Daito Jima
(Rasa Id.)

VOLCANO ISLANDS

+25°N 140°E

Kita Iwo Jima
(S. Alessandro)
Iwo Jima (Sulphur)
Minami Iwo Jima
(S. Augustine)

PESCADORES ISLANDS

FORMOSA

MAP 3

face. The most northerly is the Parry group, of which Muko and Nakando Shima are important ornithologically. Next to the southward, the Peel or Beechey group (Chichi) is composed of three islands, lying only a few yards apart. They are Buckland (Ototo), Stapleton (Ani), and Peel (Chichi). The last has attracted most attention, for the best harbor is there. It is the largest island, though only $9\frac{1}{2}$ miles long by $4\frac{1}{2}$ wide. It is also the type locality for all peculiar native species. The Baily Islands (Haha Jima), south of Peel, were apparently never visited by naturalists who recorded birds until 1889.

[1] *Horornis cantans panafidinicus* Momiyama, Bull. Biogeogr. Soc. Japan, 1, no. 3, 1930, p. 175 (note) (Tori-Shima, Seven Islands of Izu). Oliver L. Austin, Jr., has kindly given me the information on the status of this bird.

About a hundred miles southeast of the Baily Islands are three small islands separated from each other by 50 to 60 miles. They are the Volcano or Sulphur (Iwo or Kazan Retto) Islands: Kita Iwo Jima, Nako Iwo Jima (*"Iwojima"*), and Minami Iwo Jima. A small rail (*Poliolimnas cinereus brevipes*) once occurred on the two southern islands but was extirpated by rats and cats introduced by man.

Both Bonin and Sulphur archipelagoes resemble the Seven Islands geographically in that they are of volcanic origin but differ in having a much warmer climate and in their slightly larger land area. Furthermore, they are farther from Japan, and their faunas are therefore more isolated.

The bird fauna of the Bonin Islands is like that of the Seven Islands in the random and fortuitous nature of the distribution of unrelated species characteristic of all oceanic islands but differs in being much richer in distinct species, presumably because of the distance from the larger land masses, and in the greater number of habitats for birds. Of the 32 forms of 14 species that are known to breed there, no less than four are very distinct species known only from the Bonins. Specialized as they are in form, they were presumably so in habit. Three of them are extinct. The two subspecies bring the total to five forms, or 18¾ percent of the bird fauna.

EXTINCT

IWO-JIMA RAIL: *Poliolimnas cinereus brevipes.* Naka and Minami Iwo-Jima
BONIN NIGHT HERON: *Nycticorax caledonicus crassirostris.* Nakondo and Chichi Shima (Peel Island)
BONIN PIGEON: *Columba versicolor.* Nakondo and Chichi Shima (Peel Island)
BONIN THRUSH: *Zoothera [Aegithocicla] terrestris.* Chichi Shima (Peel Island)
BONIN GROSBEAK: *Chaunoproctus ferreirostris.* Chichi Shima (Peel Island)

The history of the islands that concerns the birds is summarized as follows:

June 1827: H.M.S. *Blossom,* Captain Beechey, landed personnel on the central island of the central group. Beechey named the island Peel and the harbor Lloyd's. On no other island of the Bonins was landing made. He thought the island had never had permanent human habitation.

The party found two shipwrecked sailors who had built a hut of ship timbers. It is possible that rats came ashore with this wreck, although no mention is made of them. Beechey wrote: "Of birds we saw some handsome brown herons with white crests [*Nycticorax*], plovers, rails, snipes, wood pigeons [*Columba versicolor* and *C. janthina nitens*], the common black crow [*Corvus macrorhynchus* subsp.], a small bird 'resembling a canary' [*Hapalopteron familiare* Kittlitz] and a grosbeak [*Chaunoproctus*]." They were all very tame and "until alarmed at the noise of a gun

suffered themselves to be approached." The plover and snipe were migrants, and perhaps they saw migrant crakes (*Porzana pusilla*) as well.
In addition, the grosbeak was also collected.

June 1828: The naturalist J. H. von Kittlitz landed at Port Lloyd, Peel
Island. He found the two sailors mentioned by Beechey and remarked
that pigs, escaped from the wreck, were very numerous in the woods. In
addition to the birds noted by Beechey's party, Kittlitz saw the ground
thrush repeatedly near Port Lloyd. This is the last time any of the four
native species were recorded on Peel. The grosbeak and thrush probably
disappeared from the earth a few years afterward. Kittlitz visited Peel
Island only.

1830: The first permanent settlement by man was made on Peel Island.
A party of colonists consisting of American and British sailors together
with Polynesians of both sexes arrived, and this event had the most serious
consequences for the birds.

1853: A small squadron of American naval vessels, commanded by Com-
modore M. C. Perry, en route to Japan, anchored in Lloyd's Harbor for
four days. Both Peel and nearby Stapleton Island were explored and the
Baily group was visited, but apparently no birds were collected. Except
that all birds were very rare, little mention was made of them, although
the naturalist Heine and other expedition personnel had made collections
on the Riu Kiu Islands shortly before.[1] Deer, goats, pigs, sheep, and
"innumerable dogs and cats" were then to be found where no mammals,
except a bat, existed before. The human population was estimated at
only 31.

There were strong indications that whalers had been using the southern
islands. The name "Coffin" for that group (called "Baily" by Captain
Beechey, R.N.) indicates that American whalers had been in the habit
of stopping there. It was the custom of these vessels not only to replenish
stores of wood and water, but also to supplement a meagre diet with
whatever could be shot or caught. A custom of "heaving down" ships on
suitable beaches for cleaning and repair must often have introduced rats
by accident.

1854: A second American naval expedition arrived under the command
of Captain Rodgers. William Stimpson, naturalist, collected many mollusks and crustaceans, but very few birds, and none of the native land
birds (Cassin, 1862).

1875: The Bonin Islands were annexed by the Japanese Government.
In that year there were no more than a hundred people on the whole

---

[1] Cassin reported these. See Perry, 1856.

group; 20 years later there were 5,000. A descendant of the colonists of 1830 estimated in 1930 that one-half of the original forests had been destroyed.

1889: A. P. Holst collected for Seebohm. He found the pigeon and night heron on Nakondo Shima in the Parry group, but not on Peel Island.

1900–1929: Collectors for Allan Owston of Yokohama, the Marquess Yamashina, and others searched for the rare birds from time to time but without success.

The following table shows the known distribution of four peculiar native species. X means that the bird has been recorded and E that it is presumably extinct.

	Parry Islands (Muko)	Peel or Beechey Islands (Chichi)	Baily or Coffin Islands (Haha)
Columba	X E	X E	—
Hapalopteron	X	X E	X
Zoothera (Aegithocichla)	—	X E	—
Chaunoproctus	—	X E	—

So little of the true ornithological history of the Bonins is known that it is now impossible to reconstruct with certainty the events leading to the disappearance of all these forms. By inference it is probable that men and animals introduced by them were directly responsible, for the most characteristic and most common of the distinct native species, the small, yellow *Hapalopteron familiare*,[1] disappeared from Peel Island shortly after permanent settlement by man, but still exists on less civilized islands. A tree-dwelling species, it was at an advantage, but it could not overcome the pressure of predators on the most heavily inhabited island. Probably the two other species, the finch and the thrush, were extirpated even earlier, for according to Kittlitz they were usually to be found on the ground, and no doubt nested on or near the ground; they were more vulnerable to the cats, dogs, and hogs we know were present after 1830, and the rats which had almost certainly come ashore with wrecks and from whaling ships.

So little is known of the southern group before the coming of introduced animals that it cannot be said that the grosbeak or the thrush

---

[1] *Ixos familiare* Kittlitz, Acad. Imp. Sci. St. Pétersbourg, Méem divers savants, 1, 1830, p. 235 (Bonin Islands, Peel Island). This species is thought to belong to the family Timeliidae (Delacour and Mayr *in verb.*). Peel Island (Chichi Shima) is the type locality for all species described by Vigors and by Kittlitz (see Beechey's "narrative" and Kittlitz, 1858, pp. 164, 190).

certainly never occurred there. It is quite possible that they did. If so, they were probably extirpated there also by introduced animals.

Because the pigeons nested in trees they were, like *Hapalopteron,* at an advantage. Perhaps they were attacked at nesting time by black rats, although we have no direct evidence of this. No doubt they were shot and trapped by the settlers who suffered sometimes from famine after typhoons. In 1828 they were already becoming rare because of shipwrecked sailors, according to Kittlitz. The settlers were excellent shots—so good that the sea-otter hunters sent especially for them from Japan. Austin and Kuroda (1953) remark of the related *Columba janthina* that "were it more prolific it would be an excellent game bird, but it cannot withstand hunting pressure and disappears the moment it is subject to persecution." This would apply equally well, if not better, to the long-isolated Bonin species. No doubt the same fate came to the night heron on Peel Island soon after 1830. The little yellow *Hapalopteron,* not such a tempting target, remains to this day.

Borodino Islands (lat. 25° 50′ N., long. 131° 14′ E.). Three small islands lie almost midway between Okinawa, in Riu-Kiu chain, and the Bonin Islands, 320 miles east of the former, about 650 miles west of the latter. They are called Kita Daito Jima or North Borodino, Minami Daito Jima or South Borodino, and Okino Daito Jima or Rasa Island. Kita, the largest, has only about 7 square miles of land surface; the others are slightly smaller. The highest hill (on Rasa) is no more than 107 feet. Both North Borodino (Kita) and Rasa have valuable deposits of guano; phosphate beds have been worked there for many years. Although exact information is lacking, it is probable that much original vegetation was destroyed before any ornithological exploration was accomplished and several subspecies were extirpated. South Borodino (Minami) is the only island upon which a considerable number of human beings has not been settled, and it is the only one except Kita upon which native and peculiar subspecies have been found. Of five such forms, only one was found on Kita in 1922, when a Japanese collector explored the place. It may well be that all five are now extinct, for a large part of the bush of Minami (South) was cleared for the construction of an airfield and other military installations during the war 1942–1945.

The forms listed below are all confined to Minami Daito Shima (South Borodino) except *Parus varius orii,* which occurred also on Kita or North Borodino. All are extremely rare if not already extinct. According to Hachisuka (*in litt.*), both *Cettia diphone restrictus* and *Parus varius orii* are probably extinct.

NATIVE LAND BIRDS

Zosterops palebrosa daitoensis Kuroda
Parus varius orii (Kuroda)
Microscelis amaurotis borodinensis Kuroda
Cettia (*Horeites*) *diphone restrictus* (Kuroda)
Troglodytes troglodytes orii Yamashina

In addition, although no specimens have been recorded, a colony of Steller's albatross (*Diomedea albatrus*) is said [1] to have been seen on South Borodino until about 1919. Probably fishermen turned feather hunters destroyed the birds.

The *Marianas Islands* (lat. 20° 32′ N. south to 13° 27′ N., long. 144° E.) are apparently an extension of the long arc of volcanic mountains, although the most northern island is more than 300 miles south of the Sulphur (Iwo) Islands and depths of 2,000 fathoms separate them. There are 15 in all; three of the most northerly are little more than the peaks of active volcanoes, and the fourth is a reef with almost no dry land. The remaining islands are of volcanic origin, some having the magmatic base covered with limestone. Such are the five most southerly: Saipan, Tinian, Aguigan, Rota, and Guam. They are larger than the Bonins. The land area of Guam is 200 square miles; that of Saipan is 72; Tinian is a little smaller; and the land area of the smaller islands less than 10 square miles each. The larger islands were all heavily forested at one time. Much of this has been destroyed and replaced by sugarcane on Saipan and Tinian during the past 40 years, but on Guam there remain large tracts of tropical forest.

Eighteen species are known to inhabit these islands. Six of them are distinct species which occur nowhere else in the world. The table below shows their distribution, including that of the peculiar ground-

DISTRIBUTION OF THE ENDEMIC SPECIES OF THE MARIANAS ISLANDS

Species	Date of discovery	Asuncion	Agriban	Pagan	Almagan	Saipan	Tinian	Rota	Guam
Anas oustaleti	1894	—	—	—	—	X	X	X E	X
Megapodius l. lapérouse..	1823	X	X	X	X	X E	X E	X E	X E
Rallus owstoni	1867	—	—	—	—	—	—	—	X
Ptilinopus roseicapillus ...	1831	—	—	—	—	X	X	X	X
Monarcha taka-tsukasae ..	1931	—	—	—	—	—	X	—	—
Cleptornis marchei	1889	—	—	—	—	X	—	—	—
Corvus kubaryi	1836	—	—	—	—	—	—	X	X

X indicates occurrence and E that the bird is extinct on that island.

[1] Yamanari as quoted by Hutchinson, 1950.

nesting *Megapodius,* subspecies of a species that occurs also in the Palau Islands.

The remarkable fact is that no species or subspecies is known to have been totally extirpated in spite of permanent habitation by Europeans for almost 300 years. A Spanish mission was established on Guam in 1668. The Germans took over the government of the islands in 1899, and the Japanese in 1919. No account of the natural history of those times is available, but when Lord Anson landed on Tinian in August 1742, he found that the island was being used as a range for cattle and hogs introduced by the Spaniards to feed the missions of Saipan and Guam. No doubt the stores of galleons of the annual Manila plate armada were replenished there also.

Where attempts have been made to supply ships and settlers from small islands in other parts of the world, the result has been rapid extirpation of numbers of species. Examples are the Bonin, Caroline, and Society Islands in the Pacific and the Mascarene Islands in the Indian Ocean.

Two explanations are possible to account for the apparent survival. First, so few Europeans inhabited the islands, and their efforts to develop them were so relatively harmless, that their presence was not so disastrous as on other smaller islands. This hypothesis involves particularly the size of the islands, for in general it may be said that larger islands afford better chance of survival than smaller.

As a corollary of the first hypothesis, there is a possibility that the Spaniards made no attempt to live upon native animals, but only upon the cattle and pigs they introduced. Certainly they were in the habit of naturalizing pigs on uninhabited islands all over the world, as they did on Bermuda.

Alternatively, it is possible to assume that a certain number of species disappeared from the islands between 1668 and 1822, when first Kubary, a naturalist who recorded what he saw, arrived on the Marianas. The fact that there are only 20 distinct species of land and fresh-water birds known from these relatively large islands, whereas there are 32 on the Palaus and 18 from the island of Ponape, lends credence to this last alternative.

The mound-builder (*Megapodius*) has been extirpated from the more heavily populated islands of the group during the past 50 years. It is significant that in 1870 the human population of the Marianas was about 8,000; in 1930 it was estimated at about 88,000.

The *Palau* (Pelew) and *Caroline Islands* [1] emerge from coral reefs ex-

---

[1] See Baker (1951) for a complete account of the birds.

tending east to west for 2,000 miles, thus crossing a huge inverted "T," the upright portion of which is the Marianas.

There are 26 islands in the Palau group as well as a vast number of tiny islets. None of the islands are large; Babelthuap, the largest, is 23 miles long and rises 1,968 feet above sea level. Others are much smaller, the next largest being Peleliu, only about 5 miles long, one-fourth of a mile in width, and rather flat.

As far as is known no form of bird has been extirpated from these islands. However, of a total of 31 species to be found on the Palaus, five (an owl, a warbler, a flycatcher, a shrike, and a zosterops) are peculiar to the islands, occurring nowhere else. Their isolation and small numbers make them particularly vulnerable.

The Palau representative of the Nicobar pigeon (*Caloenas nicobarica pelewensis*) has been mentioned as possibly extinct because there was no report of the bird between 1880 and 1948. However, naturalists of the U.S. Navy saw birds on Garakao on five occasions among the bushes and vines on the high coral cliffs in 1949. Although none were collected, there can be little doubt that the record is accurate.

Ships long avoided these islands because of the reputation of the natives for ferocity during the early part of the nineteenth century. Civilization had little or no influence until after World War I.

DISTRIBUTION OF SPECIES CONFINED TO THE PALAU ISLANDS

Species	Babelthuap	Koror	Arakabesan	Urukthapel	Eil Malk	Garakao	Ngesebus	Peleliu	Angaur
Megapodius l. senex	X	X	—	—	—	X	X	X	
Otus podarginus	X	X	—	—	—	—	—	—	X
Psamathia annae	X	X	—	—	—	X	—	X	—
Rhipidura lepida	X	X	—	—	—	X	—	X	—
Colluricincla tenebrosa	X	X	—	—	—	X	—	X	—
Rukia palauensis	X	—	—	—	—	—	—	—	X

It is possible that the white-eye (*Rukia*) has been extirpated from some islands.

Few birds have been collected on Arakabesan, Urukthapel, and Eil Malk; perhaps these species may yet be found on them.

The *Caroline Islands* have fewer native species than have the Palaus— 28 as against 31—but perhaps because of the greater distance to the continents and the consequently enforced isolation of their faunas, they possess nine peculiar native species that occur nowhere else. It should be said parenthetically that it is also possible that because the Palaus are close to the continent of Asia, competing species arriving from time to time during centuries may have extirpated older immigrants. Of a total of 500 islands and islets, only four are of the mountainous, volcanic type that support land and fresh-water birds. These are Yap (80 sq. mi.), Ponape (134 sq. mi., and with a mountain 3,000 feet in height), Truk (50 sq. mi.), and Kusaie (42 sq. mi., with a peak 2,064 feet in height).

Two species, a rail and a starling, were extirpated from Kusaie Island between 1828 and 1880. Both were confined to that island, and the rail was specialized in form and most probably also in its habits; indeed it appears to have had very poor powers of flight. The starling was less obviously specialized but most probably was a relict, for a related species still exists on Ponape.

EXTINCT

Kusaie—*Aphanolimnas monasa; Aplonis [Kittlitzia] corvina*

Reasons for the extinction of the birds of Kusaie and not any of the other islands are probably many and complex, and certainly not known.

That both were long established on the smallest island of the group and that both were specialized in form cannot be doubted. Neither can it be doubted that populations so situated are handicapped. But Kusaie is not much smaller than other islands of the group. In fact, Truk is a group of islands smaller than Kusaie and forming a lagoon: distances between them are not great but probably are partial barriers to these small birds.

A more obvious, even though not quite satisfactory, explanation lies in the fact that Kusaie harbors enormous populations of rats, probably both "jungle rats" (*R. rattus*) and "house rats" (*R. norvegicus*). The difficulty with this explanation is the question whether such rats do not occur on Yap, Ponape, and Truk also. The answer is that they almost certainly do. Possibly the difference lies in the fact that Kusaie was the favored resort of whalers during the early years of the nineteenth century, and their vessels, "hove down" on the beaches for cleaning and repair, were the means of introduction and reintroduction, to the end that enormous populations overran the island. No reason can be given for occasional pestilential infestation of certain islands by rats, whereas others apparently escape such calamity, even though rats be present.

The following are rather small populations of somewhat specialized birds. Since a species of similar attributes has been extirpated from Kusaie, it may well be that they would be in danger should further changes in their habitat take place.

SMALL POPULATIONS

Yap
  Zosterops (*Rukia oleagina*)
  Flycatcher (*Monarcha godeffroyi*)
Ponape
  Parrot (*Trichoglossus rubiginosus*)
  Starling (*Aplonis pelzelni*)
  Zosterops (*Rukia sanfordi*)[1]
Truk
  Thrush (*Metabolus rugensis*)
  Zosterops (*Rukia ruki*)

Arc III. Central Pacific (Wake south through Fiji and Samoa and islands off new Zealand).

Wake Island (lat. 19° 18′ N., long. 166° 35′ E.) is isolated 200 miles north of the northernmost of the Marshall Islands, itself a low atoll without refuge for any but sea birds. Wake is no more than 20 feet above seal level, but its islands are covered by "scrub" about 20 feet high. The only native land bird was a small rail (*Rallus wakensis*), which is now probably extinct. It has not been seen since the Japanese occupation of 1942-1945.

The *Marshall Islands* (lat. 15° to 4° 30′ N., long. 160° 30′ to 172° E.) curve in a semicircle to the northeast and eastward of the Carolines, at 400 to 500 miles distance, and separated from them by a 3,000-fathom channel. There are 29 coral-limestone atolls and 5 islets in the group, none of them more than 20 feet above the sea. On some are small forests of great trees (*Messerschmidia* and *Pisobia*) where terns (*Gygis*), boobies (*Sula sula*), and frigatebirds (*Phaethon rubricauda*) nest; others are covered only by low brush, principally *Scaevola* and *Pandanus*. The only land bird is a large pigeon (*Ducula oceanica*), and that has been recorded from only four atolls: Jaluit, Elmore, Arno, and Wotje. J. T. Marshall (1951) records that these pigeons are sometimes kept as pets by the natives, which observation casts a little light on the record of the fruit pigeon, (*Ptilinopus*) from Ebon Atoll,[2] claimed to have been a population of the Marshalls but now extinct. The only known specimen does not differ from Caroline Island birds, and was probably a captive.

The *Gilbert and Ellice Islands* (lat. 0° 2′ 45″ to 11° S., long. 172° 30′ to 180° E.) are virtually a southern continuation of the Marshalls. There are 25 atolls in the two groups. Small, low islands, no more than 8 or 10

---

[1] Perhaps this bird should be called *longirostra* Takatsukasa.

[2] *Ptilinopus marshallianus* Peters and Griscom (1929). I quite agree with Ripley and Birckhead (1942) that this is a synonym of *P. porphyraceus hernsheimi*. See also Baker, 1951, p. 184.

miles long, emerge from the reefs. Rainfall is said to be less than that of the Marshalls. No resident land or fresh-water birds have been recorded.[1]

Nauru (Nawado, Pleasant) Island and *Ocean* (Paanopa) Island are usually included in these groups although they are isolated 450 to 350 miles, respectively, west of the Gilberts. Both are raised atolls about 100 feet above the sea. Both were covered with vegetation. Because of the considerable destruction by guano diggers and by military operations during the years 1942–1945, a reed warbler (*Acrocephalus luscinia rehsei*) has been thought to be in danger. Much vegetation remains on Nauru, however, and although it is a very small bird population, it has apparently survived.

South of the Ellice Islands are hundreds of square miles of shoal water, of coral reefs and small islands. The Horne Islands, together with Wallis and Rotumah, are largest, with mountains rising to 2,500 feet on the Horne Islands. There is only one native species confined to the group, a mound-builder (*Megapodius pritchardii*). This population is small and to be found only on Niuafau, between the Horne Islands and Fiji. Other populations are rather small also, but appear to be in no great danger, unless war, industrial operations, or some other catastrophe should overtake them.

The *Fiji Islands* (lat. 15° to 21° S., long. 178° to 176° E.) lie to the southeast of the Horne group, with high, volcanic, though small islands spread no more than 150 miles apart, between them. These may be likened to landing fields for an occasional wind-borne bird. There are five large, high islands and thousands of islets and rocks in the archipelago. The most northerly is Vanua Levu, 2,130 square miles, with jagged mountains of 2,420 feet. Close by is Taviuni, which is 26 miles long and about 10 miles wide. Viti Levu, the largest, has 4,053 square miles of land surface, its heavily forested mountains rise to more than 4,000 feet. It is 40 miles southwest of Vanua Levu. Approximately the same distance to the southward lies the long narrow island of Kandavu. Close to the east coast of Viti Levu are the small islands of Ngau and Ovalau. Both are mountainous and covered with trees.

Of the 54 species of birds, 22 are confined to the archipelago. Thirteen specialized native species, living in restricted habitats and now few in numbers, are listed below, for they will be in danger as civilization advances. Two are listed as possibly extinct because collectors of the Whitney Expedition were not able to find them in 1928. The islands are so large that it may be that a very small population of a warbler, and even (al-

---

[1] *Tricholimnas conditicius* Peters and Griscom is a synonym of *T. sylvestris. See* Greenway, 1952.

though less probably) the rail may be found. It was specialized to the extent that its powers of flight were impaired, and it was also a small population, quite restricted in its range and habitat and may have fallen prey to the introduced mongoose, which was perhaps responsible for the great reduction in numbers of three other species of rails as well. Reasons for the extreme rarity, or possibly even the extinction, of the warbler (*Trichocichla rufa*) are unknown. Nothing whatever is known of the habits or habitat of the species.

POSSIBLY EXTINCT

RAIL: *Nesoclopeus poecilopterus*. Viti Levu and Ovalau
WARBLER: *Trichocichla rufa*. Viti Levu

SMALL POPULATIONS

PIGEONS:
  Ptilinopus (*Chrysoena*) *layardi*. Kandavu and nearby small islands
  Ptilinopus (*Chrysoena*) *victor*. Vanua Levu, Taviuni, and nearby small islands
  Ptilinopus (*Chrysoena*) *luteoventris*. Viti Levu, Ovalau, Mhengha, Waia, and
    Ngau
PARROTS:
  Vini amabilis. Viti Levu, Ovalau, Taviuni
  Phygis solitarius. All large and many small islands (genus confined to Fiji)
  Prosopeia personata. Viti Levu and perhaps formerly Ovalau
FLYCATCHERS:
  Rhipidura personata. Kandavu
  Mayrornis versicolor. Ongea Levu
HONEYEATER: *Xanthotis* [*Meliphacator*] *provocator*. Kandavu
FAMILY?:
  Lamprolia victoriae kleinschmidti. Vanua Levu
  L. v. victoriae. Taviuni

Compared to the rate of extinction on smaller Pacific islands in the past 200 years, this loss is small. Size of the islands themselves is perhaps a partial explanation, greater size offering more and larger habitats and actually more room to hide from introduced predators, such as tree-dwelling black rats (*R. rattus*). It is possible that the introduction of such rats was less often accomplished and less long-continued, and therefore less effective as a destructive force, because the fierce reputation of the people prevented whalers from using the beaches to repair their ships. No sailor wanted to be cooked and served as "long pig."

The East India Marine Society of Salem, Mass., reported that of 12 ships of Salem, trading in *bêche-de-mer* and shells in the Fijis between 1820 and 1830, three were wrecked and lost among the islands, two were severely damaged, and two were attacked by natives, losing part of the crew in one case and all but one in another. A whaling ship in calm weather near the shore was boarded by natives, who killed the unsuspecting captain and two mates with sharp spades used to cut up whales, and it was only by the courage and presence of mind of the third mate that the ship was saved.

Stories like those of Captain Peter Dillon, who discovered the grisly fate of Pérouse and the crews of the frigates *Astrolobe* and *Boussole* 30 years after their mysterious disappearance, must have been discouraging to those who thought to trade in the Fijis. Dillon was ambushed at Wailea on the northwest coast of Vanua Levu in 1812. Fourteen of his companions were killed, two being cooked and eaten before his eyes. He (1829) wrote: "These people sometimes, but not very often, torture their prisoners in the following manner. They skin the soles of the feet and then torment their victims with firebrands so as to make them jump about in that wretched state. . . ."

A wry poetic justice is interjected in these situations. When a ship was wrecked on these islands, she piled up on fringing reefs where rough seas and hungry fish prevented many rats from swimming ashore. On the other hand, if natives captured the ship they brought her into a harbor and beached her, and then the rats escaped and often brought havoc to their island.

The *Tonga Archipelago,* about 200 miles east of Fiji, is a chain about 150 relatively small islands. The two largest are Vavau, 9 miles long by 6½ wide, with hills 600 feet in height; and Tonga Tabu, in the south, 24 miles long by 10 miles wide, but nowhere more than 60 feet above sea level. The only native land bird peculiar to this group is a whistler (*Pachycephala melanops*), which is confined to Vavau, and a few islands nearby. Possibly there were once more specialized residents or earlier immigrants which were extirpated without our knowledge, although no such are known. Certainly Forster in his "Descriptiones" says that a bird he called *"Muscicapa atra,"* which was probably a *Pomarea,* occurred on Tonga Tabu. No such is to be found now.

The *Samoan Islands* have probably lost a rail (*Pareudiastes pacificus*), which, like the barred-wing rail of Fiji, was so specialized as to have been almost, if not quite, incapable of flight. Like all such rails it nested on the ground and was entirely terrestrial. It was also a small population confined to Savaii. Probably it was extirpated by imported rats and cats, which are thought to have been introduced by whaling ships. It is remarkable that the most peculiar tooth-billed pigeon (*Didunculus strigirostris*) was able to survive. Because it was only partially terrestrial and a good flier it was certainly at an advantage compared to the rail. It is not known to have been found on any islands other than Savaii and

Opolu. When first discovered this was a common species, but 40 years later it was said to be so rare as to be on the verge of extinction. Not improbably, the species would have been extirpated had the islands been smaller or the birds confined to a restricted habitat.

Five hundred miles farther to the eastward the Samoan Archipelago is more isolated than the Fijis. The islands are also smaller. Savaii is 703 square miles in total area and Opolu is 430 square miles. Both are mountainous with quiescent volcano peaks 3,000 to 5,400 feet high and heavily forested. Smaller (40.2 square miles) is Tutuila Island, lying 36 miles to the southwest, and the nearby, even smaller, islands, Tau, Ofu, and Olosinga.

The *Tokelau* (Union) and the *Phoenix* groups, northeast of Samoa, are all low coral atolls. No endemic species are to be found on them and no populations are known to have been extirpated. They have had little contact with European or American civilization.

The *Kermadec Islands* (lat. 30° 12′ S., long. 178° 28′ E.) are about 550 miles north-northeast of New Zealand, the nearest land, and more than a thousand from any other. There are three of them. Curtis and Macaulay are very small and are not forested. Volcanic activity has been reported on Curtis. Raoul (Sunday) Island is a little larger with a total area of 11.2 square miles, and its forested peak is 1,723 feet high.

Untimely reports of the extirpation of the native parrot (*Cyanoramphus*) have agitated the public. All native birds, however, are said to be present in numbers. In fact, there appears to have been little interference of civilization, although there are now nine settlers on Raoul (Sunday) Island.

Only five species of native land birds are known to breed on the islands, and only one is a passerine bird: it is of an Asiatic family called honeyeaters, the tui of New Zealand (*Prosthemadera novaeseelandiae novaeseelandiae*). Seven song birds have arrived and occupied habitats on this small island during the past 50 years. These include six European birds—two thrushes, three finches, and the ubiquitous starlings that were artificially introduced into New Zealand but naturally extended their ranges. The seventh is an Australian white-eye, which has colonized New Zealand and some of the islands without assistance from men. It may well be that the introduced blackbird and starling may compete directly with the native tui; indeed they have been seen to kill tuis on the mainland of New Zealand.

The *Chatham Islands* (lat. 44° S., long. 176° 30′ W.) are 370 miles east of New Zealand, whence they have received all their native birds and plants, and from which general direction winds blow almost constantly and sometimes with great force. These are the largest of the islands that lie on the edge of the New Zealand continental shelf, having a total land area of 372 square miles. Chatham (Whare Kouri) is the largest, being about 31 miles long, with a hill 938 feet high. Next in size is Pitt Island, about 8 miles long by 3 in width, and Mangare, a large hill 940 feet high but only about a mile in diameter. Close to its cliffs is Little Mangare, also a high rock with a acre of low trees at its top. A couple of miles off the east coast of Pitt is South East Island with only about 2 square miles of land. Both of these islets are the last refuges of two quite specialized native forms, all four of which were once more widely distributed among the islands.

Representatives of 32 superspecies are known to have bred on the group within the past 154 years. It should be said parenthetically that 10 of these European birds introduced into New Zealand by man late in the nineteenth century. At least three, and perhaps more, have extended their range naturally to the islands (Williams, 1953, p. 681). Four species and five subspecies, quite specialized in form, have been known only from this island group (endemic). Of these, three species and one subspecies— two rails, a warbler, and a honey-eater—are extinct.

EXTINCT

RAILS (Rallidae):
  Rallus [Nesolimnas] dieffenbachii, Chatham Island only. Extinct 1840–1845
  Rallus (Cabalus) modestus. Chatham Island, Mangare Island. Extinct 1895–1900
FERNBIRD (Sylviidae): *Bowdleria rufescens.* Pitt Island, Mangare Island. Extinct 1890–1894
BELLBIRD (Meliphagidae): *Anthornis melanura melanocephala.* All islands. Last seen 1906 on Little Mangare

Extirpation of these birds was no doubt due to many complex causes now quite unknown. However, two facts indicate that the major causes were directly related to the advent of man. First, the two rails and the fernbird nested and fed on the ground. Second, these three disappeared rapidly as men and their cats, dogs, pigs, and particularly rats, infested the birds' habitat. As early as 1840 Dieffenbach recorded black rats (*R. rattus*) on Chatham Island. These were probably introduced by parties of sealers or whalers, or accidentally from a wreck before 1843, when the islands were first permanently settled by Europeans. Fires after that destroyed much habitat.

The disappearance of the bellbirds (*Anthornis*), like that of other members of the family on the Hawaiian Islands, cannot be explained.

Evidence of the role of introduced predators in reducing (if not actually extirpating) populations is afforded by the distribution of four surviving native forms. Once with a more extensive range, they are now confined to islets free from rats and cats.

SMALL POPULATIONS

SHOREBIRDS (Scolopacidae):

Thinornis novaeseelandiae. Now confined to South East Island; formerly North and South Islands of New Zealand and Chatham Islands: Pitt and Mangare.

Coenocorypha aucklandica pusilla. Now confined to South East Island; formerly Mangare and perhaps other islands. Other subspecies occur on the Auckland, Snares, Antipodes Islands.

PARROT (Psittacidae): *Cyanoramphus auriceps forbesi.* Now confined to Little Mangare. Former distribution unknown but perhaps also Pitt and Mangare (see Fleming, 1939, p. 503).

FERNBIRD or WARBLER (Sylviidae): *Petroica (Miro) traversi.* Now confined to Little Mangare. Formerly Mangare and perhaps other islands, but original range unknown (see Fleming, 1939, p. 508).

Three local Chatham Island populations, representing subspecies still to be found elsewhere, have been extirpated from the large lakes of Chatham Island. They are two species of ducks and a bittern (*Anas castanea chlorotis, Anas rhynchotis,* and *Botaurus stellaris poecilopterus*).

Here, as on other Pacific Islands, the history of the impact of Europeans and their animals is obscure. We do not know how much the occupation of sealing parties affected the fauna, when rats were introduced, or any other details. There were no naturalists to observe the birds when the islands were discovered in 1791.

Even though the evidence of the last century for extirpation by Europeans and their introduced animals may be scanty, there can be little doubt that these have been recently the major causes. There is also even more convincing evidence here that the processes of extinction were in force before their coming. The following species have been recorded not only from bones found in the dunes of Chatham Island, but also (in the cases of the large rail and the raven) in a kitchen midden of the natives of Polynesian stock called Morioris, which people are themselves almost extinct.

KNOWN ONLY BY BONES

Chenopsis sumnerensis, a swan, said by Forbes (1892) to be larger than the Australian black swan, once was to be found on Chatham Island and South Island, New Zealand. According to Oliver (1945) it was close to *Cygnus.*

Coenocorypha chathamica was larger than *C. aucklandica pusilla,* which still exists on South East Island.

Diaphorapteryx hawkinsi, a large rail, was apparently related to the wood rails (*Tricholimnas*) of New Caledonia and Lord Howe Island and to the weka of New Zealand. Forbes found bones in a kitchen midden.

Palaeolimnas chathamensis, a large coot which perhaps should be placed in the genus *Fulica.*

Nestor meridionalis, the kaka, is a species of large parrot still to be found on New Zealand and only recently extirpated from Norfolk Island.

Paleocorax moriorum was close to the ravens and was not a good flier, according to Pycraft (1909). Forbes found bones in a kitchen midden on Chatham Island with those of the rail. The species occurred also on both North and South Islands.

The *Bounty Islands* (lat. 47° 41′ S., long. 179° 03′ E.) are a group of islets lying 360 miles east of South Island, New Zealand, and about the same distance southwest of the Chatham Islands. No land birds nest on them; there is little or no vegetation. No population of sea bird is in danger.

The *Antipodes Islands* (lat. 49° 40′ S., long. 178° 50′ E.), 400 miles from the coast of New Zealand and 100 miles south of Bounty, consist of an island 5 miles long by about 3 in width and two high rocky islets close by. A low shrub (*Coprosoma*) and a coarse tussock grass grow thickly in crevices and depressions in the hills, where the force of fierce gales is less. Known to breed here are two forms of parrots (*Cyanoramphus unicolor* and *Cyanoramphus novaezelandiae erythrotis*[1]), the New Zealand pipit (*Anthus n. novaeseelandiae*), and the British goldfinch (*Carduelis c. britannica*), which was introduced by man in New Zealand but extended its range naturally to this island. There are no human beings there, and attempted introductions of domestic animals have failed.

The *Auckland Islands* (lat. 50° 20′ S. long. 166° 18′ E.) are next largest in size to the Chatham Islands, having a total land area of 234 square miles. Eight islands form the group, of which Auckland and nearby Adams Island are the largest, the former being 25 miles long by about 15 in width. Four smaller islands, Enderby, Ewing, Ocean, and Rose, are just north of these, and Disappointment lies 5 miles to the westward. The nearest land is Stewart Island, 200 miles north, itself close to the southern tip of New Zealand.

The group lies well within the region of gales called "the roaring forties." These winds have much affected the trees of the forest, which lie prone for two thirds of their length even in protected places. Exposed ridges support only tussock grasses and moss.

Such gales blow almost all the time from westward; it is remarkable that the 20 forms that found their way to the Aucklands to breed did so from New Zealand. Of these one species and three subspecies are to be found nowhere else. Two, a duck and a rail, are extinct, and these two were somewhat specialized in form, particularly the rail, which was probably not at all a good flier.

---

[1] Recorded from Antipodes by Waite (1909) but not by the Ornithological Society of New Zealand (1953). Occurred on Macquarie Island also.

EXTINCT

Mergus australis
Rallus muelleri

Settlement of the islands has been only of a semipermanent nature. Parties of seal hunters were put ashore on most of the islands soon after 1810, and some of them remained over periods of years until the seals were all but gone. This period probably ended about 1830, for Captain Morrell (1832) says that there were almost none there during his visit, although a few seals were killed from time to time. A whaling company, under the leadership of the well-known whaling firm of Charles Enderby, established a small colony on Enderby Island, but apparently it only lasted during the two years 1847–49. Numbers of ships have been wrecked, some within Carnleys Harbor, between Auckland and Adams Islands. It would be remarkable if rats had not been introduced accidentally, but there is no mention of them either in Captain Musgrave's account of his long stay after the wreck of his ship or by any of the more scientific accounts.

Apparently the earliest introduction of domestic animals was in 1807, when pigs were put ashore. It may be that these were at least partly responsible for the extirpation of the rail, a ground-nesting bird, as Waite (1909) implies. Captain Musgrave wrote in 1863 of "a small animal like a cat which lives in holes and we frequently find feathers and egg shells about their holes." No one has determined what this animal may have been, but it was obviously destructive.

After 1850 cattle were introduced on Enderby and Rose, and goats on Ewing, Enderby, and Ocean. The latter have been particularly destructive to the plant life, according to Cockayne (1904). When the dogs, mentioned by the "Grafton" castaways, came ashore is not known.

The confiding nature of the birds, characteristic of so many island species, appears in the account of marooned sailors. Musgrave wrote in his log: "Amongst the birds with which the woods abound there are three kinds of songsters some of whom are so tame that they come and feed out of our hands. . . . They also come into the house in flocks when the hawks [presumably *Falco novaeseelandiae*] come after them . . . and then there is the green parrot [*Cyanoramphus*] and the Robin Red Breast [*Petroica*]. . . . We could put out our hands and take hold of them but we do not disturb them. . . . "

In fact all these birds have survived. A list of the small populations confined to these islands follows:

SMALL POPULATIONS

Coenocorypha aucklandica aucklandica. Range: Formerly throughout the group. Now confined to Ewing and Adams Island.
Petroica macrocephala marrineri. Now confined to Adams, Enderby, and Rose Islands.

Eight British song birds, introduced by man on New Zealand, have flown to the Aucklands and are known to breed there (Williams, 1953, p. 679).

Campbell Island, 148 miles southeast of the Aucklands and 318 miles from New Zealand, has about 43 square miles of land surface, resembling Antipodes in having but little vegetation. Only four native birds breed there. Of these the only form confined to the island is a subspecies of the gray duck of New Zealand and outlying islands. It is *Anas castanea nesiotis* Fleming 1935, so rare that a small party of men with guns could probably wipe it out.

Macquarie Island (lat. 54° 45′ S., long. 158° 40′ E.) is the southwesternmost of New Zealand Islands, being 340 miles south of the Auckland Islands and 540 miles from New Zealand. The land area is about 100 square miles with hills rising to 1,421 feet. Lying as they do in the region of powerful westerly gales, the vegetation is limited to tussock grasses and mosses. Nevertheless, five populations of land birds certainly have bred there. One of these, the weka (*Gallirallus australis*), was introduced. Two (*Zosterops lateralis* and *Sturnus vulgaris*) have arrived naturally within the past 40 years. Local populations of two characteristic New Zealand species, the oldest inhabitants of the island, are extinct. *Rallus philippensis assimilis,*[1] still to be found on New Zealand and nearby islands, has not been recorded since 1880. Since then six or more scientific expeditions have visited the group; one party from the Mawson Expedition of 1911 remained for two years, but no one has seen these birds. Perhaps both disappeared between 1880, when J. H. Scott visited the island and recorded both as common, and 1894 when Hamilton was unable to find them in spite of search. A sealing party was questioned but had no information.

A little green parrot has been extirpated.

EXTINCT

Cyanoramphus novaezelandiae erythrotis

Falla (1937) sums up the situation: ". . . no full account of the fauna in its pristine condition was ever written, and seventy years of occupation

---

[1] *R. p. macquariensis* Hutton, 1879, is a synonym, according to Falla, *in litt.*

by sealers and the depredations of imported vermin elapsed before any such survey was attempted." The vermin included "innumerable" wild dogs which were present as early as 1815. Rats, as is so often the case, were not reported until an observing naturalist saw them much later, but because so many ships were wrecked here it is probable that they were present. Certainly large and wild "domestic" cats increased between 1880 and 1894 as had the bellicose weka. Members of the Mawson expeditions believed that these birds were interfering with the nesting petrels. It is not improbable that they contributed to the extirpation of the ground nesting rails and parrot.

Arc IV. Eastern Pacific (Line Islands south through Tuamotus).

The *Line Islands* (so called because some of them are near the Equator) are not properly a group at all. Rather they are a chain of nine small atolls separated by great distances and great depths of water, extending from Palmyra (lat. 5° 53′ N.), 960 miles south of Honolulu, south to Flint Island (lat. 11° 26′ S.), which is 390 miles north of Tahiti. The four most northern are Palmyra, Washington, Fanning, and Christmas, all within 150–200 miles of each other. Malden and Starbuck are isolated about equidistant (450 miles) between the northern group and the three southern islands, Caroline, Vostok, and Flint. Howland and Baker Islands (lat. 0° 49′ N., long. 176° 43′ W.), although they are nearer the Phoenix group and 600 miles west of Christmas Island, should be included here.

Certainly such distances as these are strong barriers to the passage of land birds. The islands themselves offer but little encouragement to immigrants and there probably have not been many. However, history indicates that there has been much human interference, and there is direct evidence that some damage has been done to their fauna.

A subspecies of the gadwall of North America is known to have been extirpated from Washington Island (lat. 40° 43′ N., long. 160° 25′ W.). This island is extraordinary in that it is an atoll raised above the sea and having in its center a fresh-water lake rather than a lagoon. Whether because of introduced predators or overshooting or both, the birds are now extinct. According to Hutchinson (1950), the peat bog that has accumulated in and replaced the western part of the lake "cannot be many centuries old." The subspecies therefore must have evolved in a very short time indeed.

A population of birds, which may be identified with the bird Latham called the "barred phalarope" (*Aechmorhynchus cancellatus*), once lived on Christmas Island (lat. 1° 55′ N., long. 157° 20′ W.) but is now extinct there.

A representative of Polynesian warbler (*Acrocephalus aequinoctialis*), inhabitant of so many dry Pacific atolls, is still to be found in numbers. Other than the beautiful little parrot (*Vini kuhli*), which was probably introduced by the Polynesians, there are no land birds.

Naturalists of Captain Cook's last voyage noted only 13, or perhaps 14, birds on Christmas Island in December 1777; except for the resident shorebird *Aechmorhynchus*, this is more or less a fair sample of what may be seen on the Line Islands. According to Stresemann (1950) they saw a large petrel (*Pterodroma alba*), which was no doubt nesting there, and four species of terns (*Procelsterna australis, Gygis alba candida, Anous stolidus pileatus, Sterna fuscata oahensis*). The last named was breeding. A resident shorebird (*Aechmorhynchus cancellatus*) and the migrant golden plover from North America (*Pluvialis dominicus fulvus*), as well as the bristle-thighed curlew, which breeds in Alaska, were present at that time. Two ground-nesting boobies, the common brown one (*Sula leucogaster plotus*) and the blue-faced (*Sula dactylatra personata*), are mentioned. No doubt raiding the tern colonies, and perhaps their own, were the ubiquitous frigatebirds (*Fregata minor palmerstoni*). Bo'sn-birds (*Phaethon rubricauda melanorhynchus*) were to be seen, and perhaps in those low latitudes they had begun to breed. The only land bird was the little gray warbler (*Acrocephalus aequinoctialis*). Land crabs, little lizards, and small rats were all probably accidental introductions by Polynesians, who must have carried in their great canoes coconuts and stems of bananas in which such things could hide.

As their names indicate, these islands were almost all made known to the world by American traders and whaling captains. Certainly the Polynesians had landed on them long before, but only temporarily in their search for larger and more hospitable places. They introduced the small Polynesian rat (*R. exulans*). What damage these may have done during the years when populations had attained numbers above the point of subsistence cannot be determined. Where they now occur they are said to be quite harmless to birds.

Within the past century small colonies of men have been established on all these islands either for the purpose of digging guano or drying copra. Because guano supplies are almost exhausted and substitutes have been found for copra, it may be that even small bird populations will be safe from depredation; those that have survived the depredations of the enemies imported by men will now probably continue until some more formidable danger appears.

Hutchinson's (1950) work on the guano deposits of these islands indicates strongly that there have been great changes in the rainfall and climate of the central Pacific. It follows logically that there have been great changes in the sea-bird populations.

Summarizing this work, perhaps rather roughly, it may be said that there are two types of Central Pacific islands, the wet and the dry. The wet are well wooded and harbor comparatively few sea birds. The absence of accumulation of much recent guano indicates that large colonies of terns and boobies have not occupied them for thousands of years. Ancient guano deposits prove that such birds were formerly present in great numbers. Dry islands have no woody plants, only a stunted and scattered scrub. Populations of terns and boobies are large and consequently accumulations of recent guano are considerable, although varying.

The dry islands are Howland, Baker, and five of the Phoenix group as well as Jarvis and Malden in the Line Islands. They are all close to the equator, between lat. 13° N. and 4° S.

Wet islands lie both northward and southward of these. To the north are Palmyra, Washington, and perhaps Fanning; to the south Starbuck, Caroline, and Flint.

Christmas Island, situated on the border between wet and dry belts, is intermediate. But little guano has been deposited, although large colonies of terns have nested there since at least 1777.

Obviously the rainfall of the wet islands, lying relatively far from the Equator, must once have been much less. At that remote time no trees grew on them; their ground was covered with nesting terns and boobies during a great part of the year. In view of the extremely strong predilection of terns to return to known nesting sites, and the enormous distances separating the islands, it is almost a certainty that many populations were extirpated by this change. Probably two ground-nesting boobies (*Sula leucogaster* and *S. dactylatra*) and the wideawake or sooty tern (*Sterna fuscata*) were the largest colonies of species involved.

Hutchinson offers an explanation for this considerable change in rainfall which he says is in accordance with evidence offered by the Chincha Islands off the coast of Peru. It involves another factor governing sea-bird populations. This hypothesis holds that there has been in the past a "shifting of the climatic and oceanographic zonation." Because, as he says, the islands richest in recent guano are closest to the upwelling of the south equatorial current, a change in the current would obviously cause a change in bird populations, for food supply would be altered.

The *Marquesas Islands* (lat. 7° 40′ to 10° 40′ S., long. 138° 20′ to 140° 55′ W.) are, except for the Hawaiian Islands, the remote group in the Pacific. They are also rather small and widely separated. Thirteen islands have a total land area of 480 square miles. The two largest are Hiva-oa (Dominica), with 95 square miles of mountainous and beautifully forested country, and Nuku Hiva (Madison), whose mountains are almost 4,000 feet in height. All who have seen it say that it has strikingly beautiful scenery. Hatutu, farthest north, and Fatu Hiva in the south are isolated 30 and 25 miles from their nearest neighbors. Others are 10 to 20 miles apart; marked subspecific variation in the flycatchers and warblers reflect this isolation. Both Hatutu and Fatu Huku, last resort of the rare ground dove (*Gallicolumba*), are uninhabited, although it is said that the Polynesians sometimes visit them. Ten of these harbor populations of land birds, and these are listed in the table below.

DISTRIBUTION OF ENDEMIC SPECIES

Species	Hatutu	Eiao	Nuku Hiva	Huahuna	Hua Pu	Fatu Huku	Hiva-oa	Tahahuku	Motane	Fatu Hiva	Number of subspecies
Ptilinopus mercieri	—	—	X	—	—	—	X	—	—	—	2
Ptilinopus dupetithouarsi	—	—	X	X	X	—	X	X	X	X	2
Ducula galeata	—	—	X	—	—	—	—	—	—	—	—
Gallicolumba rubescens	X	—	—	—	—	X	—	—	—	—	—
Coriphilus ultramarinus	—	—	X	—	X	—	—	—	—	—	—
Halcyon [*Todirhamphus*] *godeffroyi*	—	—	—	—	—	—	—	X	X	—	—
Pomarea mendozae	—	X	X	X	X	—	X	X	X	X	4
Pomarea whitneyi	—	—	—	—	—	—	—	—	—	X	—
Acrocephalus caffra	X	X	X	X	X	—	—	—	X	X	7

No form is known to have been extirpated from the Marquesas. Bird populations were never large, the birds confiding and the habitat somewhat limited. Because the islands were discovered as early as 1598 and rediscovered in 1774 by Europeans, this apparent immunity is surprising.

No completely satisfactory explanation can be given, but perhaps a part of the reason lies in the fact that the islands were not as popular a resort of whalers as was Tahiti. Both Edmund Fanning and Herman Melville, who visited the islands in 1798 and 1842, respectively, reported the Marquesans to have had a reputation for ferocity, as did the Fiji Islanders. A whaling captain would have gone to Tahiti to repair his ship, and thus the Marquesas were spared the many introductions of rats and other predators. Even after 1842 when the French took the islands, they appear to have had less contact with civilization than the Society Islands. An alternative is possible: that species have been extirpated without having been recorded and even more possible that populations have disappeared from islands without our knowing.

The *Society Islands* are about 150 miles southwest of the Tuamotus. Their importance to the birds lies in their relatively large size and the number of habitats of Tahiti, which has a total area of 600 square miles, and its neighbor Moorea (Eimeo), only 7½ miles west and with a total area of 50 square miles. Four smaller islands lie 60 miles to the westward. These are the Leeward Islands. Their names are Bora-Bora, Raiatea, and Huahine. Largest, Raiatea, has about 150 miles of land surface; others have about 5 to 14. All these islands are extinct volcanoes—1,000 feet in height on smaller islands to 7,321 feet on Tahiti itself; all are beautifully forested.

Twenty-three miles farther to the westward are three isolated atolls, Mopiti, Mopelia (Mopihaa), and Funua Ura (Scilly). Providing only the most restricted habitat of coconuts, *Scaevola,* and other low-bush characteristic of such islands, they have only one species of land bird.

The following table shows that of the 12 species of land birds confined to the Society Islands, 6 are quite extinct and 5 have disappeared from the island of Tahiti, although they still exist on other, smaller islands. A sea bird, species of petrel (*Pterodroma rostrata rostrata*), which still nests on other Pacific islands, has also disappeared from Tahiti. Derscheid (1939) called attention to a note in the diary of James Morrison, boatswain's mate of the *Bounty,* who remarked that there was a bird about the size of a goose which nested in the mountains and was good to eat. Probably he referred either to this species or to *Puffinus l'herminieri polynesiae.* Certainly three (a shorebird, a rail, and a parrot) of a total of 17 species of land birds are extinct.

EXTINCT

Prosobonia leucoptera
Rallus pacificus
Cyanoramphus zealandicus
Cyanoramphus ulietanus
Turdus ulietensis
Aplonis mavornata

These had in common that they were quite distinct in form and apparently were long-establishd residents. This is particularly true of the rail and the shorebird, whose exact relationships cannot now be traced. As far as is known they occurred on this group only, but we cannot be sure that their relations on other islands did not disappear before the coming of Europeans. Both the shorebird and the rail were terrestrial; they probably fell prey to rats very quickly. Certainly they were not found by Lesson and Garnot, naturalists of the French frigate *Coquille* in 1823, or by subsequent expeditions. The parrot was known to exist until 1844, two years after the islands were annexed by France.

Exactly when the ground dove (*Gallicolumba*) and the blue parakeet (*Vini*) disappeared from Tahiti and Moorea is not known. Probably the kingfisher (*Halcyon tuta*) never occurred on Tahiti, and the records of Latham are in error, according to Stresemann (1950).

There has been some doubt in the past as to the actual provenance of *Cyanoramphus ulietanus, Turdus ulietensis,* and *Aplonis mavornata,* but there can be little doubt that they came from the Society Islands. Certainly the first two were recorded by Forster, naturalist on Captain Cook's voyage, in his diary (Stresemann, 1950). Nothing resembling them has been found since on Ulieta (Raiatea) or Tahiti.

DISTRIBUTION OF CERTAIN LAND BIRDS CONFINED TO THE TUAMOTU, SOCIETY, AND AUSTRAL ISLANDS

Species	Tuamotu	Tahiti	Moorea	Huahine	Raiatea	Bora Bora	Mopelia	Mopiti	Henderson
Prosobonia leucoptera	—	E	E	—	—	—	—	—	—
Aechmorhynchus parvirostris	X	—	—	—	—	—	—	—	—
Rallus pacificus	—	E	E	—	—	—	—	—	—
Nesophylax ater	—	—	—	—	—	—	—	—	X
Ptilinopus purpuratus	X	X	X	—	—	—	—	—	—
Ptilinopus huttoni	—	—	—	—	—	—	—	—	X
Gallicolumba erythroptera	X	E	X	—	—	—	—	—	—
Vini stepheni	—	—	—	—	—	—	—	—	X
Vini peruviana	—	E	E	—	—	—	—	X	—
Cyanoramphus zealandicus	—	E	—	—	—	—	—	—	—
Halcyon tuta	—	—	—	X	X	X	—	—	—
Halcyon venerata	X	X	—	—	—	—	—	—	—
Pomarea nigra	—	X	—	—	—	—	—	—	—
Acrocephalus caffra	—	X	X	—	—	—	—	—	—
Acrocephalus atypha	X	—	—	—	—	—	—	—	—
Acrocephalus vaughni	—	—	—	—	—	—	—	—	X

"E" indicates extinct. Such restricted populations on isolated islands are potentially in danger of extinction.

The advent of Polynesians probably brought considerable change to the fauna for they brought with them the small rat, pigs, dogs, and, almost without question, birds in cages. This may account for the peculiar distribution of certain species. Certainly it is also probable that bird diseases were disseminated in this way, although it must be said that we know nothing specific of these. The Polynesian rat (*R. exulans*) appears now to be perfectly harmless to bird populations, but when first introduced it may not have been. This cannot be said of the rats of European ships; these were always deadly when introduced on islands and they are said to have been found soon after Cook landed. Certainly Peale found them in 1840.

The important difference between Tahiti and the Fiji Islands lies in the attitude of natives toward Europeans, and the consequent use of Tahiti by whalers and traders before permanent settlement in about the year 1830. The introduction of rats was disastrous. After permanent settlement, other effects of civilization began to be apparent. More of the lowland country along the coasts was cultivated, and introduced plants began to force out the indigenous ones. Bennett in 1840 speaks of the imported guava bush which had taken over much of this land, and no doubt bird habitats were preempted by many introductions. This same traveler lists the animals thus: "limited to swine, dogs and rats indigenous [these were introduced by Polynesians and were not indigenous] and horses, oxen, goats, asses and cats. Imported rabbits have been introduced upon the coral islets, but they have not increased, and it is suspected that the rats destroy their young."

Older editions of the British Admiralty Sailing Directions, as well as Bennett, speak of facilities for "heaving down" ships for cleaning and repair at Papeete in earlier days. Every time a ship was beached in this way hundreds of rats escaped to the land.

The *Tuamotu* or *Low Archipelago,* together with the *Mangareva* or *Gambier* group and the *Southeastern* Islands (lat. 15° to 27° S., long. 124° to 149° W.), complete the most easterly and final chain of Pacific islands. There are 80 islands in the groups, as well as hundreds of small islets and rocks, lying in the general direction northwest to southeast. Only five of these are of the high, volcanic, and forested type, affording habitat for land bird species: in the northwestern Tuamotus, Makatea, and Tikei, and in the extreme southeast, Managareva, and the distant Henderson and Pitcairn Islands.

The endemic shorebird *Aechmorhynchus,* unlike all other birds of the Pacific islands (except Hawaii), is probably a very old colonist from North America.

About 350 miles southeast of Tahiti are the *Cook Islands* (lat. 18° 48' to 22° S., long. 157° 20' to 160° W.). There are seven high, volcanic, and heavily forested islands 20 to 50 miles apart. Curiously, there are only two or three species confined to the group and to be found nowhere else. A representative of the widespread fruit pigeon (*Ptilinopus rarotongensis*) and the South Sea flycatcher (*Pomarea dimidiata*) are both confined to the Rarotonga. Both are said to be fairly common. The island is 20 miles in circumference and its volcanic peak is 2,920 feet high.

Komandorskie and Aleutian Islands

The *Komandorskie* and *Aleutian Islands* form a 3,420-mile loop from across the North Pacific Kamchatka to the Alaska Peninsula (long. 160° E. to 163° W., lat. 55° to 51° N.). The Komandorskies are somewhat more isolated than the others, being about equidistant (200 miles) between the Russian coast and Attu, the nearest Aleutian island. There are two of these barren, rocky and cold islands called Bering and Copper, chiefly known for having been the place where the explorer Witus Bering died in 1734, and for its fur-seal nurseries.

According to Stejneger the all-but-flightless cormorant *Phalacrocorax perspicillatus,* which was confined to rocks off Bering Island, was extirpated about 1850 by ruthless hunting.

It is probable a Canada goose (*Branta canadensis asiatica*)[1] has been extirpated from Bering Island. On the Aleutian Islands the race *leucopareia* nested in great numbers on the grassy hills as well as on outlying rocks in 1887–1889, but were shot and otherwise persecuted by the Aleuts, according to Dall, Nelson, and Turner. Bent (1912) recorded "a few" seen on Kiska, Adak, Atka, and Attu Islands in the summer of 1911, and one was collected as she flew from the nest. None were seen by Laing during a cruise through the islands in the spring and summer of 1924, nor did Wilson (1948) during a stay of 13 months on Attu in 1944 and 1945 see a single one. They are now nesting on Buldir Island only (1966). Austin and Kuroda (1953) report that Canada geese are no longer to be seen in Japan in autumn and winter. The last certain record was February 7, 1929. It is not known with certainty where those birds came from, but quite probably they nested in the Kuriles, Komandorskies.

New Zealand

New Zealand (between lat. 34° 15′ and 47° 30′ S.; long. 166° 30′ and 178° 45′ E.) is the antipode of the coast of the Bay of Biscay and central Spain. It is composed of two large islands, separated by a strait only 12 miles wide at its narrowest point. The area of North Island is about 44,131 square miles, that of South Island 58,120, and Stewart Island, 15 miles south, is 665 square miles. Except for three volcanic peaks, the mountains of North Island are no more than 4,000 feet, but Mount Ruapehu, a snow-capped active volcano, is 9,008 feet high. On the South

---

[1] The taxonomic status of these birds has not been settled. See Aldrich, 1946b, p. 95 (*Branta hutchinsii asiatica* described from Bering Island); Delacour, 1954, p. 173, who recognizes this form without doubt.

Island a great chain of mountains rising to more than 10,000 feet occupies almost the entire western and central areas. The scenery, characterized by mountain torrents fed by melting glaciers, pouring in great cascades into deep-cut fiords, is wild and awesome. At sea level the climate as a rule is mild although of course more so in the north. The mean annual temperature of New Zealand is 63° F. in summer and 48° F. in winter, 57° on the North Island and 52° on the South.

Before the coming of the Polynesians at an unknown date, highly endemic, unique forests are said to have covered almost the entire country below 5,000 feet. It is probable, however, that extensive grasslands existed, even on the North Island, but more extensively on the South Island. These were the habitats of the great flightless birds, the moas. There can be no doubt that very much forest has been destroyed, especially since the coming of Europeans. Because the Maoris' fields were small and the populations not large, it is doubtful that much bird habitat was destroyed by them, although they no doubt did burn forests in places. According to Cumberland (1941) forests have been reduced from 97,200 square miles to 53,000 square miles. This author is of the opinion that 25,000 square miles, almost half of the South Island, was covered by tussock grass in 1840, when European settlement began there. Statistics of this sort vary greatly.

In comparison with New Guinea and other of the larger islands of the Asiatic continental shelf, New Zealand's actual land area is smaller and its avifauna is likewise smaller. It has only 57 species of land and freshwater birds (excluding those introduced by man), as against 260 in New Guinea. It is separated from Australia, the nearest continental land-mass, by 1,400 miles, and in at least partial consequence the proportion of endemic species is high; there are 34 of these. In all probability New Zealand has received all birds over water, whereas New Guinea was apparently connected to Australia within recent geological times. Two facts summarize this situation. First, the size of the island naturally restricts habitat and consequently the populations of birds. Second, its relative isolation from the nearest continent has permitted fewer species to colonize and forced those that have done so into reproductive isolation that has handicapped them in survival. This comparison leads to exactly the same conclusions as that of the smaller islands of the Pacific Ocean, not only in these respects but also in point of extinction. Whereas the larger islands nearer the coasts have lost no bird forms, New Zealand has lost certainly two and perhaps three.

EXTINCT [1]

Coturnix novae-zelandiae novae-zelandiae
Sceloglaux albifacies rufifacies
Xenicus lyalli
Heteralocha acutirostris

A distinct subspecies of quail disappeared about 1868 from North Island and 1871 from South Island. It was a representative of a species to be found also in Australia (*Coturnix novae-zelandiae pectoralis*). Australian birds are thought to have been introduced on the North Island a year or so before the final extirpation of the native birds there. Introductions on the South Island took place about the same time, but because the native birds were to be found only in remote places by that time, it is unlikely that there was ever opportunity for the two subspecies to interbreed.

It has been suggested by Falla that the Australian brown quail [2] (*Synoicus*) may once have had a native New Zealand form, which is now extinct. The evidence is much confused by introduction and subsequent interbreeding of two subspecies (*Synoicus ypsilophorus* and *australis*).

The second was a subspecies of an owl that occurs now on South Island only, the last reported sighting on the North Island having been in 1886.

The third was the small but famous Stephen Island "wren," which was discovered and extirpated by the lighthouse keeper's cat. It was apparently quite specialized and confined to a tiny island in Cook Strait.

The fourth was an extremely peculiar endemic genus. Described as "curious" and "confiding" and flying but little, it was peculiarly easy prey, like so many such species confined to islands. Much of the forest habitat had been destroyed before the last bird died.

Here as elsewhere the exact reasons for extirpation cannot be assigned with any certainty, but it appears that they are directly related, in these known cases, to interference brought about by the coming of men. Ten populations have been much reduced in numbers since 1840, when Dieffenbach and other naturalists knew the country. These also are species confined to the islands, showing again that endemism is in many cases a predisposing factor to extinction and that the following populations are in danger.

[1] The weka (*Gallirallus australis*) disappeared from South Island, but was reintroduced from Chatham Island, a stock which had been introduced on Chatham from South Island in 1905 (*Oliver, 1955, p. 370).

[2] See Turbott and Buddle, 1948, p. 327, and Checklist of New Zealand Birds, 1953, p. 37.

SMALL POPULATIONS

TAKAHE (Rallidae): *Porphyrio* (*Notornis*) *mantelli hochstsetteri*
LAUGHING OWL (Strigidae): *Sceloglaux albifacies albifacies*
KAKAPO (Owl): *Strigops habroptilus*
BUSH "WREN" (Xenicidae): *Xenicus longipes stokesii*
SADDLEBACK (Eulabatidae): *Philesturnus carunculatus refusater; P. c. carunculatus*
WATTLEBIRDS (Eulabatidae): *Callaeas cinerea wilsoni   C. c. cinerea*
NEW ZEALAND "THRUSHES" (Turnagridae): *Turnagra capensis tanagra; T. c. capensis*
STITCHBIRD (Meliphagidae): *Notiomystis cincta*

The Moas

The extinct family of moas, known only from New Zealand, well illustrates the theory, as indeed do all fossil forms, that in the long view of nature extinction is the result of certain phases of evolution. In this case, however, we have the outline of a history not to be seen elsewhere or in other groups of birds, for the fossil evidence is more complete.

New Zealand is thought to have been habitable for birds as long ago as Eocene time. The first evidence that we have of the moa is Miocene or early Pliocene. This earliest population (*Anomalapteryx antiquus*) is known only from two leg bones, but none like them have been found in later deposits, and it may be assumed that a population had become extinct long before the arrival of man in New Zealand. The same may be said of other populations which disappeared during Pliocene and early Pleistocene. According to Archey (1941) the last of the moas were probably exterminated by natives before the great immigration of Maoris in A.D. 1350. Oliver (1949) was of the opinion that one species at least (*Megalapteryx didinus*) was to be found on the South Island as recently as 300 years ago. He cautiously refrained from assigning an immediate cause for extinction but pointed out that the birds disappeared shortly (in geological time) after a wet climatic period, during which much of the birds' grassland habitats were largely overgrown forests. This was thought to have taken place to a greater extent on North Island, and both authors agree that the birds disappeared from there first. Oliver implies that men may have had some influence on the populations, for he indicated that the birds were all dead within approximately 300 years after the arrival of the Maoris.

Certainly heads and necks of two specimens of *Megalapteryx didinus,* with ligaments, skin, eyes, parts of trachea, and in one case feathers, are preserved in the British Museum and in the Dominion Museum, Wellington, where there is also a huge leg with skin and feathers attached. "They look," wrote Oliver, "as though they had died only a few years ago." These remains were found in relatively dry limestone caves near Queenstown and Cromwell in the Otago District of South Island. There can be no question that humans killed and ate moas, as the investigations of Teviotdale (1924) at Shag River, South Island, have shown. Whether those humans were the people now known as Maoris is not yet known. The only certain date has been determined by radio-carbon analysis of seeds and twigs found in a moa's stomach. They were eaten 670 years ago.[1] Bones have been found throughout kitchen-midden refuse, in one place with the skull of a dog, in another just above a mass of fish scales. Eggs have been found punctured artificially. Bones of six species have been discovered in such places.

Fires that may possibly have been set by the Maori apparently destroyed breeding colonies of the birds among grasses and brush near the beaches.

The fact is that no good conclusive reasons for the extinction of the moas can be given. As in the case of most modern birds, it may be said only that the time of the occurrence of certain events is significant. In this case a long wet era which destroyed habitat, and the later advent of man, may have been fateful phenomena.

The following arrangement, the most recent, is that of Oliver (1949), who admitted 2 families, 7 genera, and 27 species. Archey (1941) admitted also 2 families, but only 6 genera and 20 species by suppressing those names that represented bones of intermediate size. Probably the number of species will eventually be considerably reduced. It is a curious coincidence that Amadon (1950), in the latest revision of the endemic family Drepanididae of the Hawaiian Islands, admits 2 subfamilies, 10 genera, and 22 species.

Pachyornis (Mauiornis) septentrionalis Oliver, 1949: North Island
Pachyornis (Mauiornis) mappini Archey, 1941: North Island
Pachyornis (Pounamua) murihiku Oliver, 1949: Known only from the type locality in southern South Island
Pachyornis (Pachyornis) australis Oliver, 1949: South Island, known only from type locality
Pachyornis (Pachyornis) elephantopus (Owen, 1865): South Island, found in kitchen middens
Emeus huttonii (Owen, 1879): South Island, found in Maori kitchen middens
Emeus crassus (Owen, 1846): South Island, found in Maori kitchen middens
Eurapteryx curtus (Owen, 1846): North Island
?*Eurapteryx tane* Oliver, 1949: North Island
Eurapteryx geranoides (Owen, 1848): North Island
Eurapteryx gravis (Owen, 1876): North Island and South Island
Zelornis exilis Hutton, 1897: Known only from type locality Wangaehu, North Island
Zelornis haasti (Rothschild, 1907): South Island

[1] Deevey, 1954. Apparently he doubts that moa remains have actually been found in kitchen middens.

Anomalopteryx parvus (Owen, 1883): North Island and South Island

Anomalopteryx oweni (Haast, 1896): North Island, found in kitchen middens

Anomalopteryx didiformis (Owen, 1844): North Island and South Island

Anomalopteryx antiquus Hutton, 1892: Known only by a tibia and fragment of metatarsus from Miocene or lower Pliocene formation of Eleniti Valley, near Timaru, southern Canterbury, South Island

Megalapteryx hectori Haast, 1886: South Island, found in Maori kitchen middens

Megalapteryx didinus (Owen, 1883): South Island, found in Maori kitchen middens

Megalapteryx benhami Archey, 1941: South Island

Dinornis gazella Oliver, 1949: North Island

Dinornis novaezealandiae Owen, 1843: North Island and South Island

Dinornis torosus Hutton, 1891: South Island

Dinornis ingens Owen, 1844: North Island, South Island

Dinornis robustus Owen, 1846: South Island. A foot and leg with skin still adhering are preserved in the Otago Museum.

Dinornis hercules Oliver, 1949: North Island

Dinornis giganteus Owen, 1844: North Island

Dinornis maximus Haast, 1869: South Island

(?)*Dinornis queenslandiae* DeVis: Known only by the type, a fragment of the head of a femur, found at King's Creek, Queensland, Australia; said to be "post-tertiary"

Oliver (1949), who compared this type with moa bones, asserted that it belonged to the genus *Pachyornis*. However, he writes: "Such of the bone as was discovered is imperfect in that a small portion of the head and a large part of the great trochanter are missing. It is thus impossible to make a satisfactory comparison with other femora." In view of this fact, and because McDowell (in Mayr and Amadon, 1951) is of the opinion "that the development of the trochanter is so different that it seems impossible that *queenslandiae* belongs in the *Dinornithidae*," the true identity of this fragment must remain in question.

Islands of the Indian Ocean

The Mascarene Islands

In the Indian Ocean, between 400 and 500 miles east off the coast of Madagascar, are three small islands named for a Portuguese navigator, Mascarenhas, who came upon them in 1505. Whether or not the Arabs already knew of their existence, they were uninhabited by men at that time.

Because of their theatrically beautiful mountains, clear streams, and mild temperatures, they were compared favorably with the Garden of Eden by the Dutch, French, and English travelers and settlers since the sixteenth century. Little of Portuguese comment has been recorded, and apparently no effort was made by them to settle the islands. The Dutch in 1598 introduced a few convicts on Mauritius, but no permanent settlement was established by them either. However, the position of the islands had become generally known to sailors, and they became a port of call for fresh provisions. When game became scarce on Réunion early in the eighteenth century, hunting parties were sent to Rodriguez to replenish larders of a growing human population. It was due largely to this commercialization of wildlife that the extirpation of the birds was accomplished.

Excluding those named forms which are known only from single bones, and the "dreamed-up birds" listed by authors, a conservative estimate of land and fresh-water species (zoogeographical species) of the group is 28, of which all but one are, or were, species to be found nowhere else (endemic). In all, 24 are certainly extinct, and perhaps 25, if the collared parakeet of Rodriguez is added.

The largest island, Réunion or Bourbon, has 970 square miles of land surface. Its geologically recent peaks rise to 10,069 feet above the sea and so steeply that there can be little cultivation on their sides, but the plateau is highly cultivated.

Mountains of Mauritius (Île de France or Cerne) are less steep and much cultivated now. In spite of the fact that it is a little farther from Madagascar, the nearest large land mass, and also a little smaller (720 square miles), 19 species of birds certainly were once to be found there, as against 13 on Réunion. However, the steep slopes of Réunion offer less habitat for birds, and although there may have been a few species of which no satisfactory information remains, it is theoretically quite as possible that Réunion never possessed as many species.

If the populations of Mauritius are compared with those of the Society Islands, which are comparable in size and relative isolation, no great difference in numbers of zoogeographical or superspecies is apparent, although percentage of extinct species is slightly higher. The striking difference is in degree of endemicity, in degree of specialization of the forms, exemplified by the distinct family Raphidae, the dodos.

The island of Rodriguez, 400 miles east of Mauritius, is the smallest (43 square miles of land surface) and most isolated of the group. It is characterized by low hills of volcanic origin, the highest point being about 1,300 feet above sea level. A reef, from which islets emerge, almost surrounds it. When the little colony of French settlers, whose doings were chronicled by François Leguat, arrived in 1691 the entire island was covered with forest, but this was gradually destroyed and remains are found only on the cays of the reef. Most of the main island is cultivated in coffee and tobacco. Only nine species of land and fresh-water birds are known to have inhabited it.

BIRDS OF THE MASCARENE ISLANDS

Species and island	Extinct	Known by bones and/or travelers' stories only
Anas theodori: Mauritius	X	X
Sarcidiornis mauritianus: Mauritius	X	X
Anhinga nanus: Mauritius	X	X
Falco punctatus: Mauritius	—	—
Phaenicopterus sp.: Mauritius	X	X
Dryolimnas c. cuvieri: Mauritius	—	—
Aphanapteryx bonasia: Mauritius	X	X
Aphanapteryx leguati: Rodriguez	X	X
Fulica newtoni: Réunion	X	X
Ardea duboisi: Réunion	X	X
Ardea mauritiana: Mauritius	X	X
Ardea megacephala: Rodriguez	X	X
Raphus cucullatus: Mauritius	X	X
Raphus solitarius: Réunion	X	X
Pezophaps solitaria: Rodriguez	X	X
Alectroenas nitidissima: Mauritius	X	—
Nesoenas mayeri: Mauritius	—	—
Columba rodericana: Rodriguez	X	X
?*Streptopelia picturata* subsp.: Rodriguez (see p. 303)	—	—
Psittacula subsp: Réunion	X	X
Psittacula echo: Mauritius	—	—
Psittacula exsul: Rodriguez	X?	—
Lophopsittacus mauritianus: Mauritius	X	X
Necropsittacus rodericanus: Rodriguez	X	X
Mascarinus mascarinus: Réunion	X	—
Tyto sauzieri: Réunion	X	X
Athene murivora: Rodriguez	X	X
Collocalia f. francica: Mauritius	—	—
Fregilupus varius: Réunion	X	—
Necropsar rodericanus: Rodriguez	X	X
Tchitrea b. bourbonnensis: Réunion	—	—
Tchitrea b. desolata: Mauritius	—	—
Saxicola tectes: Réunion	—	—
Microscelis b. borbonica: Réunion	—	—
Microscelis b. olivacea: Mauritius	—	—
Coquus typicus: Mauritius	—	—
Coquus newtoni: Réunion	—	—
Malacacirops b. borbonicus: Réunion	—	—
Malacacirops b. mauritianus: Mauritius	—	—
Bebrornis rodericanus: Rodriguez	—	—
Zosterops c. curvirostris: Mauritius	—	—
Zosterops c. haesitata: Réunion	—	—

Of the many animals introduced during the past 250 years, no doubt the ubiquitous black rats (*R. rattus*) and brown rats (*R. norvegicus*) have been the greatest enemies of the birds. According to Carié (1916), who published an account of known introduced animals, they came from

Bombay. Bernardin de St. Pierre spoke of them as being common, even in the trees, in 1768. Swine, extremely dangerous for ground-nesting birds, are said to have been introduced by the Portuguese soon after their discovery of the islands. Possibly they were at least partly responsible for the extirpation of the native land tortoises, for they were fond of the eggs. The first mention of the mischievous macaque monkeys (*Macaca cynomalogus*) is 1606. They are said to live in the hilly, rocky, and heavily wooded parts of the interior and to destroy eggs and young birds. According to M. de Charmoy and M. Antelme (*in litt.*) they are powerful limiting factors on bird populations. Cats probably ran wild soon after settlement by human beings but, according to Carié the first written reference is 1803, when they were said to be abundant.

Other than these known enemies of birds, 11 mammal species are known to have been acclimatized. These include goats, deer, rabbits, and several forms of wild animals from Madagascar.

SMALL POPULATIONS

HAWKS

Falco punctatus Temminck

    Falco punctatus TEMMINCK, Planches Col., Livr. **8**, 1821, pl. 45.

Range: Known only from Mauritius.

Status: Still to be found in small numbers. M. Georges Antelme (*in litt.*) believes that the species is gradually becoming extinct. The implication of its local name, "mangeur-de-poules," is the probable reason. It is a close relative of *Falco newtoni,* the kestrel of Madagascar.

PIGEON

Nesoenas mayeri (Prévost)

    Range: Mauritius.
    Status: Rare.
    See page 299.

PARROTS

Psittacula exsul (A. Newton)

    Palaeornis exsul NEWTON, Ibis, 1872, p. 33, pl. 7.

Range: Rodriguez.

Status: An extremely small population, if not actually extinct. No information about the species has been forthcoming since 1875 (Ibis), when it was apparently a rare bird. Inquiries are not answered. Perhaps they persist on the small outlying islets, as is often the case when birds are persecuted.

Description: A medium-sized parrot (total length about 12 inches) of a pale bluish-green color with a grayish cast.

No specimen of the adult male has been collected, if the anonymous author of *Relation de l'Île de Rodrigue* is believed,[1] for he speaks of a parrot with "green plumage like the preceding [presumably *Necropsittacus*], a little bluer and on top of the wing a little red, as well as the bill." No known specimen has red on the wing, but the mature male of presumably related species is so marked.

The bill of a supposedly immature male is reddish. The upper mandible of the female is black.

Location of specimens: I know of specimens only in Cambridge, England.

Psittacula krameri echo (A. and E. Newton).

   Palaeornis echo NEWTON, Ibis, 1876, p. 284 (Mauritius).

Common names: Cato, ring-necked parakeet.
Range: Mauritius.
Status: A small population, according to M. Georges Antelme, of Les Fougères, Mauritius, but still exists in the forests. Apparently the birds do not visit gardens habitually, which has probably saved them from destruction.

Description: A medium-sized (total length about 12 inches) parrot, generally green with a rose collar.

The head is green, tinged with blue on top. A narrow black line runs from eye to nostril and a wider black stripe from the base of the bill to a pale rose-colored collar, which extends from the throat to the sides of the neck. Body plumage is green. The under wing coverts are tinged with yellow. The central tail feathers are bluish. Bill is red, the lower mandible darker.

The female is all green, with a black bill.

Location of specimens: I know of specimens in Bremen; Cambridge, Mass.; London; New York; Paris; and Port Louis, Mauritius.

M. René Guerin, of Curepipe, Mauritius, kindly writes me that in addition to the species listed above, the "merle de Maurice" (*Microscelis borbonica olivacea*) and the "coq des bois" (*Tchitrea bourbonnensis desolata*) have become quite rare, and he fears for their survival.

---

[1] Manuscript in Ministry of Marine in Paris. This account has generally the ring of truth.

HYPOTHETICAL BIRDS

Podiceps "gadowi" HACHISUKA, Extinct Birds Mascarene Island, 1953, p. 124.
Podiceps sp. NEWTON AND GADOW, Trans. Zool. Soc. London, **13**, 1893, p. 289.

Whether there was ever a distinct species of grebe on these islands is not known. A single right ulna was found on Mauritius. The present repository is unknown. Newton and Gadow identified it as *"Podiceps* sp."* Hachisuka's description of *"gadowi"* adds little to their remarks and those of Lambrecht (1933, p. 262), who also refrained from a formal description of a species. According to them the bone measures 82 mm. in length, within the range of variation of the African species *auritus*.

Astur "alphonsi" NEWTON AND GADOW, Trans. Zool. Soc. London, **13**, 1893, p. 285.

Neither Newton and Gadow in their original description nor more recent investigators have been able to point out any difference between the bones upon which a conception of *"alphonsi"* is based and the bones of *A. melanoleucus* of Africa.

Tibiotarsal and metacarpals in Port Louis, Mauritius.

Leguatia "gigantea" SCHLEGEL, Vers. Med. Akad. Wetensch. Amsterdam, **7**, 1858, p. 142.

The conception of a large long-legged and flightless bird, once resident on Mauritius, is based upon observations of François Leguat in the year 1700, and a picture which accompanied them. It is most probable that Leguat existed and many of his observations were correct, in spite of suspicions to the contrary of Atkinson (1922) and others. However, Leguat was no naturalist, and no other travelers of those days described anything resembling what he called *Géant*.

If we could be sure that the illustration by Adriaan Collaert, called "Avis Indica," is accurate (it does not accord at all with Leguat's description) we might suppose that there was once indeed a giant gallinule or waterhen to be found on Mauritius. The elaborate color plates that Rothschild (1907) and Hachisuka (1953) have caused to be painted might also have some basis in fact, even though color patterns of the birds depicted are not those of Collaert's figure. However, the conclusion that Leguat saw a waterhen has been reached by many authors. Schlegel (1858) said, "Seems to represent the Crane-type among Waterhens," which is perhaps a reference to size. Rothschild (1907) agreed, as did Hachisuka (1953).[1]

---

[1] The accuracy of even modern plates is illustrated by Hachisuka's remarks (p. 165): "I am at a loss to know upon what basis Lord Rothschild has coloured the tip of the wings black." A footnote to this sentence reads, "Through the carelessness of my supervision the wing tip [in my own plate] was coloured black; it should be entirely white."

Because seventeenth-century travelers spoke of flamands, flamencos, flamends, and flamingos, and indeed the bones of flamingos have actually been found on Mauritius,[1] Strickland and Melville (1848) were of the opinion that Leguat referred to these, and Carié (1930) came to the same conclusion.

In my opinion quite properly, Lambrecht (1933) called *Leguatia "ein erdachter Vogel"*—a dreamed-up bird (see also Stresemann, 1923c, p. 511).

Can a concept such as this be placed on a level of credibility with species based on specimens or, at least, evidence adduced by those dedicated to the discovery of truth?

"Kuina mundyi" HACHISUKA, Bull. Brit. Orn. Club, **57**, 1937, p. 156.

This conception of a small, flightless rail is based solely upon the account of Peter Mundy (1914), who was on the island from April 14 to April 18, 1638. No other traveler described a bird exactly like this. Mundy mentions two birds, a dodo and a "hen," which it must be assumed he actually saw, for his diaries evidence a truth-telling man, as Sclater (1915) observes. But he was not a trained observer of birds, and details of his descriptions should not be singled out to the end that we conceive it necessary to add a new species to the fauna of the island.

He wrote: "A Mauritius henne, a Fowle as bigge as our English hennes, of a yellowi[s]h wheaten coullour, of which we got only one. It hath a bigge long Crooked sharpe pointed bill, Feathered all over, butt on their wings they are soe Few and smaller thatt they cannot with them raise themselves From the ground.

"There is a pretty way of taking of them with a redde Cappe, butt this was strucke with a stick. They be very good meat. And are allsoe Cloven Footed, soe thatt the can neyther fly nor swymme . . ."[2]

Mundy's notions of color and form are illustrated by his description of petrels: "Pittrells, the smallest Sea Fowle that useth the Ocean, seldome seene neare land, like unto a Swallow near off the Coullour and not much bigger." Would it be wise to assume that Mundy saw a petrel that no one else has ever seen?

With so little evidence it appears probable that he saw the otherwise undescribed young of some other bird, probably *Aphanapteryx bonasia.*

"Victoriornis imperialis" HACHISUKA, Proc. Biol. Soc. Washington, **50**, 1937, p. 71.

This concept is based on discrepancies between descriptions of travelers of the seventeenth and early eighteenth centuries who were not trained

---

[1] Metacarpals in Tring. See Clark, 1866, p. 144; Newton and Gadow, 1893, p. 282.
[2] Mundy, 1914, p. 352. Hakluyt Society transcript of manuscript at Oxford.

naturalists. Evidence of this sort must be approached with care and gentleness. The descriptions in question are vague and unquestionably written for the general reader. It is unthinkable, and indeed improper, to interpret them as the statements of writers pretending to a scientific approach.

In describing *Victoriornis imperialis,* Hachisuka has compared the accounts of Tatton, the mate of a vessel, and Bontekoe, a Dutch traveler, with those of Carré and Dubois, missionaries. They are here summarized.

Tatton (in Castleton): (1) Bigness of a turkey, (2) cannot fly, (3) white in color.

Bontekoe: (1) Fat, so fat they can scarcely walk, (2) cannot fly.

Upon these two accounts rests the hypothesis that there was once a species that was distinct from the solitaire of Réunion, and which Hachisuka calls the Réunion dodo.

There are in addition two aquarelle paintings of fat, white dodos, but, assuming Pieter Witthoos, the artist, had a living model, we still have no assurance that his captive bird came from Réunion. With the fragmentary evidence available to us it is quite possible to assume that differences in descriptions are due to inaccuracy, or to actual variations according to age and sex of birds, rather than to specific variation. To assume that a seventeenth-century drawing, representing a fat bird rather than a thin one, must have had as model a bird of a given species from a certain place is circular reasoning, leading nowhere.

Indeed in the De Bry edition of van Neck's voyage there is a woodcut showing activities of the ship's company near the beach at Mauritius. In the background is a long-legged dodo, much like the solitaire in general form. Shall we assume that there was once another species on Mauritius as well as Réunion?

The following accounts are thought by Hachisuka to refer to a second species, generally referred to as the solitaire of Réunion, as distinct from his "White Dodo of Reunion" or "Victoriornis."

Carré: (1) They are solitary, (2) of a changeable color, which merges upon yellow.

Dubois: (1) They go alone, (2) white plumage, with tips of wings and tail black, (3) tail like an ostrich, (4) legs and feet like a turkey.

"Apterornis coerulescens" SELYS, Rev. Zool., 1848, p. 294.

Berlioz (1946) suggests that the *Oiseaux Bleus* that Dubois described on Réunion may well have been *Poules Sultanes,* but with the remark that identification of the following description must remain in question. Dubois (1674) wrote, "[They are] large as the Solitaires, have the plumage all blue, the bill and feet red, made like chickens' feet, they do not fly but run very fast so that a dog can scarcely catch them. They are very good." This description was given the name *Apterornis coerulescens* by Selys. It may be that the bones of a *Porphyrio* will some day be uncovered, but until then these descriptions must remain mysterious.

Nesoenas "duboisi" ROTHSCHILD, Extinct Birds, 1907, p. 166.

Former range: Réunion.

Known only by the description of Dubois, who mentions two pigeons on Réunion. Probably Rothschild (1907) was correct in assigning this to the genus *Alectroenas*. Dubois wrote: "Of a russet red color, a little longer than European Pigeons, with a beak longer, red at the base near the head, the eyes surrounded by a fiery color as in Pheasants . . ."

Necropsittacus "francicus" ROTHSCHILD, Proc. 4th Int. Orn. Congr., 1905 (1907), p. 197.

Range: Mauritius (?).

In his original description Rothschild writes, "which we only know from Dubois's description." Actually there is no reason to believe that Dubois wrote of any parrots other than of Réunion. In his "Extinct Birds" Rothschild says, "We only know this bird from the various 'voyages' to Mauritius in the 17th and early 18th centuries." I cannot find any such reference.

Hachisuka (1953) repeats this statement with the remark that although he hardly believes that "nomenclatorial separation is necessary," he still hopes that evidence in the form of contemporary accounts and drawings will be found in Portugal or Holland.

"Necropsittacus ?borbonicus" ROTHSCHILD, Extinct Birds, 1907, p. 62, pl. 8.

Former range: Réunion.

Known only by the short description of Dubois, who saw, among other birds, "a green parrot of the same size [as presumably *Psiattacula krameri eques*] having the head, the upper part of the wings the color of fire." No other traveler described such a bird.

Could it have perhaps been an escaped pet lory that he saw? Dutch paintings of the sixteenth century show the somewhat similar East Indian *Domicella garrula* perched in parlors.

"Bubo (?) leguati" ROTHSCHILD, Extinct Birds, 1907, p. 71, *ex* Milne-Edwards, Ann. Sci. Nat. Zool. (5), **19**, 1874, p. 13, "small Bubo."

Range: Rodriguez Island.

Known only from a single bone (tibiotarsus in Tring). Because of this single bone, which cannot be referred to that of *Athene murivora* in his opinion, Rothschild described a new species of Rodriguez. It is now impossible even to assign it to a given genus.

Hachisuka (1953) correctly points out that Leguat did not mention such a bird.

Scops commersoni OUSTALET, Ann. Sci. Nat. Zool. (8), **3**, 1896, p. 35, fig. 3.

Range: Island of Mauritius.

Known by unsatisfactory drawing and description only. A drawing by Jossigny and a description by Desjardins (1837), contemporary naturalists of that island, show it to have been a large owl with prominent ear tufts. It was called a *Petit Duc* or Scops owl at the time. Rothschild doubted that the bird had been in reality a Scops owl because of its great size, its approximate total length having been 594 mm. or 23½ inches.[2] Indeed, Desjardins' description applies better to some form of *Asio,* and it is not impossible that it was related to the Madagascar long-eared owl (*A. madagascariensis*). No species of this genus has unfeathered tarsi, however, and Desjardins makes it clear that the tarsi were devoid of feathers. Certainly the tropical species of Scops owls have naked toes and scantily feathered legs.

All we can know is that there was a very large owl which is extinct. Desjardins' notes (1837) say: "In September 1837 many residents of La Savane told me that they had seen owls in their forests; Doctor Dobson, of the 99th regiment, assured me that he had killed one in the wood of Curipipe." This is the last information we have of it.

"Strix newtoni" ROTHSCHILD, Extinct Birds, 1907, p. 79.

Range: Mauritius.

This supposed form is based on a remark of Newton and Gadow (1893, p. 287) that two metatarsal bones in their possession were smaller than those of *sauzieri.* It is quite probable that if enough bones were available it would be apparent that individual variation, rather than specific, was responsible for the difference of 6 mm. recorded.

"Testudophaga bicolor" HACHISUKA, Proc. Biol. Soc. Washington, **50**, 1937, p. 213 (based on the testimony of an anonymous author *teste* Hachisuka).

Range: Rodriguez.

In the belief that no starling could eat flesh as described in "Relation de l'Ile Rodrigue," Hachisuka supposed that the bird was truly a chough, or small crow. No bones of this nature have been found on Rodriguez or any other Mascarene Island. Although the description could be that of

---

[2] Rothschild's measurement is a misprint.

a chough, it could also be that of a starling, and since there is real evidence of the existence of such a bird, another hypothesis is unnecessary. It is no more improbable that a starling should peck decayed flesh and eat eggs than a rail, and we know that *Porzanula palmeri* did so on Laysan Island, as does *Telespiza*.

"Foudia bruante" (Müller)
Fringilla bruante P. L. S. MÜLLER, Natursyst., suppl., 1776, p. 164 (Réunion).

Range: Réunion.

This conception is based on a picture in Daubenton's "Planches Enluminées" (no. 321, fig. 2), which resembles the common foudia of Madagascar, except that the back is entirely pale pink, not crimson and not streaked with black and brownish. There are, however, other plates in Daubenton's work that indicate that he may not have painted accurately always.

Montbeillard (1778) described a ruddy bird with wings and tail apparently of the same color. This he called *Le Mordoré,* and he thought it came from Réunion, but we have no assurance of this whatever.

In short, it is quite as likely that the foudias of Réunion were introduced from Madagascar as it is that a species was confined to the island.

EXTINCT

DUCKS AND GEESE (ANATIDAE)

Anas theodori NEWTON AND GADOW, Trans. Zool. Soc. London, **13**, 1893, p. 282, 291.

Range: Known only from bones found on Mauritius.

A fragment of sternum, coracoid, humerus, and tarsometatarsus are in Cambridge University. A source (George Pine, "The Isle of Pines") quoted by Hachisuka (1953) was almost certainly written as a hoax. It contains numerous untrue and even egregious statements.

Sarcidiornis mauritianus NEWTON AND GADOW, Trans. Zool. Soc. London, **13**, 1893, p. 290.

Range: Known only from bones found on Mauritius and perhaps a traveler's account of Réunion.

Metacarpal and a fragment of pelvis are in Cambridge University.

Père Dubois saw geese on Réunion about 1670, which he said were smaller than European geese. They had red feet and bills, he said. Hachisuka (1953) quotes J. Marshall ("Memorandum concerning India," 1668), who noticed geese on Mauritius with wings black at the tip and with the basal half black.

Anhinga nanus (Newton and Gadow)

Plotus nanus NEWTON AND GADOW, Trans. Zool. Soc. London, **13**, 1893, p. 288.

Range: Known only from bones found on Mauritius (Pleistocene, ?Recent).

A humerus, pelvis, and tibiotarsus are in Cambridge University and humerus in the British Museum. They indicate a smaller bird than exists in Madagascar, according to Newton and Gadow.

The following species have been described. There must still be considerable doubt about their relationships.

?Ardea (?Nycticorax) mauritiana (Newton and Gadow)

Butorides mauritianus NEWTON AND GADOW, Trans. Zool. Soc. London, **13**, 1893, p. 289, pl. 24, figs. 6–8.

Former range: Mauritius.
Status: Extinct, known by bones found in subfossil state.

?Ardea (?Nycticorax) megacephala Milne-Edwards

?Ardea megacephala MILNE-EDWARDS, Ann. Sci. Nat. Zool., **19**, art. 3, 1874, p. 10, pl. 14.

Former range: Rodriguez.
Status: Extinct, known only by bones.

?Ardea duboisi Rothschild

Ardea duboisi ROTHSCHILD, Extinct Birds, 1907, p. 114.

Former range: Réunion.
Status: Known only by the seventeenth-century description of the Abbé Dubois.

That there were herons on Mauritius, Réunion, and Rodriguez which no longer exist cannot be doubted. Two of them are known by bones found in subfossil deposits. Curiously enough, no bones of any of the three herons that live there now have been found. Since the extinct birds are thought to have been night herons, the chance of their having been confused with *Butorides striatus, Bubulcus ibis,* or *Egretta dimorpha,* herons and egrets that have been recorded, is not probable.

The extinct Mauritian herons are known only by bones found in the Mare aux Songes.[1] They were originally described and named *Butorides mauritianus* by Newton and Gadow (1893), but there has been no clear

---

[1] Cranium, pelvis, coracoid, ulna, radius, and tarsometatarsus in Cambridge University.

reason given for placing the form in that genus. In fact, the authors seemed to have felt that the relationships might be with the bitterns or the night herons because of a peculiarity of the coracoid. Rothschild (1907), although he listed the bird as true heron (*Ardea*), said that he believed that all these forms were night herons (*Nycticorax*).

Unlike those of Réunion and Rodriguez, no description of this bird has been found. It was smaller than *megacephala* of Rodriguez.

MEASUREMENTS

(Following various authors cited)

	Tarsometatarsus mm.
A. *mauritiana*	81–87
A. *megacephala*	93
Bubulcus	80–82
Butorides	50–58
Egretta	100–120

Ardea megacephala (or perhaps *Nycticorax*) Milne-Edwards (1874) is known not only by its bones,[1] found on Rodriguez, but also presumably by contemporary descriptions. It appears that it was about the size of a night heron but differed in having a large, strong bill and in having almost lost the power of flight (Gunther and Newton, 1879). Both Leguat and an anonymous writer [2] apparently described it as a bittern, but Milne-Edwards, Gunther and Newton, and Rothschild consider it unlikely that it was truly a bittern because of the skull, which is constricted at the temporal region, and the shape of the bill, which is wider.

Leguat said of it only: "We had bitterns as big and fat as capons. They are tamer and more easily caught than the 'gelinotes' [3] . . . the lizards often serve as prey for the birds, especially for the bitterns. . . ."

It was written in *Relation de l'Ile Rodrigue* before 1730, "There are not a few Bitterns which are birds which fly only a very little, and run uncommonly well when they are chased. They are the size of an Egret and something like them."

Dubois, who visited only Réunion (Bourbon), described a heron, but there is no other evidence of it. He said, "Bitterns or Great Egrets, large

---

[1] Cranium, coracoid, sternum, humerus, femur, metacarpal, tibiotarsus, and metatarsus in Paris; vertebrae, pelvis, radius, ulna, and phalanges in Cambridge University; humeri, femora, tibiae, and metatarsi at Tring. Gunther and Newton thought this to be a night heron.

[2] Anonymous manuscript in Ministry of Marine, Paris, "Relation de l'Ile Rodrigue." Written before 1730. See Milne-Edwards (1875); Newton, 1875a.

[3] A rail, presumably of the genus *Porphyrio*.

as capons, but very fat and good. They have grey plumage, each feather spotted with white, the neck and beak like a heron, and the feet green, made like the feet of Poullets d'Inde. This bird lives on fish."

This description is reminiscent of *Egretta dimorpha* of Madagascar, which is resident, although rare on Mauritius and Réunion. These birds have been imported into the islands from time to time (Meinertzhagen, 1912).

Rothschild (1907) gave it the name *Ardea duboisi*. On the theory that a representative of these lived once on Réunion, Berlioz (1946) suggests that Dubois described the immature plumage of the supposed night heron of Mauritius and Rodriguez.

RAILS

There were without doubt rails and coots on Mauritius, Réunion, and Rodriguez when men first landed in 1505. They were extirpated in about 150 years and are known to us by bones found in swamps and caves and by pictures and descriptions of travelers only.

The following species have been described. They are known only from bones:

Aphanapteryx bonasia Selys: Island of Mauritius
Aphanapteryx leguati Milne-Edwards: Rodriguez Island

Identification of these contemporary accounts with the bones has been the subject of much speculation. Descriptions and figures of that time leave much to be desired always. Any of them may be discredited under the microscopic eye of modern scholarship, or, on the contrary, given an importance of detail that they do not deserve, but we cannot doubt that some such bird once existed. The bones prove that there was a large rail on Mauritius.

This species has been called *Aphanapteryx bonasia* Selys.[1] With this long-legged bird [2] several drawings, paintings, and descriptions have been identified, and from them the concept has been built of a generally red-

---

[1] Rev. Zool., 1848, p. 294. The name is based on seventeenth-century descriptions (see Strickland). Mandible, tibiotarsus, and tarsometatarsus in Paris (see Milne-Edwards, 1868). Pelvis, sacrum, femur, humerus, sternum, and mandibles in Cambridge University (see Newton and Gadow). Pelvis, humerus, tibiotarsus, and tarsometatarsi in Tring. *A. broeki* Schlegel (1854) and *A. imperialis* Frauenfeld (1863) are thought to be synonyms.

Hachisuka (1953, p. 138) has used the name *Aphanapteryx broeki* Schlegel 1854 (not 1848 as he states) for the reason that he believed *A. bonasia* Selys (1848) to be a composite species, including as it did the "hen" of Sir Thomas Herbert. I find it impossible to believe that Herbert has given us enough evidence for more than an hypothetical species (*Pezocrex herberti*).

[2] Tarsometatarsus 70–76 mm. as against 78–56 mm. for the American *Rallus longirostris*.

dish-brown bird, with plumage fluffy and decomposed, and short, useless wings and tail. The bones themselves show that there was considerable individual variation in size and amount of curvature in the bill. It may well be assumed that there was variation of plumage due to age and sex about which we can know nothing, and unless more bones are found to prove that another species of rail existed, there is no need for any such theory.

When he gave the name *bonasia* to this species, Selys had grave doubts as to whether there had been more than one species, because of differences in the drawings and descriptions of seventeenth century travelers. Those men were not experts, however. Students recently (Sclater, 1915; Lambrecht, 1933) have identified the drawings of Van den Broeck and Sir Thomas Herbert's "hen" with this species in spite of the difference in the shape of the bill. It appears probable to them that the "Velt-hoendt" Cornelisz saw in 1602 and the "Red Hens with Woodcock's beaks as big as our hens and excellent eating" seen by François Cauche in 1638 were one and the same as the "Mauritius Hen" of Peter Mundy.

Milne-Edwards has suggested that some of these travelers confused young dodos with the rail, which may well have been so.

Besides these evidences there are three paintings, attributed to Roelandt Savery, that have been assumed to be this bird. The models for these were captive birds in European zoos of the seventeenth century. All show a large rail, generally reddish brown in color, with long legs and short wings. There is some variation in detail. One has a gray belly, one has bare skin about the eye, and the bill is rather more straight in one than in the others.

No record or other evidence remains of such a bird ever having existed on Réunion, but this may be due to the few bones that may remain, the lack of careful search for them, and the scarcity of records of the seventeenth-century travelers. There are bones to prove that a relative once lived on Rodriguez. These show it to have been a smaller bird with longer wings than *bonasia* of Mauritius. It is *Aphanapteryx leguati*.[1]

It has sometimes been assumed that the "gelinotes" described by Leguat were these birds, although it seems more probable that they were either a species of *Porphyrio,* of which we have no evidence on Rodriguez save this, or perhaps *Porphyrio madagascariensis* imported from Madagascar. Leguat wrote: "Our 'gelinotes' are fat all the year round and of most delicate taste. Their colour is always of bright grey,[2] and there is very

---

[1] *Erythromachus Leguati* Milne-Edwards (1874). See also Gunther and Newton, 1879. The bill, a scapula, coracoid, humerus, metacarpal, pelvis, femur, and tibiotarsus are in Cambridge University.

[2] Gray and blue are often synonymous in these descriptions.

little difference in plumage between the two sexes. They hide their nests so well that we could not find them, and consequently did not taste their eggs. They have a red naked area around their eyes, their beaks are straight [1] and pointed and is red also. They cannot fly, their fat makes them far too heavy for it. If you offer them anything red, they will fly

FIG. 1.—*Aphanapteryx*. (After Rothschild.)

at you to catch it our of your hand, and in the heat of the combat we had an opportunity to take them with ease."

Fulica newtoni Milne-Edwards, 1867

Bones of a large coot have been found in the Mare aux Songes on Mauritius. They are now in Cambridge University and at Tring. No

---

[1] Beaks found in a subfossil deposit are curved like the Mauritian *bonasia*.

contemporary description or other clue to its appearance remains. It was a large bird; the tibiotarsi measure 77–86 mm. as against 52–64 for the European coot (*F. atra*). Because of the "double or basally divided spina externa of the sternum," Hachisuka has given the new generic name *Paludiphilus* to this species.

Probably there was a coot also on Réunion. Du Bois mentions "Poules d'Eau" as large as hens. "They are all black and have a large white crest on the head . . .," he wrote.

THE DODOS

The Dodo. *Raphus cucullatus* Gmelin

Didus ineptus is a synonym.
Range: Island of Mauritius.
Status: Extinct since the end of the seventeenth century. Known by osseous remains, a foot and beak, in the British Museum and at Oxford.

The Solitaire. *Raphus solitarius* (Selys-Longchamps)

Didus borbonicus is a synonym.
Range: Island of Réunion (Bourbon).
Status: Extinct since perhaps the end of the seventeenth century. Known only by pictures and accounts of travelers of that time.

Rodriguez Solitaire. *Pezophaps solitaria* (Gmelin)

Range: Island of Rodriguez.
Status: Extinct since the latter half of the eighteenth century. Known by bones and accounts of contemporary travelers.

These three stout and flightless species inhabited the Mascarene Islands in the Indian Ocean, off the coast of Madagascar, before the arrival of men, pigs, dogs, and rats. Presumably because they had no enemies, they had become the archetype for specialization and even degeneration of form. Dutch sailors called them dod-aarse,[1] whence perhaps dodo.

It was these, and men from French and British ships following them, together with the pigs, monkeys, cats, and rats, that extirpated the birds. They disappeared from Mauritius soon after 1680, from Réunion about 1750, and from Rodriguez perhaps as late as 1790–1800.

All three species must have been isolated for many thousands of years

---

[1] Some dictionaries say this is not to be said. The derivation Doudo, Portuguese for simpleton, has been questioned because the Portuguese did not remain in the islands long after discovering them and no reference to dodo birds has been found in Portuguese.

before these horrid events. We know but little of the bird of Réunion, but those of Mauritius and Rodriguez were very distinct from each other.

Raphus cucullatus Gmelin,[1] the bird of Mauritius, is the dodo. Since the island was discovered about 1507 by Portuguese Mascarenhas, it required 174 years for the birds' enemies to make an end of them.

It is remarkable that they were able to reproduce themselves for this length of time, for pigs and the mischievous macaque (*Macacus cynomologus* Linnaeus) were introduced by the Portuguese during the sixteenth century.[2] Pigs, when they are not fed but allowed to roam in the forest, are always a great danger to ground-nesting birds, and the macaque

FIG. 2.—Mauritius dodo (*Raphus cucullatus*). (After Van den Broecke.)

is notoriously fond of eggs. Certainly Dutch ships provisioned in part with dodos as early as 1598,[3] as no doubt the Portuguese had done from time to time for nearly a century. The diary of a Dutch ship's captain in the fleet of Admiral Schuurmans in 1602 has this to say of the practice:[4]

They also caught birds which some name Dod-aarsen, others Dronten when Jacob Van Neck was here, these birds were called Wallich-Vogels, because even long boiling would scarcely make them tender, but they remained tough and hard, with the exception of the breast and belly, which were very good . . . on the 25th of July, Willem and his sailors brought some Dodos which were very fat; the whole crew made an ample meal from three or four of them, and a portion remained over . . . on the 4th of August Willem's men brought 50 large birds on board the

---

[1] This is undoubtedly the oldest name. Lambrecht (1933) made plea for *Didus ineptus* Linnaeus as a *nomen conservandum*.

[2] Carié, 1916. Deer also were brought from Batavia by the Dutch, but there is no reason to believe that they were dangerous.

[3] Voyage of Jacob Cornelius Van Neck. See Strickland and Melville.

[4] Translation from Strickland and Melville, p. 14.

'Buyn-vis'; among them were 24 or 25 Dodos . . . another day, Hogeneen set out from the tent with four seamen provided with sticks, nets, muskets, and other necessaries for hunting . . . they captured another half hundred of birds, including a matter of 20 Dodos, all which they brought on board and salted.

It has often been repeated, although on questioned source,[1] that the birds laid a single white egg on a mass of grass in the forest, and that the cry was like that of a young goose. There is no reason to doubt that large pebbles were taken into the crop. (Herbert, 1638.)

According to descriptions and pictures, the bird was extraordinarily gross, weighing as much as 50 pounds. The bill was 9 inches long, strongly hooked at the tip, "blackish with a little redness above the tip on its arched part." The forehead, region about the eyes, and cheeks were bare, and the skin was light ash color. The body plumage was ashy gray, paler, almost white on the lower breast, and blackish on the thighs. Primaries were yellowish white and the coverts white with black tips, feathers of the tail loose and "curly," placed very high (were these not probably upper tail coverts?). The feet and legs were yellow and the irides "whitish."

A great store of bones has been found in a marsh, the Mare-aux-Songes, in Mauritius, and these are in the British Museum, Paris, Leyden, Brussels, Darmstadt, Berlin, and New York. A head and foot are preserved in the Ashmolean Museum at Oxford University. The specimen of which these are remnants was in the Tradescant collection called "the Ark." In 1662 this came into the hands of Ashmole, who in turn left it to Oxford. It is thought by Renshaw (1931) to be the specimen described by Sir Hamon Lestrange as "coloured before like the breast of a young cock pheasant, and on the back of a dunne or deare colour," and that it may have been a young female. The British Museum possesses a foot, and there is a head in Copenhagen, and a small fragment in Prague.

THE SOLITAIRE OF RÉUNION

The solitaire or white dodo of the island of Réunion (*Raphus solitarius*)[2] is known only from descriptions and pictures. These, however, were done with specimens as models. Accounts of travelers substantially agree.[3]

---

[1] François Cauche. A manuscript document in Paris says that *Pezophaps* of Rodriguez laid but a single egg, however.

[2] *Apterornis solitarius* Selys-Longchamps, Rev. Zool., 1848, p. 293 (Bourbon, now called Réunion). This is an older name than *Didus borbonicus*.

[3] Picture by Pieter Witthoos was drawn from a specimen brought to Amsterdam in 1670 (A. Newton, 1867). Accounts to travelers are Tatton in 1613, Bontekoe in 1618, Carré in 1668, Du Bois in 1669–71.

There can be little doubt therefore that there once lived on Réunion a bird very like the dodo of Mauritius, but because of certain discrepancies in these descriptions and pictures there has been disagreement in the conclusions drawn from them.

The observation of Newton (1867) that the bird of Réunion differed from that of Mauritius in being white with some admixture of yellow is well substantiated. The statement that the first four primaries of the wings were not directed backward but downward and forward accords

Fig 3.—Solitaire of Réunion (*Raphus solitarius*). (After Frohawk.)

well with Witthoos's painting, but seems likely to have been the result of some accident.

Oudemans (1917) and, following him, Rothschild (1919) are of the opinion that the many discrepancies in the paintings of the birds are to be explained by differences between the sexes, as follows:

Male. The horny sheath of the upper mandible was hooked and sharp; its distal end black, its proximal half was yellow with transverse black stripes; the rest of the bill was white . . . a ball-shaped tail of ostrich-like feathers gradually passed into the subcaudal coverts and circumanal feathers.

Female. The horny sheath of the upper mandible was not hooked, but obtuse, sometimes ending in a blunt point, sometimes rounded; it was greyish or light fawn coloured, the rest of the bill being greyish or greenish. . . . The tail consisted of at least six white rectrices, which resemble in shape those of a Silver Pheasant.

These are unusual sexual variations. There are too many unknown factors involved in this problem to solve it satisfactorily. We do not know how interested these artists were in accuracy. Assuming that they were accurate, we do not know without question the sex or actual provenance of the specimens and we have no way to determine their age or the character and extent of individual variation.

Berlioz (1946) says gently:

> One may conceive that these descriptions, of which the test seems even purposefully to exaggerate the strangeness of the bird, have induced the authors to consider this as different enough from the true Dronte of Mauritius, even to the point of separating them generically [Hachisuka, in 1937, has named the species of Bourbon *Victoriornis imperialis*]. In this respect such thoughts cannot fail to be very hazardous for, if the bird of Mauritius, of which a quantity of bones of all sorts have been found, is now well defined, it is not the same in the case of Réunion: because absolutely nothing remains of the latter, according to Rothschild, outside of old descriptions, except four drawings or painting executed in Holland nearly at the end of the 17th century.

In fact the dodo of Réunion may simply have been an albinistic form. Whatever may be thought of the distinctness of the population, it did once exist. When Bory de Saint Vincent visited the island in 1801 the birds had been extirpated for some time, probably by the same agencies that destroyed their neighbors in Mauritius.

THE SOLITAIRE OF RODRIGUEZ

Perhaps the birds had disappeared considerably earlier from Réunion, for we know that the neighboring island of Rodriguez, where there were fewer human beings than on the other islands, was used as food store as early as 1730. Small ships were sent there from Réunion (Bourbon) to collect tortoises and birds, and 30,000 of these land tortoises were taken in less than 18 months.[1] Probably as a result of this traffic the bird of Rodriguez was so rare in 1761 that the Abbé Pingué, an astronomer, never saw one, although he was assured that they still existed.

The number of bones found in fissures and caves indicate that the population was quite large. Enough have been found to reconstruct several skeletons, and it is largely upon osteological grounds that this solitaire of Rodriguez has been separated generically from its relations.

It is presently called *Pezophaps solitaria* (Gmelin), but it might better be called *Raphus,* for it was clearly a related species. It differed obviously

---

[1] Milne-Edwards, 1875, from "Relation de l'Isle de Rodriguez," MS. in Ministère de la Marine, Paris.

in having a shorter bill, without the great hook at the tip so characteristic of the dodo.

Other osteological differences are here summarized.[1]

The plane of the foramen magnum forms an angle with the base of the skull of 58° in *Pezophaps,* 55° in *Raphus,* and 35° in *Didunculus.*

The foramen of *Pezophaps* is like that of *Didunculus* but is laterally compressed in *Raphus.*

The unique structure of the occiput, steeply pitched, flat and elevated at the expense of the parietal, is common to both *Pezophaps* and *Raphus* and is only somewhat less apparent in the former. The ratio of height to breadth of the occipital condyl in most pigeons is ¾; in *Goura* and *Pezophaps* it rises to 1 and exceeds 1 in *Raphus.*

The frontal processes and premaxial palatine converge in *Pezophaps,* whereas they are parallel in *Raphus;* the palatine is more reduced in the latter.

Contemporary accounts are fragmentary; the most quoted is that of François Leguat.[2] He described a brownish gray "male" and a "female" "the color of fair hair." The Abbé Dubois described a white bird. It seems possible that there existed a white phase with intermediates, or there may have been sexual or age variations. It is not necessary to assume that a second species existed. An anonymous author [3] of about 1730 agrees fairly well with the others except in the description of color. This is as follows:

Description: "The Solitaire is a large bird which weighs 40–50 pounds. They have a large head with a sort of bandeau which one might say was of black velvet. Their feathers are neither feathers nor hairs, they are grayish white, the upper back somewhat blackish . . . they have a very short bill of the width of nearly an inch (pouce), which is sharp . . . they have a little lump in the wings which has a sort of musket ball at the end, and this serves them in defense."

Habits (same source). "They have scaly and very strong feet, and run fast, principally among the rocks, where a man, agile as he may be, will have trouble in catching them. They never fly, having no feathers in their wings, but they fight with them and make a great noise with their wings when they are angry and the noise is quite like thunder . . . they do not lay, according to what I think, but once a year, and lay but one egg; not that I have seen their eggs, for I have not been able to discover where

---

[1] See Lambrecht, 1921, 1933.

[2] Atkinson, 1922; Swaen, 1940; Mortensen, 1934, Vivielle (1926) and Dehérain (1926) discuss Leguat and his "Voyage et Avantures."

[3] "Relation de l'Isle de Rodriguez," in the Ministère de la Marine, Paris. See Newton, 1875a, and Milne-Edwards, 1875.

they lay, but I have seen only a single young one with them; and when one wishes to endanger himself by approaching them they would bite very hard. These birds live on seeds and leaves of trees that they collect on the ground. They have a gizzard larger than a fist, and what is striking is that a stone is to be found in it large as a hen's eggs, oval and a little flat, whereas this animal cannot gulp (anything) as large as a little cherry. I have eaten them; they are fairly good tasting. . . . ."

PIGEONS

?*Columba rodericana* Milne-Edwards

Columba rodericana MILNE-EDWARDS, Ann. Sci. Nat. Zool., **19**, art. 3, 1874, p. 16.

Range: Known only from bones (Recent) found on Rodriguez.

A sternum, humerus, femur, tibiotarsus, and tarsometatarsus are in Tring, according to Hachisuka (1953).

This has been assumed to be a relative of *Alectroenas nitidissima* (a form known by skins taken on Mauritius) by Shelley (1883), Rothschild (1907), and Hachisuka (1953). Differences found by Milne-Edwards are reprinted by Lambrecht (1933) and Hachisuka (1953). Relationships of the bird are obscure. *Alectroenas nitidissima* see page 292.

PARROTS

Lophopsittacus mauritianus (Owen)

Psittacus mauritianus OWEN, Ibis, 1866, p. 168.

Range: Mauritius. Known only from bones and rather unsatisfactory accounts of travelers.

The bones prove that it was a large parrot with an enormous bill. Sir Thomas Herbert's sketch of 1638, which has been identified with this birds, shows a distinct crest. Renshaw (1933a) suggests that the strange crested bird to be seen sitting in a tree in an illustration of van Neck's voyage represents it also.

The color, in spite of the plates in Rothschild (1907) and Hachisuka (1953), is a debatable matter. Either they were gray or blue. Admiral van Neck said he saw a blue parrot ("Caerulei quoque psittaci ibi frequentes sunt"). Captain van der Hagen in 1607 saw a gray parrot ("on reçut . . . de perroquets gris"), and Verhuffen in 1611 spoke of "grawe Papagayen." [1] However, since the only other parrots known to have occurred on Mauritius were green, it may well have been this to which they referred.

---

[1] See Strickland and Melville, pp. 9, 17. These sources not seen by me.

Newton and Gadow (1893) and Berlioz (1940) suggest that this species was closely related to *Necropsittacus* of Rodriguez.

Necropsittacus rodericanus (Milne-Edwards)

Psittacus rodericanus MILNE-EDWARDS, Ann. Sci. Nat. Zool. (5), **8**, 1867, p. 151 (Rodriguez).

Range: Rodriguez.

Known from bones and the description of an anonymous contemporary observer.[1] He wrote: "The Perroquets are of three kinds and in quantity. The largest are larger than a pigeon and have a very long tail, the head large as is the bill. Most of them live on the islets that lie to the southward of the island, where they eat a little black seed which produces a small tree (arbrisseau), the leaves of which have an odor of lemon [leaves], and come to the large island to drink water. The others stay on the large island, where they are to be found in small trees (arbrisseaux)."

A cranium, tibiotarsus, coracoid, humerus, femur, and tarsometatarsus are in Cambridge, *fide* Lambrecht.

They were very large birds, approximately the size of a large cockatoo, and had huge bills. They were probably related to forms of the islands of the southwestern Pacific and to *Psittacula,* according to Milne-Edwards (1874), Gunther and Newton (1879), and Berlioz (1940).

Mascarinus mascarinus (Linnaeus)

Psittacus mascarin LINNAEUS, Mantissa Plant., app., 1771, p. 524 ("Mascarina" = Réunion).

Range: Réunion. Because Peter Mundy records having seen "Russet Parrats" on Mauritius in 1638, Hachisuka (1953) suggests that they occurred there also.

Status: Extinct. Known only by two specimens in Paris and Vienna. The last known living bird was in the garden of the King of Bavaria in 1834.

Description: A medium-sized parrot (total length about 14 inches). The bill was strong and smaller at the base; it was red, as was also probably the naked cere. A circlet of black feathers extended all around the front of the head forward of the eye. The entire head behind the eye was grayish lilac. The back and wings were grayish brown, the underparts paler. Tail brown, white at base. The feet were reddish brown.

Relationships are obscure.

Location of specimens: I know of specimens in Paris and in Vienna.

---

[1] "Relation de l'Isle de Rodriguez," in the Ministère de la Marine, Paris (Isle de France correspondence, vol. 12, 1760). See Newton (1875a); Milne-Edwards (1875).

Psittacula krameri eques (Boddaert)

> *Psittacus eques* Boddaert, Tabl. Pl. Enl., 1783, p. 13 (based on Perruche à collier d'Isle de Bourbon of Daubenton, Pl. Enl., pl. 215, 1783).

Former range: Réunion.

Status: Extinct. Known only from Daubenton's plate (see above) and from the statement of Dubois, who no doubt saw them about the year 1700, that there were "Perroquets Verts . . . ayant un collier noir" on the island.

It is probable that there was such a parrot on Réunion therefore, and it is also probable that, unless it had been introduced, it differed from *echo* of Mauritius. How it may have differed is not known.

Oustalet (1893b) states that the birds disappeared at the beginning of the nineteenth century, but gives no reason for this belief.

OWLS

Athene murivora Milne-Edwards

> *Athene murivora* Milne-Edwards, Ann. Sci. Nat. Zool. (5), **19**, art. 3, 1874, p. 13 (Rodriguez).

Range: Rodriguez Island.

Status: Extinct. Known from a tibiotarsus and mandibular in the British Museum, and probably by a description in the anonymous manuscript "Relation de l'Ile Rodrigue," in the Ministry of Marine in Paris, as follows: "A bird is to be seen which is very like the brown owl, and which eats little birds and small lizards. They almost always live in trees, and when they think the weather will be fine, the utter always the same cry at night. On the other hand, when they think the weather will be bad, they are not heard."

Tyto sauzieri (Newton and Gadow)

> *Strix sauzieri* Newton and Gadow, Trans Zool. Soc. London, **13**, 1893, p. 287 (Mauritius).
> *?Tyto newtoni* Rothschild (Extinct Birds, 1907, p. 79) is considered to be a synonym.

Range: Island of Mauritius, Indian Ocean.

Status: Extinct. Known only by bones (Recent).

There can be no doubt that owls were present in Pleistocene time, and perhaps they persisted until the beginning of the eighteenth century, but that there were two species on Mauritius is questionable.

Of seven tarsometatarsi found in the marsh called Mare-aux-Songes, a pair are shorter; they are 56 mm. long as against 63–66 mm. The longer birds were named *Strix* [1] *sauzieri* by Newton and Gadow (1893), and the

---

[1] Now *Tyto*, after Peters.

supposedly small one they differentiated tentatively as *"Strix* sp." They remarked that the range of individual variation is considerable in the barn owls and that it seemed unlikely that two species existed on Mauritius. Rothschild (1907) assumed that no doubt there were two species on Mauritius and named the smaller one *Strix newtoni,* pointing out that *Tyto alba affinis* (a barn owl) and *Tyto capensis* (grass owl), two distinct species with different nesting habits, live in the same parts of Africa. The tarsometatarsus of the former is about 81 mm. and the latter 89 mm. This was quite possibly the case, but there are not enough bones to prove it. The few that have been found are in Port Louis, Mauritius. No bones have been identified with the supposed smaller species save the tarsometatarsi; several tibiotarsi (90–93 mm.) and a humerus (71 mm.) are said to belong to *sauzieri.* The larger cnemial process of the tibiotarsus and the shorter humerus [1] are supposed to distinguish this species.

STARLINGS

Fregilupus varius (Boddaert)

> *Upupa varia* BODDAERT, Tabl. Pl. Enl., 1783, p. 43, *ex* Daubenton, pl. 697 (no locality = Réunion Island). Fig. Rothschild, 1907, pl. 1.
> *Necropsar rodericanus* SCLATER, Philos. Trans. Roy. Soc. (London), **168**, 1879, p. 427, pl. 42 (Rodriguez Island).
> ?*Necropsar leguati* FORBES, Bull. Liverpool Mus., **1**, 1893, p. 34, pl. 1. Type locality unknown, thought to be Île-au-Mât, near Rodriguez. It is probably a synonym of *N. rodericanus.*

We know without doubt that there were once two species of long-isolated and aberrant starlings on the islands of Réunion (Bourbon) and Rodriguez. They were called huppe or calandre (*Fregilupus varius*) on Réunion and disappeared about midnineteenth century [2] from that island. The last specimens were collected between 1835 and 1840. A total of 23 have been preserved.

They were large grayish-brown starlings and differed from any others in having a crest of pale, decomposed feathers.

A mounted skin, thought to represent a mature male, is described by Sharpe (1890) as having the head and hind neck light ashy gray. Feathers of the crest with white shafts. Lores and a line above the eye white. Black ashy brown, lower back, rump and upper tail coverts washed with rufous. Wings and tail blackish brown, with an external wash of gray. Primary coverts white for the terminal half, with brown tips forming an alar speculum. Cheeks, throat, and underparts white.

---

[1] Humerus of the Madagascan *Tyto soumagnii* measures 75 mm.

[2] Berlioz (1946) quotes d'A. Legros, Album de la Réunion, 4, 1883, p. 151, to the effect that a few were seen not long before that date.

Females are thought to be like males but smaller (Rothschild, 1907), and the young are said to be browner.

The bird of Rodriguez is known by bones[1] and, perhaps, a unique skin

FIG. 4.—*Fregilupus varius* of Réunion.

which is in Liverpool. The bones have been named *Necropsar rodericanus* Gunther and Newton. They prove that this form was closely related to

---

[1] Cranium, sternum, coracoid, humerus, metacarpal, femur, tibiotarsus, and tarsometatarsus in Cambridge University.

Fregilupus but considerably smaller, and no doubt should be called *Fregilupus,* as all descriptions indicate.

A description of the bird in life, written before 1730, also shows it to have been a close relative, although it lacked the distinctive crest.[1] A translation follows:

A little bird is found which is not very common, for it is not found on the mainland. One sees it on the *Islet au Mât,* which is to the south of the main island, and I believe it keeps to that islet on account of the birds of prey which are on the mainland, as also to feed with more facility on the eggs, or some turtles dead of hunger, which they well know how to tear out of their shells. These birds are a little larger than a blackbird, and have white plumage, part of the wings and tail black, the beak yellow, as well as the feet, and make a wonderful warbling. I say a warbling since they have many and altogether different notes. We brought up some with cooked meat, cut up very small, which they eat in preference to seeds.

Whether the specimen in the Liverpool Museum represents this species cannot now be determined, for its true provenance is not known. Forbes (1898) made it the type of *Necropsar leguati,* upon the assumption that it was confined to the tiny Île-au-Mât near the coast of Rodriguez. This hypothesis was based on the fact that the leg bones of this specimen are shorter than those found on Rodriguez itself. According to Forbes the measurements (in mm.) of these bones are as follows:

	Tarsometatarsus	Tibiotarsus
varius	45	65
rodericanus	36	52–59
leguati	31.5	46

Neither sexual nor individual variations were considered, since nothing was known of these. Rothschild (1907) quotes Hartert to the effect that females of *varius* appear to be considerably smaller than males. It is quite possible that if enough bones of *rodericanus* were collected the difference of 4 or 5 mm. would prove to fall within the range of variation of that species, especially since the difference between the length of the smallest female and the largest male tarsus of *Sturnus vulgaris* is 4 mm.

There is little doubt that this specimen is a starling,[2] the first primary is much reduced, 38 mm. according to Forbes. His figure of the bill is that of a starling. The plumage is white, "except for a lighter or darker ferruginous wash on the external webs of the distal half of the primaries and secondaries, as also on the outer webs of the newly moulted, and on both webs of the unmoulted rectrices. Bill, legs and feet yellow." Wing 109 mm.

---

[1] Anonymous, "Relation de l'Ile Rodrigue," in the Ministère de la Marine, Paris, Isle de France Correspondence, vol. 12, 1760. See Newton (1875a); Milne-Edwards (1875).

[2] See also Amadon, Amer. Mus. Novit., no. 1496, 1951, p. 30.

Because of the difference in size, Rothschild recognized *leguati* as a species distinct from that of Rodriguez, with a remark that since it is improbable that Île-au-Mât harbored a distinct species, the specimen probably came from Mauritius. However, there is no evidence that such a bird ever existed on that island.

Seychelles Islands

North of Mauritius and Madagascar is a series of banks in 300 fathoms of water, and from them hundreds of small, dry cays and islands appear. They are inhabited by sea birds and the rather common white-throated rail (*Dryolimnas cuvieri*). In the middle of the most northern bank, 600 miles northeast of Madagascar, are the Seychelles—four rather larger islands, both hilly and well watered. Mahé is the largest, being 17 miles long by 4 to 7 miles in width, and with two peaks rising from the central ridge, the highest being 2,390 feet. Praslin is 8 miles long by 1 to 3 miles, and La Digue has about 4 square miles of land surface, but their hills are 1,260 and 1,175 feet in height. Silhouette has 8 square miles of land surface and its mountain is 2,473 feet.

So small and isolated are they that it is not surprising that only 17 representatives of superspecies are known to breed on them. Fourteen of these are not to be found elsewhere (endemic).

According to Vesey-Fitzgerald (1940) and P. V. Hunt of Mahé (*in litt.*) the following are extinct.

EXTINCT

Psittacula eupatria wardi
Otus insularis
Zosterops semiflava

SMALL POPULATIONS

Tchitrea corvina
Bebrornis sechellarum

South Atlantic Islands

Between Africa and South America a submarine ridge extends down the middle of the South Atlantic Ocean. There are depths of water ranging to 2,000 fathoms over this ridge, but there are also four islands at long intervals; these are huge, inactive volcanoes. They are all steep, high, and rocky, with but limited life.

The most northerly, Ascension, is just south of the Equator at lat. 7° 55′ S., the Tristan da Cunha group at 37°, Diego Alvarez or Gough Island, 200 miles south of Tristan, and Bouvet, the southernmost, at 55°

Although St. Helena rises from much greater depths and lies a little closer to the African coast than the others it is like them in that there were no mammals or reptiles on it when men first landed.

All support colonies of most interesting sea birds, but of native land birds there are very few. None have been recorded from Ascension, and St. Helena possesses but one, a plover (*Charadrius sanctae-helenae*) of African origin. Curiously enough, on the Tristan group and Gough Island, seven species have evolved; three of these have representative subspecies in the islands, and one represents a superspecies. Bouvet is covered by an ice cap and possesses no land birds.

Bones found on St. Helena indicate that there has been some change in the character of the population of sea birds, but no evidence of the former presence of land birds has been discovered, nor has any such evidence come to light on any of the other islands of the mid-Atlantic plateau.

In fact, the islands are nothing but deeply eroded cinder cones towering from the sea, in the case of Tristan to a height of 6,760 feet above the beaches. Tristan is the most interesting island, for it is the only one upon which speciation has occurred within the limits of the group itself. These islands form a rough triangle with Tristan on the north and Inaccessible and Nightingale at the other angles. The sides are about 20 miles in length, and the base, between the two latter islands, is about 15 miles long. There are no inlets or harbors. Mountain torrents have scored the cliffs with gulches, and at the foot of these ravines are rocky beaches. There are occasional sandy beaches also, and the largest of these, on the northwest side, is usually used by the islanders as a landing place. Above this is a narrow plateau, no more than a mile in width, upon which the single small village is huddled at the foot of the mountain.

High winds, generally from the northwest, beat upon the mountain in savage gusts that blow sand and spray to thousands of feet. At sea level the climate is mild, about 33°—48° F., but snow and sleet are the rule on the mountain during June, July, and August, and the temperatures fall much lower. Fog and rain occur frequently.

Vegetation is luxuriant in sheltered places, but even there the trees are stunted and in more exposed areas are reduced to creepers or are prone for two-thirds of their length. The only woody tree is *Phylica arborea,* a buckthorn. This is the only wood serviceable to man. There are also tree ferns. The most peculiar vegetation is the native tussock grass (*Spartina arundinacea*), which grows in tangled thickets about 12 feet high that are extremely difficult to penetrate and are the habitat of most land birds.

A great deal of the vegetation has been cleared from inhabited areas. The brush in some other areas has been burned.

Following is a list of the land birds of the Tristan group:

A rail—*Atlantisea rogersi*, Inaccessible.
Two gallinules—*Gallinula n. nesiotis*, Tristan; *G. n. comeri*, Gough.
Two thrushes—*Nesocichla e. eremita*, Tristan; *N. e. gordoni*, Inaccessible.
Four buntings—*Nesospiza a. acunhae*, Tristan and Inaccessible; *N. a. questi*, Inaccessible; *N. wilkinsi* Nightingale; *Melanodera goughensis*,[1] Gough.

Of these the only species the origin of which is to be traced is *Melanodera goughensis*, and this is clearly South American. In all likelihood the others came also from South America, since the prevailing winds come from that quarter with great strength. However, we must conclude that either the stock from which they sprang has become extinct on the mainland, or that evolution and mutation have obscured the characters the ancestors once possessed. Both Africa and South America have gallinules similar to the island forms.

One of these birds, the Tristan coot (*Gallinula n. nesiotis*), is extinct. "Starchy," the Tristan thrush (*Nesocichla*), is much reduced in numbers and must be considered to be in great danger because of the cats, which, in spite of reports that they had been extirpated by the rats, are still living in the bush on the mountain sides (Christophersen, 1940). The rails that nest on the ground on Inaccessible (*Atlantisea*) and Gough Island (*Gallinula*) will be in very great danger should mammals, and particularly rats, be introduced.

It is probable that from time to time many birds have blown out to Tristan from South America with the "roaring forties" and perhaps sometimes they were in numbers, but no evidence of their being has ever been discovered. We may only speculate about the distribution of living birds; there is no evidence upon which to make conclusions; there are no fossil remains. It is possible that *Atlantisea* once occurred on Tristan da Cunha itself and that it was extirpated by the coot, a bellicose bird, and very probably a much later immigrant. But there remains the presently unanswerable problem of why *Atlantisea* does not occur on Nightingale Island and why the gallinule was confined to Gough and Tristan and has never been recorded from Inaccessible and Nightingale. The fact is that these problems are inherent in the study of insular faunas; there is seldom an answer. The only obvious difference between the islands upon

---

[1] Lowe (1923, pp. 512–513) proposes *Rowettia* as a new generic name for this bird. He points out that it resembles *Melanodera* of the Falkland Islands and mainland of South America. The only character he mentions that is not due to adaptation to island environment is the acuminate tail feathers in the young. I can find no difference, the tail feathers of the streaked young of *Melanodera* being also acuminate. Lowe (p. 525) points out also that *Nesocichla* is, superficially, simply a *Turdus* modified to an island environment, differing in its peculiar tongue and its wing formula, however.

which the gallinule does occur and those on which it does not is the lack of ponds and streams on the latter. Other water rails, however, have adapted themselves to drier islands than these; for example, the Laysan rail and the Sulphur Island ashy crake.

FIG. 5.—Starchy (*Nesocichla eremita*).

The problem of the preservation of species here is dissimilar to the problem in more civilized places in that the people have but little to eat. Aside from the motive for killing, however, the underlying factors are similar: an increase in the population of human beings and their intro- duced animals, and a consequent decrease in the native fauna. Tristan was first settled permanently in 1817. In 1811 there were three Americans on the island engaged in killing seals. During Napoleon's captivity, and the existence of H.M.S. *St. Helena,* a garrision was maintained on Tristan. In 1829, 27 women were imported, and by 1852 the population had increased to 85. In March 1938 there were 188, and all able-bodied persons were interested in killing the birds for food and for profit, but most often for food it must be said.

Of the 14 [1] species of albatrosses and petrels recorded to have bred on Tristan and Gough there are three which are locally considered to be extremely rare. It is doubtful that the Tristan populations will long survive. Of these, two are probably distinct races confined to the group, and the third breeds on islands all around the edges of Antarctica. These are: The Tristan wandering albatross (*Diomedea exulans dabbena*), the great-winged petrel (*Pterodroma macroptera macroptera*), called "black eaglet" by the people of Tristan, and the giant fulmar (*Macronectes giganteus*), familiarly named Nellie, stinker, and stink pot.

The albatross used to nest on the sides of the mountain at Tristan but is now confined to the mountain plateau of Inaccessible, where there are only a few remaining (Hagen, 1952), and to Gough Island, where they probably survive in numbers, for there are no enemies there. The bird lays but a single egg a year, and the young are about the nest for 10 months without being able to fly. Furthermore, this period of helplessness is during the winter when much snow lies on the mountains where they nest. Add to this their natural enemies, the skua and the giant fulmar, and you have powerful natural checks on the population, but introduced enemies mean death for all. The birds have disappeared from Tristan within 50 years. Men have moved in on them and literally devoured them.

Another albatross, the "Molly" of the islanders (*Diomedea chloroptera*), has disappeared from its former nesting grounds within walking distance of the settlement but still occurs in considerable numbers on the opposite, and much less accessible, side of the island and on the uninhabited islands. Thousands of young and eggs are taken each year. A third (*Phaebetria fusca*), the "pee-o," nests in such rugged cliffs that it is in less danger.

The second of the sea birds that has almost disappeared is the great-winged petrel (*Pterodroma m. macroptera*). Murphy (1936, pp. 218, 689) is of the opinion that this is probably a form peculiar to the Tristan group. It is in great danger; Hagen of the Norwegian Expedition of 1937–38 says that only an occasional stray pair could be found on the mountain side. In this case again, meat-hungry human beings have eaten them almost every one. One man alone is said to have brought in 1,000 chicks

---

[1] Mathews (1932, pp. 19–20) and Murphy (1936, p. 213) think that probably more birds will be found to breed on Gough Island.

during a single winter, and this in the memory of the present generation. Inhabited nesting burrows were located by dogs trained to stand and bark until their owners came up to dig the birds out of the burrows. The young are considered to be toothsome, and the fat is used for cooking grease. Their nesting season, like that of the wandering albatross, begins in autumn; the young in their burrows are, of course, less exposed.

The giant fulmars are also gone from Tristan. They were all eaten.

Hagen records details of death in the Rockhopper penguin colonies.

With man came his domestic animals. As early as 1790, when sealers came to the islands, goats and pigs were found. It is not recorded who introduced them, but no doubt the Spaniards or Portuguese did so. It is well known that their early voyagers left animals on lonely islands for shipwrecked mariners. At present cattle, donkeys, feral house cats, rats, and mice are present on the island of Tristan. There was a single pig on Inaccessible in 1938 (Hagen, l.c.), which incidentally is recorded as gorging himself on the young of sea birds, but no other mammals were seen there or on Nightingale. Presumably none have been introduced on Gough Island.

Little hope for the preservation of the fauna of these islands can be held. A growing population of human beings, without adequate food, certainly cannot be denied. It can only be a matter of time until the people move to Inaccessible and Nightingale Islands. The chances then of the introduction of enemies will be greatly increased. With the coming of cats, pigs, and rats the birds will disappear.

Accounts of Extinct or Vanishing Forms

The Ostriches

There can be no doubt that ostriches have disappeared from a large part of their range and have been disappearing for perhaps a million years. No reason can be given for this, for there is no evidence. There are no more ostriches in southern Europe, in India, or in China, where fossil remains have been found. Presently they are confined to Arabia and the drier parts of Africa, which excludes the forested Congo region. They are nowhere common and have disappeared from the neighborhood of human habitation over the range of the species.

The following forms have been described [1] from fossil remains. These were all somewhat larger than existing birds.

1. *Struthio novorossicus* Alexeiev, 1916: Described from a tarsometatarsus from the lower Pliocene beds near Odessa, west coast of the Black Sea. Other remains have been found in that vicinity. An egg found near Cherson, north coast of the Black Sea, has been named *S. chersonensis* Brandt, 1873.
2. *S. karatheodoris* Forsyth Major, 1888: Described from bones of Lower Pliocene found on the island of Samos.
3. Remains have been found in Algeria (*Struthio* sp. Thomas, 1884) and in Egypt (*Struthio?* sp. Stromer, 1902). The horizon of this first is thought to be the Tertiary and the second late-Pleistocene or Recent.
   S. oldawayi Lowe 1933: Tanganyika Territory, East Africa. Lower Pleistocene. How this differs from other Pleistocene forms is not explained; Lowe says, "I have distinguished this . . . solely for convenience of reference."
4. *S. asiaticus* Milne-Edwards, 1871: Siwalik Hills, India. Lower Pliocene.
   S. indicus Bidwell, 1904: Nullas, Kain River, Banda, United Provinces, India. Known from egg fragments.
5. *S. wimani* Lowe, 1931: Described from a pelvis found in Lower Pliocene Ponticum, Hipparion beds) at T'ai Chia Kau in Pao Te Hsien, North West Shansi, on the Yellow River, China.
   A fossil egg was found 75 miles south-southwest of Kalgan, on the border of Shansi and Inner Mongolia. This was named *S. anderssoni* Lowe, 1931. Considerably smaller eggs found at Ertemte, 130 km. north-northwest of Kalgan, and other localities of Inner Mongolia were named *mongolicus* Lowe, 1931. [2]

Six subspecies have been described from populations still living but becoming rare with the advance of civilization. They are *S. c. camelus*

---

[1] The question of names has been solved arbitrarily by using those which were given with a description of bones. Where an egg was described, even though it might be distinguished from those of other forms, the name has been considered to be a synonym.

[2] There are persistent rumors of a population of small birds in North Africa.

Linnaeus, which occurs sparingly in the Sahara desert, Egyptian Sudan, and the west coast of the Red Sea; *S. c. spatzi* Stresemann, 1926, which is confined to the Río de Oro region of northwestern Africa; *S. c. syriacus* Rothschild, 1919 (see below); *S. c. molybdophanes* Reichenow, 1883, which is still to be found in Ethiopia and Somaliland; *S. c. massaicus* Neumann, 1898, a rare bird of eastern Kenya and Tanganyika; and *S. c. australis* Gurney, 1868, still to be seen in the wilder parts of South Africa south of the Zambesi and Cunane Rivers.

The following key to the subspecies of *Struthio camelus* is after Rothschild, 1919b. The subspecies *spatzi* was described from eggs which have pores shaped like "a short, straight comma," rather than a needle prick. Few young birds were available, but apparently they did not differ from *camelus* since no mention has been made of any differences.

1. Neck and legs red.. 2
   Neck and legs leaden blue.................................. 4
2. Head on top hairy, size large.............................. *massaicus*
   Head with naked shield.....................................
3. Size larger .... *camelus*
   Size smaller .... *syriacus*
4. Size larger, top of head hairy; red on front tarsus reaching well
   on to larger toe.. *australis*
   Size smaller; top of head with horny shield: red on front of
   tarsus only reaching end of tarsus........................ *molybdophanes*

KEY TO EGGS

1. Eggs smooth, not pitted.................................... 2
   Eggs pitted 3
2. Eggs larger, not so highly polished....................... *camelus*
   Eggs smaller, highly polished............................. *syriacus*
3. Eggs closely pitted, pits smaller......................... *australis*
   Eggs closely pitted, pits large........................... *molybdophanes*
   Eggs with pitting more scattered, pits large.............. *massaicus*

Arabian Ostrich

Struthio camelus syriacus Rothschild

> *Struthio camelus syriacus* ROTHSCHILD, Bull. Brit. Orn. Club, **39**, 1919, p. 93 (Syrian Desert).

Common names: Zazilhum (male), reidah (female); na-ama.

Status: Now much reduced in numbers and probably confined to the southern part of its range in Saudi Arabia. Last specimen killed and eaten by Arabs near the oil pipe line north of Bahrein, Hasa Province, between 1940 and 1945.[1] An individual was seen near Tallul Basatin (lat. 33°2N.,

---

[1] S. D. Ripley *in litt.*, 1950.

long. 38° 50′ E., approximate), western Iraq, in March 1928 (Henry Field, *in verb.*).

Range: Formerly perhaps Sinai and in Syrian and Arabian deserts from the Euphrates Valley at about latitude 34° N. on the north, and on the west from the Jordan-Saudi Arabian border at Wadi Sirhan (lat. 31° N., long. 38° E.) and the western edge of the great central Nafud (desert) eastward, south of the Euphrates, to the Persian Gulf, and southward along the coast to the vicinity of Bahrein at about latitude 25° N., and, in the more remote past, south to 22° and perhaps farther.[1]

Habitat: "According to Arab information the ostrich is found most plentifully about 50 miles southwest of Jauf . . . also, to a certain extent, north of Jauf at the heads of the Wadis which drain into Mesopotamia . . ."[2]

The former locality is close to the northwestern edge of the terrible central desert of sand dunes called Nafud, where Carruthers found fragments of eggshell of the ostrich in 1909. He (1910) says of the plateau just west of this desert:

> It was as if the whole country had been swept away by the continual westerly winds that always blow here with great violence. The ground was covered with worn rocks that stood up in hundreds like giant mushrooms.
> The sand dune area, curiously enough, supports a considerable amount of vegetation, a heavier growth, indeed, than the surrounding desert. The Nafud is covered with the ghada bush. . . . I have measured roots of the ghada . . . as 50 feet in length. I have also found a curious parasite (*Cynomonium coccineum*) growing on the roots of the ghada bush, which produced a long spadix . . . these are juicy . . . the antelope . . . dig in the sand in order to get at them. It is worthy of notice that the Arabian Oryx seemed to be most numerous where these parasites were growing.

It has been said that the ostrich is to be seen most often in the general vicinity of antelope and oryx.

It is possible that this great, central dune area, as well as smaller ones, may be the last resort of the ostrich. D. Ripley (*in litt.* 1950) says, ". . . might occur in the Nafud southwest of Riyadh (pure speculation)."

Habits: "This bird breeds in the middle of winter and lays from twelve to one and twenty eggs. The nest (madhab) is generally made at the foot of some isolated hill. The eggs are placed close together in a circle, half buried in sand, to protect them from rain, and a narrow trench is drawn

---

[1] Prater, 1921; Carruthers, 1922; Cheesman, 1923. These are almost all casual observations of travelers. The range may once have been more extensive. A fragment of eggshell collected at about latitude 22° N. in Central Arabia is thought to be that of *Psammornis* (an extinct genus with a similar range) by Lowe, 1933.

[2] Cheesman was given two live specimens from the vicinity of Jauf in 1923, according to Leachman (in Prater, 1921). Wadi means riverbed. Carruthers (1922) says a chick was brought to Captain Shakespear 2 marches east of Jauf in April 1914.

around, whereby the water runs off. At ten or twelve feet from this circle, the female places two or three other eggs which she does not hatch, but leaves for the young ones to feed upon immediately after they are hatched. The parent birds sit on the eggs in turn; and while one is so employed, the other stands keeping watch on the summit of the adjacent hill." [1]

The Emus

Dromaeius novaehollandiae diemenensis Le Souef

Dromaeius diemenensis LE SOUEF, Bull. Brit. Orn. Club, **21**, 1907, p. 13 (Tasmania).

Status: Extinct.
Range: Tasmania.

Dromaeius novaehollandiae minor Spencer

Dromaeius novaehollandiae minor SPENCER, Victorian Nat. (Melbourne), **23**, 1906, p. 140 (King Island).

Status: Extinct. Known only from bones (post-Pliocene, ? Pleistocene).
Range: King Island, Bass Strait, Australia.

Dromaeius novaehollandiae diemenianus (Jennings)

Casuarius diemenianus JENNINGS, Ornithologia, 1828, p. 382 (Kangaroo Island).

Status: Extinct.
Range: Kangaroo Island, Bass Strait, Australia.

Emus (*Dromaeius novaehollandiae*) are still to be found throughout most of their range, the semiarid regions of Australia excluding the deserts. They are much reduced in numbers, and the subspecies that once lived on Tasmania and Kangaroo and King Islands of Bass Strait are extinct.

These large flightless "ostriches of Australia" appear to be ill fitted for adaptation to human beings. The birds were removed from the protected list in parts of Western Australia because of their inroads on the wheat crops as long ago as 1923. At that time a bounty was paid in certain districts. So great was the outcry against the birds in 1932 that soldiers were sent against them armed with machine guns. Although this adventure met with no great success the bounties have continued; as much as 4 shillings per beak plus sixpence for an egg was being paid in 1948 by local "vermin boards." The great rabbit fences stretching from north to

---

[1] Burckhart *in* Prater (1921).

south are a hindrance, and many birds are found dead on the desert side of them.

It seems unlikely that the emus can survive once men have actually declared war upon them. We may hope that sanctuaries will be adequate. (See Serventy and Whittell, 1951.)

Fig. 6.—Kangaroo Island emu (*Dromaeius novaehollandiae diemenianus*).

The Tasmanian birds (*Dromaeius novaehollandiae diemenensis*) became extinct in the early part of the 19th century. The exact date cannot be determined because of the introduction of *novaehollandiae* from Australia. Two specimens upon which the subspecies is based were presented to the British Museum in 1838 however, and there is a third in the museum at Frankfurt, Germany. These lack the black feathers on the fore neck of *novaehollandiae*.

Bones have been found (H. H. Scott, 1924, 1932), but they do not reveal any size difference between this form and its mainland relative.

Dromaeius n. minor[1] of King Island, between Tasmania and Australia in Bass Strait, is known only by bones found in post-Pliocene (? Pleistocene) sand dunes. Most of these are smaller than those of other forms, but there is more individual variation than was at first suspected. Perhaps it is not a recognizable subspecies. However that may be, the birds of that island are gone and with them a population of wombats (*Vombatus u. ursinus*), which, however, survives on Tasmania, where it was introduced. There is no explanation for the disappearance of these species. Although men made their way to the islands in Bass Strait in late Pleistocene, there is no reason to believe that these few negritos killed all the birds. It is possible that fires started by men or by lightning on so small an island may have been responsible.

Forest fires were probably the cause of the extirpation of the Kangaroo Island birds (*D. n. diemenianus*),[2] in the opinion of the Australian naturalists Morgan and Sutton (1928). The island was discovered in 1802 by Flinders, who saw many. Except for a few seal hunters, there were no inhabitants until 1836, when a few settlers arrived from England. The birds were gone by that time. It is unlikely that a few men could have extirpated the population from an island 90 miles long by 35 miles wide, even though they depended on the country for food, according to those authorities.

Only one specimen is known and that is in Paris. François Péron and other naturalists of the naval expedition captured three birds on Kangaroo (Decrés) Island in 1803 and brought them alive to Paris, where they lived for a time. The skin in the Muséum National, Paris, is one of these. It is the type of *D. ater* Vieillot (see Berlioz, 1929, 1935). It is darker and smaller than the other subspecies. Skeletal material is in Paris, Florence, and Adelaide, South Australia.

These allied forms[3] have been described from bones found in post-Pliocene beds in Queensland and New South Wales. In the opinion of DeVis, one form (*patricius*)[3] had shorter, stouter legs than the living birds and the coracoid is said to differ. A second (*graciliceps*)[4] had longer and more slender legs. A third species (*Dromaeius* (?) *queens-*

---

[1] See also Spencer and Kershaw, 1910. *D. bassi* Legge (1907) and *D. spenceri* Mathews (1912) are synonyms.

[2] *D. peroni* Rothschild (1907), *D. parvulus* Mathews (1910), and *D. spenceri* Mathews (1912) are synonyms; *D. ater* Vieillot, 1817, is a synonym of *novaehollandiae. D. ater* of many later authors is a synonym of *diemenianus*.

[3] De Vis, 1888, pp. 1290–1292. Tibiotarsus in British Museum; coracoid, pelvis, tarsometatarsus, femur, and tibiotarsus in Sydney.

[4] De Vis, 1892, pp. 445–448. Tarsometatarsus fragment in Sydney.

MEASUREMENTS *

(In mm.)

	novae-hollandiae Australia	diemenensis Tasmania	diemenianus Kangaroo Island	minor King Island
Femur	217–243	225	173–182	130–186
Tibiotarsus	415–446	422–446	329–336	266–354
Tarsometatarsus	335–411	348–349	271–284	199–292

* After Scott, Spencer, Rothschild, Morgan, and Sutton.

landiae)[1] is known only by a fragment, but so large that De Vis identified it as a moa (Dinornis). Hutton (1894) pointed out that this resembles both the emu and a larger species, Dromornis australis Owen, which also inhabited Queensland in post-Pliocene.

Recently (C. Anderson, 1937) a sternum had been found in Wellington Caves, Queensland, but no other such bone has been identified as a member of the genus, and apparently no other bird bones were found with it. It was a larger bird than novaehollandiae.

Because of the considerable individual variation in the size of the leg bones of the living emus, it seems probable that there were three species in Pleistocene times (F. S. Colliver, in Morrison, 1942): D. patricius, the possible ancestor of the living novaehollandiae; Dromornis australis Owen, 1872; and Genyornis newtoni Sterling and Zeitz, 1894. Both of the latter were large, robust species, and it is likely that because of their relatively large heads their feeding habits were specialized. Any change in their environment would jeopardize their existence. Undoubtedly men were present at that time, but it is doubtful that they contributed to the extinction of these species.

Location of specimens: I know of bones of Dromaeius diemenianus in Adelaide, Paris (where a skin is preserved), and Sydney, and of Dromaeius novaehollandiae diemenensis in Augsburg, Florence, Frankfurt, Gotenburg, Launceston, London, and Nijmegen.

Steller's Albatross[2]

Diomedea albatrus Pallas

Diomedea albatrus PALLAS, Spic. Zool., 1, fasc. 5, 1769, p. 28 (off Kamchatka).

Common names: Short-tailed albatross, Steller's albatross.

Status: Extremely rare. A very small population is again to be found

---

[1] De Vis, 1884, pp. 23–28. Femur fragment in Sydney.

[2] For an excellent account see Austin, 1949, whose sources of first-hand information are his own, as well as those recorded in the original Japanese. I am obliged to him for the most recent information.

breeding on Torishima, Seven Islands of Izu, according to Masaji Yama-moto (*fide* Austin, Sokkojiho, 21(8) 1954, p. 232), a member of the staff of the weather station on the island. His pictures are unmistakably of birds of this species.

On May 17, 1951, Comdr. G. S. Ritchie and other officers and men of the British research vessel H.M.S. *Challenger* saw what is believed to have been a Steller's albatross in the southwestern Pacific, north of New Guinea and south of the Caroline Islands (lat. 4° 04′ N., long. 147° 55′ E.).[1]

A bird, thought to have been this species, was reported to have been seen off the coast of California in 1946 (Traylor, 1950). Last unquestionable previous record was June 30, 1933, when five bands taken from birds banded on Torishima in 1932 were returned to Tokyo. Possibly small colonies still exist on small islands off the coast of Formosa, but it is doubtful.

There can be no doubt that this population had been reduced to the point of extinction by the depredations of men motivated by the feather trade. It is true that the largest colony, that of Torishima, was destroyed by the volcanic eruptions of 1933 and 1941, but had it not been for the raids of fishermen and feather hunters on smaller colonies, the birds would be still as numerous as formerly.

The history of Torishima since its settlement in 1887 is short and dis-graceful. There were anywhere from 10 to 50 human inhabitants until 1900. These knocked the birds on the head as they sat on the nests, and it was easy for one man to kill 100 to 200 birds a day according to Hat-tori (in August, 1949), who saw it being done. In 1903 an eruption killed all the feather hunters, but others took their places. When next an ornithologist visited the colony, in 1929, there were about 2,000 Steller's albatrosses. In 1933 there were less than 100. Yamashina (in Austin, 1949) says: "This may have been partly the result of pasturing cattle on the breeding grounds after 1932, but the fatal cause was the last great mas-sacre perpetrated by the inhabitants in December 1932. . . . The elemen-tary school teacher told Yamada that . . . over 3,000 Albatrosses had been killed."

The Japanese Government declared Torishima a "Kinryoku," or no hunting area, in 1906, and it was about this time that Californians began to miss the birds off their coasts. The sanctuary was reaffirmed in 1933, and in that year Yamashina in Tokyo persuaded the marine laboratory, as well as a large fishing industry, to collect albatrosses at sea; no Steller's

---

[1] See Ibis, 1952, p. 536. Unfortunately the photograph does not show diagnostic characters.

albatross was found.[1] In 1941 the volcano erupted again. The whole island, save the steep northwest slope, is now covered with new lava, Austin reports. He sailed close to the island in 1949 during the breeding season but saw no Steller's albatrosses. At this time he was able to circumnavigate islands of the Bonin group also, but no albatrosses of this species were seen.

The only enemies of these birds, other than men, to be recorded are parasitic insects and crows.[2] Of the latter Hattori says: "Two or three first attack a chick at the hip, and killing it, finally devour it. The chicks that perish by these accidents will amount almost to one-third. . . ."

Range: Torishima in the Seven Islands of Izu, and formerly in the Bonin Islands on Kita-no-shima,[3] and Nishi-no-shima (or Rosario) about 80 miles southwest of Torishima,[4] as well as (possibly) on Mukoshima and Yomeshima,[5] and on the southern Borodino Islands (Yamanari quoted by Hutchinson). In the southern Riukiu group (Liu-Kiu or Nansei-Shoto), on Kobisho Island.[6] Possibly the birds bred also on Agincourt Island, off the northern tip of Formosa, and perhaps on the Pescadores Islands in the Formosa Channel.[7]

In winter, birds were to be seen often at sea from Kamchatka south to the coast of China, and eastward to the coasts of Alaska and California.[8]

Habitat: Nested on the ground on small islands. Hattori (in Austin, 1949) says they nested "in open places with some tender grasses, and the reed fields are avoided." Yamashina (*l.c.*) mentions a nesting colony at the top of the crater of the volcano on Torishima where the ground was covered with pebbles. During a great part of the year the birds were to be found only at sea.

Description: A large albatross (wing spread about 7 feet). White, with dark primaries and tail tip. To be distinguished from the Laysan albatross by its white, not black, back and larger size.

---

[1] The record published by Yamashina proves to be an error.

[2] Probably the Oriental raven or thick-billed crow, *Corvus macrorhynchus* subsp.?, a close relative of which is said to rob nests and raid hen roosts in China (Caldwell and Caldwell, 1931).

[3] *Shima* and *jima* mean island. Kita is the northernmost of the Parry Group or Muko-Jima Retto, lat. 27° 43′ N., long. 142° 06′ E. Five eggs in Yamashina collection in Tokyo. Details of this range from Austin, 1949, *q.v.* The oft-repeated record for Wake Island, based on Peale, is quite probably an error. He saw only *nigripes* and *immutabilis,* according to his descriptions. There are no specimens.

[4] An egg taken December 1, 1924, and an adult bird taken October 20, 1924, by fishermen and given to Momiyama.

[5] Reported by a local resident. No recent record of sighting.

[6] Lat. 25° 56′ N., long. 123° 42′ E. An egg recorded by Kobayashi, 1930.

[7] Based on report of a traveler. No specimen or other record extant.

[8] Turner, 1886; Grinnell and Miller, 1944.

The immature may be distinguished from the mature black-footed alba-tross by the flesh-colored bill and feet.

Mature birds have the head and neck white, often tinged with buff above; back and underparts pure white; primaries dark brown; tail white with a narrow dark brown tip; bill pink, and feet and legs pale blue.

Fig. 7.—Steller's albatross (*Diomedea albatrus*).

The immature is dark brown.

"Chicks are covered with pale brown down, thinner on the head, with black feet and bills" (Hattori, in Austin, 1949).

Habits: The birds appeared on Torishima in September and October after having been at sea since the previous June. No doubt the males fought, and there was a curious and stylized dance or drill conducted with

clashing of bills, stamping of feet, stretching, and groaning, as among
other albatrosses. A single egg was laid in a cup-shaped mound of earth
as in most other related species, but whether there was a second laid if
the first was stolen is not known. The chicks hatched in January.[1] Since
the sexes are so similar, it is not known whether both parents brooded and
fed the young, as other members of the family are known to do. They
were fed in the same manner: the parent regurgitates more or less digested
food, and this is poured into the bill of the chick.

By mid-July both parents and young had left the breeding colonies.
The account of Hattori, the only one with details of habits, says: "They
[the young] remain and feed freely near the shore for the first week or
so.[2] But after the first windy night with rough seas, you will not find a
single bird remaining the next morning."

The staples are said to have been squid, shrimp, and fish, as with other
albatrosses.

Newell's Hawaiian Shearwater

Puffinus puffinus newelli Henshaw

> *Puffinus puffinus newelli* Henshaw, Auk, **17**, 1900, p. 246. (Waihee Valley,
> Ulani=Maui Island, Hawaiian Islands).

Common name: Newell's shearwater, Ao.

Status: Breeds on cliffs of Kauai, Hawaiian Islands, in considerable
numbers. Sixty banded in 1964 by *W. King (Bu. Sp. Fish. Wildl.).
Many sight records at sea indicate that probably breeding populations
exist on Niihau, off which island three were seen in 1947 (Fisher, 1951).
Munro (1944) says, "It has most likely been killed out by the mongooses on Hawaii, Maui and Molokai." It is possible that if the birds nest
in the high cliffs of the north side of Molokai, where the mongoose cannot go, they may still survive there. Certainly the mongoose would destroy
eggs, young, and old in any accessible burrow. Fortunately there are as yet
none on Kauai or Niihau. Two specimens are in the Bishop Museum,
Honolulu, and one in the American Museum, New York. Five have apparently been lost from other collections; they cannot now be found.

Range: Hawaiian Islands: ?Niihau, Kauai, Molokai, Maui, Hawaii
(sight).

Habitat: Nested in burrows at the foot of cliffs near the sea at from
500 to 1,000 feet elevation (Fisher, 1951). In this the bird resembles its

---

[1] Incubation for albatrosses varies from two to three months.

[2] In Austin (1949). He writes also of a high mortality of immature birds after the parents
have left. In some other albatrosses this period is much longer.

close relatives in Lower California and the islands off northwestern Mexico, which deposit also their eggs on the ground in caves in the cliffs which tower to 3,000 feet.

Description: Entire upper parts, including top of the head and upper surface of wings and tail, black. Below pure white. Feet and tarsi yellow, black on outer, posterior side of tarsus; outer toe and half the middle toe black. Maxilla and edge and tip of mandible black; rest of maxilla light brown. (Henshaw, 1902.) According to Ridgway (in Rothschild, 1893–1900), "It comes nearest to *P. auricularis* Townsend, of Clarion Island (Revillagigedo Group, N.W. Mexico), but differs in blacker color of upper parts, wholly white malar region, more extensive, more uniform, and more abruptly white anterior and central under tail coverts, more extensive and 'solid' blackish border to under wing covert region, and especially in the very abrupt line of demarcation along the sides of the neck between the black of the upper parts and white of underparts. *P. auricularis* also has the bill entirely black and also stouter."

This form is treated as a subspecies of the Manx shearwater of the North Atlantic by Murphy (1952). He remarked that it shows a closer resemblance to typical *puffinus* of the Atlantic than to its nearer neighbor *auricularis* of the Pacific, but that further study is required.

Habits (Bent, 1922): Townsend's shearwater (*P. auricularis*), to which this bird is closely related, is said to breed in May, but the season probably begins earlier, for young were seen in May; and the black-vented shearwater (*P. opisthomelas*) is reported to begin in March, the season is at its height at Natividad on April 10. A single egg is laid, and this resembles very closely the eggs of other members of the genus, the eggs of which are all much alike. The young remain in the burrows until they are fully grown and fledged.

The note is described as "a gasping, wheezy cry, somewhat resembling the escape of steam from a partly clogged pipe."

Their migrations are thought to take the form of wanderings in search of food rather than definite seasonal movements.

Black-capped Petrel

Pterodroma hasitata [1] (Kuhl)

> *Procellaria hasitata* KUHL, Beitr. Zoöl., 1820, Abth. **1**, p. 142 (*ex* Forster). (Dominica.)

---

[1] Murphy, 1936, p. 697, states that *hasitata, phaeopygia, externa,* and *cahow* form a natural formenkreis. They are here treated as a superspecies.

Pterodroma caribbea CARTE, Proc. Zool. Soc. London, 1866, p. 93, pl. 10 (Blue Mountains, Jamaica). This is the black phase.

Procellaria jamaicensis BANCROFT, Zool. Journ., **5**, 1829, p. 81 (summit of Blue Mountain Peak, Jamaica) is a *nomen nudum*.

Procellaria diabolica L'HERMINIER IN LAFRESNAYE, Rev. Zool., 1844, p. 168, and Lawrence, Proc. U.S. Nat. Mus., **1**, 1879, p. 450, is a *nomen nudum*.

FIG. 8.—Black-capped petrel (*Pterodroma hasitata*).

Common names: West Indian petrel, black-capped petrel, diablotin, blue mountain duck (*P. "caribbea,"* the black phase).

Status: Breeds in considerable numbers on the cliffs of Morne La Selle, Haiti, where they were found in February 1963 by David B. Wingate (*Wingate, 1964, pp. 147–158).

Bond (1936) records a sighting of the black phase (*P. "caribbea"*) at sea west of the Bimini Group, Bahama Islands.

Man and the animals he has introduced are directly responsible for the almost complete disappearance of this form. A great earthquake and landslide is reported to have destroyed a great many birds and an entire breeding area on Martinique in 1847 (Noble, 1916). However, the young had been taken from the nests, even in great numbers, and from the highest cliffs for many years before that. It is probable that the birds were almost extinct by the time the Burmese mongoose was imported to Jamaica sometime after 1872.

On Hispaniola the mongoose has penetrated into the mountains called Morne La Selle (Bond, 1942), and as we know the diablotin nests somewhere in this vicinity, it must be considered to be in great danger on that part of its range.

It is questionable whether the bird nests on Dominica, but it would be in less danger there, since the mongoose has not been introduced.

In former times, when the French islands and Jamaica were probably supporting larger populations of human beings than was physically possible, and the people were meat hungry, these birds were considered a great delicacy and were hunted constantly during the breeding season. When possible this was done with dogs. When the burrows were found, long sticks were poked in, the birds seized these and were pulled out (Labat, 1722). Even the difficult and dangerous cliffs were scaled. The young in down, called "cottons" on the French Islands of Guadeloupe and Martinique, were thought to be particularly succulent.

The Burmese mongoose may or may not have contributed to the extirpation of the bird on the French Islands. In any case the process of extinction was well on its way by the time the mongoose was imported. The exact date of the first importation is not known, but it was without doubt after 1872. Certainly, if birds were still nesting at lower altitudes in the mountains in the last part of the nineteenth century, they would have been in great danger from the mongoose. Noble (1916) reports that an old Negro, who had once hunted the diablotin, told him that the birds had not been seen since the time of a great earthquake that tumbled the cliff in which the birds nested into the valley.

Range: Recorded recently from the islands of Hispaniola and Dominica. Formerly in Jamaica and Guadeloupe (Bond, 1950) and perhaps Martinique (Lawrence, 1879). Recorded as an accidental stray, probably as a result of hurricanes, from New York State and New England.

Habitat: Breeds in burrows, probably above 1500 feet, and probably in small caves in cliffs. Feeds at sea.

Description: In life, most likely to be confused with the greater shear-water, but the white rump distinguishes it.

Above, dark brown except for prominent white spot on hind neck, rump, and base of tail. Below, white. The sides, gray.

Downy young are said to be yellow.

The black phase (*P. "caribbea"*) lacks the white spots on neck and rump.

Habits: The birds probably appear on the breeding grounds in September (L'Herminier, as quoted by Lafresnaye, 1844). Here they breed in December and raise their young in burrows or rocky crevices during January, February, and March (Noble, 1916), after which they return to sea, where they remain until the following autumn. Since they have never been seen far from the Caribbean area it is reasonable to suppose that they do not stray far, except through stress of weather.

All through the year they search for food at sea. During the breeding season one or the other of the parents remains on the nest during daylight hours. It is not known which parent remains or whether they relieve each other, but by inference we suppose that they do relieve each other or return at night to feed their families, for they are recorded by Labat as having cried out over the land at night in the West Indies. During the evening the majority of the birds fly out to sea, search for food, and return just before daybreak. At this time the strange cries that caused the Negroes long ago to call them "little devil" are heard. It is probable that the diablotin, like his close relative the Galápagos petrel, towers in spirals up and up and swoops back to the earth or the sea, screaming all the while (Loomis, 1918).

At sea they are gliding constantly but with very little movement of wings, searching for food. Their close relative the Galápagos petrel is known to eat "sea butterflies," tiny shellfish with wings on their heads and very fragile shells, that swarm in certain parts of the warm seas, as well as small jellyfish (Loomis, 1918). No doubt many other minuscule marine creatures are eaten. R. H. Beck and others have successfully attracted related forms by scattering bait judiciously at sea, and the birds have been seen to gobble bits of fat.

Location of specimens: I know of specimens in Berlin; Cambridge, England; Cambridge, Mass.; Chicago; Kingston, Jamaica; London; Milan; New York; Oslo; Philadelphia; and Washington.

Bermuda Petrel or Cahow

Pterodroma cahow (Nichols and Mowbray)

Aestrelata cahow NICHOLS AND MOWBRAY, Auk, **33**, 1916, p. 194 (Gurnet Rock, southeast side of Castle Island, Bermuda).
Aestrelata vociferans SHUFELDT (1916) is a synonym.

Common names: Cahow, cowhaw, cowkoe (in imitation of the cry).

Status (Murphy and Mowbray, 1951): Extremely rare. Breeding range confined to a few small, rocky islets. Probably no more than a hundred individuals remain. At least a part of the population appears to be endangered by the invasion of nesting places by rats, which have reached even the rocks offshore. No living birds are to be found on those that rats occupy or have occupied. All possible nesting areas are now protected from human interference by law. However, Murphy and Mowbray say further:

Islets offshore . . . have evidently fallen within the wartime target area, for the tops of some of them are peppered with green, corroded, and distorted bullets of small arms. So far as could be judged, this has had no deterrent effect upon the return of the Cahows. Even more remarkable is the fact that the petrels still come back seasonally to their ancestral homes despite the blinding glare of beacons on the nearby airfield and the roar of planes that pass overhead day and night from one year's end to another.

There can be no doubt at all that the birds were very common between 1609 and 1622, for although contemporary accounts may be somewhat confused, both the "pimlico," a shearwater (*Puffinus l'herminieri*), and the cahow were found in great numbers according to letters and journals written on the spot.[1]

It is also quite likely that the famine of the winter 1614–15, which so distressed the English colonists, reduced the numbers of birds almost to the point of extinction. Governor Butler of Bermuda at that time wrote: "150 persons of the most ancient sick and weake wer sent into Coopers Iland, ther to be relieved by the comeigne in of the sea birds, especially the Cahowes."

Furthermore, in 1616, "a proclamation . . . was published . . . against the spoyle and havock of the Cahowes, and other birds, which already wer almost all of them killed and scared away very improvidently by fire, diggeing, stoneing, and all kinds of murtherings." And in 1621–22 a law was enacted to protect the birds on their nesting grounds. By this time the populations were no doubt so reduced on Coopers Island that they could no longer feed any number of people and the few that were taken

---

[1] Verrill (1902) contains a bibliography of contemporary sources.

on St. Davids and less accessible islands attracted little or no attention. Even the identity of the bird once called cahow became a mystery, and naturalists during the nineteenth century often confused it with Audubon's shearwater.

The specimen found by Louis A. Mowbray at Gurnet Rock in 1916, as well as bones found in caves, answered the question of identity, but where the few remaining birds bred remained a mystery except to all but a few. The occasional discovery of a dead bird made it apparent that some remained. William Beebe recorded the death of a fledging in 1935, and a second bird, apparently mature, was found dead in 1941.[1] It remained for Louis S. Mowbray, son of the man who found a specimen on Gurnet Rock in 1916 and described it as new to science, and Robert Cushman Murphy to tell of the few birds remaining in 1951.

Range: Now confined to small islets near Castle Roads, close to the southeastern coast of Bermuda. On only three are the birds known to nest, but their exact locations have been withheld—the better to protect the birds (Murphy and Mowbray).

No doubt cahows nested in suitable places over all the islands of Bermuda before the coming of man, for bones and eggs have been found many feet below the surface in the limestone hills that were once sand dunes (Shufeldt, 1916, 1922). The account of the visit of Diego Ramirez in 1603 indicates very large populations.[2]

Perhaps these colonies were exterminated by semi-wild hogs, which are known to have been present in 1593.[3] It would appear that the rats and cats came later (*id.*), but they may have contributed.

Possibly the cahow was once to be found in the Bahama Islands. Bones found on Crooked Island, in one of the few undisturbed caves remaining, have been identified, although without doubt, as belonging to this species.[4] No eyewitness account from the Bahamas has been found.

Habitat: Seventeenth-century accounts differ, but most have it that the birds nested in burrows in the sand dunes, not in rock crevices. Murphy and Mowbray (*l.c.*) point out that even now, on rocky islands, they burrow into soil or decomposed rock, often beneath what is, in effect, a roof of rock. Available nesting places on the rocky islands are said to be limited, and the competition for them fierce. If rats could be controlled on the

---

[1] *Time,* June 23, 1941. It is doubtful that this specimen was preserved.

[2] Beebe, 1935. I have not found the source.

[3] Verrill (1902) cites Clarke, J. S., Naufragia or Historical Memoirs of Shipwrecks, London, 1805 (account of Henry May).

[4] Wetmore, 1938. Size accords well with *cahow,* larger than *hasitata,* but this authority thinks they may belong to an undescribed species. Metacarpal, ulna, and radius in Peabody Museum, Yale University.

larger islands, where sandy dunes and grassy turf make better nesting sites, no doubt the bird populations would increase.

Description: A large petrel (total length about 16 inches), brown above and white below. Its *white forehead* distinguishes it from the the greater shearwater and Audubon's shearwater; from its closest relative, the

FIG. 9.—Cahow (*Pterodroma cahow*).

diablotin of the West Indies (*Pterodroma hasitata*), in lacking the white (or grayish) collar and in having a dark, not a white, rump. There is a narrow band of white on the upper tail coverts, as in the greater shearwater.

The wings are shorter than the diablotin's (260–262 as against 275–295 mm.).

Habits: The facts known of the annual life cycle are as follows (from Beebe, 1935, and Murphy and Mowbray, 1951): An egg containing an embryo about half incubated was found on January 29; a chick hatched on February 25; a fledgling with down on its belly was found June 6.

These, together with seventeenth-century accounts of the known cycles of related species, indicate that the cahows arrive on the breeding grounds in late October and early November. During November and December pairs are formed and burrows dug. Eggs are laid in late December and early January. The incubation period is about 56 days and young are hatched by early March. The old birds have left the nesting site for the open sea by late May, and all the young have gone by the second week in June.

Like other petrels the parents take turns in incubating eggs. One parent or the other returns to the nest after dark, relieves its mate of the duty, and he or she, apparently it is not known which, departs within a few hours to feed at sea. They spend a short time together in the burrow. Their performance has been described by Murphy and Mowbray as follows: "They could be seen nibbling each other's beaks and necks and then launching into an 'ecstatic' turn-around in the small chamber, puffing up their plumage and shuffling about each other on flat tarsi." This lasted for about 3 hours, and then one put out to sea alone.

Murphy and Mowbray (*l.c.*) remarked that when the birds were handled they did not eject oil from the nostrils, nor was the peculiar, musky, "petrel" smell to be detected.

The strange, wild cry apparently is heard only in October and November.

Remarks: Man himself primarily, and the hogs and rats that he has purposely or inadvertently landed on Bermuda, are obviously the cause for the small size of the population and extreme localization of breeding habitat. This is true not only of the cahow but also of Audubon's shearwater, the "pimlico," and perhaps also the Manx shearwater (*Puffinus p. puffinus*),[1] which apparently breeds occasionally on Bermuda. Murphy and Mowbray suggest that the rats should be extirpated from the larger and more suitable islands in order to increase the size of the bird's habitat. In the meantime Mowbray is experimenting with artificial burrows, breaking up the hard limestone soil surface near occupied nesting sites.

Location of specimens: I know of specimens only in New York.

---

[1] See Bradlee and Mowbray, 1931. *Puffinus mcgalli* Shufeldt (1916), described from a sternum, is a synonym according to Wetmore, 1931, as is *bermudae* Nichols and Mowbray, 1916.

Guadalupe Island Petrel

Oceanodroma macrodactyla Bryant

Oceanodroma leucorhoa macrodactyla W. E. BRYANT, Bull. California Acad. Sci., **2**, 1887, p. 450 (Guadalupe Island, N.W. coast Mexico, Pacific).

Common name: Guadalupe petrel.

Status: Probably extinct. No specimen has been seen alive since 1911.[1] About a dozen qualified naturalists have visited the island since then but no one has seen the birds. Anthony (1925, p. 286) reports as follows: "In former years there was a considerable colony along the ridge in the pine growth at the north end of the island . . . in July of the current year (1922) the same ridge was explored and but little was seen to indicate the recent occupation of the nesting ground. A few burrows were seen but they seemed to be very old." There has been no further word.

Imported domestic cats are reported to have preyed upon the colony, and it is probable that these were at least partly responsible for the disappearance of the population. Dozens of birds were in fact found to have been torn to pieces by cats (Anthony, 1925; Thayer and Bangs, 1908).

Range: Bred formerly on Guadalupe Island off the coast of northwestern Mexico and was to be seen at sea off the coast of California.

Habitat: Nested in burrows among the pines and oaks about 2,500 feet above the sea.

Description: Black above and brown below, with a forked tail. Can only be distinguished from Leach's stormy petrel by the paler under side of the wings. Leach's petrel also breeds on Guadalupe Island, as well as on the coast of California. It would probably be impossible to distinguish the two on the wing.

Habits: The breeding season came rather earlier in the year for this bird than it did for other birds of the genus that bred nearby (*melania, socorroensis*). It began in early or mid March and ended early in May (Anthony, 1898; Kaeding, 1905). Thayer and Bangs (1908) record eggs as late as June 17; but the breeding season was over for most birds at that time.

An attempt, at least, to line the burrows with leaves and pine needles was made (Anthony, 1898). Only a single egg was laid. It is white, with a wreath of minute spots, of a faint reddish-brown color and lavender about the larger end; some specimens have fine dots of pale lavender mixed with reddish spots (Bent, 1922).

Location of specimens: I know of specimens in Berkeley, Calif.; Cambridge, Mass.; London; Los Angeles; Pittsburgh; Providence, R. I.; San Francisco; Chicago; New York; Philadelphia; and Washington.

---

[1] Reported by Townsend, 1923.

Fig. 10.—Guadalupe Island petrel (*Oceanodroma macrodactyla*).

Spectacled Cormorant

Phalacrocorax perspicillatus Pallas

Phalacrocorax perspicillatus PALLAS, Zoögr. Rosso-Asiat., **2**, 1811, p. 305 (Bering Island).

Common names: Pallas's cormorant; spectacled cormorant.

Status: Extinct. All known specimens were given away or sold by Governor Kuprianof of the Sitka district (which at the time comprised Bering Island), during the decade 1840–50. The species must have disappeared about this time, for natives told Stejneger in 1882 that cormorants had been seen (and also eaten) about 30 years before. They said that Aij Kamen, a small island off the coast, was the last refuge of the birds.

In spite of the fact that Steller said that the birds were common on the rocky shores of the island in 1741, they were so specialized, isolated, and relieved of competition that they could not adjust themselves enough to escape extirpation in a period of 100 years.

There can be no doubt that the immediate cause of the extirpation of this species was man. The bird was good to eat. Aleuts, imported by the Russian-American Co. in 1826, considered the bird a delicacy (Stejneger, 1886). No doubt the seal and sea-otter hunters who visited the island at intervals also killed them for food. Steller, the only naturalist who ever saw the bird alive, says that it was eaten by Bering's party.

Later it was seen to be prepared by the Kamchatkan natives, who buried the bird, feathers and all, encased in a large lump of clay in hot coals.

It is probable that the Arctic blue foxes, which seem to have bred up to and perhaps beyond the point of subsistence by 1871, were a natural enemy of this large, stupid, clumsy, and almost flightless creature.

Range: Bering Island, Komandorskie (Commander) Islands, North Pacific. Stejneger (1889) suggests that they may once have existed on other islands in the north Pacific.

Habitat: It is quite possible that the birds never bred on the Bering Island itself but only on small islands off the shore, for there was a very large, hungry population of Arctic blue foxes on the main island when Bering's expedition was wrecked there in 1741. Bering Island had never, or at any rate very seldom, been visited before that time, for the foxes and sea otters were completely fearless (Stejneger, 1936).

Description: Very large (14–15 pounds) and with small wings. Body plumage, wings, and tail deep bronze-green with steel-blue reflections on neck; head, except forehead (which is naked), dark greenish blue. Two greenish-blue crests. Head and upper part of neck with long, hairlike, pale-yellow feathers. A large patch of white on flanks. Tail black.

Habits: There is no first-hand information on the habits of this bird. It may be assumed that it fed on small fish and pelagic forms. Probably the nesting season was early in summer. It is said to have been a stupid bird, by which we may suppose it was not at all afraid and was easily killed.

Location of specimens: I know of specimens in Dresden, Helsinki, Leyden, Leningrad, and London.

Double-crested Cormorant

Phalacrocorax punctatus featherstoni Buller

  Phalacrocorax punctatus featherstoni BULLER, Ibis, 1873, p. 90 (Chatham Island). Fig.: Buller, Birds New Zealand (ed. 1), pl. 32.

Status and range: Chatham and Pitt Islands off the coast of New Zealand. Said to be fairly common on the south coast of Chatham from Waitangi to Ouenga, and at Kaingaroa (breeding), but even more common on Pitt Island.

The "Double Crested Cormorant," as it is called locally, has survived many years of predation by introduced animals. During these years the shorebird *Thinornis* has disappeared from the larger islands. Unless there should be a marked increase in the human population or some accidental introductions of predators, there is no reason to think that this population is in any great danger.

Harris's Galápagos Cormorant

Phalacrocorax harrisi Rothschild

  Phalacrocorax [1] *harrisi* ROTHSCHILD, Bull. Brit. Orn. Club, **7**, 1898, p. 52 (Narborough Island, Galápagos Islands).

Status: A very small population, in danger should any change take place in its habitat, or if enemies should be imported, for the birds are tame and cannot fly.

Wild dogs are said to have been imported into the Galápagos, and Beebe records having seen one. It is conceivable that these might be a danger to nesting birds and to the young.

---

[1] Reduction of the length of wing and keel of sternum is apparently the only reason for retaining a monotypic genus *Nannopterum*. The bird is otherwise close to *carbo*.

If the Ecuadorian Government should decide to establish colonies on these islands the danger would, of course, be greatly increased.

Range: Galápagos Islands: Narborough and adjacent shore of Albermarle.

Habitat: Nests on the rocky shores of the large islands, apparently not on the outlying islets. The nests are constructed of bits of wood and seaweed.

Description: A large bird with very small wings. The general color is brown, but the wings have a purplish gloss. The sides of the head and neck are covered with long, hairlike plumes.

Habits[1]: Known to feed on small fishes, eels, and octopuses and probably other small pelagic creatures. Nests are constructed of dried seaweed, starfish, and other jetsam, or just a few sticks in a depression in the sand. The nests are about 3 to 15 feet apart. Unlike some isolated populations of cormorants, they resent intrusion and bite savagely. In captivity they are quarrelsome.

Bonin Night Heron

Nycticorax caledonicus crassirostris Vigors

> *Nycticorax caledonicus crassirostris* VIGORS, Voy. *Blossom,* Zool., 1839, p. 27 (Peel Island or Chichi Jima, Bonin Islands). Fig.: Bull. Biogeogr. Soc. Japan, **1**, 1930.

Status: Extinct. Last specimen, taken in 1889 on Nakondo Shima in the northernmost group, is in the British Museum. Subsequent Japanese expeditions have been unable to find the birds (Yamashina, *in litt.*). Human interference is probably responsible.

Range: Peel Island (Chichi) and Nakondo Shima in the Bonin Islands, western Pacific.

Habitat: Nested in low trees near the coast. Frequented beaches and marshes.

Description: Total length about 2 feet. The crown is black, ornamented with two long white plumes extending over the hind neck; back cinnamon-brown and underparts white. Bill black, feet and legs orange.

This subspecies differs from all others of the species *caledonicus,* which is distributed from the Celebes, the Palaus, New Caledonia, and the Philippines north to the Bonin Islands, by its thicker straighter bill.

---

[1] See Gifford, 1919; Rothschild and Hartert, 1899.

Hawaiian Goose

Branta [*Nesochen*] *sandvicensis* [1] (Vigors)

Anser sandvicensis VIGORS, List Anim. Garden Zoöl. Soc., ed. 3, 1883, p. 4 (Hawaiian Islands).

Common names: Néné, Hawaiian goose.

Status: About 30 birds remained in 1951.[2] In 1964 an estimated 285 were on Hawaii and Maui, and 200 in captivity over the world.

The Pacific Science Association kindly sent the following information in November 1951:

"In an attempt to save the néné from extinction there are at present three projects of raising the geese under wire. Two of these are located on the island of Hawaii and the third in England. As early as 1823, néné were successfully raised in England, and in the next 75 years became quite common in zoos and menageries in Europe. By 1900 they had become scarce, and the last néné in Europe disappeared from Clères, France, in 1940. The Territory of Hawaii Board of Agriculture and Forestry began raising néné in 1927. Its flock grew to 42 birds but was broken up in 1935, and the geese were distributed to various persons in the islands. All but one of these néné had vanished by 1949, when the Board started a new farm.

"The Board's new flock was started from two pairs of geese from the flock of Herbert Shipman, of Hilo, Hawaii, and increased by a third pair composed of a gander lent by the Honolulu zoo and a wild-caught goose. The main objective of the Board's farm is to produce 50 geese a year to be released into the wild, but so far only five goslings have been produced in two years of operation. In the 1951–52 breeding season it is hoped to increase the production rate by removing the first clutch to set under chickens and leaving the second clutch for the goose to incubate.

"The English néné restocking project is on the ground of the Severn Wildfowl Trust at Slimbridge. This flock was started in 1951 with a pair of geese from Mr. Shipman's flock. This English flock will provide an insurance against the entire loss of breeding stock should the Hawaii flocks be subject to a natural disaster.

"At present there are 24 néné in captivity in addition to an estimated 30 wild geese, and from the captive geese something is being learned of the life history of these birds. A limited study of the néné in the wild was authorized by the Board of Agriculture and Forestry in May 1951. This will consist of investigations by a ground party of néné observations

---

[1] For reason for abandoning the name *Nesochen,* see Delacour and Mayr, 1945, p. 9.
[2] Baldwin, 1945; Board of Agriculture of Hawaii, 1951 (*in litt.*).

reported to the local office of the Board. The problem in the past was a consistent and discouraging inability to find the geese again after they were reported, and an important part of the new study will be directed toward setting up a good reporting system and a procedure whereby the ground party of biologists can be on the way in the field within four hours of receiving sight reports. It is hoped that eventually there will be money available for a thorough ecological investigation of néné in the wild."

Baldwin (1945) says—

The causes of the decline [in numbers] were no doubt multiple. Direct influences were brought to bear upon the Nènè by white men directly and indirectly through altered environment. The following list gives a few of the events occurring after civilization of the western world penetrated the Hawaiian Islands:

Activities of man: exploration, hunting with firearms, probable increase in capture of live birds and eggs, flushing and frightening of birds from nests and foraging grounds, sandalwood gathering in uplands, ranching developments and activities, and building of beach resort homes and of military roads in uplands.

Indirect agents of the white man's activities: introduced animals such as the rat, goat, sheep, cattle, horse, pig (new stock), ass, feral dog and cat, mongoose, game birds (pheasant, quail, guinea hen, jungle fowl, turkey, peafowl), the mynah, and the ant. Introduced plants such as pasture weeds and grasses, mesquite, thimbleberry, and pampas grass.

Range: West coast of the island of Hawaii. "The present range encircles Mauna Loa on Hawaii, between roughly 5,000 and 7,500 feet elevation, and extends to or near the seashore in parts of Kau, north Kona and south Kohala" (Baldwin, 1947). Formerly Maui, where the range "apparently included lowlands as well as upper slopes of Haleakala and the crater itself . . . complete proof of [its] former existence . . . on Maui is lacking" (Baldwin, 1945).

It is the opinion of all recent writers that records from Kauai, Niihau, and other islands represent accidental visits.

Habitat: In grasslands and in open forests from sea level up to the edge of the rain forest, in the open country from 5,000 to 7,500 feet; often recorded as breeding and being seen on lava flows, but these are, as a rule, the older lava flows where *Vaccinium,* a favored fruit, has begun to grow. The birds apparently avoid the closed, humid forest of lowland and upland (Baldwin, 1947).

Description: Similar to the cackling geese of northwestern America (*Branta hutchinsii*) but with larger, stronger feet and legs and peculiar long, pointed feathers of buffy color, with black bases on throat and sides of the neck. Total length about 25 inches.

Head and hind neck black. Back brown, edges of feathers paler, lower back and tail, darker, almost black. Wings dark brown. Below, like back but sometimes paler and grayer. Belly and under tail coverts white.

Habits: Nests in grass or low brush near the edge of the rain forests at various altitudes from about 1,500–2,000 feet, and sometimes at altitudes of 5,000 feet and above, where Baldwin (1941) records nest and young. Formerly it nested at sea level. There is a mass movement of the great majority of the birds from lower to higher altitudes in the spring, and return in the autumn. Henshaw (1902) wrote:

The breeding season . . . is a very protracted one, and Mr. Eben Low informs us that some pairs begin to lay the latter part of August, and the nesting season is not at all over and the young able to shift for themselves till April or even later. . . . Mr. Palmer Wood . . . tells me that they lay from three to six eggs, the former being doubtless the usual number. The nest of a wild bird which he found in a lava flow was placed among low bushes and was made by scraping the surrounding dirt into a hollow pile. The eggs are laid directly upon the earth, but finally are surrounded with down plucked from the breast of the old birds.

Fig. 11.—Néné (*Branta sandvicensis*).

Baldwin (1947) has made a careful study of food habits. He has identified 29 plants from droppings. It appears that grasses are the most important food item, but there is no particular food plant that especially attracts the birds in all regions and none plays the role of limiting factor. He believes that there is probably a lack of green food at lower altitudes during the drier summer months. It seems possible that this might account for the movement of birds to the uplands.

In captivity: The following is kindly contributed by J. Delacour:

The Hawaiian goose first arrived in Europe about 1832 and bred very soon after (1834) at Lord Derby's menagerie at Knowsley in England. The birds bred regularly for many years in a number of zoos and in private collections and became fairly common in Europe, but they had become scarce again after 1900 and eventually disappeared entirely. Intelligent management and planning could have kept it going forever. The last specimen lived in my collection at Clères until 1940, when it vanished at the time of the invasion of France. It had come from the late F. E. Blaauw's collection in Holland, where it had been reared 42 years before. Blaauw had a great deal of experience with the species and was the last breeder to rear it successfully. He reports as follows in the Ibis, 1904, p. 68:

The next birds to lay eggs in 1903 were a pair of Sandwich-Island Geese (*Neochen* [*sic*] *sandvicensis*), a species which has become extremely rare in Europe of late years. These birds are kept at Gooilust in a small grass-grown enclosure, with plenty of shrubs and a wooden shed in it. As they are not happy in the frost and snow, I have them shut up in the shed every night after winter has set in. One good result of this arrangement is that the birds usually build a nest in the hay which covers the floor, and this makes it possible to protect them and their eggs from the cold weather that often prevails at the early season when they are accustomed to lay.

Last February five eggs were laid. These were all hatched, and during the whole time of incubation the male was constantly on the watch beside the female, running with great fury at everyone who came near.

The chicks are of a dark olive-green, darker on the head and back and whitish on the under parts. The tips of the wings are nearly white, the bill and legs black.

These birds grow very quickly, so that at the age of about nine weeks the wings have to be cut to prevent them from flying away.

The immature dress is very much like that of the adults, but the general tone is more grey than yellow, and all the black or brown markings are less clearly defined. The curious spiral ridges in the neck feathering are already visible in the young bird. The yellow of the neck is greyish-black, and is darker in the young males, so far as my experience goes.

At the end of July the young birds began to moult, and in about six weeks had acquired the adult plumage, the flight and tail feathers being retained, as in other young waterfowl, until the following year.

Of this particular brood of five, I had the misfortune to lose three when they were only half-grown. A very cold night seems to have been the cause of their death.

Hawaiian geese require the same treatment as most other geese. In Europe, the only difficulties were their limited resistance to cold winters and early laying, and the bad temper of the ganders; some were inclined to attack their mates suddenly, even after years of peaceful life, and to kill them. Otherwise, they were easy to keep and to breed.

As with all geese, the first condition of success is an abundance of grazing. The birds feed mostly on grass, to which they must have free access throughout the year. Hawaiian geese are quarrelsome, and it is necessary to separate each adult pair in a vermin-proof pen, large enough never to become overgrazed. At Clères each pair of geese needed a minimum space of 3,000 square feet, but on poorer soil more was necessary (Normandy is one of the richest grass countries in the world). It is advisable to surround the pen with a thick live hedge to ensure privacy and shelter and to plant a few clumps of shrubs and small trees for shade.

There should be sufficient water for free swimming and mating, a pool 6 by 10 feet being a minimum. The old Hawaiian gander at Clères swam freely on the ponds at times. If rain and wind are strong, an open shelter' is useful. The geese should be pinioned or wing-clipped. Besides grass and various other greens, a daily ration of grain and rich protein mash are all they need.

The geese will take care of the goslings, but if any predators are around, it is safer to incubate the eggs under a broody hen (or in an incubator) and to rear the young in safe coops with a sufficient run attached which is moved as often as necessary on a good fresh lawn, usually three or four times a day. When the young are two weeks old, they are let out during the day on good grass, under constant watching if any enemy is likely to attack them. Besides grass, lettuce, and other greens, mash, soaked biscuits, and custard (egg and milk cooked together) and plenty of duckweed, if available, form the best diet.

Hawaiian Duck

Anas platyrhynchos wyvilliana Sclater

> *Anas wyvilliana* SCLATER, Proc. Zool. Soc. London, 1878, p. 350 (Hawaiian Islands). Fig.: Rothschild, 1893–1900, p. 271; Phillips, 1922–1926, pl. 21.

Common names: Hawaiian duck, kaola maoli, kaola.

Status: Rare and local. This species should, without doubt, be protected from gunners. It has too small a population to withstand for any length of time any open season, considering the rapidly increasing population of the Hawaiian Islands and improved methods of transportation. As Munro (1944) remarks, Mokulua Islet, off Lanikai, northeastern coast of Oahu, should be set aside as a refuge, and it is probable that Kawainui Swamp at Kailua and the neighboring Kaelepulu Pond, where Munro records their bringing their young and feeding, should be guarded as well.

Range: Hawaiian Islands. Recorded from Niihau, Kauai (where common), Oahu (breeding recently on Mokulua Islets, northeastern coast [Munro, 1944] but extirpated by 1964), Molokai, Maui, and Hawaii.

Habitat: Ponds, lakes, and swamps along the coast and mountain streams and swamps to 8,000 feet.

Description: The only small brown duck to be seen on the Hawaiian

Islands. It resembles a small, dark female mallard; a brown bird with a deep purple mirror (which is bordered with white) on the wing. Sexes alike, except that males sometimes have a greenish tinge on the head and upturned tail feathers as in the mallard.

Habits: Generally those of the mallard. Nests on the ground, well concealed, the nest itself lined with down and containing 8–10 eggs, usually 8 (Munro, 1944, p. 43). Henshaw (1902) found stomachs filled with fresh water mollusks, and Munro (1944) records some as having eaten earthworms. They are excellent fliers.

Enemies: Man and his introduced mammals, the Norway rat, the cat, and the mongoose.

Remarks: This is without doubt a species which was originally a small population, probably closely related to the mallard, which arrived fortuitously and remained on the Hawaiian Islands to breed. The mallard has been recorded as an accidental visitor in winter to the islands (Perkins, 1903). They ceased to acquire what is in the male mallard a mature plumage; in this they resemble many tropical ducks.

Location of specimens: I know of specimens in Bremen; Cambridge, Mass.; Genoa; London; New York; Pittsburgh; Oslo; and Honolulu.

Laysan Island Duck

Anas platyrhynchos laysanensis Rothschild

> *Anas laysanensis* ROTHSCHILD, Bull. Brit. Orn. Club, **1**, 1892, p. 17 (Laysan Island). Fig.: Rothschild, 1893–1900, p. 19; Phillips, 1922–1926, **2**, pl. 2.

Common names: Laysan teal; Laysan Island duck.

Status: Extremely rare; must be considered in grave danger of extinction. In 1950, Vernon Brock (H. J. Coolidge, *in litt.*) counted 26 adults and 7 young of the Laysan teal. One duck had a brood of three and another of four. Nine to 11 birds were seen in 1936 (Munro, 1944); 500 in the wild in 1964 (*Bu. Sp. Fish. Wildl.).

Laysan is now a Federal bird sanctuary, but, as Munro remarks (1944), a bird sanctuary must have a measure of supervision to be an effective protection. This bird, in spite of rats and cats, which must have made terrible enemies for such an unwary creature, and rabbits, which made a desert of the island, has somehow managed to survive. When we consider the shipwrecked sailors, the Japanese plume hunters, and the sealing parties that visited this island before it was inhabited and after these the guano diggers, it is truly remarkable that this population has survived.

Range: Laysan Island. Record from Lisiansky (Kittlitz, 1834, p. 124) has never been confirmed.

Habitat: Fresh-water ponds and surrounding shores as a rule, but apparently roaming about the island. "The wild ones ran across the buildings in the evenings and early mornings. . . . In the daytime several might he seen sitting along the top rail of a fence around a vegetable patch at a brackish water seep." (Munro, 1944.)

Description: The only small, dark-brown duck to be found on Laysan. Resembles a small, dark female mallard, but with white feathers sur-

Fig. 12.—Laysan teal (*Anas laysanensis*).

rounding the eye. It is by these characters that it differs also from the Hawaiian duck (*A. wyvilliana*). Females have slightly smaller white eye ring than males. Males sometimes have a greenish head and the turned up tail feathers of the mallard and the Hawaiian duck.

Downy young are darker than the Hawaiian teal.

Habits: ". . . Though strong on its feet . . ., weak on the wing and swims but little. It has difficulty in rising and generally flies but a short distance, but one I chased to test its flight 'rose pretty high and flew a long distance.'" (Munro, 1944.)

Location of specimens: I know of specimens in Berlin and Bremen;

Cambridge, Mass.; Chicago; Honolulu; London; Los Angeles; New York; Paris; Pittsburgh; and Washington.

Marianas Island Duck

Anas oustaleti Salvadori

> *Anas oustaleti* SALVADORI, Bull. Brit. Orn. Club, **4**, 1894, p. 1 (Marianne Islands). Fig.: Phillips, 1922–1926, pl. 21.

Common names: Oustalet's gray duck, Marianas Island duck, Ladrone Island duck, naga's, n'gang, or ngaanga (various spellings of Chamorro name).

Status: This is an extremely rare bird in great danger of extinction. It has not been seen on Guam [1] for many years in spite of the fact that naturalists have searched for it. Twelve birds are reported to have been seen in Susupe marsh, Saipan, and Lake Hagoi, Tinian, during 1945 (Moran, 1946; Marshall, 1949). Marshall remarks that it is probable that the same birds fly from one island to the other. It has been suggested that birds may be found on Rota (Baker, 1948), but it is not likely that the populations, if they exist, are large. It is a certainty that should these marshes be drained, or that if much shooting be permitted, this species will disappear.

The encroachments of the works of man are the principal dangers to this population. The human populations of Saipan and Tinian are relatively very large and will increase rapidly in the future.

Range: Marianas Islands: Guam, Saipan, Tinian, ?Rota (Baker).

Habitat: Fresh-water marshes, ponds, and streams.

Description: A very dark brown bird, appearing almost black in certain lights. The darkest duck to be seen on Saipan, Tinian, and Guam. The top of the head sometimes green, and the tail feathers in males sometimes upturned as in the mallard. The speculum is dark purple, bordered with a black line followed by a white band. Sides of the head and neck are buff, streaked with brown. A dark stripe extends from the base of the bill through the eyes. Sexes do not differ.

Following is a quotation from Yamashina (1948):

> *Anas oustaleti* has two types of plumage, one resembling *Anas platyrhynchos* and the other *Anas poecilorhyncha*. Therefore, certain earlier investigators, who saw only the *platyrhynchos* type, were convinced that *Anas oustaleti* is a near relative of *Anas platyrhynchos* (Salvadori, 1894), while others, who saw only the *poecilo-rhyncha* type, thought it nothing but a subspecies of *Anas superciliosa* (= *Anas*

---

[1] Stophlet, 1946, remarks that there is much uninhabited and practically unexplored country on Guam.

poecilorhyncha superciliosa) (Hartert, 1930). . . . Kuroda (1941–42) later proved this conjecture by observing the moult of living specimens.

Yamashina's descriptions of both types are as follows:

A. PLATYRHYNCHOS type: *Adult male in nuptial plumage:* Whole head is dark green, except at the sides where buff feathers are plentifully intermingled, a dark brown streak through the eye, and faint white ring on the lower neck. Feathers on scapulars and sides of body are as those of *Anas poecilorhyncha.* Sides of body are vermiculated but some brown feathers are found even in the full nuptial plumage. Upper breast is dark reddish chestnut with dusky spots. Upper and under tail coverts are as in *Anas platyrhynchos.* Speculum is as that of *Anas platyrhynchos,* but the tips of the greater coverts are buff instead of white. Central tail feathers are more or less curled upward. Base of bill is black, tip is olive color. Iris is dark brown. Feet, reddish orange, webs darker.

Adult male in eclipse plumage: Resembles the eclipse plumage of *Anas platyrhynchos.*

A. POECILORHYNCHA type: *Adult male in nuptial plumage:* Resembles *Anas poecilorhyncha pelewensis* from the Palau Islands and Truk Island, but sides of head are browner, superciliary stripes and ground color of cheeks are more buffy. Feathers on upper breast and sides of body are more broadly edged with brown. Speculum is usually violet-purple, as in the *platyrhynchos* type, but in two specimens from Saipan and Tinian, respectively, it is dark green as in *Anas poecilorhyncha pelewensis.* Tips of the secondaries are usually white, but sometimes very faint as in *Anas poecilorhyncha pelewensis,* and in one specimen from Saipan they are buffy. Bill is olive color with a black spot in the center of the upper mandible. Iris, dark brown. Feet, dark orange, darker in joints and webs.

Adult male in Eclipse plumage: Same as the nuptial plumage.

Thus it is apparent that the *platyrhynchos* type is composed of characteristics 90 per cent peculiar to *Anas platyrhynchos,* and 10 per cent peculiar to *Anas poecilorhyncha,* while the *poecilorhyncha* type reverses the percentages, being 90 per cent similar to *poecilorhyncha* and 10 per cent to *platyrhynchos.* Therefore, *Anas oustaleti* may conceivably have originated from a compound of *Anas platyrhynchos* and *Anas poecilorhyncha* stock.

Habits: Eggs and young have actually been taken in June and July (Kuroda, 1942), and ducklings have been seen in April (Marshall, 1949). Adult males in breeding condition have been collected in April and October (*idem*). They are said also to breed in January and February on Guam and Saipan (Phillips, 1923; Yamashina, 1948).

In general the habits resemble those of *Anas poecilorhyncha* (Yamashina, 1948). "They are frequently found in pairs, not out in the open water but in little ponds completely surrounded by tules. When flushed, they circle the lake and land in concealed ponds. . . . This mallard feeds on green vegetation and seeds by plucking away at the grass in very shallow water. No 'tip-up' feeding was seen." (Marshall, 1949.)

"Breeds in tall grass along the river . . ." [on Guam] (Phillips, 1923).

"Nests in swamps and streams" (Seale, 1901).

Remarks: It is possible that if the populations, or at least a few pairs, could be removed from Saipan and Tinian to Guam, the chance of survival for the species might be increased.

Location of specimens: I know of specimens in Cambridge, Mass.; Lisbon; London; New York; Paris; and Tokyo.

Washington Island Gadwall

Anas strepera couesi (Streets)

> *Chaulelasmus couesi* STREETS, Bull. Nuttall Orn. Club, **1**, 1876, p. 46 (Washington Island, Line Islands, Pacific). Fig.: Phillips, 1922–1926, pl. 27.

Status: Extinct.[1] No specimen has been procured since Streets obtained the types, nor has the bird been reported as seen, although an expedition from the B. P. Bishop Museum, Honolulu, visited the island in 1924; as well as R. H. Beck for the American Museum of Natural History.

The cause of the extirpation of this population is not known. Man and the cats, rats, and other animals he has introduced on this small island are presumably responsible. Duck shooting by the few residents of these islands was a common sport (Wetmore, 1925b). It is quite likely that any shooting of this small, aberrant population would cause its extirpation (Phillips, 1926, p. 314).

Range: Formerly Washington Island, Line Islands, Pacific.

Habitat: Fresh-water lakes and peat bogs (Streets, 1876). Washington Island is an atoll that has been raised above sea level, so that the lagoon has been drained of salt water and been filled with fresh on the eastern side. The rainfall in this group is sometimes more than 100 inches per year.

Description: The only two existing specimens are approximately the size of a teal (although slightly smaller, wing 199 mm.), otherwise resembling the young of the gadwall except that the bill and feet are black and the number of lamellae in the bill is greater.

Habits: Nothing recorded; presumably same as the gadwall.

Remarks: It is most probable that this population arrived at Washington Island as a stray or accidental visitor. The gadwall is recorded as an accidental visitor from the Hawaiian Islands (Munro, 1944). On the assumption that the types are immature birds, there is little difference between these and young gadwalls. The population then was perhaps nothing more or less than a number of aberrant gadwalls.

We know that gadwalls come occasionally to these islands as accidental visitors, but do they breed there occasionally? They are recorded as acci-

---

[1] *Fide* Wetmore, Condor, 27, 1925, p. 36. This information came to Wetmore from an associate who had lived on the island for many years.

dental visitors to the Hawaiian Islands where they have been seen in winter only (Munro, 1944). But, on the other hand, Dr. S. Dillon Ripley tells me that a duckling taken on the Tuamotu Islands was raised by Charles Nordhoff at Tahiti. When it reached maturity it turned out to be a gadwall. Perhaps the population called *couesi* was never anything more than the offspring of a pair or two of wounded ducks which remained, recovered, and bred.

Location of specimens: I know of specimens in Washington.

Pink-headed Duck

Rhodonessa caryophyllacea (Latham)[1]

> *Anas caryophyllacea* LATHAM, Ind. Orn., **2**, 1790, p. 866 (India). Fig.: Phillips, 1922–1926, pl. 8.

Common names: Pink-headed duck; saknal (Bengali); lal-seera, golab-lal-seer (Hindustani); dumrar (Nepalese); umar (Tirhoot).

Status: Certainly rare and local, if not extinct. Last specimen was said to have been taken in the wild in 1922 (Bucknill, 1924), but probably it was taken between 1925 and 1935, since they were living in captivity at Foxwarren Park, England, in 1932 (Seth-Smith, 1932; Ezra, 1942) and at Clères, France (Delacour, *in verb.*), and probably in Calcutta. The last captive birds in Europe died during the years 1940–44.

During the last 20 years of the nineteenth century these birds appeared from time to time in the markets of Calcutta, and during these years, as well as during the early part of the twentieth century, a few were shot by gunners. Stuart-Baker estimates that it was not unusual to bag six in a day. It has seldom been recorded as a common bird and almost always as rare.

Range: India: formerly in scattered localities from Bhutan south and west to central Bihar, along the Ganges, and south along the coast to the vicinity of Madras.

The center of distribution, or area in which they were to be found most commonly, was north of the Ganges and west of the Brahmaputra, including Purneah, Maldah, Purnlia, Bhagalpur, and Tirhoot (Baker, 1908; Finn, 1915; Phillips, 1926).

Habitat: Small ponds, surrounded by bushes and high grass. The country in which they were to be found most commonly was a vast plain, crossed by deep streams inhabited by crocodiles, and often inundated by floods. It was sparsely inhabited by human beings but heavily by tigers

---

[1] See *Humphrey and Ripley (1962)

(Simson, 1884; Baker, 1908). Sometimes, in the cold weather, they resorted to rivers. This district is said to be more heavily populated of late
years, and to be drained and cultivated.

Description: Unlike any other duck in having the head and neck pink.

Male: Entire head, sides and back of neck pink. Throat dark brown.
Back and underparts darker brown, paler below. Upper back, sides, and
breast with small spots of pinkish white. Wings appear buff (the outer
webs of primaries are brown); the speculum is dull salmon-pink. Tail
dark brown. Bill reddish pink, eyes orange-red, feet dark with a tinge of
red (Baker, 1908).

Female: Differs in having the head paler pink and the throat pink (not
brown). The rest of the plumage is somewhat paler.

Immature: Head and neck white with only a tinge of pink. Rest of
plumage paler.

Habits: The breeding season begins in April. The nest is built in the
center of tufts of high grass not far from the water. The nest is made of
grass with a few feathers but is not lined. The egg is pure white or pale
yellow and is said to differ from all other ducks' eggs in being almost
perfectly spherical (about 45.7 by 42.2 mm.) (Hume and Marshall, 1879;
Finn, 1915). Both male and female have been found in the vicinity of
the nest.

Although as many as 10 have been kept at Foxwarren Park in England
in semicaptivity, under ideal conditions, and individuals have lived for
10 years, they never attempted to nest (Ezra, 1942).

Labrador Duck

Camptorhynchus labradorius (Gmelin)

> *Anas labradoria* GMELIN, Syst. Nat., **1**, pt. 2, 1789, p. 537 (Arctic America,
> Connecticut, and Labrador). Fig.: Rothschild, 1907, pl. 36; Phillips, 1922–1926,
> p. 58.

Common names: Labrador duck, pied duck, skunk duck (Long Island,
New York), sand shoal duck (New Jersey).

Status: Extinct. Last recorded specimen shot in the autumn of 1875 on
Long Island, N. Y.; it is No. 77126 in the United States National Museum.
The last one previous to this was taken at or near Grand Manan Island,
Canada, in 1871. There is a later record the authenticity of which is
questioned because of the loss of the specimen, which is said to have been
shot by a boy in the overflow of the Chemung River, Elmira, N. Y.
(Gregg, 1879; Phillips, 1926). The mature male was always rare, even

during the first half of the nineteenth century when the birds could sometimes be found in the markets of New York.

Range: Breeding range unknown. The only known eggs were in the
Staatliche Museen für Tierkunde und Völkerkunde, Dresden, Germany.
These were supposed to have come from Labrador. Audubon's son John
was shown nests in Labrador, which were thought possibly to be those
of the Labrador duck, but it may be doubted that they really were
(Audubon, 1838, p. 271).

It cannot be doubted that birds were taken near Grand Manan Island,
New Brunswick, and in winter occasionally on the coast of New England

Fig. 13.—Labrador duck (*Camptorhynchus labradorius*).

(Townsend, 1905) and more commonly on Long Island and New Jersey
(Giraud, 1844) south at least to the Delaware River and perhaps to
Chesapeake Bay, where Audubon claimed to have seen them "near the
influx of the James River."

Habitat: In winter the birds were found in sandy bays and estuaries;
character of nesting site unknown.

Description: Male: Head white, except for a black line through the top
of the head; neck white with a velvety black collar; back, rump, and upper
tail coverts and tail black; wing coverts white, primaries black; underparts
black: Bill black or brownish, base yellow or orange. Legs and feet: "fore
part of legs and ridges of toes pale whitish ash, hind part the same be-
spattered with blackish, web black" (Wilson). Eyes reddish hazel to
yellow.

Female: Brownish gray with a whitish line behind the eye, and whitish throat. A pure white speculum on the wing. Bill like the male.

Immature male: Probably somewhat darker than female.

Downy young: Unknown.

Enemies: Nothing recorded except that the bird was shot and appeared in markets in New York, New Jersey, Baltimore, and probably other cities on the eastern coast of the United States.

Remarks: What is known about this bird is as follows:

1. Relatively small populations appeared in the Bay of Fundy, on Long Island, and on the coast of New Jersey in autumn and winter.

2. It preferred sandy coasts, bays, and inlets.

3. It was shot in winter and appeared irregularly in the markets, but it was not good to eat and often rotted before it could be sold.

4. It was said to be a wary bird and difficult to shoot.

5. During a period of about 20 years (1850–70) it disappeared gradually from its winter range and from the markets.

6. It was never thought to be a common duck, and as early as 1844 (Giraud) it was rare on its winter range. Mature males were latterly very rare indeed.

The reasons for its extinction are not known. Outram Bangs (Phillips, 1926) suggested that its peculiar bill might be due to peculiar feeding habits. The peculiarity lies in the extraordinary number of lamellae and in the swollen cere. Certainly the population was never a large one, even as early as the seventeenth century. It may by that time have been restricted in its breeding range to small islands or special habitat. The true facts can never be discovered, in all probability, but the loss illustrates the great danger to any population of small numbers.

Location of specimens: I know of specimens in Albany, N. Y.; Amiens, France; Berlin; Boston, Mass.; Brussels; Brooklyn, N. Y.; Cambridge, Mass.; Chicago; Dresden; Dublin; Frankfort; Halberstadt; Leningrad; Leyden; Liverpool; London; Montreal; Munich; New York; Paris; Philadelphia; Poughkeepsie, N. Y.; Tring, England; Vienna; Washington; and Burlington, Vt.

Crested Sheldrake

Tadorna (*Pseudotadorna*) *cristata* (Kuroda)

Pseudotadorna cristata KURODA, Tori, **1**, 1917, p. 1, fig. 1 (Naktung River, near Fusan, Korea). Fig.: Phillips, 1922–1926, **4**, p. 310.

Status: Most probably extinct. The last specimen was taken early in December 1916 near Fusan, Korea. It is a female, the type of the species, and is in the Kuroda collection in Tokyo. In spite of search, none have been seen since. The birds were described and drawn by Japanese artists about 120 years ago and are thought to have been seen near Hakodate, Japan, in those days (Kuroda, 1940; Austin, 1948).

Range: Exact range unknown. Probably bred formerly in eastern Siberia (perhaps also in Korea and Japan) and migrated south to Korea and Japan.

FIG. 14.—Crested sheldrake (*Tadorna cristata*).

Other than the type, only two specimens are known. A female taken near Vladivostok in April 1877, is in Copenhagen. A male was shot at the mouth of the Kun-Kiang River, Cholla Pakto, Korea, late in November or early in December 1913 or 1914; this specimen is also in the Kuroda collection in Tokyo. The collector, a Mr. Nakamura, is said to have given away a specimen from the same locality; where it is now is not known.

Description: Slightly larger than the mallard, about the size of the common sheldrake (*Tadorna tadorna*), its particolored and crested head (green on top with pale gray face and neck in the male and, in the female, black on top with a white face and neck) should distinguish it.

Male: Top of head dark metallic green separated from the face by a black line extending from the forehead and below the eye to the middle of the hind neck. Metallic green feathers cover the upper breast in a wide band which extends, narrowing, to the hind neck. The back, lower breast, and belly are generally dark gray with thin, wavering white lines. In general the wing resembles that of the common sheldrake. Median coverts are gray, vermiculated with white like the black, shading to brown. The inner secondaries are chestnut. The greater wing coverts are white and the outer secondaries green. Primaries and tail are greenish black. Under tail coverts pale brown shading into white.

Female: Differs conspicuously from the male in having the forehead and region around the eyes white; a narrow circle of black feathers surrounding the eyes like spectacles. Back, median coverts, flanks, and underparts pale grayish brown with narrow white lines or vermiculations.

Remarks: The first specimen sent to Europe was thought by P. L. Sclater to be a hybrid between the ruddy sheldrake and falcated teal (*Casarca ferruginea* × *Anas falcata*). Phillips (1926) found no reason to believe it to be a hybrid and quotes Sushkin as being of the same opinion. In this they agreed with Hartert (1924, p. 46) who cautiously advised breeding experiments before final judgment.

Location of specimens: I know of specimens in Tokyo and in Copenhagen.

Brazilian Merganser

Mergus octosetaceus Vieillot

Mergus octosetaceus VIEILLOT, Nouv. Dict. Hist. Nat., **14**, 1817, p. 222 (Brazil). Fig.: Phillips 1922–1926, **4**, p. 302.

Common name: Brazilian merganser.

Status: Apparently always a small and scattered population, which may be in danger in years to come when the human population over its range is larger. Recorded 1948 by Giai (1950).

Range: In small, scattered populations recorded from southeastern Brazil in the states of São Paulo, Santa Catarina, Goyaz; eastern Paraguay; and Misiones in Argentina.

Habitat: Small, secluded brooks.

Description: Head and neck black with green reflections. Back brown with greenish reflections. Below, irregularly barred brown and white; the barring is less distinct on the upper breast and sides, giving a gray appearance.

Habits: Bertoni (1901) writes as follows:

It lives in the brooks which run silently under the thickness of virgin and deserted woods, far from human dwellings, preferring the mouth of the brooks, and it is probable that from time to time it comes to the coast of the River. They go by couples, or in small bands. I have seen them in winter. . . . Their flight is very rapid but seemingly not far distant or sustained; the habits look to me to be those of a sedentary bird. On the ground it runs very quickly and it is hard to overtake it among trees and vines, which easily explains the position of the legs inserted not so far back as in other genera. Nevertheless it is in the water that its movements are livelier and it is in this environment that it spends most of its time. It swims very fast and when necessary it can dive with great ease, running after the fish under the surface for some seconds. It is very voracious and I believe it eats only fish. If it is taken alive, it defends itself with courage and tries to escape and it is not possible to tame it in any way. It continues to be wild and untamed. One that I caught refused any food and let itself die at the end of ten days . . . nevertheless it showed plenty of intelligence. Its meat does not smell badly. It is very rare.

Auckland Island Merganser

Mergus australis Hombron and Jacquinot

> *Mergus australis* HOMBRON AND JACQUINOT, Ann. Sci. Nat. Zoöl., (2), **16**, 1841, p. 320 (Auckland Islands). Fig.: Phillips, 1922–1926, **4**, p. 302.

Common name: Auckland Island merganser.

Status: Dr. R. A. Falla reports [1] this bird to be extinct, in all probability; it has not been recorded as having been seen since 1901.

Pigs were imported in 1806. Rats and mice came ashore with the many wrecks that have occurred. Sealing and whaling stations have been established on these islands during the first part of the nineteenth century.

Range: Auckland Islands.

Habitat: Fresh and brackish waters of creeks and upper estuaries (Waite, 1909; Phillips, 1926). One of the specimens Lord Ranfurly sent to the British Museum in 1901 was shot in Carnley's Harbour, which is salt, indicating that they have wandered from fresh water in the past.

Description: Head (crested) brownish red, rusty on throat and foreneck; back blackish brown, feathers edged with gray; sides of body gray; underparts barred irregularly gray and white; a double white patch on the gray wing.

The thin, "toothed" bill should distinguish this duck from any other on Auckland Island.

The sexes were alike.

Habits: Food, at least in part, fresh-water shrimps (Waite, 1909); nothing more recorded. Other members of the family nest in holes in trees or sometimes on the ground.

---

[1] *In litt.* He lived from December 1942 to January 1944 on the island.

Location of specimens: I know of specimens in Christchurch; Dresden; London; New York; Otago; Vienna; and Wellington.

California Condor

Gymnogyps californianus (Shaw)

> *Vultur californianus* SHAW, Nat. Misc., **9**, 1797, p. [1], pl. 301 (Coast of California).

Common names: California condor, California vulture.

Status: Rare, local and in great danger of extinction. Species is reduced to between 50 and 100 individuals [1]; Koford's estimate was 60 in 1950, and only 40 in 1964 (*Bu. Sp. Fish. Wildl.). The United States Government has established a refuge to protect these on the breeding ground.

No enemies of the California condor are recorded save man alone. The vast increase in human population and the increased range and efficiency of the rifle are the reasons (at least apparently) for the diminution in numbers of birds. During the same period great cattle ranches have been transformed into fruit farms, and it seems probable that the consequent reduction of available food has been a powerful contributing factor.

It has been said that the birds were poisoned by cattle ranchers and others who made a practice of exterminating coyotes and squirrels by means of poisoned meat. In view of the fact that there is only one eye-witness account of birds so killed, and also their high tolerance for the several poisons used, there is some doubt about poison as a limiting factor, in spite of many indignant statements to the contrary. Koford (1953) cites records of several birds caught in mammal traps and suffering other accidents due usually to interference by men.

Range: Breeding confined to coastal range of Santa Barbara and Ventura Counties in southern California.

Birds have been seen occasionally within the past 20 years approximately (200 miles north of the probable breeding area) as far north as Madera County, not far from Stockton (Cohen, 1951). They are also seen on the western slopes of Sierra Nevada, Fresno County (Grinnell and Miller, 1944); in northeastern San Luis Obispo County (Johnson, 1945); in Kern County between Kern Canyon and Bakersfield (Lofberg, 1936); and in northern Los Angeles County (Anderson, 1935). The southernmost recent record in California is from the vicinity of Palomar Mountain, above San Luis Rey River in San Diego County, but birds were seen in 1937

---

[1] American Ornithologists' Union Committee, 1939; Grinnell and Miller, 1944. A good general account of the history to 1900 is Harris, 1941. A definitive work on the species by Koford (1953) should be consulted (reprinted by Dover Publications in 1966).

near La Encetada, Baja California (Meadows, 1933; Koford, 1953). These sight records indicate only that the birds range far in search of food.

It is probable that the birds bred formerly (within the past 100 years) in all available territory in the mountains of California.

In their wanderings in search of food they have been seen within the past 120 years from the Columbia River in Oregon on the north to Lower (Baja) California in the south. In 1826 they are recorded as having fed near the banks of the Columbia River.[1]

As lately as the last decade of the nineteenth century they were seen in Humboldt County in extreme northern California (Smith, 1916), where they had been common 40 years before (Newberry, 1857). One hundred years ago birds were sometimes seen on the beach (Gamble, 1847). In 1861 they were one of the most characteristic birds in the neighborhood of Monterey, but during the next 10 years they gradually disappeared from the lowlands (Cooper, 1871). During the intervening years to the end of the century they were seen as far east as the eastern slopes of the Sierra Nevada, and there is a questionable sighting recorded from Utah (Henshaw, 1874). The southernmost records are Arizona on the east (Coues, 1866) and Laguna and San Pedro Mártir Mountains of Lower California (Anthony, 1893). They are thought to have been extirpated in that region (C. D. Scott, 1936).

No doubt a closely related form once ranged widely over North America, for bones (Pleistocene)[2] have been found in Florida (Wetmore, 1931b), and in the state of Neuvo León, northeastern Mexico (Miller, 1943). They have also been found in kitchen middens in Nevada and New Mexico.

Habitat: Nests in caves on the sides of the mountain canyons in the region of pines and oaks, and once at least in a cavity in a large redwood tree near Springville, south of the Sequoia National Park.[3]

Description: Considerably larger than any other land bird in North America. If seen on the wing it is possible that the white under wing coverts, which extend from the base of the wing almost to the tip, will distinguish it. Immature golden eagles have some white near the tip. If the bird is seen feeding on a carcass, as is sometimes the case, the orange or reddish-orange head and great size will at once distinguish it from the turkey vulture.

---

[1] D. Douglas, Journal, reprint, 1904, p. 252. An ill-defined migration has been postulated to explain the occurrence of the birds so far north (Grinnell and Miller, 1944).

[2] According to H. I. Fisher (1947) and Wetmore (1956, p. 37) the Pleistocene birds should be called *amplus* because of skull differences.

[3] *Los Angeles Times,* September 24, 1950.

Adult: Sexes alike. Head and neck bare, the skin being red or orange. The forehead is covered sparsely with black, stiff feathers. There is a prominent ruff of lanceolate feathers around the neck. Back, wings, and

Fig. 15.—California condor (*Gymnogyps californianus*). (After Allan Brooks.)

underparts black with brown edges to the feathers, except for the outer secondaries and a few feathers on the back which are grayish. The under wing coverts and axillaries are white. Primaries and tail are black. Bill and feet are gray horn color with a small patch of red at the knee. The color of the bare head and neck is reported to change from yellow to red during a morning display (Davis, 1946).

Young: When hatched, the bird is covered with gray down, which is retained for at least two years. The plumage is then much like the adult, except the under wing coverts are gray, edged with white, and the head and neck are darker with patches of small, gray feathers. Subadult plumage is retained for four or five years.

Habits: Condors probably do not breed before the age of six, according to Koford (1953), and sometimes even at greater ages. No nest is built. The single egg is usually deposited on the ground within small caves or crevices or under protected ledges on the walls of canyons, and occasionally near rocks or in the more or less protected places on mountain slopes.

Accounts of the wariness of the birds have often referred to them as being "shy." They have been approached quite closely, however, particularly when the birds were gorged with food and when they were nesting. Finley (1906) and others found that the parent birds showed no fear of human intruders.

Ordinarily one egg is laid in a season, but a second may be laid if the first is destroyed, and probably they do not lay every year. The period of incubation is approximately 40 days, and the eggs have been found between February 17 and May 28, though the majority are laid between March 23 and April 25. Koford observed sexual display on 30 occasions between October 1 and April 13. Twenty-two were in March. (Bent, 1937; Taylor in Harris, 1941; Koford, 1953.)

Territory is well defined where other condors are concerned, although, at least on one occasion, the parent birds took no notice of the human intruders, except to move a short distance (Finley, 1908).

The birds are definitely carrion eaters but they prefer fresh meat (Finley, 1908; Clyman, 1928). Great distances are covered in a search for food and this search is made with minute care which probably is the cause of an apparent inquisitiveness (Webb, 1939).

In the past, a hundred and more years ago, there may have been a seasonal movement of population. This would account for the appearance of the birds near the Columbia River in Oregon in certain seasons (Grinnell and Miller, 1944). There had never been any record of a movement of birds approaching the extent or regularity of migration; there have been recorded great congregations of birds in the autumn, particularly where food was plentiful (Scott, in Bent, 1937).

Specimens are to be found in all large museums.

Guadalupe Island Caracara

Polyborus lutosus Ridgway

> *Polyborus* [1] *lutosus* RIDGWAY, Bull. U.S. Geol. Geogr. Surv. Terr., **1**, 1876, p. 459
> (Guadalupe Island off the coast of Lower California).

Common names: Quelili, Guadalupe caracara.

Status: Extinct. Last seen alive and last specimens collected on December 1, 1900, by R. H. Beck (Abbott, 1933, p. 14). At least six visits to Guadalupe Island have been made by naturalists since that time; no one has found the bird.

This is one of the few cases of a bird population having been purposefully extirpated by man. The goat herders, whatever the facts may actually have been, were convinced that the great brown quelili killed the kids. These men are said to have poisoned and shot almost the entire population (Bryant, 1887).

Range: Confined to Guadalupe Island, off the west coast of Baja (Lower) California, Pacific Ocean.

Habitat: Nest was a "large affair of sticks on top of a pile of rubbish and cacti" (Swann, 1925). Mexicans reported in 1885 that the birds nested in the cliffs (Bryant, 1887).

They were observed about the bodies of slaughtered goats and were also to be seen in numbers near springs and pools of fresh water (Bryant, 1887).

Description: A large, brown hawk, larger than any other to be seen on Guadalupe Island, except the osprey, which is white below.

The head is black, the back and underparts dark brown, irregularly barred with buff. Wings dark brown above, but show a wide stripe of gray (the primaries being gray indistinctly barred with brown for two-thirds of their length). Tail gray, indistinctly barred for two-thirds of its length, and dark brown at the tip.

Young birds are paler brown with wide buffy or whitish streaks through the middle of the feathers; wings and tail more brownish, less gray, but are otherwise like the adult.

These birds differ from their congeners on the mainland by having less white on the upper breast and in having the breast and belly alike. In *cheriway* and *plancus* the lower breast and belly are black. The differences are of degree and not of kind. These represent a superspecies.

---

[1] See Hellmayr and Conover, Publ. Field Mus. Nat. Hist., **13**, pt. 1, No. 4, 1949, p. 281 (note), and Ridgway and Friedmann, U.S. Nat. Mus. Bull. 50, pt. 11, 1950, p. 595 (note) for remarks on the status of Vieillot's name *Polyborus*. I prefer to retain the name for the reason that Vieillot misidentified Buffon's "Caracara," now thought to be *Circus buffoni*. He certainly indicated the type of the genus as *Polyborus vulgaris* Vieillot in his Nouvelle Dictionnaire, **5**, Dec. 1816, p. 257.

Habits: Breeding season was early in spring, the nest having been constructed and eggs deposited by April 17. The nest was built of sticks and other coarse material. The egg had a whitish ground, obscured by heavy blotching of reddish brown (Swann, 1925, p. 73). A local informant told Palmer (*in* Ridgway, 1876), collector of the original type series, that three were laid. He reported that the birds fed extensively on live kid goats, as well as small birds, mice, shellfish, worms, and insects. "To

Fig. 16.—Guadalupe caracara (*Polyborus lutosus*).

procure the latter they resort to plowed fields where they scratch the ground almost like domestic fowls," said Bryant (1887), who also records their feeding upon the carcasses of skinned goats (he did not see an attack upon any live animal) and the carcasses of petrels. Further he records that they ate caterpillars in season, as well as beetles and other insects.

Of the call and other sounds, Palmer (*l.c.*) says, "In fighting among themselves they make a curious gabbling noise; and under any special excitement the same sounds . . . with an odd motion of the head, the neck being first stretched out to its full length and then bent backward till the head almost rests upon the back. The same odd motions are made and similar noises emitted when the birds are about to make an attack upon a kid. When surprised or wounded they emit a loud, harsh scream something like that of a Bald Eagle."

Remarks: The fossil evidence of the history of this form is more complete than one has a right to expect. Its ancestor on the mainland is *Polyborus prelutosus* Howard, 1938 (*q.v.*). This species was distributed pretty generally over North America, having been found in California and Mexico (*grinnelli*), Florida, and Puerto Rico (*latebrosus*). It is probable that this common species of the Pleistocene migrated to Guadalupe Island and its ancestor, the recently extirpated quelili, resembles it more nearly than do its ancestors on the mainland (Howard, 1938). It illustrates the theory that the forms that remain near the point of dispersal of a species are more highly specialized or, at any rate, more different from their ancestors than those which are at the periphery of the population. In other words, a section of the population has removed itself from the main current of evolution, and when the competition for existence diminished (perhaps because of the lack of enemies in this special backwater) this population of the periphery did not change to the same extent as those nearer the original center of distribution. This theory is not always applicable; see for example the case of the great auk, and the spectacled as well as Harris's cormorant.

That we can expect nothing but constant change in nature is beautifully clear here. For when *Polyborus cheriway* and its insular counterpart the Guadalupe caracara were described, they were thought to be a fixed quantity, a more or less stable unity. It now becomes apparent that the ancestor population of the Pleistocene has bred itself out of existence, so to say.

Location of specimens: I know of specimens in Cambridge, Mass.; Chicago; London; New York; Norwich; Pittsburgh; Providence, R. I.; San Diego, Calif.; and Washington.

Marianas Island Megapode

Megapodius lapérouse lapérouse Gaimard

> *Megapodius lapérouse lapérouse* GAIMARD, Bull. Gén. Univ. Annon. Nouv. Sci., **2**, 1823, p. 451 (Tinian Island, Marianne Islands).

Common names: Marianas megapode, Marianas tsukatsukuri (Jap.), sasuga-chō (Jap. in Marianas), sassenat (variously spelled) (Chamorro).

Status: Rare, local and presumably in danger of extinction. No birds were found by collecting parties of United States Naval medical units on Guam, Rota, Saipan, or Tinian. Presumably the birds have disappeared from Saipan, Tinian, Rota, Guam (Baker, 1948, 1951). Though they appear never to have been common on those islands and to have preferred outlying islands, there can be no doubt that they have diminished greatly. It may be assumed that destruction of habitat and the importation of rats, dogs, and cats are responsible. Practically the whole of Saipan and Tinian was planted in sugarcane by the Japanese during their tenancy (1919–45). Certainly construction of airfields and other military installations has not added to the natural habitat of megapodes.

By designating suitable small islands as refuges for these birds, it is probable that they could be saved. On the larger islands an increasing human population will sooner or later put an end to them.

Range: Marianas Islands: Recorded from Asunçion (Assongsong), Agrihan, Pagan, Almagan, Saipan, Tinian, Agiguan, Rota, and Guam.

Habitat: Generally to be found in small flocks in the thickets just behind the beaches on most tropical islands, though they do occur in the hills (Taka-Tsukasa, 1932, p. 16, note). Here they may also build the large mound of sticks, leaves, and sandy soil where the eggs are laid. They are to be found more commonly on the low, sandy, small outlying islands.

Description: A dark bird, about the size of a small hen, generally to be seen skulking in the shadows of thick vegetation near beaches.

Back and underparts are black; wings, tail, and rump dark brown. A small, inconspicuous crest is gray; the feathers are paler on the sides of the head and on the neck, and through them the skin may show bright orange or brown. The bill may be yellow or orange; feet and legs yellow; claws and joints of the toes black. Eyes orange.

The chick has the head and back brown. Wings and lower back barred with paler. The face, throat, and abdomen pale yellowish buff. Bill and feet are dark brown. This plumage is succeeded by a first year plumage. The body plumage is dark brown, paler on the throat and underparts. Feathers of the wing coverts and secondaries are edged with rufous. How long this plumage is retained has not been recorded, but the gray head of maturity is apparently acquired before the brown plumage (with rufous bars on wings and buffy under tail coverts) is lost.[1]

Habits: Like other members of the genus, this is reported to be a gregarious bird. It is often to be seen skulking on the ground in deep

---

[1] For notes on the taxonomy of the group see Mayr (1938a) and Amadon (1942a). They consider *lapérouse* and *senex* to be subspecies of one another, the two forms being specifically distinct from other members of the genus.

shadow, but may be seen also in bushes and low trees. The birds are fast on foot but appear to be rather slow and clumsy on the wing, though their close relatives *senex,* of the Palau Islands, are reported to be good fliers, able to fly short distances from one island to another (Finsch, 1875).

Also, like other members of the genus, they deposit the eggs in a large mound made of sand, leaves, and grass (Schnee, 1912). In the case of *senex* this mound may be as large as 4 feet high and 24 feet in diameter

Fig. 17.—Marianas Island megapode (*Megapodius lapérouse lapérouse*)

(Taka-Tsukasa, 1932). Both sexes are said to collaborate in nest construction.

The breeding season is reported to be from January or February through June by Oustalet (1896a), but a chick has been found as late as July (Hartert, 1898) in the Marianas Islands. In the Palau Islands the birds are said to breed all the year round, though eggs are to be found more commonly during the southeast monsoon (Kubary, *in* Finsch, 1875). This question requires clarification, for many females are reported to lay in the same nest, but they do not use the same mound more than once (Taka-Tsukasa, 1932).

It is probable that the period of incubation is a long one, for the birds are well developed when they hatch out. Taka-Tsukasa (1932) suggests that the embryo hatches at a lower temperature than that of birds which incubate their eggs themselves.

Food is insects and tender shoots.

The egg is extremely large, of a dirty brownish color, and always stained by nest material.

When disturbed they sometimes set up a loud clamor.

Location of specimens: I know of specimens in Bordeaux; Boulogne; Cambridge, Mass.; Delmenhorst; Frankfurt; New York; Paris; St. Omer; Toronto; Vernon, France; and Washington.

Palau Island Megapode

Megapodius lapérouse senex Hartlaub

> *Megapodius lapérouse senex* HARTLAUB, Proc. Zool. Soc. London, 1867, p. 83 (Pelew or Palau Islands).

Common names: Pelew megapode, apagaj, bagai or bakai (local), tsu-katsukuri (Jap.).

Status: Rare and local. A United States Naval medical unit obtained two chicks on Garakayo and a female on Peleliu in 1945 (Baker, 1951, p. 107). Now absent from Koror, Babelthuap (Dillon Ripley, *in litt.,* 1946; Taka-Tsukasa, 1932).

Range: Palau (Pelew) Islands: Recorded from Kajangle, Babelthuap, Koror, Auron (Aulong), Peleliu ("Arumidiu and Gayangas off Peleliu").[1]

For known details see *Megapodius l. lapérouse,* the preceding form.

Heath-hens and Prairie Chickens

In the seventeenth century, before Europeans came to America, the heath-hens and prairie chickens inhabited infertile plains of the northern Atlantic coasts, where stunted trees and bushes grew sparsely, as well as the vast grassy plains of the West, from the edge of the forests beyond the Mississippi River and almost to the Rocky Mountains. Now the birds are no longer to be found on Atlantic coasts and occur in much reduced and more or less isolated populations in the western part of their range.

Several factors, some of them but little understood, complicate the problem of both present and past distribution as well as the sizes of living populations. In early records no distinction was made between the prairie

---

[1] Cannot be found on any chart or coast pilot available; see Hand List Birds Japan, ed. 3, 1942, p. 223, and Taka-Tsukasa, 1932.

chicken and the sharp-tailed grouse (*Pedioecetes phasianellus*), which occur together and occasionally hybridize. In some places the ruffed grouse (*Bonasa umbellus*) was also confused with them. Especially, cyclical increase and decrease in numbers, so characteristic of grouse all over the world, have made it extremely difficult for accurate estimates of populations to be made. Furthermore, irregular and partial migrations of northern birds, as well as wanderings of smaller groups (Hamerstrom and Hamerstrom, 1949), make complications. Added to this is the shifting of birds from place to place by man during times of real or imagined scarcity. Countless introductions of western birds were made on the eastern coasts, beginning perhaps as early as 1824, when laws were passed on Marthas Vineyard Island, Mass., to protect the birds. No doubt the majority of these are unrecorded, but magazines devoted to shooting refer to importations into New Jersey and New England all during the latter half of the nineteenth century (Phillips, 1928).

Because of these facts, only a somewhat vague history of the species can be given.[1] They were once to be found in great numbers on the grassy prairies and in suitable places at the eastern and western edges of the forests, where trees grew sparsely; in years when diseases plagued them less and predators were fewer, thousands were seen feeding together. As Europeans advanced into their territories and cultivated the land, destroying favored nesting places and breeding territories, prairie chickens became fewer. To be sure, when trees were cut to make farms on the forests' edge, there was a temporary increase in numbers in those places, but farmers and their sons shot birds all during the year through the nineteenth century. Professionals shot for the cities' markets, so that between the years 1900 to 1910 birds had become so few that the spectre of the heath-hen caused laws to be passed for their protection.

All recent authorities have recognized a superspecies containing two species, one of three forms and a second of one only. However, ranges of the eastern subspecies and, consequently, the applicability of names have been questioned (Poole, 1949). When (and if) more specimens from the eastern mainland are found, a more satisfactory disposition may be made. In the meantime the only possible arrangement conforms to the conventional one, for it is quite possible that all the known specimens from the eastern mainland were imported from the west by man.[2]

---

[1] See Schorger (1943) for a good one.

[2] If more material will prove that birds from New Jersey and Pennsylvania are referable to the western subspecies, *pinnatus*, then the name *brewsteri* Coues (Key to North Amer. Birds, ed. 3, 1887, p. 884) will be available for populations of New England as represented presently only by skins from Marthas Vineyard Island.

The Heath-hen

Tympanuchus cupido cupido (Linnaeus)

> *Tetrao cupido* LINNAEUS, Syst. Nat., **1**, 1758, p. 160 (Virginia, *ex* Catesby).

Status: Extinct.

Range: Eastern coast of North America, perhaps from Maine and New Hampshire, but certainly from Massachusetts to the Potomac River at Washington, and possibly southward to the Carolinas.

Greater Prairie Chicken or Pinnated Grouse

Tympanuchus cupido pinnatus (Brewster)

> *Cupidonia pinnata* BREWSTER, Auk, **2**, Jan. 1885, p. 82 (Vermillion, South Dakota).
> *Tympanuchus americana* (of authors, not Reichenbach) is a synonym.

Status: Common near the center of populations, but somewhat localized and relatively rare on the periphery.

Range: The center of population is between the Missouri and Mississippi Rivers at about lat. 35° to 45° N. The northern limit of the birds is south central Alberta, southern Saskatchewan and southern Manitoba, Minnesota, Wisconsin, and Michigan. Formerly the birds occurred in eastern Ohio and Kentucky, but now the approximate eastern and southern boundaries run through southern Indiana, Illinois, Missouri, northern Arkansas, and northern Texas. The western limits roughly parallel the Rocky Mountains, from central Alberta through eastern Montana, Wyoming, Colorado, New Mexico, and northern Texas, where presumably it breeds within the range of the following species.

Lesser Prairie Chicken

Tympanuchus pallidicinctus (Ridgway)

> *Cupidonia cupido* var. *pallidicincta* Ridgway, Bull. Essex Inst., **5**, no. 12, 1873, p. 199 (prairies of Texas).

Status: As far as is known this was never a large population; it was always confined to a relatively small range within the memory of man, but with a much wider range in the Pleistocene, for bones have been found in Oregon.

Range: Formerly from the Arkansas River in southeastern Colorado eastward to south-central Kansas and southward through northeastern Oklahoma and the "Panhandle" of Texas. Recorded as a straggler in Missouri and Kansas. Now confined to restricted ranges in eastern New Mexico, northern Texas (Panhandle), and northeastern Oklahoma.

Attwater's Prairie Chicken

Tympanuchus cupido attwateri Bendire

Tympanuchus cupido attwateri Bendire, Forest and Stream, **40**, no. 20, 1893, p. 425 (Refugio County, Tex.).

Status: Range, and populations within it, much reduced.

Range: Formerly from Bayou Teche on the coasts of Louisiana (but now confined to the extreme southwestern corner in Cameron and Calcasieu Parishes) westward along the Texas coast almost to the Rio Grande, and north to the vicinity of Austin, where it is rare (Oberholser, 1938; Ridgway and Friedmann, 1946).

These forms may be distinguished by the following key [1]:

1. Axillaries usually banded or heavily spotted (pinnae pointed).... *cupido*
   Axillaries usually pure white (pinnae rounded)................. 2
2. Back of legs (tarsi) usually feathered........................... 3
   Back of legs unfeathered....................................... *attwateri*
3. Smaller (wing ♂ 207–220; ♀ 195–201); paler (breast feathers with
   narrow pale brown bars)..................................... *pallidicinctus*
   Larger (wing ♂ 217–241; ♀ 208–220); darker (breast feathers with
   wider chocolate-brown bars)................................. *pinnatus*

The last HEATH-HEN (*cupido*) was seen on March 11, 1932, on Marthas Vineyard Island. It was a male. On April 1, 1931, it had been trapped and banded and at that time was 7 years old and in good condition. Apparently it had been alone since December 1928 (Auk, **49**, 1932, p. 524; Gross, 1931). None have been seen since.

It has been suggested by Gross (1928, p. 500) that perhaps the heath-hen was not originally indigenous to Marthas Vineyard. However, suitable habitat existed in 1603, as recorded by Captain Gosnold's historian. Whether or not they were imported by man, birds were present in 1824, for a law was enacted to protect them in that year. By accounts of gunners during the nineteenth century it appears that, in years that birds were plentiful, 15 or 20 could be started in a day, indicating that they were quite common. In 1890 a count revealed 120–200. This census was made every year after 1906 for 20 years. Cycles of population were marked; there were only 77 in 1906 and 60 in 1908. Numbers increased from 300 in 1910 to 1,000 in 1914, and almost 2,000 in 1916. In that year a fire devastated the habitat and, although numbers increased from 150 to 600 between 1918 and 1920, the population never recovered from the following cycle of decrease; only three were counted (25 estimated) in 1925 (Phillips, 1926; Gross, 1928). In that year the Federation of New England Bird

---

[1] These characters are subject to individual variations. Not every specimen can be identified by a single character. See Ridgway and Friedmann, 1946; Poole, 1949.

Clubs employed special wardens to control predators and otherwise protect the heath-hens. During the following year 120 cats were killed in the reservation, and there was indirect evidence that poachers had been at work. Efforts to protect the few remaining birds were in vain. As Dr. Gross (1928) remarked, "The problem of saving the Heath Hen is not the simple one of providing protection against hawks and cats and sup-

FIG. 18.—Heath-hen (*Tympanuchus cupido cupido*).

plying food when needed, but is much more complex. There are other factors such as the inadaptability of the species, excessive inbreeding, excess male ratio and disease which are most important in the decline of the Heath Hen." By that time the population was so small that no wild species could have maintained itself.

It is recorded that birds were shot on Naushon and Nashawena, small islands lying close by to the north of Marthas Vineyard. Probably they were imported or they may have spread naturally from the mainland after the trees were felled and the land was cultivated. Such may have been the origin of the population of Marthas Vineyard as well.

Whether the birds of Marthas Vineyard differed from those of New Jersey and Pennsylvania has been questioned, for the only six specimens known from the eastern mainland differ from those of Marthas Vineyard (and perhaps New England[1]) and resemble western prairie chickens (*pinnatus*) in having the axillaries barred, the pinnae rounded, and the breast darker (Poole, 1949). However, the probability is great that the record is confused by long-forgotten introductions of western birds, or simply by errors in recording the true provenance of specimens collected in the mid-nineteenth century. Furthermore, the individual variations are no doubt greater than has been suspected.[2]

Although there is some suitable habitat in the State of Maine, all records have been suspect, having been based on the account of Audubon who probably never saw the birds there.[3] The spruce grouse (*Canachites canadensis*) has been called heath-hen by residents of eastern Maine (Palmer, 1949). Prairie chickens were brought into Maine from the west probably between 1870 and 1880, and no doubt it was these that were protected by law in 1876 and 1878 (Grinnell, 1888; Palmer, 1949).

A similar confusion of names and species is quite likely to have occurred in New Hampshire. A single eighteenth-century record for "Growse" (Belknap, 1792, p. 171) has been believed by some to refer to heath-hens, but the record is not convincing.

To the southward and westward, in Massachusetts, there can be little doubt that the birds were to be found at the edges of the forests not far from the coasts as well as in the valley of the Connecticut River, for they are mentioned by travelers of the seventeenth century,[4] and residents of the eighteenth and nineteenth, in unmistakable fashion. The ornithologist Thomas Nuttall (1840) says that Governor Winthrop told him that the birds had been "so common on the ancient brushy site of Boston, that . . . servants stipulated with their employers not to have Heath Hen brought to the table oftener than a few times a week." This state of affairs could not have lasted long, for in 1855 the indefatigable collector Dr. Samuel Cabot stated that he knew of no specimen taken on the main-

[1] No specimens have been preserved.

[2] There are six specimens in the Museum of Comparative Zoology that probably came from Marthas Vineyard before 1860 but that resemble *pinnatus* in the characters mentioned.

[3] Audubon, Orn. Biogr., 2, 1834, p. 491, says they occurred on Mount Desert Island and at Mar's Hill (near Houlton), but since he says himself that "they have been confounded with the Willow Grous" at Mar's Hill it is likely that they were also so confused on heavily forested Mount Desert Island.

[4] Wood, 1635; Forbush, 1912; and Gross, 1928, are good accounts.

land of Massachusetts except one on Cape Cod, which might have flown from Marthas Vineyard. In the western part of the state the last bird is said to have been shot in 1830,[1] and it was probably about that time that they disappeared from New England. To be sure, the State of Rhode Island passed a law to protect them in 1846, but this is the only evidence that they were ever to be found there at all, and perhaps men imported those for sport.

Since they were to be found near the Connecticut River, it is not surprising that Nuttall found them on bushy plains near Westford, Connecticut. But between 1830 and 1840 they were no longer to be found anywhere in the State (Linsley, 1843).

Certain enlightened men, even as early as 1791 began to see the necessity for protecting the birds on Long Island, and in that year a bill was introduced in New York legislature.[2] The bill protected the bird during spring and summer and provided for a fine for illegal shooting, but it seems to have been generally disregarded, as such laws were elsewhere. A sharp focus is brought to bear on the state of mind of the people of those days when conservation was discussed. It is recorded that when the chairman read the name of the bill "An Act for the preservation of the heath-hen and other game," the northern members were astonished and could not see the propriety of preserving "Indians or any other heathen" (Wilson, 1811, vol. 3, p. 106, note). At this time they were said to be most common in the Long Island towns of Oyster Bay, Huntington, Islip, Smithtown, and Brookhaven, "though it would be incorrect to say that they were not met with sometimes in Riverhead and Southampton," according to Alexander Wilson's informant. Wilson goes on to say that a brace of grouse had increased in price during the first 20 years of the nineteenth century from $1 to $5. Gradually the bird decreased in numbers until it was thought to be almost, if not quite, extinct in 1844. It is said that its popularity as a game bird and the high prices it commanded in the markets were responsible for this sad state of affairs (Giraud, 1844; DeKay, 1844, p. 205). A bone (sternum) in the American Museum of Natural History is the only tangible evidence of the birds' having once lived in New York State.

Specimens from Monmouth and Burlington Counties in coastal New Jersey,[3] added to many records of time,[4] prove that a pinnated grouse was

---

[1] Gross quotes a note in the magazine *Forest and Stream* (1882), p. 344.

[2] By Cornelius J. Bogert, Esq., of New York City.

[3] In the Academy of Natural Sciences of Philadelphia, the Reading (Pa.) Public Museum, and the private collection of J. J. Gillin of Ambler, Pa. The specimen at Reading is thought to have been taken before 1860, the others before 1892 (Poole, 1949).

[4] Wilson, 1832; Audubon, 1842; Turnbull, 1869; Chapman, 1903; Huber, 1938.

common on those sandy plains before 1850, but these few specimens resemble the western race, and we cannot now be altogether certain that they are representative of the original stock. A very similar bird was to be found on those same plains in the Pleistocene, as a bone discovered at Hornerstown, Monmouth County, attests.[1]

Perhaps it was not until 20 years later that the birds disappeared from the less settled parts of Pennsylvania; the only three remaining specimens were supposedly taken in mid-nineteenth century.[2] Birds are said to have been seen near Broadhead's Creek, on the "huckleberry barrens" of Monroe County in 1869 (Weygandt, 1906, p. 9), and this is the last notice of them in the East.

A specimen recorded in the catalogues of the U.S. National Museum confirms its former occurrence near Chesapeake Bay. Although this skin was burned in a Chicago fire, there can be no doubt that it once existed. It seems probable that if the birds occurred in Maryland they would also occur in Virginia, but there is no record. The type locality "Virginia," as well as early statements that the bird occurred in "Carolina," was based on the fact that Catesby (1743, app.) included mention of them in his book. But Catesby never saw the birds in this country, nor does he record having seen anyone who had seen them.

The GREATER PRAIRIE CHICKEN (*T. c. pinnatus*). In the Canadian provinces of Alberta, Saskatchewan, and southern Manitoba, northern limit of the species, the birds are not uncommon in certain regions; indeed Alberta allowed a short shooting season in 1951. Elsewhere on the north-western prairies the birds are relatively rare (certainly in comparison to the imported pheasants), and only in South Dakota have they been found commonly in recent years.

Before Europeans destroyed the forest, few were to be found in the heavily wooded country bordering the Great Lakes. It is quite probable that they bred only in the southern counties of Wisconsin,[3] where travelers of the early nineteenth century found them commonly. It is said that bags of 20 to 30 birds a day were not uncommon between 1830 and 1850. Soon afterward great scarcity was noticed, and this particularly during certain years when the effects of the unknown factors that cause periodic reductions in populations were intensified by overshooting. Laws were passed during this period, but apparently they were not consistently enforced.

[1] *Tympanuchus lulli* Shufeldt, Trans. Connecticut Acad. Arts. Sci., **19**, 1915, p. 69. Humerus, in the Peabody Museum, Yale University, is said to be longer than that of *pinnatus*.

[2] Three from Monroe County in collection of Col. Henry W. Shoemaker, taken about 1860; one from Lancaster County taken before 1850 in the Franklin and Marshall College collection, and one in the Reading Public Museum from Carbon County.

[3] Schorger, 1943, p. 2. See also Gross, 1930, and Hamerstrom and Hamerstrom, 1949.

More recently there has been a rapid, although local, increase in numbers in the deforested territory in Michigan (N. A. Wood, 1951) where a 19-day open season was allowed in 1951.

In the one-time prairie States to the southward, relatively small and isolated populations maintain themselves on uncultivated land. An estimate of a total population of 32,000 birds was made for the entire State of Illinois in 1943. The birds were to be found on a continuous range several hundred miles in extent in the southeastern part of the State as well as a smaller area in the northwest, but elsewhere only in scattered groups near streams.[1] A melancholy history records that 20,000 birds were sent to market from Lake County, Ind., in 1850, but only 6,000 were thought still to exist throughout the whole State in 1951.[2] Perhaps a few remain in Iowa, but in 1935 the bird was said to be extinct (Hendrickson, *in* Bennitt and Nagle, 1937, p. 47).

To the eastward, in Ohio, the original populations confined to the northern and western extremities are thought to have been extirpated not later than 1900 (Trautman, 1935), and although introductions have been made the birds have not been successful enough to afford sport.

Perhaps the birds once occurred in unforested Kentucky "barrens," for Wilson as well as Audubon [3] recorded them there, but they were probably extirpated soon afterward.

West of the river, in Missouri, local histories record great flocks on the prairies and clearings on the edges of the great forest. Now small groups are to be found, but numbers are thought generally to be declining. In 1935 only 5,100 birds were thought to exist, but this was due to cyclical phenomena, for by 1938 their numbers had risen to 10,000 and by 1941 to 13,692. In the spring of 1952 the Missouri Conservation Commission estimated 9,380.[4]

Populations in certain parts of Kansas (Topeka and Baldwin) as well as in north-central Oklahoma were reported to have increased in 1950 and 1951.[5] In October of that year a short shooting season was allowed in Kansas. The greater prairie chicken disappeared as a breeding population in Texas between 1900 and 1912.[6] Only stragglers have been seen in

[1] Yeatter, 1943; Audubon Field Notes, **6**, no. 1, 1952, p. 20.

[2] Schorger, 1943; Audubon Field Notes, **6**, no. 1, 1952, p. 20. Perhaps the count of 6,000 is low.

[3] Amer. Orn., **3**, 1811, p. 114; Orn. Biogr., **2**, 1834, p. 490.

[4] Bennitt and Nagle, 1937; Schwartz, 1945; Pittman-Robertson Quarterly, **11**, no. 4, 1951, p. 425; the 1952 Prairie Chicken Census: Missouri Conservation Commission, Pittman-Robertson Program Project no. 13-R-6 (1952).

[5] Audubon Field Notes, **5**, no. 3, p. 209; no. 4, p. 264.

[6] Texas Game Fish and Oyster Commission, 1945. I can find no authentic breeding records for Texas. In fact, the only records of occurrence are contained in Ridgway and Friedmann

part of the range since then. However, on the tall-grass prairies of north-eastern and north-central Oklahoma, populations numbering 12,655 birds were estimated by State biologists in 1944.[1]

ATTWATER'S PRAIRIE CHICKEN: Far to the southward, on the coasts of Texas and Louisiana, *Tympanuchus c. attwateri* once occupied coastal prairies that extend, with little interruption, behind the beaches bordering the Gulf of Mexico, from south-central Louisiana almost to the Rio Grande on the border of Mexico. As elsewhere, populations are much isolated and reduced, only one small flock remaining in extreme southwestern Louisiana.[2] In Texas these flocks are also isolated and the total number was thought to be no greater than 8,000 in 1937.[3]

THE LESSER PRAIRIE CHICKEN (*pallidicinctus*), once extremely common on the lower or "rolling" plains of northwestern Oklahoma, northern Texas (and southeastern New Mexico, is now much reduced in numbers. The bird's habitat, characterized by sage grasses and shinnery oaks (*Quercus harvardii*), has been changed or denied to it by agriculture, by sheep, and by cattle. About 12,000 birds were estimated to remain in Texas in 1937.[4] Perhaps that figure represents a low point, due to factors re-sponsible for periodic rise and fall in size of populations, for 14,914 were estimated to remain in the northern Oklahoma counties, bordering Kansas and Texas, in 1944.[5] The fact that in the late nineteenth century they bred in southeastern Colorado, southwestern Kansas, and central Okla-homa illustrates the great reduction of range in the past 50 years. Unless some great change in their environment is brought about, they should be able to survive.

Location of specimens: I know of specimens of *Tympanuchus c. cupido* in Ann Arbor, Mich.; Berkeley, Calif.; Basle; Brussels; Cambridge, Mass.; Charleston, S. C.; Chicago; Edinburgh; Exeter; Geneva; Gotenburg; London; Los Angeles; New York; Nijmegen; Oslo; Paris; Philadelphia; Pemberton, N. J.; Pittsfield, Mass.; Providence, R. I.; Quebec; Rouen;

(1946): namely, Gainsville, in northeastern Texas on the border of Oklahoma, and Tascosa, Oldham County, in the northwestern "panhandle."

[1] Prof. F. M. Baumgartner, *in litt.* See also Duck, L. C., and Fletcher, J. B., "A Survey of the Game and Furbearing Animals of Oklahoma," Oklahoma Game and Fish Commission, Pittman-Robertson Series no. 2, State Bull. no. 3, Norman Okla., undated [1944]. Birds were found in the following counties: Pawnee, Osage, Nowata, Craig, Ottawa, Payne, Tulsa, Rogers, Mayes, Wagoner, and Muskogee.

[2] In Cameron and Calcasieu Parishes. McIlhenny, 1943; Oberholser, 1938.

[3] Texas Game, Fish and Oyster Commission, 1945.

[4] Texas Game, Fish and Oyster Commission, 1945. This authority believes that the popu-lation numbered perhaps 2,000,000 before 1900.

[5] Prof. F. M. Baumgartner, *in litt.* Birds were found in the following counties: Cimarron, Texas, Beaver, Harper, Woods, Alfalfa, Grant, Ellis, Woodward, Major, Dewey, Roger Mills, Beckham, Greer, Harmon, Jackson.

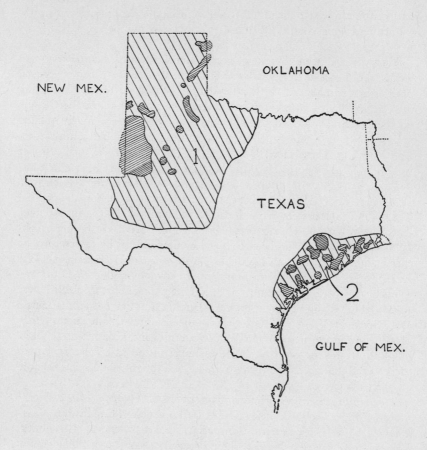

MAP 4.—Distribution of prairie chickens, southwestern United States.

San Francisco; Springfield, Ill.; Springfield, Mass.; Trenton, N. J.; Vienna; and Winnipeg, Canada.

New Zealand Quail

Coturnix novae-zelandiae novae-zelandiae Quoy and Gaimard

Coturnix novae-zelandiae novae-zelandiae [1] Quoy and Gaimard, Voy. de l' Astrolabe, Zool., **1**, 1830, p. 242 (Baie Chouraki=Hauraki Gulf, North Island, New Zealand). Fig.: Buller, 1887–1888, p. 23.

Common name: Koreke (Maori).

Status: Extinct. Last specimens taken at Blueskin Bay in 1867 or 1868 (Buller, 1887–88). Although thought still to exist on South Island in 1870, there are no specimens taken as late as that.

The disappearance of this bird seems to have been so rapid that it is difficult to see how it could have been saved. Buller (1887–88) has the following to say:

Sir Edward Stafford . . . about the year 1848 . . . went out to his own estate, about thirty miles from Nelson . . . and in the course of a few hours the party bagged 29½ brace. In the hope of preserving the game, he prohibited any shooting over the ground during the following year; but in the ensuing season, when he naturally looked for some good sport, there was not a single quail to be found. Sir Frederick Weld . . . about the same period . . . tried a similar experiment . . . but with no better success.

It was still numerous as late as 1861 in South Island, even on the east coast, on the open down of Waikonaiti, north of Dunedin, and in Nelson Province, northwestern South Island. Buller thought it still existed in the remote parts of South Island until 1875, but the records are considerably confused by the importation of *Coturnix pectoralis,*[2] *Synoicus australis,* and *Synoicus ypsilophorus.*

These birds were not brought into New Zealand before 1866, but many importations of pheasants were made between 1842 and 1869 (Oliver, 1930). Of course, it can never be proved that diseases imported with the pheasants were responsible for the extermination of the native quail, but it does seem probable that they were at least partly responsible.

Certain it seems that the imported rats, cats, ferrets, stoats, and other enemies must have preyed on the imported *Synoicus ypsilophorus,* which has thrived in New Zealand and even has extended its range to Three Kings Island, 32 miles off Cape van Dieman. Why should these enemies extirpate the New Zealand quail within a period of 30 years but allow

---

[1] See Ornithological Soc. New Zealand (1953).

[2] Recorded by Hutton (1871) and by Buller (1887–1888) but not by Oliver (1930) or Checklist (1953).

the imported quail to thrive during the following 30 years? It seems more
probable that the astonishing diminution in numbers of *novae-zelandiae*
noticed at the end of the 1840's was due to an imported disease. Fires,
although they must have been a serious contributing cause on sheep
ranches, cannot be cited as a danger on the estates of Sir Walter Buller's
friends, where an attempt was made to preserve the birds, although they
may have been very destructive to the birds elsewhere. Three bird diseases
have been proved to be shared by pheasants (*Phasianus colchicus*) and
American bobwhite quail (*Colinus virginianus*) as well as other gallinaceous birds, and to be lethal to that quail (Tyzzer, 1929; Stoddard, 1931).
Whether these diseases would have been lethal to *Coturnix novae-zelandiae* cannot, of course, be proved. At least one gapeworm has been
recorded from Australian game birds, however.

Range: New Zealand: North Island, South Island, and Great Barrier
Island.[1]

Habitat: "Abundant in all the open country, and especially in the grass
covered downs of the South Island" (Buller, 1887-88).

Description: A large, dark quail. Male: Top of the head, back, wing and
tail coverts and flanks brown, shafts of the feathers yellowish white, giving
a streaked appearance. Face, chin, and throat solid rusty brown. Feathers
of the underparts dark brown edged with paler. Bill black. Feet reddish
brown. *Female:* Differed in having the ear coverts, chin, and throat
whitish, the feathers having brown centers giving a spotted appearance.

Young male: Rusty brown throat paler than in adult male, upper parts
resembling female. Underparts washed with fulvous.[2]

The sequence of plumages and time element involved are not recorded.
It is likely that the young male acquired a plumage similar to that of
the female during molt of down and that the rusty brown throat, paler
than that of the adult, recorded by Buller, was an intermediate plumage.

This bird was apparently closely related to *C. pectoralis.*

Habits: Like other quails, this bird was to be found in open grasslands
and swampy country. Seeds constituted the major part of its diet. It made
a nest of grass in a depression in the ground. Eggs, 10 to 12 in the clutch,
were pale brown blotched with darker brown. The period of incubation
was about three weeks. Young birds are recorded to have been seen as late
as April 10 (Potts, 1870, p. 66). One call of the bird was likened to a
"low purring sound that one might suppose to proceed from an insect

---

[1] A sight record for Three Kings Island, Kermadec Islands, by a party from the Government steamer *Stella* in 1885 (Buller, 1887-1888, p. 226) must be mentioned, but may be
questioned.

[2] Buller, 1887-1888. "Similar to the adult female" (Oliver, 1930).

rather than a bird . . . it sounds something like 'Twit, twit, twit, twee-twit' repeated several times" (Potts, 1871). These meagre records are all that have come down to us of a once great game bird.

Location of specimens: I know of specimens in Auckland; Cambridge, England; Cambridge, Mass.; Christchurch; Dresden; Edinburgh; London; New York; Paris; Philadelphia; Pittsburgh; Toronto; Tring, England; Vienna; Washington; and Wellington.

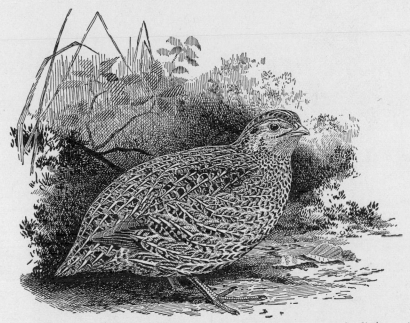

FIG. 19.—New Zealand quail (*Coturnix novae-zelandiae novae-zelandiae*).

Himalayan Mountain Quail

Ophrysia superciliosa (J. E. Gray)

> *Rollulus superciliosus* GRAY, Knowsley Menagerie, **1**, 1846, p. [8], pl. 16 ("India" = Mussoorie, Himalayan foothills, eastern Punjab).

Status: This species has not been found since June 1868. However, no very intensive search has been made; the birds are difficult to find and it may be that they will be found again. Ripley (1952, p. 905) writes of recent reports from eastern Kumaon, but he was unable to determine their accuracy.

Range: Known only from the foothills of the Himalayas, near Mussoorie in the eastern Punjab and Naini Tal in Kumaon, India. Specimens have been taken between Budraj and Benog Hills, and others close to Jerepani (5,500 feet) in the vicinity of Mussoorie and on the eastern slopes of

Fig. 20.—Himalayan mountain quail (*Ophrysia superciliosa*).

Sheir-ka-danda, near Naini Tal at 7,000 feet, according to Hume and Marshall (1879, p. 14).

Description: About the size of the European partridge (total length about 12 inches). Extremely shy and retiring, it is not likely to be seen. The long, decomposed feathers distinguish it from quails and partridges.

The male has a distinctive black-and-white head and neck, top of the

head white merging into gray, sides of the head and neck mottled black and white. The throat is black. Ground color of back is dark brown, the feathers with black edges, below grayer with black edges giving the same mottled appearance. The female has the top of the head and the throat pale brown and white feathers only around the eye. Bill bright red, feet and legs paler.

Habits: The only accounts we have are those of Captain Hutton and Kenneth Mackinnon (*in* Hume and Marshall, *l.c.*). They are said to have been found in long "seed grasses" on the steep slopes of the hills and that they were made to fly only when pressed by a dog or trod upon. Their flight was slow and heavy. A covey is said to have remained near Captain Hutton's house at Jerepani from November 1867 to June 1868, after which it disappeared. Hume assumed that they migrated north in spring. He points out that Blyth's account (Ibis, 1867) of a flock flying overhead is an error.

I know of specimens in Liverpool, London, and New York, but apparently an untraced skin exists.

Masked Bobwhite Quail

Colinus virginianus ridgwayi Brewster

> *Colinus ridgwayi* BREWSTER, Auk, **2**, 1885, p. 199 (18 miles southwest of Sasabe, Sonora, Mexico).

Status: According to Ligon (1952, pp. 48-51), this form is extremely rare. Only with the greatest trouble and after long, diligent search was he able to find a few birds in suitable habitat of Sonora, Mexico, where formerly they had been quite common. Probably the birds had disappeared from the northern part of their range in the mountains of southern Arizona between 1907 and 1914.[1] The last specimen was taken in 1885.

Range: Formerly in the Baboquivari, Whetstone, and Huachuca Mountains of southern Arizona and in Sonora, where recorded from Sasbe, Magdalena, Bacuachi, Campos, and the Yaqui Valley. Last seen 60 miles west of Hermosillo.

Introduced in Arizona and New Mexico in 1937 and 1950 (Ligon, *l.c.*).

Habitat: Grasslands above 1,000 feet.

Description: The only quail in the area (see above) with the distinct call *bobwhite*. Size of the Virginia bobwhite (total length about 8 inches). The male differs from that subspecies in having a black chin and throat and the underparts solid brown.

---

[1] See Smith, 1907, p. 196, and Swarth, 1914, p. 21.

The reason for extirpation in Arizona is said to have been the destruction of its grass habitat by drought and by introduced cattle (Brown, 1904, pp. 209–219). Cattle are thought to have destroyed habitat in Sonora also (Ligon, *l.c.*).

Manchurian Crane

Grus japonensis Müller

Grus japonensis P. L. S. MÜLLER, Natursyst. Suppl., 1776, p. 110 (Japan).

Common names: Manchurian crane; tancho (Japanese).

Status and range: Small populations are known to breed only in the marshes bordering Lake Khanka,[1] Siberia, about 120 miles north of Vladivostok, and perhaps in isolated pairs along the Ussuri River and Amur River to its lower reaches. In Japan a very few are resident in a marsh near Kushiro, in eastern Hokkaido.

Fifty years ago, according to Dementiev (1951), the birds were to be seen commonly along the rivers that flow into Lake Khanka, the Lef and Mo, and the Sugatch, which flows out to the eastward, as well as at the junction of the Ussuri and Amur Rivers, but they are now seen in numbers only on the marshes on the eastern side of the lake and at the mouth of the Lef River on the western side. Birds were found along the middle reaches of the Amur as far west as the Seja River by Shulpin, but that they were breeding is not certain.

The often repeated "Manchuria and Corea" given as part of the breeding range appears to be based on sight and hearsay records [2] or sightings of migrants. Perhaps the birds once bred over a larger part of eastern Asia, but there is no evidence.

During the last years of the nineteenth century they were seen in numbers on the marshes near Petuna on the Sungari River during September (James, 1887). These were no doubt migrants. Sowerby wrote in 1923: "Indeed the bird is so common in Manchuria during the migratory seasons that it has become known as the 'Manchurian Crane'."

Apparently they have been seen only rarely in eastern China during the past 50 years (David and Oustalet, 1877; La Touche, 1934) but much more commonly in Korea (Austin, 1948), where probably most of the population spends the winter. Probably for thousands of years Koreans have trapped them for sale in Japan and China. The presence of so many

---

[1] Also spelled Chanka and Hanka. Details of status and range from Dementiev, 1951, and Austin and Kuroda, 1953 (Japan).

[2] Taczanowski, David, Bonaparte, Hartert, and Steinbacher do not include these areas, nor does Austin, 1948.

MAP 5.—Range of Manchurian crane.

armed men puts them in great jeopardy. They are reported as rare winter visitors in Kyushu (Austin and Kuroda, 1953), but whether these breed in Siberia or in Hokkaido is not known.

In Japan the birds are extremely rare and localized; they are presumably in danger of extinction. As recently as 1949 H. Saito estimated that only 35 or 36 were still to be found breeding in a large fresh-water marsh near Kushiro, Hokkaido. This place is said to be almost inaccessible, except

FIG. 21.—Manchurian crane (*Grus japonensis*).

during the winter when the ground is frozen. The region is protected as a "natural monument."

During the Tokogawa regime, when the killing of game was strictly controlled by a feudal system and the Buddhist religion forbade meat to a large extent, the birds throve in Japan, but with the coming of the Meiji restoration their numbers declined rapidly.

Between 1924 and 1945 the population increased only by 15 or 16 birds, principally because only one egg (rarely two) is laid and of course not all eggs produce young, and but few young reach maturity under natural conditions. A few stray from the marsh when springs freeze over in exceptional winters and these usually starve (Austin and Kuroda, 1953).

Description: A very large (stands about 3½ feet long) pure-white bird with black wings and broad stripes of pale gray from the sides of the head down the neck.

The top of the head is bare and reddish in color with long coarse black hairs covering the forehead and lores. Cheeks and throat gray as are the sides of the neck. Primaries and tail white but generally covered by the long black secondaries.

Whooping Crane

Grus americana (Linnaeus)

> *Ardea americana* LINNAEUS, Syst. Nat., **1**, 1758, p. 142 ("in America septentrionali" = Hudson's Bay *ex* Edwards).

Status: Twenty-eight wild whooping cranes were thought to be alive in January 1956; in 1966 there were 44, 8 of them young. Nesting birds were found in the northeast corners of Wood Buffalo Park just south of Great Slave Lake, on the border of Alberta and Northwest Territory in June 1955. They are protected by the Canadian Government. Wild birds have been carefully counted on their winter grounds at the United States Government's Aransas Refuge on the coast of Texas for 15 years, and although it is possible that a very few birds might have escaped notice, it does not seem probable. Some hope is to be found not only in the protection on the winter grounds by the government, but also in the efforts of the National Audubon Society which retained the services of R. P. Allen in 1945. Mr. Allen (1952) has made a most complete and excellent study, "The Whooping Crane," which is largely the source for this account. He has made certain recommendations which, if followed, can at least prolong the species' existence.

It is clear from the numbers of young birds appearing each year that no more than six to eight pairs have been breeding during that time. The last known nesting place was Muddy Lake in Saskatchewan, but that was deserted in 1923, and it was not until 1955 that nesting birds were seen again.

Losses during migration during a 10-year period are about 30 percent, and more than half of this percentage were known to have been shot. To prevent recurrent catastrophes, Allen recommended that a refuge be established along 50 miles of the Platte River in Nebraska, where the birds concentrate during migration, and that the public be constantly informed of their great rarity. Although the Aransas Refuge has proved to be effective in protecting the birds during winter, it is desirable to protect them wherever they may go by establishing new refuges when necessary and to increase the number of guards, to prevent airplanes from flying too low, and to improve the public's understanding of the situation.

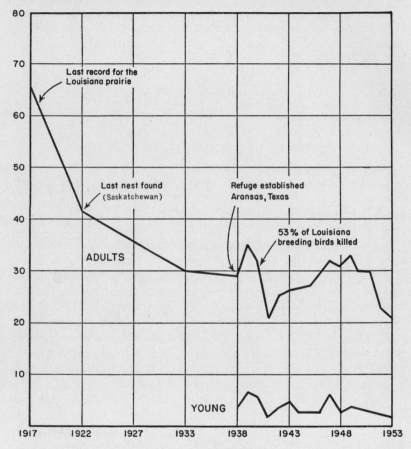

CHART 2.—Figures in vertical column indicate the known numbers of whooping cranes in the years shown in the horizontal column. After R. P. Allen (1952) and information furnished by the National Audubon Society.

Even though such protection could be given, the survival of the species would be a miracle. The accompanying graph [1] shows a slow shrinkage of the population during the past 35 years. Gradual encroachment by

[1] Figures 1918–1948 from R. P. Allen (1952, table K, p. 78; table L, p. 81). Figures 1949–1953 by the kindness of John H. Baker, president of the National Audubon Society.

civilization, draining of the great marshes where they nested, and building of roads and canals with the consequent influx of intruders and shooters both into their breeding and wintering grounds were the principal reasons. There is little doubt that the populations that bred in Louisiana were thus destroyed—those of the upland prairies in 1918 and of Whitemarsh near the coast in 1940 (see also McIlhenny, 1943).

The population was just holding its own in 1953. The breeding success is about 50 percent, the average annual gain was 0.44 and the loss 3.6 birds, Allen calculated for the years 1940–49. It might have been expected that a wild population, with what is usually a mortality rate of 80 to 90 percent during the first year of life and 50 to 75 percent in later years, might long ago have disappeared. Perhaps the same factors of mortality do not apply to small populations not occupying all available habitat. Perhaps even limited protection has permitted survival. We may hope, but perhaps we cannot expect, that the cranes will survive.

Bones found in Pleistocene deposits in Florida and the tar pits of Rancho La Brea, California, indicate that the range of the species was then much greater and the birds no doubt much more numerous at that time. However, the populations have probably not been large during the period of recent recorded history. Indeed Allen estimates a total of no more than 1,300 to 1,400 even before 1860.

Range: Within the past century breeding pairs have been recorded in scattered localities from the Slave River, which flows from Great Slave Lake at about lat. 60° 30′ N., from the "prairie states" of northwestern Canada and the United States, between the Mississippi and Missouri Rivers, and also on the coast of the Gulf of Mexico in western Louisiana.[1] The center of distribution apparently had been from Alberta southwestward to Iowa and Illinois (see map).

The two nesting records for the vicinity of Great Slave Lake of the "sixties" and MacFarland's note that he saw birds on the Mackenzie River near the shores of the Arctic Ocean every spring between 1862 and 1865 indicate that the birds may once have bred over a wide area of northern Canada. Perhaps the small breeding populations on the periphery of the range disappeared before those nearer the center of distribution, as did the small Canadian colonies of passenger pigeons.

Nests have been found in Alberta 12 miles north of Killam (1904 and 1905), 30 miles north of Edmonton (1906–09), Whitford Lake (1909), Wainwright (1914) and Wood Buffalo Park in 1955; in *Saskatchewan* east of the Moose Mountain area, Battleford on the North Saskatchewan

---

[1] Allen (1952) found no good evidence of breeding in New England or New Jersey, and I agree.

River (1884), near Yorkton (1900), 20 miles north of Davidson (1911), Bradwell (1911, 1912), 9 miles northwest of Kerrobert (1922), 7 miles south of Unity at Muddy Lake in 1922; in *Manitoba* near Winnipeg

●   indicates substantiated records of occurrence. Breeding places are surrounded by dotted lines. Dates are the last known breeding records.

⟃   indicates occurrence in Pleistocene time, substantiated by finds of bones.

○   indicates numbers of records.

MAP 6.—Former distribution of the whooping crane.

(1871), Dufferin County, southwest of Winnipeg (1906), Lake Winnipeg (1877, 1883), mouth of the Red River (1891), Lake Winnipegosis (1885), Oak Lake (1891, 1893–94); in *North Dakota* near Ina in Rolette County (1871), Lakota in Nelson County (until 1908), and perhaps other places, but the records are not beyond question; in *Minnesota* near Brainerd in Morrison County (1874), Elbow Lake in Grant County (1876), and Thief Lake in Marshall County (1889); in *Iowa* somewhere "in the northwestern part of the state" (1866–67), Sac County (1868), Dubuque (1868), in "western Iowa" (1871), Blackhawk County (1871), Lake Mills in Winnebago County (1879), Oakland Valley (1874), Cherokee County (1877), Franklin County (1880), Clear Lake in Cerro Gordo County (1880–82), Wright County (1881), Kossuth County (1881), Eagle Lake in Hancock County (1883, 1894), Midway (1883); in *Illinois* only in the "central Illinois marshes" (185?); in *Louisiana* at White Lake marshes from near Florence to the Mermenteau River in Vermilion and Cameron Parishes, where they bred in numbers before 1900 and perhaps a few until recently.

From these breeding places the birds migrated south to winter quarters, the coastal marshes of the Gulf of Mexico in Louisiana and Texas, and the plateau of Tamaulipas, Mexico, as well, the migration route being mainly through the prairie States. Records of birds that were actually shot indicate that there were great concentrations on the North Platte River in Nebraska. No doubt settlers arriving between 1865 and 1900 shot too many. The history in this respect is quite like that of the Eskimo curlew (*Numenius borealis*).

Habitat: Nests were built in the midst of great shallow marshes or sloughs once characteristic of the prairie lands as well as similar flat country lying to the north, but which is overgrown with willows and clumps of small deciduous trees, called "aspen parkland," and especially "*Salix*-Chrysomelidae communities."

In Louisiana the birds nested in the vast brackish coastal marshes among grasses (*Panicum*) and bulrushes (*Scirpus*).

Habits: As early a mid-December pairs are to be seen dancing, bowing, flapping their wings slowly and leaping into the air with neck arched over the back. The exact significance of this is not known, for they do not leave their winter quarters until late March or April, and they have not been seen to dance any more actively or often just before departure than they have earlier.[1] On their northward journey they have been recorded in Nebraska between April 1 and 18 and on the breeding ground in Saskatchewan April 24 to May 3. Probably the birds that wintered in Louisiana followed a route farther to the eastward, but a large proportion of the migration and even the whole population were concentrated during a period of two weeks every spring in an area of prairie along the Platte

---

[1] Allen (1952) and Warden Wallmo of the Aransas Refuge determined that the birds left between April 6 to 18, 1947, and April 5 to 18, 1948. Some birds (presumably those that do not breed) remain for the summer.

River no more than 200 miles wide from east to west, where most of the recorded losses of birds took place. This was the most direct route from Chicago to San Francisco.

Whooping cranes fly in pairs, families, and small bands. The flight itself is heavy and rather slow. Just before leaving the winter grounds, and perhaps at other times, birds have been seen to fly upward in great spirals, to sail and call and descend in a wild dive. The buglelike notes, *ker-loo! ker-lee-oo,* "send a shiver down your spine," says Robert Allen.

Presumably the birds dance and call in "prenuptial courtship display" on their arrival at the nesting marshes, but during incubation they are silent (p. 182), unlike sandhill cranes (*G. canadensis*). The nest is a little island of grasses heaped to a height of a foot or two above the water. Usually two eggs are laid, but sometimes one or three. One or the other of the pair is to be found almost invariably on the nest, and they relieve each other in this duty about six times during daylight hours, although sometimes the watches may be shorter or longer. During the 33 to 34 days of incubation, both parents feed the chick, although the female does most of the work. In eight months the young can fend for themselves, and before the beginning of the new breeding season, just before departure for the north, the parents turn on the young and drive them away.

The birds' food during the winter includes crayfish, snakes (even the poisonous cottonmouth moccasin), frogs, and shellfish and insects of many kinds, to the number of at least 28 species of animals and 17 of plants (see Allen, 1952, pp. 105–123). The stomach of a whooping crane killed in Manitoba was found to contain several ducklings.

Description: A very large, long-legged, white bird with bare red skin around the bill and black wings. Like all cranes it flies with the neck extended straight forward, whereas the herons (often called cranes) fly with the head drawn back. It is often confused with the sandhill crane (*G. canadensis*), with which it associates on migration, but its pure-white (not gray) plumage and larger size distinguish them. Old gunners in Nebraska called them the "big white ones."

Probably sight records for flying birds should be disregarded and none of any kind taken too seriously. It was apparently errors in identification that caused ornithologists to believe the birds were much more common than they actually were between 1923 and 1933, causing protection to be delayed too long.

FIG. 22.—Whooping crane (*Grus americana*).

Auckland Island Rail

Rallus muelleri Rothschild

Rallus muelleri ROTHSCHILD, Bull. Brit. Orn. Club, **1**, 1893, p. 40 (Auckland Island). Fig.: Proc. 4th Internat. Orn. Congr. 1905 (1907), p. 204.

Status: Extinct; known by unique type in Stuttgart.

The Auckland Islands were discovered in 1806, and for the next 40 years at least they were the resort of sealers and whalers, who camped near the shores. Pigs were introduced on the small islands near the north end in 1807 so that castaways might have food. Cattle, sheep, and rabbits were brought later for the sealing gangs. None of these animals, however, seem to have been on the islands in 1864, and it seems probable that they were all eaten, for Captain Musgrave (1865, p. 27), who spent 20 months there in that year and the next, found no trace of them. He did find cats, and he says of them: "This animal evidently lives on birds and eggs. We frequently find feathers and egg shells about their holes." Mice were very common as they had been in 1840, when a party from the U.S. Exploring Expedition landed. It seems quite possible that even a small population of cats could exterminate a ground-nesting bird, but there were dogs also to contribute to the death of this species.

It is astonishing that the whalebirds (*Pachyptila desolata altera*) that still nest in holes along the shore escaped a similar fate. Perhaps the oil these birds shoot from the nostrils deters predators.

Range: Formerly Auckland Island, off the coast of New Zealand.

Description: Resembled *Rallus pectoralis* of Australia more closely than any other form but differed from that in its tawny coloration and all others in the extraordinary development of the feathers of the back and rump.

Head and neck brownish; red faintly and irregularly striated with black; cheeks reddish gray; center of throat reddish white; back and rump bright chestnut, with centers of the feathers black; wings brownish black, faintly edged with rufous gray; lower part of throat rufous gray like breast; flanks, abdomen, and under tail coverts black, each feather tipped with pale rufous and with two white bands; tail rufous with indistinct gray bands (Rothschild, 1893).

Mathews and Iredale (1926, p. 76) note that in comparison with *R. pectoralis* the wings are much reduced and the legs and feet stronger. They propose the genus *Hyporallus* for it.

Remarks: Rothschild in his original description has only this to say of the history of the unique type: "The single specimen was sent to me for description by Count von Berlepsch. . . . It is the property of the Stuttgart Museum. It is named in honor of the famous bontanist Baron von Müller of Melbourne, who presented the specimen." This is all that is known of its history.

The supposition of Buller (1905, p. 42) that the type of *R. muelleri* was purchased by von Hügel in New Zealand in 1874[1] and collected by "the unfortunate Captain Musgrave of 'the Grafton'," which was wrecked on

Fig. 23.—Auckland Island rail (*Rallus muelleri*).

the Auckland Islands in 1864, seems not to be the fact. Actually this specimen is a *Rallus pectoralis pectoralis,* according to Mathews and Iredale (1913, p. 209), who examined the specimen at Tring. This specimen is now in the American Museum, New York. I have examined it and agree with Mathews. It may be doubted that the specimen was collected on Auckland Island.

In spite of the fact that we cannot be quite sure of the provenance of either the type or the von Hügel specimen, it is most probable that there

[1] Ibis, 1875, p. 392.

was once a rail on Auckland Island and that it is extinct now. Captain Musgrave of the *Grafton* (1865, p. 39) says that he captured a bird "something like a water hen." We may safely assume that the captain saw a ground bird resembling a rail.

Rallus muelleri is probably a representative of the superspecies *Rallus pectoralis,* which is distributed from Flores and New Guinea, eastern and southern Australia, and Tasmania. It is probable that a few arrived on Auckland Island, aided by the strong prevailing westerly winds and remained there to breed. A limited population such as this might tend to force a comparatively rapid change in the appearance of the bird, and a new species might be created within a relatively short number of generations.

It is curious that the species has never been recorded from New Zealand or any of the neighboring islands. The obvious hypothesis is that individuals have arrived on those islands but have not been able to survive, either because of their enemies or because their numbers were not great enough to propagate. Such discontinuous distribution is characteristic of distribution on oceanic island groups.

Location of specimens: I know of one specimen in Stuttgart.

Wake Island Rail

Rallus wakensis (Rothschild)

> *Hypotaenidia wakensis* ROTHSCHILD, Bull. Brit. Orn. Club, **13**, 1903, p. 78 (Wake Island).

Status: Probably extinct. Neither residents of Wake Island, to which the species is confined, nor visiting naturalists have found a bird since 1945.

Thomas D. Musson, station manager for the Civil Aeronautics Administration, wrote in 1949: "The undersigned has been familiar with this island since 1946 and has been stationed here on permanent duty since March 1, 1948. During that time no bird corresponding to the description given in your letter has been noted. Conversations with individuals who had visited Wake before the war indicate that the flightless rail was at one time quite prevalent here. Rumor has it that the birds were eaten by the Japanese forces occupying the island, and this is considered quite likely, since it is known that these forces were hard pressed for food, living on almost a starvation diet, and were forced to use every possible means of getting food. The most common opinion has it that the birds became extinct on Wake during the Japanese occupation. One or two individuals have spoken of seeing what they thought to be a flightless rail, but on further conversation it appears more likely that they were confusing that bird with the plover which is fairly common here at certain seasons of the year."

Dr. Alfred M. Bailey, director of the Denver Museum of Natural History, kindly tells me: "I am afraid the rail is extinct on Wake Island. I talked to everyone I could on the occasion of my short visit there and

Fig. 24.—Wake Island rail (*Rallus wakensis*).

none had been seen after the Americans took over after the war. Of course, there is always a possibility that there might be a few on Wilkes or Peele Islands, but certainly there has been no observation."

Range: Wake Island (lat. 19° 18′ N., long. 166° 35′ E.), Pacific Ocean.

Description: Rothschild (1903, p. 78) describes this rail as follows:

"Upper surface dark ashy brown, fading to an earthy brown; ear-coverts and lores dark brown, a pale grey superciliary line; chin and upper throat whitish, neck, grey, rest of underside ashy brown, on the breast with one, on the abdomen and flanks with two or three narrow white bars; tail uniform brown; quills and under wing coverts brown, barred with white; bill and feet brown (in skin). Bill 26, metatarsus 33, middle toe with claw 38, wing (rather worn) about 96, tail about 45 mm. Wings and tail very soft, so as to suggest little power of flight."

Hartert (1927) further describes this bird: "To the original description of this remarkable species may be added, from further skins received all from 1892: There are a number of narrow white bars, both on the sides of and across the jugulum, and the sides of breast and abdomen, also on the under tail coverts. There is a pale rufous band across the chest, indistinct in some specimens. Chin and upper throat white, middle of abdomen whitish. Wings 95–100, in two specimens (none are sexed!) only about 85 mm."

According to Mayr this is a species representative of *Rallus philippensis*. It is an extreme variant of the group.

Red-billed Rail of Tahiti

Rallus pacificus Gmelin

> *Rallus pacificus* GMELIN, Syst. Nat., **1**, pt. 2, 1879, p. 717 (Tahiti). (See Lysaght, Bull. Brit. Orn. Club, **73**, 1953, p. 74, for reasons to use this name.) Fig.: Rothschild, 1907, pl. 26; said by him to be an accurate copy of Forster's painting in the British Museum, except that the legs should be less bright red and more flesh-colored.

Common names: Tevea; red-billed rail.

Status: Extinct. Said by Polynesians to have occurred on the outlying island of Mehetia a generation ago (Anthony Curtiss, *in litt.*).

No specimens of this bird have been preserved in any museum. There seems to be no doubt that Forster saw a skin, but the relationships are not made clear.

Man and introduced cats and rats are probably responsible for the disappearance of this mysterious little bird. Even on the small, outlying islands there are now populations of rats, although there are no cats on Mehetia (Anthony Curtiss, *in litt.*).

Range: Formerly Tahiti and perhaps Mehetia and other islands.

Habitat: Nothing recorded.

Description: "Black, with white spots or bars; abdomen, throat and eyebrow white; hind neck ferruginous; breast grey; bill blood red; iris red" (Forster, 1844, p. 177).

Habits: Nothing recorded.

A Rail of Norfolk Island

Porzana tabuensis tabuensis (Gmelin), which is widely distributed through the Pacific Islands from the Marquesas and Society Islands west to the Kermadecs, New Caledonia, Norfolk, and the Philippines, probably is the bird called the dark rail by Latham and *Rallus tenebrosus* by G. R. Gray. It was probably the bird discovered by J. R. Forster (according to Amadon, 1942, p. 10). It is recorded in Cook's Second Voyage (1777, 2, p. 148): "We found the same kinds of pigeons, parrots, and parroquets as in New Zealand, rails, and some small birds. The sea fowl are white boobies, gulls, tern and c., which breed undisturbed on the shores and in the cliffs of the rocks."

There are several of these birds in the American Museum that were collected within the past 30 years for G. M. Mathews. A population is probably still to be found on Norfolk Island.

Certainly, it was a *Porzana* close to *tabuensis* that Latham described. "Length between 7 and 8 inches. Bill dusky . . . irides red, general color of head, neck and body dusky black, with a deep blue tinge; back and wings deep brown; under wing coverts cinereous with darker spots; legs pale red . . . met with at Norfolk Island in the South Seas in December— Mr. Francillon."

Inaccessible Island Rail

Atlantisea rogersi Lowe

Atlantisea rogersi Lowe, Bull. Brit. Orn. Club, **43**, 1923, p. 175 (Inaccessible Island). See also Lowe, Ibis, 1928, pp. 99–131, pl. 8.

Common names: Island cock; island hen; roger.

Status: Abundant in 1948 (Broekhuysen and Macnae, 1949, p. 1–2). Hagen (1952) estimated that there might have been in the neighborhood of 1,200 birds in 1938. There is no reason to believe that any change has taken place. Should rats or other predators be introduced, the species would be in great danger at once.

Range: Inaccessible Island, Tristan da Cunha group, South Atlantic Ocean.

Habitat: The birds make runways and burrows among tangled and matted roots of the tussock grass (*Spartina arundinacea*), which covers most of Inaccessible Island. This representative of the common marsh grass grows here to a height of 10 feet. The stems are stiff and the sharp-edged leaves have been forced down by the wind and matted into a roof. Areas of undergrowth-fern are favored.

Description: A very small (total length about 5 inches) blackish-brown rail—flightless and very fleet footed.

The plumage is decomposed and hairlike. Upper parts dark, reddish brown, except the wing coverts, which are irregularly banded with white. Forehead, crown, and sides of the head are dark gray. Underparts dark gray except for the flanks, which are banded irregularly with white and the belly and under tail coverts, which are tawny with indistinct bars of black. The bill is blackish gray with sometimes a reddish tinge on the lower mandible; legs may be blackish brown or dark grayish brown. Eyes are brown in juvenile birds and bright red in adults.

FIG. 25.—Inaccessible Island rail (*Atlantisea rogersi*).

Habits: The islanders assured Hagen (1952) that eggs are usually laid in November. He is of the opinion that this is usually true but that sometimes birds lay later. Broekhuysen and Macnae found chicks estimated to be 10 days old on February 18.

As a rule two eggs are laid in a nest within the tussocks which form a natural roof.

The birds are active during the day but not at night, according to Hagen, who observed that a captive bird slept at night and those at liberty were usually silent.

Seeds, berries, and insects were found in the three stomachs examined by Hagen. He writes of them: "In the habitat of the most dense undergrowth and actual subterranean burrows . . . the individual birds have usually no visual contact, but use their voices to keep together. Of the calls, the first to mention is a 'tick . . . tick . . .,' sometimes a deeper: 'Tjupp.' Frequently a single bird can be heard . . . to call with a mournful 'tee'ap' incessantly repeated. . . . Two birds meeting in the undergrowth frequently burst into a delightful exultation, a high whistling trill."

Enemies: Fortunately there are no introduced predators on the island. There can be no doubt that the skua (*Catharacta skua hamiltoni* Hagen) is a natural enemy, but the rail population is in no danger of extinction from this source. Hagen records an old hog a resident in 1938. He was of the opinion that it did no harm. It is fortunate, however, that it was not a gravid sow.

Remarks: Dr. J. C. Bequaert tells me that a hypoboscid fly (*Ornithomyia remota* Walker) is found on *Atlantisea.* This fly is identical with Chilean and Patagonian flies of 12 species found in those countries. It is the only species recorded from Tristan and quite different from any found in Africa.

Lowe (1928) theorized that *Atlantisea* and other rails isolated in this way had never the power of flight. I agree with Stresemann (1932), Hagen (1952), and others that they more probably did have that power. Specimens are to be found in all larger museums.

Banded Rail of Chatham Island

Rallus dieffenbachii Gray

> *Rallus dieffenbachii* G. R. GRAY, in Dieffenbach's Travels New Zealand, **2**, 1843, p. 197 (Chatham Island). Fig.: Rothschild, 1907, pl. 27.

Common name: Moeriki.

Status: Extinct. The type in the British Museum was collected in 1840, and although many naturalists and others have searched for it over its restricted range, not one has been found. These were ground-nesting birds; they were probably extirpated by introduced rats and cats, as well as by bush fires.

Range: Chatham Island, 370 miles east of New Zealand.

Habitat: Nothing recorded; perhaps like that of *Rallus philippensis,* in fresh and salt-water marshes.

Description: A medium-sized (total length about 11 inches) rail resembling *Rallus philippensis,* except that the throat is banded black and white; the breast is barred irregularly with brown and black. The bill is stouter and more curved. Crown and hind neck reddish brown. Stripe from bill to eye to hind neck chestnut. A line above the eye gray. Chin and throat pale gray. A band of black feathers, tipped with whitish, separates throat and breast. The belly is barred black and white. Upper back is barred black and brown, and the lower back is very indistinctly barred with darker brown.

Location of specimens: I know of specimens in Bremen and London.

FIG. 26.—Banded rail of Chatham Island (*Rallus dieffenbachii*).

Chatham Island Rail

Rallus (Cabalus) modestus Hutton

> *Rallus (Cabalus) modestus* HUTTON, Ibis, 1872, p. 247 (Chatham Islands). Fig.: Rothschild, 1907, pl. 28.

Common names: Matirakahu; Chatham Island rail.

Status: Extinct. Last specimen, now in the British Museum (Natural History), collected between 1895 and 1900.

The combined work of men, who cut and burned, and of goats and rabbits imported by men had completely cleaned the island of Mangare of its original "bush" or forest by about 1900. Before that cats had been introduced to kill the rabbits (Fleming, 1939). Whatever birds the cats left, therefore, the destruction of the habitat prevented from breeding.

It seems doubtful that collectors contributed materially to the extinction of this form, Oliver (1930, p. 326) to the contrary. The bird was gone from Chatham long before any were collected, and the population on Mangare could not survive the destruction of its natural habitat. It certainly cannot be listed with the forms that have been purposefully extirpated. I know of only 26 specimens in museums, and it seems very doubtful that more than half a dozen have been lost.

No doubt the native hawk and skua (*Catharacta antarctica*) were enemies of individual birds when they met, but neither one nor both could have been responsible in any way for the extinction of the species.

Range: Chatham Islands: Recorded from Chatham Island, though known from bones only, and from Mangare Island. A downy chick in the Dominion Museum, Wellington, N.Z., and an adult in the Canterbury Museum, Christchurch, N.Z., are recorded from Pitt Island on original labels (Fleming, 1939).

Habitat: "Nest in holes in the ground, and when the young are hatched they get into fallen hollow trees. There is no sand at all on the island (Mangare) where the birds are" (Hawkins, *in* Forbes, 1893, p. 532).

Description: A small (total length about 7 inches) brown rail; a flightless, or almost flightless bird.

Above olivaceous-brown, chin gray; throat, underparts, and tail grayish brown, each feather faintly banded with whitish; wings brown, the outermost primaries banded with whitish fulvous. The bill is long and curved, and the genys is swollen; the nostril is placed far forward. Eyes, bill, and legs light brown.

"The young ones are always the same color as the old" (Hawkins, *l.c.*). Young uniform brownish black (Forbes, 1893).

The relationship of this bird is obscure.[1] It resembles no other, probably representing an ancestral type.

Habits: "They live on insects, principally the sand hoppers which travel into the bush here a long way" (Forbes, 1893).

"Stomach contained only the legs and elytra of beetles" (Buller, 1905, p. 45).

Remarks: Without doubt the Chatham Island rail is one of the earliest warm-blooded settlers of the islands. No bones that point to its relationships have been found on other islands; in its isolation it appears to have retained ancient characters,[2] and there is no striking similarity to any

---

[1] A committee of the New Zealand Ornithological Society records this bird as *Rallus modestus* (1953).

[2] This does not refer to reduction in size of wing and loss of flight.

living form remaining save those that illustrate the homogeneity of the world-wide family of rails.

There are certain similarities of habit between *Cabalus* and *Tricholimnas,* a form of New Caledonia and Lord Howe Islands. Both were nocturnal and frequented dry forests; both fed on insects. Their eggs, as described, are quite alike. In their nest-building habits they differed: *Cabalus* nested in holes, whereas *Tricholimnas,* it is said, lays its eggs in a depression in the ground. *Atlantisea rogersi* of Inaccessible Island, in

Fig. 27.—Chatham Island rail (*Rallus modestus*).

the Tristan da Cunha group in the far-away South Atlantic, nests in holes, but this coincidence only illustrates the fact that there is parallelism in nature.

No doubt the bird was flightless or at most was able only to flutter feebly. The outer and inner webs of the primaries are almost equal in width, and examination under the microscope shows the hamuli (barbicels) of the outer webs of the primaries to be weak, almost straight, and few in number. This condition obtains to a greater extent than in *Tricholimnas,* which has apparently gone less far along the road to retrogressive evolution. *Atlantisea* appears to have proceeded further along this road, although hamuli are still visible (Stresemann, 1932, to the contrary).

Location of specimens: I know of specimens in Auckland; Cambridge, England; Cambridge, Mass.; Christchurch; London; Liverpool; New York; Paris; Philadelphia; Pittsburgh; and Wellington.

Island Wood Rails of Lord Howe and New Caledonia

Two species so similar that it is quite possible that they would breed freely if brought together have nevertheless long been isolated on two southwestern Pacific islands and differ from other rails so strongly that they belong without question to a distinct genus. No fossil evidence has been found, and their relationships cannot be traced. It is possible, in spite of this negative evidence, that they are relics of a once widely distributed, reproductively powerful species now all but extinct.

Tricholimnas sylvestris (Sclater)

> *Ocydromus sylvestris* SCLATER, Proc. Zool. Soc. London, 1869, p. 472, pl. 35 (Lord Howe Island). Photo: Emu, **40**, 1940, p. 32.
> *Tricholimnas conditicius* PETERS AND GRISCOM, 1929, is a synonym.

Common name: Lord Howe Island wood rail.

Status: A small island population, though common on the southern and unsettled portion of the island, a region so inaccessible as to be seldom visited. "On Nov. 8, 1936 . . . I climbed to the summit of Mt. Gower . . . on the higher slopes several were observed and others heard" (Hindwood, 1940, p. 31).

The tenacity of this species to life is astonishing. In view of the disappearance of so many flightless or almost flightless forms from other islands, the ability of individuals to defend themselves and the species to procreate in spite of all the, to them, novel dangers of the past 150 years is in itself a tremendous feat.

"The haunts of the Woodhen are infested with rats, and pigs roam the mountains with but little disturbance from the islanders. . . . The occasional birds that visit the lowlands are usually set upon by dogs which seem to have a particular dislike to their . . . call note. They are looked upon with disfavor by the islanders because of their propensity for taking the eggs and chicks of domestic fowls" (Hindwood, *l.c.*).

Range: Confined to Lord Howe Island, 400 miles east of Australia.

Habitat: Dry second growth, as well as the damp forests at 2,800 feet.

Description: A large (total length about 18 inches) brown rail with greatly developed legs and feet, underdeveloped wings and a long bill.

Top of head and upper parts olive-brown; sides of head, ear coverts, cheeks, chin, and throat whitish gray. Underparts olive-brown, the under tail coverts barred with dull chestnut. Bill and feet gray; eyes red. Female is slightly smaller but otherwise resembles male.

The immature bird is paler and more tawny; primaries and their coverts barred with black (Mathews, 1928, p. 3).

Habits: According to Hindwood (1940) the bird lays a clutch of two to four large pinkish-cream eggs splotched with reddish brown or purplish in a depression in the leaves under low foliage. The breeding season is

Fig. 28.—Lord Howe Island wood rail (*Tricholimnas sylvestris*).

said to be "October, November and January" (Mathews, 1928, p. 4). Island people informed Hindwood (1940) that earthworms (no doubt imported) now constitute a large part of their diet. Beetles and other insects are no doubt their diet, but they are said to eat eggs and chicks of domestic fowl. Perhaps they are almost omnivorous like some other insular forms. In searching for food they are said to turn over debris with their bills, never using their feet.

"They are very alert and move quickly when disturbed, but they can always be attracted to within a few feet by any unusual noise. Their note is high pitched and staccato" (Hindwood, 1940).

Tricholimnas lafresnayanus (J. Verreaux and Des Murs)

Gallirallus lafresnayanus J. Verreaux and Des Murs, Rev. et Mag. Zool. (2), **12**, 1860, p. 437 (New Caledonia). Fig.: pl. 21 opp. p. 394 in Brenchley, 1873.

Common names: New Caledonian wood rail; n'dini (Layard); n'diou (Jouan).

Status: Very rare and possibly extinct; it has not been found for 50 years, although competent collectors have searched for it. In view of the extraordinary powers of survival of *sylvestris* on Lord Howe Island, and the continued existence of the kagu, quite as vulnerable a species, it is possible that some may yet be found.

All the enemies that have endangered the existence of the Lord Howe Island wood rail have also presumably preyed upon that of New Caledonia. The latter is a larger island, however, and parts of it are extremely steep and difficult to reach because of thick forests. It seems probable that in some part of its range birds survive, away from the cats, dogs, and rats that have exterminated them in their former haunts, nearer human habitations. It is possible, of course, that a disease unknown to us may have exterminated the birds.

Range: New Caledonia. The few specimens known were taken on both the east and west coasts. It is probable that they were also to be found in the hills to 3,000 feet.

Habitat: Probably only in original forests of the interior, formerly in forested river valleys near the coasts; ". . . not a swamp bird" (Layard and Layard, 1882).[1]

Description: A large brown wood rail (total length about 18 inches), with strongly developed feet and legs, underdeveloped wings and a long bill.

This bird differs from *T. sylvestris* of Lord Howe Island in having the upper parts considerably grayer, less brown; the primaries unbarred; and the wing coverts and flanks barred even more irregularly and with white, not brown.

Top of head dark gray becoming browner on the neck, chocolate-brown on hind neck; lores are dark brown and there is a paler brown stripe above them. Cheeks and throat are light gray; ear coverts are brown. Back is brown washed with gray more olivaceous on scapulars, becoming blackish brown on lower back. Underparts are dark gray washed with brown on chest. Flanks and under tail coverts chocolate-brown. Wing 185 mm.; tail 109 mm.; bill 52 mm.; tarsus 54 mm. Female is smaller than male. Young differs from adult in being almost entirely black, with a

---

[1] The legend, often repeated, that the n'dini is a marsh bird perhaps was begun by M. Bouquet de la Grye, a hydrographer. Certainly all specimens collected were from forests.

shade of chocolate-brown on the back and sides of the neck; the under-surface being more slaty black; head and throat more slaty gray, with a patch of orange-brown on foreneck (Sharpe, 1894, p. 52).

Habits: Almost nothing recorded. "We have kept it in confinement, feeding it on . . . raw meat, and garbage. It is nocturnal [1] and runs with great rapidity. In walking it elevates the tail with the peculiar flip common to the rails, and it can climb and jump like a cat. If alarmed it will squeeze itself into the smallest holes and crevices and lie 'perdue' and motionless, feigning death. . . . We have never seen it in its native haunts. . . ." (Layard and Layard, 1882, p. 535).

Location of specimens: I know of specimens of *Tricholimnas sylvestris* in Cambridge, Mass.; Dresden; London; New York; and of *T. lafresnayanus* in London and New York.

Jamaican Wood Rail

Amaurolimnas concolor concolor (Gosse)

   Rallus concolor GOSSE, Birds Jamaica, 1847, p. 369 (Jamaica).

Common names: Red rail; water partridge; Jamaican wood rail.

Status: Extinct on Jamaica since about 1890; the last specimen was taken in 1881. A closely related form (*A. c. guatemalensis*) still exists in certain localities from Guatemala south to São Paulo, Brazil, but is nowhere common, unless perhaps in the Amazon Basin. Probably the subspecies has been completely exterminated by the mongoose (Bond, 1936). Not only the mongoose but every other type of enemy, including cats and rats, were resisted for many generations. At last, after perhaps three difficult years for nesting, no doubt they died off one by one.

Range: Formerly Jamaica.

Habitat: Gosse records it as frequenting fresh-water morasses and secluded streams, rather than saline swamps, and to be found even on the mountain acclivities. Bond (1936) says: "Apparently more of a 'land rail' than other West Indian species."

Description: A medium-sized rail (total length 10 inches). Plumage generally reddish brown; the crown and nape washed with olivaceous; chin and throat whitish; primaries grayish. Bill yellowish green; feet red. Males and females are alike.

Habits: Like most rails a shy, skulking bird, preferring to run rather than fly. Its flight is said to have been heavy and labored.

Gosse said of it: "It is sometimes seen perched on a low tree by the

---

[1] It seems to me to be doubtful that the species is completely nocturnal in its habits.

road-side, at which time it seems to have lost its usual shyness, and sits looking at the sportsman until he nearly comes up to it.

"I have shot it skulking among the aquatic weeds at Basin Spring. As it roams, it utters at intervals of a few seconds, a cluck, like a hen."

Fig. 29.—Jamaican wood rail (*Amaurolimnas concolor concolor*).

The egg of the closely related subspecies of Central and South America is said to be ashy gray with a reddish tinge, speckled with dusky.

Location of specimens: I know of specimens in Cambridge, Mass.; Chicago; London; and New York.

Barred-wing Rail of Fiji

Rallina (*Nesoclopeus*) *poeciloptera* Hartlaub

Rallina poeciloptera HARTLAUB, Ibis, 1866, p. 171 (Ovalau,[1] Fiji Islands). Fig.: Finsch and Hartlaub, 1867, xii, 1.

Common names: Barred-wing rail; mbidi (Layard, 1875).[2]

Status: Probably extinct. Although naturalists, including those of the Whitney Expedition, and natives have searched carefully for it, it has not been found for more than 50 years (Mayr, 1945a). Last specimen collected not later than 1890.

The mongoose is thought to have been responsible for the extermination of this bird (Mayr, 1945a). This is quite probable. The mystery is how other ground-nesting birds, such as the friendly ground dove (*Gallicolumba stairii*), have survived this terrible enemy.

Range: Viti Levu and Ovalau, Fiji group.

Habitat: Inhabited thick taro beds and swamps (Layard, 1875).

Description: A large (total length about 14 inches) brown and gray rail.

Color above generally brown. The head, particularly on the sides, washed with gray. The lores brown and almost devoid of feathers. Primaries and wing coverts chestnut, barred for most of their length with black and paler brown. Under wing coverts black barred with white. Chin and throat whitish, under parts gray, darker on belly and flanks, feathers of these regions lightly tipped with white. Under tail coverts black, irregularly and indistinctly barred with brown, bill orange and yellow; feet yellow; iris light brown (Layard, *in* Sharpe, 1894).

Habits: "Very shy and rarely seen. . . . It lays six eggs (in a nest made of sedges), of raspberry-and-cream-coloured ground, speckled chiefly at the obtuse end with light purplish and dark blood coloured spots. . . . They nest in November and December, and I think also in March" (Layard, 1875, p. 438).

"Four eggs brought to me on October 9th" (Layard, 1876, p. 155).

"Does not fly. (Storck, *in* Reichenow, 1891). "Natives say this bird never takes wing" (Layard, 1875). Primaries are short, but the outer webs are narrow and barbicels (hamuli) are well developed.

Remarks: The nearest relative of this bird is to be found in the Solomon Islands. These two species are perhaps closest to *Habroptila* of Halmahera

---

[1] See Bolau, 1898, p. 70.

[2] Mayr (1945a) suggests that this was the mysterious "sasa," described by natives as a "wingless bird that lived in holes in the ground in the mountains"; see W. H. Walker, *Wanderings among South Sea Savages,* 1909, p. 20.

Island. They are probably all remnants of a much larger and more wide-spread population now long since gone.

Location of specimens: I know of specimens in Cambridge, Mass.; Berlin; Hamburg; London; and New York.

FIG. 30.—Barred-wing rail of Fiji (*Rallina poeciloptera*).

Laysan Island Rail

Porzanula palmeri Frohawk

> *Porzanula palmeri* FROHAWK, Ann. Mag. Nat. Hist. (6), **9**, 1892, p. 247 (Laysan Island). Fig.: Rothschild, 1893–1900, p. 9.

Common names: Laysan rail; Laysan crake.

Status and range: Probably extinct.[1] Last specimen seen (and to have been recorded) alive on Eastern Island, Midway, in June 1944 (Fisher and Baldwin, 1946). The last specimen on Sand Island was reported to be there on November, 1943 (Munro, 1945). Although Munro was told that the bird had been on Midway in June 1945, Fisher and Baldwin, both competent naturalists, had not been able to find any birds that spring, nor could anyone on the island give them a reliable report of having seen any.[2] They searched available cover on hands and knees, as well as other

---

[1] Baldwin (1949) repeats Munro's (1945) hope that the bird survived.
[2] Reports of sight records proved actually to be ruddy turnstones upon investigation.

accustomed haunts of the bird where flies collected on the carcasses of sea birds. Such small [1] islands with so little available cover could not well conceal the birds from trained men.

The birds were reintroduced into their original range on Laysan Island by naturalists from the U.S.S. *Tanager* expedition in 1923, at a time when the species was almost entirely gone from that island (Wetmore, 1925). It is not at all probable that any survive there at present, for G. P. Wilder, custodian of the island, together with W. F. Coultas, naturalists, could find none in 1944.

Other attempts at introduction, one on a large island of the main group of the Hawaiian Islands and another on Pearl and Hermes Reef, east of Laysan, have failed. It was hoped for a time that the latter attempt at introduction by Captain Anderson in 1929 might be successful, but the island was swept by the sea during a gale shortly afterward. Searches made in 1930, when G. O. Kaufmann spent three weeks there, and later (Galstoff, 1933) were all in vain.

The accidental introduction of rats into both islands and the destruction of natural habitat are thought to have been the primary and secondary causes of the death of this species (Fisher and Baldwin, 1946). Rats were introduced in 1942 or early in 1943 and apparently bred up to and above the point of subsistence, as might well be expected. The gray or roof rat (*R. rattus*) and the black rat (*R. norvegicus*) were both present in great numbers in 1944; 100 rats per acre were estimated, but the gray rat constituted 80 percent of them. Since the rats lived in the burrows of the "small moaning bird" or Bonin Island petrel (*Pterodroma leucoptera* "*hypoleuca*"), these were thought to have suffered most, though at that time the effect of predation was not obvious (Alsatt, in Munro, 1945). The rats ate eggs and young and probably, when possible, adults as well.

There are still places for the birds to live and breed—patches of brush (*Scaevola*) and imported grass (*Ammophila*)—but these have been cleared away in many places and cut into small, isolated sections in others, and this did not improve the island from the birds' point of view. This was done, perhaps, as a mosquito-control measure. This cutting and consequent reduction of natural habitat are thought to have been a secondary cause for the extinction of the bird (Fisher and Baldwin, 1946).

Munro (1947) [2] has pointed out that the frigatebirds were enemies of the rail and that, if cover were available, it might easily escape, "but on bare sand . . . the rail would have little chance with half a dozen yearling Frigates after him."

---

[1] Sand Island, 1½ by 1 mile; Eastern Island 1¼ by ¾ mile.

[2] Munro doubts that a supply of fresh drinking water was necessary for the bird, as has been suggested, and so do I.

Habitat: Scaevola thickets and coarse grass (*Ammophila arenaria*) for the most part, although when they were very numerous they could be found near the houses and on beaches, or generally in the open, where they fed on the flies to be found in the carcasses of sea birds.

Description: A very small (total length about 6 inches), pale, completely flightless rail. The sexes were alike.

Fig. 31.—Laysan Island rail (*Porzanula palmeri*).

Above pale, whitish, brown streaked with darker. A line over the eye, sides of the head and all underparts gray, flanks and belly flecked with white. Bill pale green, darker at the tip; feet and legs grayish green; eyes red.

Young had the underparts pale buff. Chicks were black (Rothschild).

Habits: The bird was, generally speaking, diurnal. It was active—"it ran like a shadow with mouse-like speed." It was fearless and inquisitive—"when a dip net was placed on the ground the birds would sometimes walk into it to see what it could be," and again, "it would stand with one foot raised examining our shoes" (Schauinsland, Palmer, *in* Rothschild, 1893–1900). But it was also unpredictable, for sometimes it would rush away at full speed, only to be halted by the sight of an insect which it would seize and swallow, after which it would return to the scene of its fright at a saunter.

The nest was usually built on the ground among heavy grasses, but sometimes upon tussocks. It was woven of grasses mixed here and there with feathers and other materials. It was quite elaborate, covered over the top,[1] with the entrance at the side, and a runway was often noticed as an approach to the entrance. The inside was often lined with softer materials and the cup-shaped bottom covered with down. This nest is apparently more elaborate than most crakes' but resembles the American *Laterallus* closely.

Although he does not definitely say so, Fisher (1904) gives the impression that the female incubated the eggs. These were blunt or elliptical and oval, greenish or creamy buff, flecked with pale reddish brown and pale purplish gray. Two and occasionally three to five eggs were to be found in a clutch.

Two notes have been described: a *chirp* of one to three short, soft notes, and a rattle compared to the sound made by handfuls of marbles thrown and bounced on a glass roof.[2] Captive birds chirped intermittently all day as did birds in a wild state, and their "cries" are reported to have been heard from time to time during the night. A creaking chorus is said to have sounded for a few minutes just before sunset.

Little is recorded of the mating performance.

Blackman (*in* Baldwin, 1947) says: "A male and a female, apparently mates, . . . sat several minutes close together . . . and in turn held their heads close to the ground while the other pecked among the feathers at the top of the head and the back of the neck." And Fisher (1904) wrote: "Two approach each other with feathers erect and when close together begin rattling in each others face. Then they suddenly ceased and slunk away in opposite directions." The sex of the birds is not stated, and it may be that this was a stylized battle between males.

Although downy young were seen as early as March 17, the nesting season began usually in April. The first eggs were to be found in May and the latest in June.

---

[1] Baldwin (1947) points out that the "roof" was not actually built by the bird, but formed of matted grass. He does say it was lined inside.

[2] Frohawk, 1892 (in description of the species, see above).

The young soon learned to feed themselves and could run as fast as the older birds in five days.

The bird was practically omnivorous; its principal food was insects, flies, and larvae, as well as adult dermestid beetles to be found in and around the carcasses of sea birds, of which there always has been an abundance. It was also fond of a large moth, and would eat seeds. In all this it was quite like other crakes, including *Laterallus,* but it also recorded as shredding and gulping the meat from the dead birds and, when possible, devouring eggs. Only two observers (Munro, 1946b; Palmer, *in* Rothschild, 1893–1900) have seen them break eggs, but many have seen them drive the small, strong-billed "finches" (*Telespiza*) from the feast after the work has been done for them.[1]

Location of specimens: I know of specimens in Berlin; Cambridge, Mass.; Chicago; London; New York; and Philadelphia.

Hawaiian Rail

Pennula sandwichensis (Gmelin)

> *Rallus sandwichensis* GMELIN, Syst. Nat., **1**, pt. 2, 1789, p.717 (Hawaiian Islands = Hawaii fide Stresemann 1950). Fig.: Rothschild, 1893–1900, p. 243; 1907, pl. 26.
> *Pennula millei* (sic) DOLE, Hawaiian Almanac and Annual for 1879 (1878), p. 54 (Hawaii).
> *Pennula wilsoni* FINSCH, Notes Leyden Mus., **20**, 1898, p. 77 (Hawaiian Islands).

Common names: Hawaiian rail; moho.

Status: Extinct. Last specimen seen 1884 (Rothschild, 1907) or possibly as late as 1893 (Henshaw, 1902). Many competent naturalists have searched for it over its former range.

Although the mongoose has often been blamed for the extermination of this bird, it scarcely seems possible that it was wholly responsible, since the bird was almost, if not quite, extinct when the mongoose was imported in 1883. Perhaps the information that the birds took refuge in rat holes gives us a clue. Although there is little evidence that relatives of the Hawaiian rat are predaceous, it may be that some of these holes were occupied by imported rats.

Range: Formerly the island of Hawaii, certainly on the windward (eastern) side of Kilauea, perhaps also in an area of about 40 miles along this coast and in the Olaa District (Rothschild, 1893–1900; Perkins, 1903;

[1] Baldwin's (1947) summary of what is known of this bird is excellent. Other sources, Rothschild, 1893–1900, and Fisher, 1904.

Henshaw, 1902). Perkins (1903) and Munro (1944) suggest that the species may have occurred on other islands.

Habitat: Open, grassy country or low scrub, just below the heavy rain forest, or in similar country in clearings in the forest (Perkins, 1903).

Habits: There is no first-hand account of the habits. Perkins (1903) says that it "seems to have lived on friendly terms with these rats [*Rattus hawaiiensis* Stone] and is said to have even been in the habit of hiding itself in the burrows in times of danger."

Description: A small brown rail (total length about 5.5 inches), flightless but active and fleet.

Dark form: Top of the head brown with a faint grayish tinge, lores fulvescent, ear coverts gray, cheeks deep reddish brown (vinous). Back dark reddish chocolate, the lower back with darker centers to the feathers. Throat and breast deep reddish brown shading to grayish chocolate on the belly, flanks, and under tail coverts.

Paler form: Black spots on feathers of the back more apparent, giving a mottled appearance.

Perhaps the paler form represents an immature plumage. The bill of the Leyden specimen measures 0.75, that of the darker form 0.8 mm., according to Rothschild (1893–1900). The bill of the New York specimen is damaged.

Remarks: It is possible that there were once two species on the island of Hawaii, but it does not seem probable. Specimens of two forms exist, however. The paler form with blackish brown centers to the feathers has been called *sandwichensis* Gmelin. According to Stresemann (1950) it was without doubt this bird that was figured by William Ellis,[1] naturalist on Cook's third voyage. The model was most probably collected on the island of Hawaii. Furthermore, a specimen in New York is without question the paler *sandwichensis,* and it too was taken on the island of Hawaii by an old native bird collector for Mills, as the label indicates (see also Rothschild, 1893–1900, p. 241). There can therefore be little doubt that the paler *sandwichensis* came from the island of Hawaii; whether it occurred on other islands remains problematical.

Because of the differences between these two forms, it has been assumed in the past (see Rothschild, 1893–1900) that the paler form, *sandwichensis,* was once to be found on some island other than Hawaii. For this theory it may be said that when Captain King listed the birds (*in* Cook, 1784, **3**, p. 118) he did not mention specific islands upon which they were to be found. Secondly, because birds were bought from the natives certainly at Kauai and possibly at Oahu, it seemed possible that the specimen,

---

[1] It is plate 70 and is in the British Museum (Natural History).

which later became the type, was acquired in this manner. On the other hand, Captain King relates that a collecting party spent several days in the interior of Hawaii in February 1778 and probably most of the birds were collected during this trip, as Stresemann (1950) points out.

Location of specimens: I know of specimens of *Pennula "millsi"* in Cambridge, England; Honolulu; London; New York; and Vienna; and of *P. sandwichensis* in Leyden and New York.

Jamaican Black Rail

Laterallus jamaicensis jamaicensis (Gmelin)

Rallus jamaicensis GMELIN, Syst. Nat., **1**, pt. 2, 1789, p. 718 (Jamaica).

A specimen was taken near Havana on July 3, 1964, according to *Bond (1965, p. 8). He believes this form to be indigenous in Cuba and in this is in agreement with Friedmann (1941, p. 159) and the American Ornithologists' Union Committee.

Common names: Jamaican black rail; little black rail; little red-eyed crake. "The negroes in Clarendon call it Cacky-Quaw by reason of its cry . . . in Westmoreland Johnny Ho and Kitty Go for the same reason" (Robinson, *in* Gosse, 1847, p. 376).

Status: The population of Jamaica and of Puerto Rico probably have been extinct since the late 1870's (Stahl, 1887; Bond, 1950). Of the Cuban population, Barbour (1943) says: "In December 1942, Doctor Albelardo Moreno took a specimen near Havana, and he informed me (*in litt.*) that several Black Rails have come to the Zoological Garden at Havana and that they are thriving in captivity."

It is probable that the mongoose, rats, and cats introduced by man were responsible for the extermination of this bird on Jamaica and Puerto Rico. It may be that the birds recorded from Cuba are quite recent immigrants from North America, as Bond thinks.

Range: Recorded from Jamaica, Cuba, and Puerto Rico.

Habitat: Dense marshes (Wetmore, 1927). In reeds on river banks (Gosse, 1847).

Description: A small (total length about 6 inches), dark rail very similar to the black rail of eastern North America. It differs in its longer middle toe (without claw, 23 mm. in length), slightly longer and heavier bill; and somewhat duller in coloration, the white markings on the upper parts smaller and fewer (Friedmann, 1941). Sexes are alike.

The head is blackish brown, hind neck chestnut; back and wings dark brown, spotted with white. Underparts blackish brown. Bill black, feet pale green, and eyes red.

Habits: Gosse (1847) says of the bird on Jamaica: "On two or three occasions, I have seen the species. Near the end of August, pursuing a White Gaulin [presumably an immature *Florida caerulea*] in the morasses of Sweet River, several of these little Rails, one at a time, flew out from the low rushes before my feet, and fluttering along for a few yards, with very laboured flight, dropped in the dense rush again. Their manner of flight and their figure greatly resembled those of a chicken; the legs hung inertly down. . . . Robinson says 'their cry was very low, and resembled a coot, when at a great distance.' The gizzard of one that I examined contained a few hard seeds."

No other information has been recorded, but habits are presumably the same as the North American *L. j. pygmaeus.* They nested on the ground.

Kusaie Island Crake

Aphanolimnas monasa (Kittlitz)

> *Rallus monasa* KITTLITZ, Denkw. Reise russ. Amerika, **2**, 1858, p. 30 (Kusaie Island, Carolines). Fig.: Tori, **11**, 1942, pl. 111.

Common names: Kusaie crake; satamanot (Kittlitz); nay-tai-mai-not (to land in the Taro garden) (Coultas).

Status: Extinct. W. F. Coultas, of the American Museum of Natural History's Whitney Expedition, searched for this bird in the marshes at sea level, where it was found originally, and also in the mountains, from January to June 1931. With the help of a Solomon Island native he made every effort to trap the bird or catch it by any means, but without success. Kittlitz found it during his visit, December 1827 to January 1828. These two specimens are in St. Petersburg (Leningrad), Russia. Finsch, during a 9-day visit in 1880, failed to find it, nor did he hear its characteristic call (Finsch, 1881).

Coultas (Ms.) writes: "In the olden days it was a sacred bird, but since the Christian missions have been established not much attention has been paid to the older faiths . . . Several oldsters seemed to remember their forefathers' speaking of the bird, but none of them admitted having seen it, except and elderly deacon, a staunch pillar of the church, who claimed to have had it pointed out to him twenty years previous to my visit."

Both Finsch and Coultas found the island overrun with introduced rats. Both thought these responsible for the extinction of the birds. This would, of course, be the simplest explanation to cover the fact that the birds are extinct. They were not common in 1828, disappearing within the next 50 years. During this time whaling ships foregathered at this island. It is probable that the rats escaped when the ships were careened on the beach for cleaning.

Range: Formerly Kusaie Island, Caroline Islands.

Fig. 32.—Kusaie Island crake (*Aphanolimnas monasa*).

Habitat: Swamps and marshes near sea level, "continually wet, shadowy places in the forest" (Kittlitz, *in* Hartlaub, 1893).

Description: A small, black rail (total length about 7 inches). Black with bluish-gray reflections; quills and tail somewhat browner; inner wing coverts brownish with white spotting, outer edge of first primary dull brownish; chin and middle of the throat somewhat paler; bill blackish (Hartlaub, *in* Sharpe, 1894). Mayr (1945a) says: "Underwing and under-tail with a few inconspicuous white bars. Iris and feet red; bill black (?)."

Habits: "It lives singly on the ground in these continually wet, shadowy places in the forest. One hears in these places from time to time its alluring voice resounding" (Kittlitz, *in* Hartlaub, 1893).

Location of specimens: I know only of specimens in Leningrad.

Iwo Jima Rail

Poliolimnas cinereus brevipes (Ingram)

> *Porzana cinerea brevipes* INGRAM, Bull. Brit. Orn. Club, **29**, 1911, p. 21 (Sulphur Island, Bonin Group = Iwo Jima).

Common names: Sulphur Island ashy crake; mamijiro kuina.

Status: "Extinct since 1911; said to have been observed by Momiyama in 1924 and 1925" (Ornith. Soc. Japan, 1942, p. 176).

Range: Volcano Islands: Sulphur Island (Naka Iwo Jima [1]) and San Augustino Island (Minami Iwo Jima).

Habitat: Grass and thickets. In other parts of its range this bird prefers fresh-water ponds and swamps.

Description: A medium-sized rail (total length about 6 inches). It is dark brown mottled with black above and white below.

Head blackish brown, neck and upper back brown, feathers of lower back and rump with black centers and olive-brown edges. Underparts white, washed with buffy. Bill yellowish olive for two-thirds of its length and with a red base, feet and legs olive, eyes red. The sexes are alike.

Described as similar to typical *cinerea,* but with lower flanks and under tail coverts more rufous, lower back and rump "inclined to be" more ruddy, bill deeper at the base, and with a shorter tarsus.

Habits and enemies: Momiyama (1930, p. 120) writes as follows:

> Originally there were no native mammals on Iō Jima other than large bats, but the increase in the human population was accompanied by the introduction of rats. Also domestic cats similarly increased. The rats caused damage to farm crops, especially sugarcane; cats pursuing them became established in the wild. The Mamijiro Kuina ordinarily hides in the tako forest or in the luxuriant foliage of large bushes in the virgin forest,[2] but may be seen where dried grass has accumulated or where rain water is available for drinking, but in time of drought this rainwater dries up, and as they need water, they may wander into the vicinity of water tanks near houses. Recently both domestic and wild cats have taken to killing them. They have become pitifully scarce. The last living form met a similar fate four or five years ago, and at present they are completely extinct.

Location of specimens: I know of specimens only in Tokyo.

---

[1] This is Iwo Jima, captured by United States Marines in 1944. "Naka" means middle; "minami" means south.

[2] Dr. O. L. Austin, Jr., kindly translated this passage from the Japanese. There were some large trees on Iwo Jima.

Samoan Wood Rail

Pareudiastes pacificus Hartlaub and Finsch

Pareudiastes pacificus HARTLAUB AND FINSCH, Proc. Zool. Soc. London, 1871, p. 25, pl. 2 (Savaii, Samoan Islands). Fig.: Proc. Intern. Orn. Congr., 1905 (1907), p. 206.

Common names: Samoan wood rail; punaé.

Status: Possibly extinct; there is no record of its having been found since 1873. The two unsexed specimens in the British Museum probably were given to the *Challenger* Expedition in 1873, but they bear no dates. An egg, also in the British Museum, was collected in 1873. The Whitney Expedition of the American Museum of Natural History searched for the bird in 1926 but failed to find it. However, Ernst Mayr thinks it possibly may still be found. The bird survived large populations of cats and rats for many years; no doubt these reduced the populations.

Range: Savaii Island, Samoan Islands (Whitmee, 1874, was of the opinion that it occurred on Opolu also).

Habitat: Nothing recorded.

Description: A small, black, almost flightless gallinule. Plumage is black, olivaceous on the back and bluish gray on breast. Bill and feet (probably) red. According to Mayr, its closest relative is *Edithornis silvestris,* confined to the mountains of San Cristobal, Solomons.

Habits: The only record of the habits of this bird is on hearsay (Whitmee, 1874, pp. 184–185). Natives told the Rev. Mr. Whitmee in 1873 that the bird nested in burrows in the mountains. It is probable that they confused it with petrels. It may have taken refuge in burrows as the Laysan rail and the moho are recorded as having done, and *Atlantisea* is known to do. He continues: "I have received some contrary evidence . . . the Rev. George Brown . . . intimated to the natives his wish to procure specimens of the bird and of its eggs; and a few weeks ago a living bird and two eggs were taken to him The man who took them declares that he caught the bird on the nest with the two eggs under it. This nest he says was on the ground and composed of a few twigs and a little grass . . . I forward it (one egg) to you for your inspection." This egg is now in the British Museum.

It is recorded as feeding on insects, and could not be kept alive in captivity.

Hartlaub and Finsch remark upon the great size of the eyes. It may be inferred that the bird was crepuscular or nocturnal.

Location of specimens: I know of specimens in Bremen, Hamburg, London, and New York.

Tristan Island Gallinule

Gallinula nesiotis nesiotis Sclater

Gallinula nesiotis nesiotis P. L. SCLATER, Proc. Zool. Soc. London, 1861, p. 261, pl. 30 (Tristan da Cunha Island, South Atlantic).

Common names: Tristan coot; Tristan cock; island hen.[1]

Status: Extinct. The last recorded specimens were received in London (three of these were alive) on May 25, 1861 (Sclater, 1861, p. 209). The *Challenger* Expedition did not find the bird in 1873, nor have later expeditions.

Range: Formerly Tristan da Cunha Island, South Atlantic. It has never been recorded from nearby Nightingale and Inaccessible Islands, and perhaps it never inhabited them.

Habitat: Carmichael said in 1818 (p. 493): "The Fulica conceals itself in the wood"; this is the only indication of habitat by an eyewitness (but see *G. n. comeri* below, a closely related form).

Description: A rather small (total length about 10 inches) gallinule or moor-hen. It resembles the common and widespread *G. chloropus* but differs in its short wings, in its almost total inability to fly, and in having a black, not gray, neck and underparts.

Head and neck black. Back and wing coverts dark olive-brown. Wings and tail and underparts black with a few pale streaks on sides. Under tail coverts white with black feathers in the middle. Edge of wing white. Bill red with a yellow tip; feet and legs yellow.

Habits: Omnivorous and bellicose in captivity; almost flightless (see *G. n. comeri,* p. 243).

Enemies: Occasionally run down by dogs. ". . . Wild hogs secrete themselves in the deepest recesses of the wood, where it is impossible to pursue them . . . they occasionally feed on dead carcasses of seals and sea lions" (Carmichael, 1818). Can it be doubted that these would not eat eggs and young of any bird?

It seems probable that the population was considerably reduced when naturalists of the *Challenger* were on the island and that they simply failed to find the few remaining individuals. No doubt the rats, which were accidentally introduced by a wreck in 1882, as well as the pigs, were the cause of the extinction of this form. Fires are reported to have destroyed the vegetation on the plateau where the settlement is and elsewhere. It is more than probable that these killed many and deprived the birds of a good part of their range.

---

[1] Carmichael, 1818, p. 496, records it as the common moor-hen (*Fulica chloropus*). See Stresemann, 1953, p. 146, for further remarks and subsequent history of specimens.

The Gough Island gallinule (*Gallinula nesiotis comeri* Allen)[1] was common in 1922, according to Wilkins (1923); there has probably been no change in its status. As far is known, no new human or animal populations have been introduced. However, should rats accidentally come ashore, this bird would be in great peril.

FIG. 33.—Tristan Island gallinule (*Gallinula nesiotis nesiotis*).

The expedition of 1888 found an old sealers' camp with potatoes growing near it. The mice (*Mus musculus,* no doubt introduced) were troublesome (Comer, *in* Verrill, 1895, p. 468).

The birds are found in dense vegetation near the bank of the stream (Clarke, 1905) and up to about 2,000 feet on the mountain. The vegetation referred to a matted tangle of tussock grass (*Spartina arundinacea*) and buckthorn tree (*Phylica nitida*), the latter much stunted. The birds are recorded as taking refuge in holes. Its drawn-out, shrill call was heard

[1] Bull. Amer. Mus. Nat. Hist., 4, 1892, p. 57 (Gough Island, South Atlantic).

on all parts of the island that have been visited, even on the highest slopes during a driving sleet storm (Wilkins, 1923, p. 506).

They cannot fly and use only their wings to help them in running, and it is said of them that they could not get on a table 3 feet high (Comer, *in* Verrill, 1895, p. 434).

Location of specimens: I know of specimens of *Gallinula n. nesiotis* in London and New York.

The Swamp-hens and "Notornis

Populations of the blue-green, red-legged, and red-billed birds called variously purple gallinule, blue coot, blue peter, and swamp-hen are to be found from the shores of the Mediterranean Sea, eastern Africa, Persia, India, throughout southeastern Asia, in Australia and New Zealand, and the islands of the Pacific Ocean as far as Fiji and Samoa. They represent an enormous polymorphic species, the populations of which differ little from one another, even when they are isolated on islands. The extent of variation among individuals of all populations is astounding (Mayr, 1949).

From these facts it may be assumed that the species has a high reproductive capacity, and has expanded its range rapidly. Even so, three small, island populations have disappeared, no doubt through excessive predation. In the Mediterranean, the Balearics and Malta know them no more, and it is probable that there was a population peculiar to Lord Howe Island, off the coast of Australia, which was extirpated more than a hundred years ago. Two tarsometatarsi found in Pleistocene bed in Queensland, Australia, are evidence of an ancestral species there,[1] but there is no other fossil evidence.

In sharp contrast to this powerfully reproductive species is its relative *Notornis* of New Zealand, so rare and isolated that it had been thought to have been extinct for 50 years, having lost the power to escape by flying, laying but four eggs a year at most and with a high proportion of those infertile, it is in a sense a "senile" species.

Much has been written of these species. The following appears to be the best arrangement of them.

> *Porphyrio (Notornis)*[2] *mantelli hochstetteri* (MEYER): South Island, New Zealand. Extremely rare.
> *Porphyrio (Notornis) mantelli mantelli* (OWEN): Formerly North Island, New Zealand. Extinct, known by bones only.

---

[1] *Porphyrio reperta* De Vis, 1888; *P. mackintoshi* De Vis, 1892.

[2] Similarities between the species of *Porphyrio* and *Notornis* are numerous and striking whereas differences are not, when these are balanced against the dissimilarity of both the other groups of birds.

Porphyrio porphyrio melanotus TEMMINCK: Australia, New Zealand, Lord Howe Island, and Norfolk Island. Common in Australia and New Zealand. Rare on Norfolk and Lord Howe Islands.

Porphyrio porphyrio albus (WHITE): Formerly Lord Howe Island. Extinct; known by the two specimens only, one in Vienna (the type) and one in Liverpool (type of *stanleyi* Rowley, which is regarded as a synonym).

The following measurements of *Notornis* are from Owen (1848), Parker (1882, 1886), A. B. Meyer (1879–1897 [1893]); those of the types of *Porphyrio albus* and *P. "stanleyi"* from Rothschild (1907) and Mathews (1910, 1928, 1936). Those of *P. melanotus* are mine of a series of 23 from Australia, 12 from New Zealand, and 9 from Norfolk Island. The value of these would be greater if the sex of the birds had been known accurately. Dr. Kinghorn, of the Australian Museum, Sydney, has kindly sent me those of the Lord Howe Island specimens of *melanotus*.

	Bill and plate	Wing	Tarso-meta-tarsus	Femur	Tibio-tarsus	Midtoe + claw
Notornis N. Island	—	—	129	118–122	194–200	—
Notornis S. Island	82–98	230–240	98–108	103, 109	165	93, 95
Porphyrio Australia	64–77	251–292	88–106	76	130	104–110
Porphyrio New Zealand	65–76	265–290	90–103	—	—	100–115
Porphyrio Lord Howe	59, 61	250, 267	86, 86	—	—	101, 106
Porphyrio Norfolk	66–70	240–276	85–98	—	—	99–110
Type *albus*	77	225, 230	82, 89	—	—	94
Type *stanleyi*	62	226	83	—	—	—

THE TAKAHE (NOTORNIS): *Porphyrio mantelli hochstetteri* (Meyer)

Notornis Hochstetteri A. B. MEYER, Abbild. Vog. Skelett., Lief 4 and 5, 1883, p. 28, pls. 34–37 (Bare-Patch Plains, east of Lake Te Anau, South Island, New Zealand). Type, formerly in Dresden Museum, has disappeared.

Notornis parkeri FORBES, Trans. Proc. New Zealand Inst. (Wellington), **24**, 1891 (1892), p. 187 (Lake Te Anau, South Island, New Zealand). Based on skeleton in Otago Museum.

Common names: Takahe; moho.

Status: Rare and localized. About 20 breeding pairs, as well as 40–60 nonbreeders, were counted in 1949–50 in the small valley where the birds were rediscovered in 1948, after a period of 50 years, during which they were thought to be extinct. It is probable that there are more to be found. The New Zealand Government has protected them and declared a 400,000-acre sanctuary including this valley. If it is possible to preserve this region in its present condition and protect the birds from molestation, they should survive.

Falla (1949) considers that deer that pass through the nesting site and stoats, which must always be a danger to ground-nesting birds, are the principal dangers to the maintenance of a safe level of population. Both of these were introduced by man. As far as deer are concerned, the only harm they can do is to step on nests. It was estimated that only six or seven passed through the valley habitually. Introduced stoats (presumably *Mustela erminea*) are of course a great danger to nests, especially where there are few small mammals. No doubt the vigilant Wild Life Section of the Department of Internal Affairs of New Zealand, which has undertaken the protection of the birds, will be able to exterminate or at least control the population of this dangerous creature.

Range: Now apparently confined to a few small valleys in the Murchison Mountains, between 5 and 10 miles west of Lake Te Anau. So steep and difficult is the country that the little lake in the valley now called "Notornis" was not marked on maps until 1949, although it was known in Maori tradition as Kahaka-takahea (nesting place of the Wanderer).

Probably formerly in forested regions over the whole of South Island, New Zealand. Within the past hundred years living birds have been captured in four places in the southwestern part of the island. These places are: (1) Duck Cove, Resolution Island, in 1849; and (2) Secretary Island in 1851; both of these are close to the southwestern coast of South Island. The specimens are in the British Museum. (3) North of the Mararoa River, which flows into the west side of Lake Manapouri, about 45 miles east of Secretary Island. This bird, the type, was taken in 1879 and was in the Dresden Museum, Germany, but "disappeared" (*fide* Stresemann) in the war, 1939–45. (4) In 1898 a dog caught one on the south shore of middle fiord, western shore of Lake Te Anau. It is in the Otago Museum, New Zealand.

Remains have been found in caves and swamps from Nelson, in the far northwest Pyramid Valley (Falla, *in litt.*), 56 miles northwest of Christchurch; Albury and Pareora Districts (Duff, 1949); Earnsclough Cave, near Alexandra; Ngapara, about 20 miles from Oamaru; Longbeach, near Dunedin; and Castle Rock, near Lumsden, Southland (Hamilton, 1893).

Habitat: Nests in open ground, covered with grasses and low bushes, close to lake and marshes and streams, and usually to be found in such places. Falla (1949) described the present habitat:

The area in which the birds were first found by Dr. Orbell [in 1948] is the basin of a hanging valley of which the floor is some 2,000 feet above the level of Lake Te Anau (itself 684 feet above sea level). Glacial moraine has partially dammed the stream, which reaches lower levels through limestone without cutting any well-marked valley on a normal incline. Most of the basin is filled by a tranquil lake with some bog above it at the head of the valley, and well drained slopes clothed in snow-grass (*Danthonia*) along its sides. Above the grass the steeper slopes of the low ridges are covered with beech and sub-alpine forest . . . wiry scrub and coarse grasses are not generally attractive [to birds] and the beech-forest floor is thick with moss. The stream and lake also show little sign of aquatic plant or animal life.

Description: A large blue-headed, green-backed bird with a bright red bill which extends backward between the eyes in a "shield." It runs well

FIG. 34.—Notornis (*Porphyrio mantelli hochstetteri*).

but cannot fly. Differs from pukeko (*Porphyrio p. melanotus*), its closest relative, in its green, not blue, back, larger bill. It is said to have "stouter feet and legs."

Head, neck, wings, and underparts dark blue, merging on the upper back to greenish blue and to bright green on the mantle and lower back. Rump and tail are olive; under tail coverts white. Base of bill and shield scarlet, remainder of bill pink, somewhat darker at the tip. Feet and legs are pink and eyes reddish brown.

Down of the chick is black; the bill is black with a subterminal white patch; feet and legs are purplish brown. There is a conspicuous spur on the wings (Falla, 1949).

Habits [1]: Observations were made during the seasons 1948-49-50, and these are the basis for concluding that the breeding season begins in October and sometimes continues into February. We know that a bird was found to be incubating on October 25. A chick, only a few days out of the egg, and also an immature, thought to have been several weeks old, were both seen on December 6. Eggs are known to have been laid on February 7 by a female that had been under observation and had been seen feeding a chick, at that time "half grown." Apparently therefore there is sometimes a second nesting.

A connubial display was seen in December and described by Fleming (1951):

The sitting bird was called off by its mate. The bird responding darted with a crouching run to the caller and straightened up facing it with the two bill tips almost touching, and both necks upstretched. After some seconds one bird, I think it was the original caller, crouched and moved round the other with a gyratory movement which spread white under tail coverts to the other's view. Drooped wings and fluffed out flank feathers gave the general impression of a round white target ringed with blue.

Nests of grasses and leaves are built "upon a foundation of earth" among clumps of snow grass (*Danthonia*) which overhang them. Pairs have been seen to make more than one nest.

Usually four eggs are laid, although there may be only one. They are said to be ovate, dull cream colored with brown spots and faint mauve blotches, and large for the size of the bird (73.5 by 48.3 mm.).

Since the birds have been heard calling at night, it may be inferred that they are partially nocturnal, but they are also active during daylight hours.

Feeding birds are described as follows:

One was seen stripping Danthonia of flowers and seeds by running the stalks from the base to the tip through its beak. Others appeared to be reaching up into various bushes for leaves or fruit . . . *Carex* . . . is pressed over with one foot, loosened at ground level with grubbing strokes of the beak, and then held in the closed foot while the succulent bases are bitten off. Special treatment was accorded the green seed stalks of *Aciphylla* from which the spiny leaflets were first carefully removed and the stripped fleshy stalk then devoured.

For the first two or three weeks, chicks feed largely upon insect larvae. Mature birds have been seen to dig among sphagnum and other bog mosses, exposing the larvae, and it is probable that they do so for the

[1] From Falla, 1949, 1951; Fleming, 1951; Turbott, 1951.

benefit of the chicks. Droppings of birds several weeks old show that the staple by that time is vegetable.

Not only the lore of the Maoris but also more recent sightings and the actual capture of specimens (see above) indicate that at least some birds may sometimes wander in winter from the region of snow to the coasts and shores of Lake Te Anau. However, none have been seen there for many years, and the only known populations appear to remain in mountain valleys.

THE TAKAHE OF NORTH ISLAND: *Porphyrio mantelli mantelli* (Owen)

> *Notornis Mantelli* OWEN, Trans. Zool. Soc. London, **3**, 1848, pp. 347, 366 (Waingongoro, North Island, New Zealand). Type based on an almost complete skull in British Museum.

Status: Extinct. Known from subfossil remains found in kitchen middens.

Range: Formerly North Island, New Zealand. Recorded from subfossil bones at Waingongoro, Wanganui River, Martinborough, Te Aute Swamp, "tertiary beds" near Hawke's Bay, a cave between the Waikato River and Mount Tongariro, and Pataua, north of Whangarie, in the extreme north.

Description: This subspecies differs from *hochstetteri* in its longer leg bones. Available measurements show the tarsometatarsus to be 18 mm. and the femur 99 mm. longer (see measurements above). It is possible that more material will show that differences are somewhat less than indicated, but unlikely that such a great difference will be bridged by individual variation. Dr. R. A. Falla writes me: "Forbes [1] was right about some of Owen's leg bones. Undoubtedly the tibia is that of a swan, and the figure of the femur is not very good. But the metatarsus is correct. . . . None known from South Island are equal to it. The smallest I have from North Island is 10 mm. larger than the largest South Island one (Aniseed Valley, Nelson)."

Perhaps more material will show that difference in size of skull, a character used by Owen in his original description, and by later authors, will fall within the range of individual variation. Dr. Falla has most kindly sent me his measurements of the distance between the temporal fossae at the narrowest point—North Island 14.5–23.5, and South Island 13.5–16. If, however, it were possible to determine the sexual variation

---

[1] Forbes, 1892. I am obliged to Dr. Falla for his kindness. Forbes (1923) indicated that there is no difference between North and South Island populations. Oliver (1930) recorded both forms as having occurred at Martinborough, North Island. Peters (Check-list, 1934) synonymized the two.

and age variation (females are probably smaller), a difference between the races might be apparent in this respect.

WHITE SWAMP-HEN: *Porphyrio porphyrio albus* (White)

> *Fulica alba* WHITE, Journ. Voy. New South Wales, 1790, p. 235 (Lord Howe Island). Type in white phase in Vienna Museum.
> Synonyms are *Porphyrio stanleyi* ROWLEY (1875) (type in white phase in the Liverpool Museum) and *Porphyrio raperi* MATHEWS (1928) based on a drawing by George Raper (c. 1790).

Range: Formerly Lord Howe Island and possibly Norfolk Island, off the east coast of Australia.

Status: Extinct—thought to have been extirpated by sailors in the early part of the nineteenth century.

There can be no question that a population of white swamp-hens existed on Lord Howe Island before the coming of man, and that this is now extinct. Drawings by Thomas Watling and George Raper, in 1788 and 1790, now in the British Museum,[1] are crude but are unmistakably white Porphyrios. The contemporary accounts of Surgeon Bowes,[2] as well as "The Voyage of Governor Phillip" (1789), mention "white birds resembling Guinea Fowl," and the latter says further: "which were found at Norfolk Island, were found here also [Lord Howe] in great numbers. The bill of this bird is red and very strong. . . ."

However, much doubt has been thrown on this single statement that the birds occurred on Norfolk (Iredale, 1910; Hindwood, 1932). It is possible that an error was made in editing the book. Furthermore, other eighteenth-century visitors to Norfolk who mentioned birds—Lieutenant King and Captain Hunter—said nothing of them,[3] nor does Forster, who accompanied Captain Cook on his second voyage, during which Norfolk Island was discovered.

Substantiating this evidence is a pure-white bird mounted in the Vienna Museum. Because of the way in which the specimen is posed—exactly like the illustration in White's "Journal of a Voyage to New South Wales" (1790)—it has been assumed to be White's type of *alba*. This was purchased by the Royal Museum, Vienna, in 1806, at the sale of the Leverian collection in London.

A second specimen, which is in the Liverpool Museum, is pure white, except for a bluish cast on the wings, and also mounted in the likeness of the picture in White's journal (*fide* Forbes). It is known to have been sold with the Bullock collection in 1819 (Sharpe, 1906) to Lord Stanley

---

[1] See Sharpe, 1906, vol. 1, pp. 108, 149. These drawings are in the British Museum.
[2] MS. in Mitchell Library, Sydney. See Hindwood, 1932, 1940.
[3] Iredale (1910) quotes Hunter, "Historical account of Port Jackson," etc., 1793.

whose son gave it to the Free Public Museum of Liverpool. At the sale it was advertised as having been brought by Sir Joseph Banks. Apparently on this basis, Forbes (1901) stated that "it must, however, have reached this country in June 1771, the date of the return of the 'Endeavour' to England." He presented no evidence for this, but it appears to have been accepted as fact by most later authorities.[1] Actually it may well be questioned. There is no real evidence as to its origin whatever. The only reason for believing that it came from New Zealand is the fact that Sir Joseph Banks visited that country on Cook's first voyage, but he collected no birds on that his only voyage. It may well have come to England after 1790.

These two types are very similar to each other. Both differ from *melanotus* of Australia and New Zealand in having shorter wings and toes (see measurements above) and (perhaps) in having the tail feathers "soft and not so stiff" as suggested by Mathews (1936). Other characters that have been used to differentiate them, including the size and shape of the "plate" above the bill, fall within the astonishing range of individual variation of this species, not only in Australia and New Zealand but also the Solomons.[2]

Partly white individuals occur in New Zealand and certain islands of the Pacific (Mayr, 1941a). That not all the birds of Lord Howe Island were white in 1788 is quite probable, for Surgeon Bowes wrote in his journal: "Fowls or coots some white, some blue and white, others all blue wt. large red bills . . ." (Hindwood, 1932). That the types are white indicates nothing more than that albinistic individuals existed in the birds of Lord Howe Island.

They were timid and the first men to see them killed them with sticks.

Both of these types differ from *Notornis* in the scutellation of the tarsi. In *N. hochstetteri* the regular transverse series of scutes, behind and in front, are separated by a series of small irregular scutes both on the inside and outside of the leg, whereas in *Porphyrio alba*,[3] "*stanleyi*" and *melanotus* the regular posterior and anterior series of scutes are not so separated.

Sharpe (1894, vol. 23, p. 6) and Rothschild (1907) separated *Notornis* from *Porphyrio* on the basis of the length of the secondaries and wing

---

[1] Iredale (1910) doubted this, but Mathews (1910), Oliver (1930) (who nevertheless calls the New Zealand bird *melanotus*), and Peters accepted it. Rothschild (1907) guessed that it came from Lord Howe, and *alba* from Norfolk.

[2] See Mayr, 1949, pp. 22–27. My measurements of Australian and New Zealand birds substantiate this.

[3] See figures in Mathews, 1910. A photograph of the chick of *hochstetteri* in Nat. Geogr. Mag., **51**, 1952, no. 3, p. 395, also illustrates this. Plate 33 in Rothschild is misleading in this respect.

coverts in relation to the length of the primaries. On this basis they placed *alba* in *Notornis*. This character (if it exists in fact) is related naturally to the length of the primaries, which are short both in *alba* and *Notornis*. The reason for this similarity may be due to an accidental parallelism in evolution, caused by the isolation of small, accidentally selected populations on islands that lacked the predators, which normally cause the more reluctant fliers to disappear and their handicapping inheritance to be eliminated. It must be a question as to whether this difference is real or whether comparisons were made by means of inaccurate pictures, however.

On the basis of the wing coverts, Rothschild (1907), also differentiated the type of *alba* in Vienna from *stanleyi* in Liverpool. It is quite possible that the difference in this case is also only apparent, being due to the fact that the type of *stanleyi* has been remounted, according to Forbes (1901), and the humeri removed. The possibilities of accidental distortion are great. No other important difference has been pointed out.

Whether or not white swamp-hens ever inhabited Norfolk Island, there are none there now, nor are they to be found on Lord Howe Island.

Typical *melanotus* (in the blue phase) has been found quite recently on both islands,[1] and a problem presents itself at once.

Was there once a form on Lord Howe Island which we now should consider to have been a subspecies distinct from other populations of *melanotus,* or was it only the white phase which was extirpated?[2] The only tangible evidences we have are the specimens recently taken on Lord Howe, and these have longer wings than the white types of *alba* and *stanleyi*. They are typical *melanotus*. Great as is the possible error in comparing such measurements (those who measured the type of *albus* differed by 5 mm.) a difference of 20 mm. appears to be sufficient for an hypothesis that a long isolated and distinct subspecies once existed on Lord Howe.[3]

It is not impossible that occasional birds still arrive on the islands from the mainland, as the appearance of one on the deck of a ship 400 miles from New Zealand attests (Henry, 1899).

Exactly when the birds disappeared is not known. A few colonists

---

[1] Hindwood (1940, p. 34) records specimens taken in 1882, 1891, and a sighting in 1936. Measurements of the skins (see above)) furnished me by Dr. Kinghorn, of the Australian Museum, are those of *melantous*. A series from Norfolk Island in the American Museum was collected in 1912 by Roy Bell, and in 1926 by the Whitney Expedition of the American Museum.

[2] See Mayr, 1931, 1938b, 1941a, 1949; Hindwood, 1932, 1940.

[3] But see Mayr (1949, pp. 25, 26), who says of the Solomon Island populations: "from Manus and the eastern Solomons, are of rather small size. The Ysabel population is gigantic, while the birds from the other islands are intermediate. Nothing would be gained in breaking up this assemblage by naming the large-sized Ysabel population."

arrived in 1834, and the birds may have gone by that time. A Dr. Foulis who was on the island between 1844 and 1847 wrote of a few birds, but no "white fowles" of any sort. Perhaps the whalers killed the birds. It is unlikely that the colonists were enough. Such predators as rats and cats did not appear until later (Hindwood, 1932, 1940).

Location of specimens: I know of specimens of *Porphyrio mantelli hochstetteri* in Dunedin; London; and Wellington; and of *P. porphyrio albus* in Liverpool and Vienna.

Kagu

Rhynochetos jubatus Verreaux and Des Murs

Rhynochetos jubatus J. Verreaux and Des Murs, Rev. et Mag. Zool. (2), **12**, 1860, p. 440, pl. 21 (New Caledonia). Fig.: Sarasin, 1913.

Common name: Kagu.

Status: Rare. Range restricted to mountains of the eastern, or windward side, in the southern third of the island. Leach (1928) was told that the range was restricted to 40 sq. km. The birds were still to be found in reduced numbers in 1945 (Warner, 1948).

There are many possible enemies of the bird, including dogs, cats, pigs, rats, and men. It is improbable that the rats have reached the birds' habitat in numbers, although Warner says that black rats are to be found in the mountains. They undoubtedly helped to exterminate the birds from areas settled by human beings, however. According to Sarasin (1913) the critical period in the life of the species was the advent of dogs, sometime after 1853. They are used by natives and Europeans to hunt the birds. Furthermore, according to Leach (1928), there are many packs of wild dogs roaming the island, and these run down the imported deer and incidentally kill kagus. Leach records the disappearance of the birds from a part of their range since the coming of lumber men, who cut kauri pines (*Araucaria*) in the mountains.

M. Joubert wrote to Bennett (1863), only nine years after the coming of Europeans, that the natives catch them to eat by placing a noose where the birds would run into it. Layard says they usually used dogs.

For a time many skins were exported to millinery establishments, and a few have been sent every year to bird dealers in Australia, but this trade is apparently not considerable.

Warner (1948) records finding many traps in some of which were carcasses of kagus. He believes that the Japanese were particularly adept, and that this practice was one of the most important reasons for the bird's rarity. He lists also destruction of natural habitat and reduction of food supply by pigs and by man's activities, such as mining, lumbering, and burning.

Range: New Caledonia: Recorded from Mont Canala (near the east coast, about 60 miles NNW of Noumea—the axis of the island lies ap-

FIG. 35.—Kagu (*Rhynochetos jubatus*).

proximately SE–NW), in the north, from Mont Humboldt and Ouhatche, on the windward side, and from Mont Dore near Noumea on the leeward side (Sarasin, 1913). The western or leeward side of New Caledonia is comparatively dry. Almost the only tree on the hillsides is a eucalypt, called locally na-oli. The trees are spaced far apart and there is little ground cover. Only in the river valleys is there suitable territory for birds.

Habitat: Formerly occurred in comparatively rich forests bordering streams near the sea (Marié, 1870; Pouget, 1875), but then more numerous in the mountains of the interior. Recorded recently from the dense forests and brush of the steep, narrow valleys (Warner, 1948).

Description: A rather large (total length about 24 inches) heronlike bird, with long legs and a conspicuous gray crest.

Head and crest of long plumes pale gray, upper parts gray, wings black-banded and tipped with white. Chest gray, belly brownish. Bill reddish orange (paler at tip); legs, feet reddish orange; eyes orange.

Downy young are dark brown with "faun" markings. Legs and feet brownish and eyes black (Campbell, 1905b).

The following is quoted from Finckh (1915): "On the 31st day feathers began to molt in on the shoulders and seven days later the wings were fully feathered and by the end of another week the entire bird had come into its first plumage. This plumage was brown but by the eighth week the color is recorded as having changed to gray, like the adult, the bird at this time being about one half the size of the adult. There was still a little down adhering to the face at the base of the bill."

There appears to be no constant external characters by which to distinguish the sexes.

The relationships of this bird are somewhat obscure. It is presently placed in a suborder in which it is the only species. It is thought to be more nearly related to the peculiar fin-foots (Heliornithes) and sun-bitterns (Eurypygae) of South America than to any other birds, although the relationship is not close or clear. It is probable that this is a surviving remnant of an older fauna. Other forms have disappeared without leaving any trace.

Habits: The breeding season has been said to begin in August and to end generally in January (Sarasin, 1913). However, eggs have been laid in captivity in Australia between April and November (Campbell, 1905a). Nothing is recorded specifically of the behavior of the birds at the onset of the breeding season, nor has anyone recorded having seen the nest in the wild. The breeding performance must be very interesting and should be studied. Much peculiar behavior is recorded, but nothing with regard to season, activation, or motivation.

Bennett (1863), who had two birds in captivity, records the female as crouching on the ground and covering herself with her wings by throwing them over her head and body.

The male bird is described as throwing up his wings alternately, as if using them as shields, and displaying much pugnacity. However, there may be some question as to whether Bennett really knew which was the male and which the female, since he says the female was very much larger, and there appears to be no size difference between the sexes. His correspondent in New Caledonia thought two species existed, one more pugnacious and smaller than the other. Most authors record certain individuals as very pugnacious.

The only recorded nest, which was fabricated by captive birds in an aviary, was a loose affair of sticks and leaves; it was built on the ground. Three eggs were laid one at a time between April and October in one year, and four the next—the last on November 16 (Campbell, 1905a). They were buff with red-brown and gray splotches. It was 55.5 by 48.5 mm.

In captivity, according to Campbell, the male was observed to sit on the nest continuously, but it was thought that the female relieved him at night. The incubation period was 35 to 40 days (Finckh, 1915). Both parents fed the chick worms. After the third day the chick began moving a short way from the nest, and after about 19 days it ran about actively. On the 29th day it began to pick up worms, snails, and slugs for itself.

In the wild, it is recorded as being in flocks and moving to its feeding grounds in numbers and, according to Marié (1870), "Its principal diet is a crustacean (*Bulimus bavayi*)[1] and it feeds on these almost exclusively on Mt. Mou (1280 m) . . . other foods are locusts which they catch on the ground and other insects." In captivity they have been kept alive on meat, canary seed, cracked corn, and worms (Finckh, 1915).

Cries are variously described as "like that of a small dog—ma-ma" and a "growling kind of scream" and "a noise between a bark and a laugh, terminating in the repeated note of ō ō, ŏ ŏ," as well as a guttural rattling sound (Marié, 1870; Layard, 1882).

Although the bird is thought by those who have visited New Caledonia to be nocturnal, this is doubted by those who have observed it in captivity. It is particularly noisy at night, however.

Apparently it cannot, or at least does not, fly efficiently. Marié (1870) said, "When chased it runs with the speed of an arrow with wings spread, but without rising from the ground. I have never seen it rise more than a meter . . . I have seen it perch on the rail of a veranda about one meter in height."

Much unexplained "displacement activity," such as picking up rubbish with the bill and throwing it about, is recorded (Tinbergen, 1951, p. 113).

---

[1] Now *Placostylus bavayi* C. and M.; this is a large mollusk, not a crustacean.

New Zealand Shore Plover

Thinornis novae-seelandiae (Gmelin)

Charadrius novae-seelandiae GMELIN, Syst. Nat., **1**, pt. 2, 1789, p. 684 (New Zealand = Dusky or Queen Charlotte Sound, South Island). Fig.: Emu, **39**, 1939, p. 1; Buller, 1887–1888, p. 214.[1]

Common names: New Zealand shore plover; New Zealand sand plover; tuturvatu; kohutapu.

Status [2]: Rare (52 pairs were counted and 70 estimated in 1938), with a very restricted range, and in danger should any change in its habitat take place. Buller (1887–88) records *Thinornis* as "comparatively plentiful," to be seen at the mouth of the Piako River, Hauraki Gulf, Manukan Harbor, and the sand spits of Tauranga on North Island, and on South Island. It probably disappeared from this part of its range within the next 30 or 40 years.

Following is quoted (with omissions) from Fleming (1939):

Taking into account the conservative and vulnerable nature of an organism of specialized habits, a suggestion can now be hazarded as to the cause of the Shore Plover's rapid extinction on the main islands of New Zealand. The cats—which were becoming numerous in New Zealand at the time of the Shore Plover's extinction—would find their natural highways of dispersal throughout the country along the same rocky shore lands on which the Shore Plover nested, and, I believe, nested fairly exclusively. Such coasts offer to traveling vermin an ever-present food supply which is largely lacking on the sand-dune coasts and river-beds where our other endemic waders breed, and for that reason I suggest the Shore Plover fell a victim to their predation whilst the dune-dwelling Dotterels and river-bed Wrybill have survived.

It will be evident to the reader that the Shore Plover is in a healthy state at its last remaining breeding ground. . . . From the past history of the species, it is evident that a more imminent danger than that of collectors is the possibility of pests reaching Southeast Island.

It is a little short of a miracle that Southeast Island has remained free from imported vermin. During the period of whaling activity, when the island was for a time a shore station for the bay whalers, every visiting ship carried in her holds rats which, had they reached the shore would have completed the extermination of the Shore Plover. But through all its vicissitudes the island remains free—as yet. Native-owned, the 540 acres of the island home of the Shore Plover, Snipe and a great population of many other indigenous species and races, have a capital value of £620, according to the last Government valuation, and are leased for grazing sheep for a paltry sum. Fortunately, Mr. F. Nielsen, the present lessee, through whose kindness we were permitted to visit the island, is fully aware of the responsibility of his trust, and takes adequate precautions that no vermin may reach the island, but it is surely to be hoped that at some date the island may become a sanctuary. The Chatham Islands, botanically and zoologically, are one of the most important biological provinces of the New Zealand Region, and are at the present date the

[1] Type taken by Forster on Cook's second voyage.

[2] This entire account is an abridged version of the excellent article by Fleming (1939, pp. 1–15).

only such province in which no sanctuary has been set aside for the preservation of the characteristic flora and fauna.

Range: Formerly New Zealand: North and South Islands; Chatham Islands: recorded from Pitt Island, Mangare Island, Southeast Island, the Sisters (?). Now confined to Southeast Island (Rangatira). Fleming (1939) estimated that 69 percent bred on the northern and eastern coasts, 23 percent on the southern and eastern, and 8 percent inland.

Fig. 36.—New Zealand shore plover (*Thinornis novae-seelandiae*).

Habitat: Rocky shores, to which it has probably always been confined. Formerly fed on sand and mud flats (Buller, 1887–88; Oliver, 1930).

Description: A small plover (total length about 8 inches). The flight is reminiscent of a turnstone.

Forehead, face, and neck black (brownish when feathers are worn). Top of the head brownish gray, surrounded by a white line; back and wings brownish gray, the feathers more or less edged with white. Below white. A prominent naked ring around the eyes red. Bill red at base, becoming pink toward the tip which is black. Feet flesh colored, claws black, eyes brown.

"Females appear slightly heavier in the body than males in life and their cheeks and neck are not nearly so jet black, but have markedly brownish areas on the sides of the face. . . . The male's bill is slightly more brilliant orange-red and the blackish tip is slightly less extensive than the female's.

"Both sexes have deep orange eyelids, olive-brown irides, pinkish orange legs and feet, and black claws" (Fleming, 1939).

In dried skins the sexes may be distinguished by the longer wing (118–128 as against 113–127 mm).

Downy young. Head, back, wings mottled gray, buff, tawny-yellow, dark brown with blackish spots. Below tawny-white.

Habits: The breeding season begins in November, and by mid-December nesting is well underway. Males defend a small and vaguely defined territory from the other males, even though nests were sometimes only ten feet apart. A "pursuit flight" is a prominent feature of courtship; males and females flying a complicated erratic flight in close coordination. Other activity is described as "corresponding with the displays of Oyster Catchers as studied by Huxley (1925)."

The nest is "a circlet of interwoven dried roots and grass," usually constructed among grass or weeds, but sometimes among rocks or even in burrows of petrels. Fleming stresses the fact that nests almost invariably have a cover or roof, whether afforded by boulders on the beach or logs, roots, or holes inland. Two or three eggs are laid. These are pale buff or brownish buff streaked or speckled with black or dark brown. Speckled eggs are sometimes more heavily marked at the longer end. They are larger as a rule than those of the banded dotterel, but there is some overlap.

Both the male and female incubate, but the female takes a larger share of the task. "The mate of an incubating bird habitually feeds on the rock platform usually within thirty yards at most of the nest . . . and is usually quick to return to the nest in times of alarm"—except when feeding at a distance. The exact period of incubation is not known. Chicks were observed to leave the nest in about twelve hours.

Enemies: Fleming (1939) records the red-billed gull (*Larus novaehollandiae*), not normally predatory, eating chicks. He believes the skua to be an enemy also.

Thinornis rossii Gray [1] is known only by the type specimen in the British Museum. It was thought to have been collected on the Auckland Islands in November 1840, but no such bird has ever been recorded from

[1] Voyage of the *Erebus* and *Terror*, 1845, no. 12, Birds, p. 12, pl. 11 (bis). (Auckland Island = Campbell Island?)

there. Since the expedition visited Campbell Island also during that season, there may have been an error. The type differs from all known specimens of *novae-seelandiae,* having the plumage and head and back uniform dark brown and a short narrow white line behind the eye.

Although Buller thought it to be an immature plumage of *novae-seelandiae,* this does not appear to be the case. Aberrant individuals, such as albinos, are known in populations of *novae-seelandiae,* and it may be that in the past other such 'sports' have occurred—and may again. On the other hand, there may have been a sympatric species which is now extinct.

Sandpipers of the Tuamotu Archipelago

Aechmorhynchus cancellatus (Gmelin)

> *Tringa cancellata* GMELIN, Syst. Nat., **1**, pt. 2, 1789, p. 675 ("in insula Nativitatis Christi"=Christmas Island, Pacific Ocean). Based on Latham's "Barred Phalarope,' Gen. Syn., **3**, p. 274.
> *Tringa parvirostris* PEALE, U.S. Expl. Exped., **8**, 1848, p. 235 (Dog, or Honden, and Raraka Islands, Tuamotu, or Paumotu, Group). Fig.: Ibis, 1927, p. 114 (specimen from Tuamotus).

Common names: Sandpiper; kivi-kivi.

Status: Abundant on small, isolated islands situated on fringing reefs of atolls in Tuamotu (Paumotu or Low) Archipelago.[1]

This species was long thought to be extinct and is so listed in Rothschild's "Extinct Birds." Lowe (1927) remarks that "its extinction, like Mark Twain's death, had been somewhat exaggerated." Beck and Quayle were convinced that populations occur only on islands uninhabited by men, cats, and rats. Natives verified this observation. No other enemies have been recorded, and it seems unlikely that there are any that might be lethal to the entire population. Hurricanes occur frequently in the Tuamotus, but fortunately they would appear to be no threat to the existence of the species.

Extinct on Christmas Island.

Latham's description was based on a specimen collected on Christmas Island on January 1 or 2, 1778, during Captain Cook's third and last voyage. It is figured by Ellis (no. 64 in the British Museum) with the notation "W. W. Ellis ad viv: delinct. et pinxt: 1778 Christmas Isles" on the margin in Ellis's handwriting.[2] According to Stresemann (1950) and others who have examined the painting, it most probably represents an

---

[1] Field notes of Beck and Quayle in the American Museum of Natural History, New York.
[2] Dr. J. D. Mac Donald, curator of birds in the British Museum, has kindly furnished me with the exact notation on the margin of the picture.

individual almost identical to those still existing on the Tuamotu Archipelago.

Range: Confined to the Tuamotu (Paumotu or Low) Archipelago. Recorded from Puka Puka (Dog or Honden Island) and Raraka by Peale,[1] Pinaki (Whitsunday) by Townsend and Wetmore (1919); on the Mangareva Group from Makaroa, Kamaka and Manoui; from South Marueta, Maria, Tenararo, Tenarunga (Minto), Maturei-Vavao, all of which lie in what used to be called the Actaeon Group, northwest of Mangareva. All other specimens have been taken in a central group, called Raeffsky, including Fakarava (Wittgenstein Island), Tahanea (Tchitchigov Island), Rataka, Katiu (Saken Island), Kauehi (Vincennes Island), Tepato (Ofiti Island), and Tuanake (Reid Island).[2] Formerly Christmas Island (Line Islands).

Habitat: To be found in every type of habitat the Tuamotu Islands afford; less commonly in the dense pandanus thickets, and commonly on the stretches of bare gravel (Quayle Ms.).

The only nest ever recorded was found on the shore by the side of a rock.[3]

Description: A small (total length about 6.5 inches) sandpiper, resembling a small curlew (see Lowe, 1927). A dark and a pale color phase exist together on the same small islands, and these apparently interbreed, for every type of intermediate is to be found.

In the darker phase the head and upper back are dark brown; a faint cream-colored line above the eye. Lower back slightly paler. Upper tail coverts and wing coverts brown, the feathers tipped with whitish or tawny. Primaries dark brown, almost black. Throat whitish, streaked with brown. Breast and belly brown, heavily streaked with darker brown. Flanks solid brown. Sexes alike.

In the paler phase there is a clear white line above the eye, and this is wider than in the darker phase. Head streaked with white and brown. Feathers of the back, as well as those of upper tail coverts, tipped with whitish. Throat white; upper breast white, feathers tipped with brown; belly white. Tail feathers tipped with white, the outer webs marked with triangular tawny or whitish spots.

Iris brown; bill black; feet blackish (Beck and Quayle Ms.).

Habits[4]: The breeding season begins in April and perhaps occasionally

---

[1] Specimens in the U.S. National Museum.

[2] These specimens were taken by collectors for the American Museum of Natural History in 1922 and 1923.

[3] In American Museum of Natural History, New York.

[4] Based on Beck and Quayle Ms. and labels of specimens collected by Beck and Quayle and now in the American Museum of Natural History, New York.

earlier. Specimens in both plumage phases have been taken in breeding condition throughout May and early June. During the same season, however, the majority of the birds were apparently not breeding, and the plumage yields no clue as to the reason. A pair, recorded as nesting, was taken on Kauehi Island on May 30, 1923. Both birds are in the lighter, though somewhat intermediate, plumage. The one nest was found on May 5, 1923, at Tuanake Island, but the parents, though seen, were not collected.

Fig. 37.—Tuamotu sandpiper (*Aechmorhynchus cancellatus*).

This nest was found on the shore of the lagoon. It was constructed of small, dry grass stems and was placed in a slight hollow in the shingle. There were two eggs in the nest (A.M.N.H. no. 5299). They resemble the eggs of the upland plover (*Bartramia longicauda*), being heavily marked with purple and violet blotches on a white ground, but, of course, smaller.

Collectors' notes speak of the bird constantly piping in a soft high pitched voice. On June 5 at Vavao the following was noted: "The Sandpipers have quieted down to a very light 'meh'. . . ."

Little more than this is recorded of the birds. They are said to be extremely curious, following the collector and his dog constantly. One specimen was caught by hand.

I know of specimens in Cambridge, Mass., London, and New York.

Barrier Island Snipe

Coenocorypha aucklandica barrierensis (Oliver)

> *Coenocorypha aucklandica barrierensis* *OLIVER, Birds New Zealand, 2d ed., 1955, p. 275 (Barrier Id., New Zealand).

Apparently a population of the New Zealand snipe has disappeared from Little Barrier Island, off the coast of New Zealand. Because it is known only by a single specimen the validity of the subspecies may be doubted. The specimen was taken in 1870 and is in the Auckland Museum.

Tahitian Sandpiper

Prosobonia leucoptera (Gmelin)

> *Tringa leucoptera* GMELIN, Syst. Nat., **1**, pt. 2, 1789, p. 678. Based on Latham's White-winged Sandpiper, Gen. Syn., **3**, pt. 1, p. 172, pl. 82 (Tahiti and Eimeo Islands). Fig.: Rothschild, 1907, pl. 35. Latham, Gen. Syn. Birds, **9**, 1824, p. 296.
> *Prosobonia ellisi* SHARPE, Bull. Brit. Orn. Club, **16**, 1906, p. 86 (Eimeo Island).

Common names: White-winged sandpiper (Latham); toromé (on Tahiti *fide* Latham); te-te (on Moorea *fide* Latham).

Status: Extinct. The unique type, supposedly from Tahiti, is in Leyden, Holland. At least three specimens were seen by Latham, but these have apparently disappeared; these were collected during Cook's voyages by Forster and Anderson (see Stresemann, 1950) in 1773 and 1777.

It is possible that the pigs or the goats that were released by Captain Cook were responsible for the extinction of this bird, but more probably rats were the cause.

Range: Formerly Society Islands: Tahiti and Eimeo (Moorea).

Habitat: Forster says it was to be found near small brooks.

Description: A small (total length about 7 inches) sandpiper; dark brown above and rusty brown below.

"General color of upper surface blackish brown; the lower back and rump ferruginous, center tail feathers blackish, the rest rufous banded with black, less distinctly on the two next the middle pair; wing coverts blackish, with white spot near the carpal bend of the wing, formed by some of the lesser coverts; crown of head blackish, the hind neck browner, mixed with black; sides of the face brown, the lores and ear coverts slightly more reddish, behind the eye a little white spot; cheeks and under surface of body ferruginous red, the throat buffy white, 'bill black, feet greenish [Latham described a specimen with yellow feet and legs]; iris dusky blackish" (J. R. Forster, *in* Sharpe, 1896). (Sharpe's description of type in Leyden.)

Latham described a similar bird which differed only in having a pale streak over the eye instead of a white spot behind the eye, and in having a rufous, not a white, throat. A second specimen was more ferruginous in color and had a ferruginous line above the eye. Sharpe in his description of this (*ellisi*) mentions a double patch of white on the wing coverts. Rothschild (1907) suggests that these variations are due to age.

Location of specimens: I know only of a specimen in Leyden, Holland.

Eskimo Curlew

Numenius borealis (J. R. Forster)

> *Scolopax borealis* J. R. FORSTER, Philos. Trans., **62**, 1772, p. 411, 431 (Fort Albany, Hudson Bay).

Common names: Eskimo curlew; doughbird (New England); fute (Long Island, N. Y.); prairie pigeon (west-central North America); pi-pi-pi-uk or tura-tura (Eskimo); courlis du nord or corbigeau des Esquimaux (French Canada); chittering curlew (Barbados).

Status: A specimen was taken on Barbados in September, 1964. It is in Philadelphia (*Bond, 1965, pp. 8–9). A specimen was taken at Battle Harbour, just north of the Straits of Belle Isle, on August 29, 1932 (Van Tyne, 1948). Just two weeks later four birds were reported at Montauk Point, Long Island, N. Y. (Murphy, 1933). As recently as February 1937 a reliable observer saw two or three in the eastern part of the Province of Buenos Aires, Argentina, near the town of General Lavalle.

The last bird recorded to have been taken in the United States was shot at Norfolk, Nebr., on April 17, 1915, and the last in South America in Argentina on January 11, 1925. A bird was reported on the coast of Texas in 1962.

Probably the most important cause for the disastrous reduction in numbers was overshooting, but it is also possible that other quite unknown factors combined to cause the catastrophe. We do not know, however, that any disease attacked them during those years, at perhaps a low ebb in the cycle of numbers. That a band of Eskimos found them on a restricted breeding ground and put them on ice is possible also, but there is no such evidence.

It may be that occasionally hurricanes destroyed the birds at sea on migration. The half dozen specimens found in Britain after severe cyclonic storms had passed over the North Atlantic Ocean may have been survivors. The greatest danger would have been encountered in the North Atlantic from the occasional storms that retained their force and traveled rapidly northwestward over the Grand Banks south of Newfoundland. If, as is possible, the majority of birds flew a great circle course from Newfoundland or Nova Scotia to points south of Cape San Roque, they passed well to the eastward of the tracks of the majority of such storms. Of 25 hurricanes occurring in August and September 1900–21, only 8 crossed the great circle route from Nova Scotia to Cape San Roque (Tannehill, 1938).

Range: Bred formerly on the tundra, north of tree line in the northern Mackenzie District, Canada, and, perhaps, west along the coast to Norton Sound, Alaska.

The only unquestionable breeding records are for localities in the northwestern corner of Canada between the Mackenzie and the Coppermine Rivers. Nests and eggs have actually been found at Fort Anderson on the Anderson River, Rendezvous Lake, Franklin Bay, and Point Lake, headwaters of the Coppermine (MacFarlane, 1891; Bent, 1929). Although no nests or eggs have actually been collected in Alaska, there are four specimens in the United States National Museum [1] taken in Alaska between May 14 and May 26, which indicate that the birds perhaps used to breed there. Nests and eggs have been found only in the latter part of June in Canada, however.

There are numerous sight records. Nelson (1887) says they passed through St. Michael, at the mouth of the Yukon, in considerable numbers in May, though few remained there to breed. Murdoch (1885) records them as rare at Point Barrow, though to be seen between May 20 and July 6.

Taverner and Sutton (1934) suggest that they may once have bred at Hudson's Bay.

From the breeding grounds, most birds crossed the continent of North America to Labrador during late summer. Few went south via the Great Plains. From Newfoundland and Nova Scotia south to the vicinity of New York,[2] they took off over the ocean to South America. Evidence of their route over the ocean is unsatisfactory. However, records from Ber-

---

[1] Data of specimens as follows: ad. ♂, no. 62448, H. W. Elliott coll. at St. Paul Island, May 26, 1872; no. 110184, unsexed, H. D. Woolfe coll. at Cape Lisbourne, May 25, 1885; male, no. 97571, E. W. Nelson Coll. at St. Michael, May 14, 1881; male, no. 88819, J. Murdoch coll. at Point Barrow, May 21, 1882.

[2] There are no specimens from New Jersey.

muda, as well as the Windward Islands of the West Indies,[1] indicate that the birds passed close to these islands every year, but did not stop regularly. There is also evidence of large migrations of birds passing over the ocean east of Bermuda during early September,[2] although no species were actually identified among these high-flying birds.

There are so few records of sightings of the birds on the coast of North America south of New York, and these are so readily explained by pressure of on-shore gales, that it is not probable that their route was close to the coast. Furthermore, there are no recorded sightings at sea between Bermuda and the mainland.

No doubt at least part of the population flew south via Bermuda and Barbados. Feilden (1889) wrote:

Arrives [Barbados] about the end of August but passes more frequently in September. The first I obtained in 1888 was on the 5th of September, when immense numbers passed over the island, though comparatively few alighted. The same day great flights of Golden Plover (*Charadrius dominicus*) and Longlegs (*Totanus flavipes*) arrived; I saw once a hundred of each of these species shot at one stand by a single gun.

Obs. I did not observe *longirostris,* nor does it appear to be known to the sportsmen of Barbados, who would at once recognize it, if it occurred, by its great length of bill.

It may be that Feilden confused *borealis* with *hudsonicus*. He lists *hudsonicus* as arriving in early August and as being less numerous, but separated them to his own satisfaction, calling *hudsonicus* "Crook Billed Curlew" and *borealis* "Chittering Curlew."

An interesting account of the arrival of the birds in Bermuda after a hurricane is that of Drummond-Hay, 1877–78. Wedderburn (*in* Jardine, 1849, p. 84) records *borealis* as "very rare"; again (*ibid.,* 1850, p. 10) Hurdis records it as "occasionally shot on the shores of Bermuda during autumn migration." In Jones (1859, pp. 41, 80) the same authorities simply say that the one shot specimens on September 15 and 25, 1848, and the other had seen four specimens, one of which was taken on August 14, the remainder in September.

Reid (1884, pp. 240–241) records them in Bermuda as follows:

Commoner and easier to approach than the preceding (*hudsonicus*). Locally termed 'Wood Snipe.' A good number accompanied the Golden Plover on their arrival in Sept. 1874, and several were killed on the north shore. Both species of Curlew remain but a short time. The Esquimaux is easily distinguishable from the Hudsonian Curlew by its smaller size and comparatively short, weak bill.

---

[1] Reliably reported to have been seen on Puerto Rico, the Grenadines, Grenada, and Barbados in late August and September (Bond, 1950).

[2] Hurdis, *in* Jones, 1859, pp. 74–76 (records from ships between Bermuda and England).

Occasionally migrations were dispersed by high winds or tropical cyclones. On five occasions single birds have been found in Scotland and once in Ireland (Dalgleish, 1880). Where dates of capture are available, they correspond with the dates of occurrence of such storms in the Atlantic. On September 6, 1855, a bird was taken near Stonehaven, Kincardineshire, Scotland; a hurricane was noted as particularly severe at lat. 50° N., long. 40° W., south of the Grand Banks of Newfoundland, September 1–2, 1855. A specimen was purchased in Dublin on October 21, 1870; hurricanes had been severe in the West Indies on October 7–8. Again, this time at Slains, Aberdeen, a bird was taken on September 28, 1878, and the cyclone responsible was reported in the Atlantic September 26–October 6. A record of May 26, 1906, at lat. 49° N., long. 27° W. is of questionable accuracy (R. Barbour, 1906).

There is no evidence to indicate where the majority of birds made their landfall on the coast of South America. On a course from Bermuda, via Barbados, the natural landfall would be the Guianas. It does not appear that they visited this coast regularly or in great numbers,[1] however.

It seems probable that many, and perhaps the majority of the birds, flew a great circle course, which of course is a straight line, coming upon the then sparsely inhabited coast of Brazil perhaps south of Cape San Roque; or they may have flown inland, stopping rarely on the coast of Guiana and Brazil, but proceeding to the plains or the inland plateaus.

Specimens have been taken in the states of Mato Grosso and São Paulo, Brazil (Pinto, 1938).

There can be no doubt that the winter months were spent on the campos of Argentina.

Stragglers are recorded from the Falkland Islands on the east to Peru and Chile on the west (Hellmayr, 1932, p. 398).

Large flocks were reported seen at Concepción del Uruguay, on September 9, 1880, but not usually to be seen after the middle of October. Between Azul and Bahía Blanca, in southern Buenos Aires, they were abundant and to be seen on the plains in company with golden plover and Bartram's sandpiper (Barrows, 1884).

But little is recorded of halts in South and Central America during the northward migration in spring. They departed from Argentina in late February or early March (Barrows, 1884). A single specimen is recorded from Guatemala in April.[2] Owing probably to lack of ornithologi

---

[1] Penard (1908, p. 270) says: "comes to the colony much more seldom than the preceding . . ." (N. hudsonicus). The Hudsonian curlew and the golden plover visit this coast in numbers and regularly, according to Penard. There is no record for the Eskimo curlew from British or French Guiana.

[2] Ad. ♂, San Gerónimo, Vera Paz. Specimen in British Museum (Salvin, 1861, p. 356).

cal interest in Central America and Mexico, nothing more is recorded. They appeared in Texas and Louisiana in early to mid March, passing northward in great numbers via the Lower Mississippi, Missouri, and Platte River Valleys. They are recorded as appearing in the markets of St. Louis, Mo., late in March, and in Kansas throughout April. In Nebraska, immediately to the north, they are said sometimes to have been seen throughout May, by which time the majority had passed through Minnesota and the Dakotas (Swenk, 1916, p. 326).

From here their movements remain largely unknown, although they were taken at Great Slave Lake.[1] They appeared on the coast of Alaska late in May, and the nests and eggs have been actually taken late in June and only in the neighborhood of the Anderson River, far to the eastward.

Habitat: Bred and nested in treeless barren grounds near the shores of the Arctic Ocean. They were found in open country, on the tundra, in or near marshes. In winter, on the pampas of Argentina or in open fields on their northern migration.

Description: The smallest American curlew (total length 13 to 14 inches). Closely resembles *N. hudsonicus,* with small individuals of which it has often been confused. The one infallible distinguishing character is that the primaries of *hudsonicus* are distinctly barred with pale brown, but this can only be seen on the under side of the wing. When both *hudsonicus* and *borealis* are seen together, *borealis* is smaller than the smallest individuals of *hudsonicus,* body weight being invariably less. Other characters that have been suggested, such as the cinnamon-colored underwing of *borealis,* the barred head of *hudsonicus,* which contrasts with the back, and the greenish legs (not bluish gray) of *borealis* do not appear to be diagnostic in the field.

Above dark brown, edges of feathers of black edged with white. Throat whitish or very pale buff. Upper breast streaked with dark brown, feathers of breast and flanks, in unworn plumage, marked with a wide "V." Primaries brown, not barred. Axillaries and wing lining reddish brown, barred with darker brown (only slightly darker and more reddish than *hudsonicus*). Bill black, paler or sometimes yellow at base of lower mandible, eyes dark brown, legs grayish blue. Sexes alike. Neither downy young nor first plumage has ever been described. Juvenile plumage is said to differ in having the feathers of the back more broadly edged with buff and in having fewer dark markings on the underparts.

Habits[2]: It is probable that breeding and nesting activities began upon

---

[1] All verifiable records for the coast of California prove to be erroneous, referring actually to *N. hudsonicus* (Grinnell and Miller, 1944, p. 566).

[2] *Numenius minutus,* which with *borealis* forms a superspecies, or, perhaps a polytypic

arrival in the tundra regions of northern Canada late in May or early in June to early in July. Almost nothing is recorded of these.

Roderick MacFarlane (*in* Bent, 1929, p. 128) says, "Very difficult to find the nests. [The birds] get off long before our approach, while the eggs nearly resemble the grass in color."

Nests are described as being mere hollows in the ground, lined with a few decayed leaves or sometimes a thin sprinkling of hay.

Eggs are about half the size of those of *N. hudsonicus* and show marked variation in ground color from dark, somewhat brownish green to blue. They are blotched with brown, more heavily at the larger end. The clutch was three or more, usually four, eggs.

Of behavior as observed during migration more has been recorded. The formation in flight was a loose wedge shape, not unlike that of other curlews. "They associate in flocks of from three to many thousands, but they generally fly in so loose and straggling a manner that it is rare to kill more than half a dozen at a shot. When they wheel, however, in any of their many beautiful evolutions, they close together in a more compact body and offer a more favorable opportunity to the gunner" (Coues, 1874).

Swenk (1916) says, "The Eskimo Curlew had several notes. During flight they uttered a fluttering tr-tr-tr note. . . . Mr. W. A. Elwood describes this note as 'a short low whistle' continually repeated by many of the birds simultaneously while in flight. . . . Before alighting, as they descended and sailed, they gave a soft whistle."

The birds on migration, at least, are recorded as having eaten insects and snails, but preferred berries, particularly the crowberry (*Empetrium nigrum*), also called bearberry and curlewberry and diddledee. This is stated to have been the case both in Alaska (Nelson, 1887) and in Labrador (Coues, 1861). During their passage through Nebraska in spring they are said to have eaten quantities of Rocky Mountain grasshoppers (*Melanoplus spretus*) and their eggs as well. "The bird in its migrations often alighted on plowed ground to feed on the white grubs and cutworms turned up by the plow, or in meadow lands, probably feeding on ants in the latter situation" (Coues, 1861).

Small snails adhering to the rocks on the littoral of Labrador (perhaps *Littorina* and/or *Saxitilis*) were commonly a part of the birds' diet.

Location of specimens: Specimens are to be found in all large museums.

---

species, has been as little studied. The birds nest in the same type of habitat in northeastern Siberia. Tkachenka (*in* Tugarinov, 1929) says, "They made a great noise, soaring with cries over our heads, sometimes alighting on the trees or on the ground, never ceasing to call."

Jerdon's Courser

Rhinoptilus bitorquatus (Blyth)

Macrotarsius bitorquatus BLYTH, Journ. Asiat. Soc. Bengal, **17**, pt. 1, 1848, p. 254 (Eastern Ghâts [1] = valley of the Godávari River *ex* Jerdon Ms.). Fig.: Ripley, 1952.

FIG. 38.—Jerdon's courser (*Rhinoptilus bitorquatus*).

Common names: Double-banded plover; Jerdon's courser; adavi wuta-titti (jungle empty purse).

Status: Certainly rare, restricted in range if not extinct. Last specimen recorded to have been taken in 1900 (Baker, 1929, **6**, p. 88). Dillon Ripley

---

[1] This district is more properly called Northern Circars.

informs me that Salim Ali and other good naturalists have searched for it recently without success.

Range: Known only from a restricted area in eastern India, from the valley of the Godávari River (near Sironcha and Badrachalam) and from Nellore, Cuddapah, and Anantpur, in the valley of the Penner River, about 100 miles north of Madras.

Habitat: Jerdon (1864, **3**, p. 625) found the birds in hilly country above the Eastern Ghâts of Nellore and Cuddapah, frequenting rocky and undulating ground, with thin forest jungle. Blanford (1898) said, "They were in thin forest or high scrub, never in open ground and I never saw any on hills."

Habits: Almost nothing recorded. Breeding season probably in June, since a male with enlarged gonads was taken in June (Baker, 1931). They were to be seen in pairs or small flocks and flew better than *Cursorius* (Blanford, 1898). Only a plaintive cry is recorded (Jerdon, 1864).

Description: The two narrow black bands across the upper breast and the straight bill should distinguish this bird from birds of similar habitat and appearance (coursers). Baker (1929) describes the bird thus:

Forehead, supercilia and a broken central coronal streak pale buff or white; remainder of crown and hind-neck dark brown, surrounded by the pale buff; tail-coverts white; remainder of upper plumage, scapulars and inner secondaries brown; tail-feathers blackish, the outermost broadly white at the base and all the lateral feathers with white apical spots on the outer webs; median coverts paler grey-brown with broad white edges forming a conspicuous wing-bar; greater and primary coverts black; primaries black, the outermost with a broad white patch on the outer web, joining obliquely with a similar broad white subterminal patch on the inner web, the white decreasing to a small spot on the inner web of the fourth; outer secondaries black, broadly edged with white on the inner webs; chin and throat white; fore-neck rufous surrounded by a black-edged white band; breast brown with a broad white belt across the lower part; under wing-coverts black and white; axillaries, lower breast, blanks and abdomen creamy-white changing to white on the under tail-coverts.

The feathers of the upper parts are obsoletely edged paler and the wing-coverts more definitely so, a character possibly of the juvenile plumage.

Iris dark brown; bill blackish-horny at the tips of both mandibles, pale yellow from the nostrils to the gape, legs pale yellowish-white with a fleshy tinge, soles flesh-coloured, nails horny.

Great Auk

Alca impennis Linnaeus

Alca[1] *impennis* LINNAEUS, Syst. Nat., ed. 10, **1**, 1758, p. 130 ("in Europa arctica" = Norwegian seas).

Common names: Great auk; garefowl; geirfugl (Iceland); gorfuglur (Faroë); an erbhoil (St. Kilda); penguin, Apponatz (?)[2]; wobble (early

---

[1] This species is closely related to the razor-billed auk, the differences being due to loss of the power of flight, degeneration of the wing and keel. Its retention in the genus *Alca* expresses its relationship.

[2] Cartier probably met the great auk on the coasts of Newfoundland and Labrador.

winter on the coast of New England); isarukitsck (Eskimo on the coast of Greenland).

Status: Extinct. Last specimens (a pair) taken at Eldey Rock off the coast of Iceland on June 3, 1844 (Salomonsen, 1945). Last specimen taken in Europe on the Waterford coast of Ireland in 1834 (Witherby *et al.,* 1941). Doubtful sight records at St. Kilda Island, Outer Hebrides, about 1840 (Harvie-Brown and Buckley, 1888; Witherby *et al.,* 1941); Grand Banks of Newfoundland in 1852 (Drummond Hay, *in* Newton, 1861).

Range: Bred, without question in recent times, on Funk Island off the east coast of Newfoundland and on Eldey Rock (where the last specimen was taken in 1844), off Cape Reykianes, south coast of Iceland, as well as the nearby Geirfuglasker rocks, which were sunk by volcanic action in 1830.

Possibly they bred on Penguin and Wadham Islands, south coast of Newfoundland; Geirfuglasker rocks west coast of Iceland and Vestmann Islands, south coast; Faroe Islands; St. Kilda; Outer Hebrides; Holm of Papa Westra, Orkney Islands; and, questionably, at least in recent times, on Lundy Island and the Isle of Man.

Recorded from the west coast of Greenland and from the coast of New Brunswick, Maine, Massachusetts, and, probably by accident of weather, Florida. In Europe from the coasts of Norway, Sweden, and Denmark (from bones only); from Scotland and Ireland; from the Channel Islands (Jersey); the coasts of France (Morbihan), and southern Spain (Gibraltar); and recorded in Pleistocene deposits in Italy (Grotto Romanelli, Otranto).

They were commonly to be found on the Grand Bank, off Newfoundland (Shaw, 1940), and coastal waters of North America during the sixteenth and part of the seventeenth century.

Details of this range involve not only the time element but also evaluation of existing evidence.

On the western side of the Atlantic the most northerly record is in Greenland. On the west coast of the great barren island, bones of the great auk have been found among Eskimo remains at Kangamiut, and a skin from the neighborhood of Fiskaernes [1] is preserved in Copenhagen. These, the only tangible records, date from the early fifteenth century to 1815, when the unique specimen was collected. Records for the east coast are based on accounts of Icelandic seamen and fishermen of the sixteenth century. [2]

---

[1] Newton, 1861; Salomonsen, 1945 (fig. 2). It is the only skin in winter plumage. A second specimen was taken on Disko Island in 1821. It has disappeared (*fide* Salomonsen, 1951).

[2] Salomonsen (1951) gives a good summary of these records.

Evidence that birds bred on or near the coasts of Greenland is of the most questionable nature. Probably they never bred there, and there is no good evidence that they were ever at all common.

Perhaps the best evidence of their breeding comes from Fabricius (1780), who describes the bird unmistakably and says: "Nests on rocks far out at sea, most remote from men: I conclude from this that its nest is never seen in Greenland but I have seen a young one taken in the month of August, having a certain amount of gray down after this for a certain number of days." Steenstrup (1855)[1] doubts this, and Grieve points out that the chick of the great auk would probably have lost its down as late in the year as August. Salomonsen is no doubt correct in saying that there was a regular northward migration of young birds from the breeding grounds farther south. To be sure, some of the bones found in ancient Eskimo camps are those of young birds, but this cannot be offered in evidence that the species ever bred in Greenland.[2]

As for the earliest records, it is said that an Icelander named Latra Clemens in 1590 filled a boat with great auks at Gunnbjorn Rocks, on the east coast of Greenland (Hørring, 1934, p. 140; Salomonsen, 1951, p. 359). This place has been identified as Leif's Island (c. lat. 64° 40′ N., long. 37° E.) (Holm, 1918; Salomonsen, l.c.). Possibly remains may be found in this vicinity.

Careful investigation of an old Eskimo camp site at the trading post of Angmagaslik yielded no bones of auks.[3] The founder of this post has told the only recent story of a possible occurrence on that coast.[4]

In 1884-85 it was said that the grandfather of a man then living had caught an Isarukitsck (great auk), of which it was related that it was a very large bird that had quite small wings with short feathers and that it could remain just as long under water as a big seal. And further, Peterson (in Helms) writes in 1896: "According to the Greenlanders a Great Auk was seen near the island of Ingmikertok in the Angmagaslik Fjord about six years ago. The man who had seen the bird said that it was as large as a Great Northern Diver, but could not fly . . . it got away."

The country north of Davis Strait appears to have been almost barren of bird life. Investigation of old Eskimo camps at King Oscar Sound reveals no bird bones whatever, and indeed no "bird darts," characteristic of Eskimo finds in western and more southerly parts of eastern Greenland, are present (Degerbøl, 1943a; Thomsen, 1917).

---

[1] Grieve (1885, App. 1, p. 3) thought Fabricius had seen some other species. Salomonsen (1951) says Steenstrup rightly disregarded the record.

[2] Hørring, in Mathiassen, 1931. Found at Kangamiut on the west coast, about lat. 65° N.

[3] Degerbøl (1935). *Larus hyperboreus* and *Lagopus mutus* were found.

[4] Helms (1926, p. 254). G. Holm established Angmagaslik station about 1894.

Even on the west coast the birds were probably never abundant. In all the many refuse heaps of Norse farms and Eskimo camps that have been investigated, few remains of the birds have been found in the Norse farms and only 12 fragments at a single Eskimo site.[1]

The disparity in the number of bird bones found in Norse sites as compared to Eskimo has been explained by the theory that the Norsemen threw fish and sea-bird offal into the sea and that the Eskimos tied up their dogs, whereas the Europeans did not. (Vahl, 1928; Degerbøl and Hørring, 1936). The Norse had domestic animals and these survived even their masters. These men were primarily farmers and herdsmen. It is probable that they were exterminated by the Eskimos.

Even the best huntsmen apparently did not kill as many great auks as did the ancestors of the Norsemen in Norway.

Sometime during the 56 years 1785–1841 the greatest known breeding colony was wiped out. This was on small, rocky Funk Island, off the east coast of Newfoundland, about 130 miles north of St. John's. Cartwright recorded the colony in 1785, and he appears to have been last to report its existence in writing. Before that, and for some years following, fishermen and voyagers [2] visited the island, in spite of the great difficulty of landing, and killed the auks and other birds in great numbers for food.

Although local fishermen probably killed many during the last years of the colony's existence, possibly the place was all but forgotten by deep-water sailors by the time the birds had disappeared. Those were times of great trouble for men. By the time North America was free of Britain, and Napoleon had been defeated, the birds were gone.

When the Norwegian naturalist Stuwitz landed at Funk Island in 1841 he found only heaps of bones, mummies, and shell membranes of the great auk, as did others who followed him (Lucas, 1888, 1890).

It is quite possible that at some remote period great auks bred on islands other than Funk, and nearer the coasts of Newfoundland, but there have been no finds of bones, only indefinite records to prove it.

---

[1] The bones of the following species have been found on Norse farms: *Cygnus cygnus?islandicus, Uria lomvia, Lagopus mutus,* and *Falco gyrfalco.* From Eskimo camps the following: *Haliaeetus albicilla, Phalacrocorax carbo, Cygnus cygnus ? islandicus, Anas platyrhynchos, Somateria mollissima, Somateria spectabilis, Larus hyperboreus, Larus leucopterus, Plautus alle, Alca impennis, Alca torda, Uria lomvia, Uria grylle, Fratercula arctica,* and *Corvus corax.* Degerbøl, 1934a, 1936; Degerbøl and Hørring, 1936; Hørring, *in* Mathiassen, 1931.

[2] Cartier in 1534 and "Capitaine Richard Whitbourne of Exmouth in the County of Devon" (about 1620) are good accounts. It is diffcult to see how "men drive them from thenes upon a boord into their boats by hundreds at a time, as if God had made the innocency of so poor a creature to become such an admirable instrument for the sustension of man," as the latter relates, for the swell sets in heavily on this rocky shore even in calm weather.

The name "Penguin Islands" in itself suggests that great auks once bred there. They lie off the south coast of Newfoundland, and some have thought that these are mentioned in an account in Hakluyt's Voyages.[1] To naturalists who have visited the place (Lucas, 1890)[2] this account has a "false ring," and I am inclined to agree. Certainly no remains of the auk have been found there. Furthermore the latitude given as 30° in one edition appears as 50° in another, so that it is as likely that this reference is to Funk Island.

In the late eighteenth century Cartwright (1911) wrote: "This [Funk Island] is the only island they have left to breed upon, all others lying so near the shores of Newfoundland they are continually robbed." Audubon says in the "Ornithological Biography" that fishermen of the Labrador coast told him that they killed great auks for bait on an island south of Newfoundland. Finally, Allen (1876) quotes Michael Carrol, of Buenavista, Newfoundland, to the effect that between 1831 and 1836 he visited Funk Island, and that fishermen surrounded the birds in small boats and drove them ashore into pounds. This could scarcely be Funk Island and is perhaps incredible.

St. Pierre-Miquelon, French colonies lying south of Newfoundland, have been suggested as possible breeding places (Grieve, 1885, app., p. 31). This is based on an egg in the Muséum National, Paris, marked "St. Pierre-Miquelon." Perhaps the egg did come from there, but more likely it was bought there, having been laid elsewhere, collected by a fisherman or a sailor, and sold at St. Pierre.

The Virgin Rock on the Grand Bank, suggested by both Newton and Grieve as a possible breeding place, has never been reported to have appeared above the surface of the ocean.

During the eighteenth century the birds were useful to sailors in more than one way, for they were to be seen commonly on the Grand Bank of Newfoundland and were used as an indication of the ship's position. An early (1742) volume of sailing directions for the north Atlantic [3] has this to say: "None [birds] are to be minded so much as the 'Penguins', for these never go without the Bank as others do; for they are always on it . . . sometimes more and sometimes less but never less than 2 together."

Although we may believe [4] that Cartier saw great auks near Bird Is-

---

[1] Vol. 3, p. 130: "greate fowles white and grey, and as bigge as geese and an infinite number of egges."

[2] Forbush (1912) suggests that Captain Hore recorded a Penguine Island near Cape La Hune.

[3] "The English Pilot," 1st ed. 1706, London. A picture, undoubtedly of two great auks, is reproduced (see also Shaw, 1940).

[4] It is quite possible that Cartier confused many birds in his concept of "Apponatz" or

lands in the Gulf of St. Lawrence in 1534, there is no evidence for the birds having bred there.

Under the date of July 5, 1785, Cartwright (1911) records a "Penguin" in the Straits of Belle Isle, between Newfoundland. His description makes the record credible. Writing of the capture of birds on Funk Island he says: "They drive as many Penguins on board as she [the boat] will hold; for the wings of these birds are so remarkably short they cannot fly."

Although Richard Fisher, "Master Hilles man of Redrife," [1] says he saw "Pengwins" at Cape Breton Island in 1593, he does not describe the bird. Farther south the birds are recorded as having been seen off the Tusket Islands, near Cape Sable, Nova Scotia, in May 1604, but Champlain did not describe great auks, and we cannot be sure that both he and "Master Hilles Man" were not mistaken. On the other hand, they may well have been great auks, for bones have been found in kitchen middens of American Indians in many localities from the Bay of Fundy south to Cape Cod. They are thought to have been seen on these coasts until the end of the seventeenth century, since their bones are found with equipment of European origin.

Only because there are many rocky islands suitable for the birds is it probable that they bred on the coasts of North America. The numbers of bones found together in kitchen middens indicate that they were quite common on the coast of Maine in the seventeenth century.[2] But Cape Cod, Mass., is a glacial moraine, and it is doubtful that the birds bred there (Forbush, 1912, to the contrary) or anywhere on the sandy coasts south of Cape Ann, Mass.

The localities where bones have been found on the coasts of North America are as follows: Nantucket Island, Grand Manan Archipelago, New Brunswick (Herrick, 1873); in Maine, along the shores of Frenchman's Bay, on Mount Desert Island (Townsend, 1907), Calf Island, Winter Harbor, Gouldsborough, Ashville, Sorrento, Soward Island, and

---

"Aporath." The birds he saw at Funk Island on May 21, 1534, he describes as follows: "As big as jayes, black and white with beaks like unto crowes: they always be upon the sea; they cannot fly very high, because their wings are so little, no bigger than halfe one's hand, yet do they fly as swift as any birds of the air level to the water; they are also exceedingly fat; we named them Aporath" (Hakluyt).

Even though "jayes" be read magpies and "crowes" ravens, as Grieve suggests, still Cartier described rather a razor-billed than a great auk (see also Lucas, 1888, p. 281). The latest translation (Biggar, 1924) adds to this confusion by translating "Apponatz" as great auk when Cartier mentions the birds seen at Funk Island, but when seen at Bird Island "Apponatz" becomes "puffin."

[1] Hakluyt, 3, p. 192.

[2] Seven humeri, several fragments of crania and sterna, as well as entire coracoids in a single midden (Wyman, 1868).

Lamoine [1]; on the following islands in Casco Bay: Goose, White, Sawyers, and Flag (Wyman, 1868; Loomis and Young, 1912). Farther south they have been found in Massachusetts at Eagle Hill, Ipswich and Plum Island, at the mouth of the Merrimac River, Marblehead (Putnam, 1868; Bullen, 1949), at East Wareham on Cape Cod, and more recently, at Squibnocket Pond, Marthas Vineyard Island (Wyman, 1869; Byers and Johnson, 1940). The find of bones on the sandy and semitropical coast of Florida [2] is probably due to the accidental visit of a bird which was forced south by northerly gales. Quite recently great numbers of dovekies (*Alle alle*) were forced upon these coasts by stress of weather (Murphy and Vogt, 1933). Only two humeri of the great auk were found, and these under numerous layers of alternate shell and earth, indicating that the bird was eaten by Indians long before the white man came to Florida.

The discovery of bones lends strong credibility to accounts of early voyagers. John Josselyn describes [3] an auk he saw at "Black Point" (now Scarborough), Maine, late in the seventeenth century ("New England's Rarities Discovered" was published in 1672). He called it "Wobble," probably because of its awkward gait.

Two days after Captain Bartholomew Gosnold sighted and named Cape Cod, on the coast of Massachusetts, his ship's writer [4] records: "the twentieth [of May] by the ships side we there killed Penguins."

It cannot be questioned that great auks bred on small islands, skerries, and rocks off the coast of Iceland. Bones of eight individuals were found at Kyrkjuvogy and Baejsker (Newton, 1861). There is also the evidence furnished by an egg formerly in the collection of John Wolley, which almost certainly came from one of the small islands off the coast, perhaps Eldey.[5] There are more eggs in collections which quite probably came from Iceland, although their history cannot be accurately traced.[6]

---

[1] Probably existed in these regions until 1700 (*fide* Wendell P. Hadlock, *in litt.*, 1949). Bones were identified by G. M. Allen, of the Museum of Comparative Zoology, Harvard. See Robert Abbe Mus. Bull. **5**, 1939, p. 16.

[2] Bank of the Halifax River near Ormond (Hay, 1902).

[3] "An ill shaped bird, having no long feathers in their pinions which is the reason they cannot fly; not much unlike a Penquin."

[4] Mr. Archer. Courses, distances, and positions of the *Concord* are vague and contradictory, but it seems probable that the ship was anchored off the outer beach near Chatham when "Penguins" were taken in 1602 (Purchas, 1905–1907). It was close to "Gilbert's Point" (perhaps Nauset), the reported position of which was lat. 41° 40′.

[5] Newton, 1905, p. 365. The egg was marked "Geir Ei." Brandt, a dealer in Hamburg, assured Wolley that it came to him from a man named Siemsen who was "travelling for a merchant or house in Flensborg, and making annual journeys to Iceland."
Brandt told Wolley in 1858 that he had had no auks or eggs for about 20 years; then he had had five or six birds or eggs in the shop from time to time.

[6] Grieve, 1885 (App.). Eggs no. 15, 25, 44, 49, 58.

According to tradition there are four small islands where the birds either formerly bred, or habitually roosted; all are (or were) called Geirfuglasker (Gare Fowl Skerries). Except for two of these, the record is by word of mouth and quite vague.

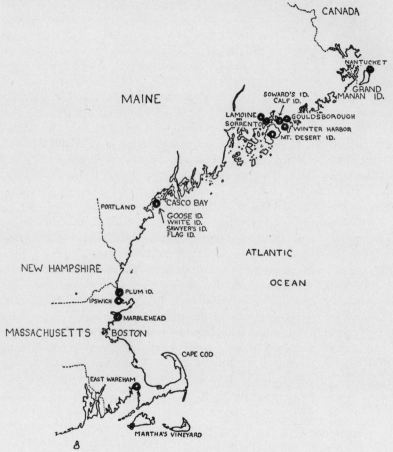

MAP 7.—Where great-auk bones have been found in New England.

The only two well-authenticated breeding places are rocks and reefs lying 15 to 25 miles off Cape Reykianes, the southwestern tip of Iceland. This is the group where Eldey Rock, the home of the last auk, was. The other rock, called traditionally Geirfuglasker, disappeared under the sea in 1830 during a volcanic cataclysm.[1] It is quite possible that many of

[1] Newton, 1861, p. 380; Grieve, 1885, pp. 18–21. The best evidence is contained in a manuscript in the public library at Reykjavik. See also Anderson (1747); Horrebow (1752).

these birds moved to nearby Eldey Rock when the ancestral home sank.

The other Geirfuglasker is in the Westmann Islands, off the south-central coast of Iceland. Evidence that the birds bred on this group comes to us through local traditions. Faber (1822), who visited Iceland in 1819, was told that the birds had occurred on these rocks 20 years before. Newton (1861) says, "We were told that some fifteen years before (1858) a young bird had been obtained thence, it was quite certain that no Great Auks resorted thither now." And in a footnote he says, "Of course it does not follow, even if the story be true, that this bird was bred there." These postmortem records are strengthened by Olafsen (1774–75), who was in Iceland 1752–57 and 1760–64, and who says, "The Geirfugl is found on one of the Westmanneyar Islands, and also on a rock off Reykjanes."

Remains of two individuals have been found at Hunafloi, north Iceland, and there is evidence that the climate of that region was warmer when those birds lived (Baroarson, 1911). However, no convincing evidence that they ever bred on the mainland of Iceland exists.

It is quite possible that the birds bred on the Faeroe Islands within historical times, but actually there is no direct evidence that they did so. Indeed evidence of any kind is scanty and lacking in convincing quality. Graba (1830) was told by an old man that he had killed an auk sitting on an egg near Westmanshavn, Sandoe Island, many years before. Svebo,[1] a resident of the Faeroes for some years prior to 1780, says that he himself never knew of a 'Gorfuglir' laying its egg in Faeroe, but that he knew of a bird having been killed with an egg in the oviduct at the island of Fugloe. There is no further evidence of its breeding on the Faeroes.

That they were to be seen there from time to time is quite likely. Olaus Wormius had one brought to him from the Faeroes about the middle of the seventeenth century.[2]

Steenstrup (1855) saw the head of a great auk preserved there, and Feilden (1872, p. 3282) was told by "an old man" that on July 1, 1808, he and some others found a 'Gorfuglir' on a ledge at the base of a cliff on Great Dimon Island (Store Dimon). If this story is to be believed, this is the last record of the great auk on the Faeroes (Salomonsen, 1935).

That great auks bred near the coast of Scotland during the Stone Age of man until about the time of the birth of Christ, and on St. Kilda until the eighteenth century, are fair assumptions.[3] On the other hand, most

---

[1] Ms. in Royal Library, Copenhagen, 1781–82, as quoted in Feilden, 1872, p. 3281.

[2] Wormius figures an auk with a white collar; this appears to have been simply an error. Although he repeats some of the confusions of Clusius, he describes his captive specimen and this was no doubt an auk (see Mullens, 1921).

[3] Witherby et al. (5, p. 150) and Rothschild (1907, p. 155). The former say "known to have bred on St. Kilda and probably in the Orkneys." The latter says both are absolutely

records are all too likely to have been the result of occasional appearances of birds that had followed schools of fish, or were forced eastward across the Atlantic by the westerly gales that prevail.

FIG. 39.—Great auk (*Alca impennis*).

The numbers of great auk bones found in low, chambered human dwellings of Orkney make it probable that birds were there in numbers in cer-

certain breeding places. Newton (Encyc. Brit., ed. 11, **11**, p. 463) may have been more nearly right in recording the bird as "an occasional visitor to certain remote Scottish islands." He says nothing of their breeding there.

tain seasons during the later Stone Age,[1] although there is no good evidence that they were common elsewhere in Scotland.

The theory that they bred there so long ago is based on the assumption of not only their seasonal arrival but also on the fact that they were difficult to catch at sea, although easily taken on the rocks during the breeding season. The men[2] of that day probably paddled hide-covered coracles. Since a six-oared boat could not overtake a swimming great auk in 1821, it is likely that it could be done but rarely in a coracle, as anyone who has ever pursued a wounded duck in a canoe can testify.

Two specimens were taken near the island of Papa Westra in the Orkneys in 1812.[3] That these were male and female may well be questioned. It was said that they were called locally the King and Queen by people who had seen them for "several years" in the vicinity. There is no evidence at all that they bred in the group. In fact the evidence is to the contrary (Newton, 1898, p. 591). According to Newton, a nearby islet called the Holm of Papa Westra would have been the most suitable breeding place.

One of the pair was a male, and is now in the British Museum. The other was not preserved (Newton, 1865).

Indeed, that it was never more than an occasional visitor to Orkneys in modern times is likely in view of the fact that Low (1813) said that he had "often enquired about the Great Auk especially, but cannot find it is ever seen here."

Undoubtedly the birds were seen from time to time during the seventeenth and eighteenth centuries on St. Kilda, the westernmost of the Hebrides. This is based on apparently good hearsay evidence of Martin (1698) and others. Martin described the bird thus: "The Gairfowl, being the stateliest as well as the largest, of all Fowls here, and above the size of a Solan Goose, of a Black Colour, Red about the Eyes, a large White Spot under each Eye . . . stands stately, his whole body erected, his wings short; he Flyeth not at all" and now comes the best evidence of the great auks' breeding on St. Kilda: "and has the Hatching spot on his breast, i. e., a bare Spot from which the Feathers have fallen off with the heat of hatching; his Egg is twice as big as that of a Solan Goose, and is variously spotted black, green and dark . . . appears on the first of May and goes away about the middle of June."[4]

---

[1] Platt, 1936, "Neolithic."
[2] Their bones show them to have been troubled by arthritis.
[3] Montagu, 1831 (2d ed.), p. 12, and other authors.
[4] There is no mention of the gairfowl or any description of it in the second edition of Martin's book (1716); we may well wonder whether Martin actually saw the bird. See also Newton (1865) and R. Gray (1871, p. 442). Cogent reasons for believing Martin are given by James Fisher (1952).

The Rev. Kenneth Macaulay, who was on the island in June 1758,[1] says that although he did not see a bird, the St. Kildians said that they saw them sometimes in July, but not every year.

The first mention[2] of the great auk is short. In 1684 Sibbald said: "There is one they call the gare-fowl which is bigger than a goose, and hath eggs as big almost as those of an ostrich."

As lately as 1821 a bird was taken alive at St. Kilda and kept alive for a few days, but it escaped (J. Fleming, 1824).

Although the latest record is by hearsay, many have believed it to be authentic.[3] This is the account of an old resident of St. Kilda, who with four others caught a bird on the rock called Stack-an-Armine. While the bird was asleep they carried it up to their "bothy," kept it alive for three days, and then killed it with a stick, "for there had been heavy gales not long before and it might have been a witch." No doubt the bird blew over from the Grand Bank of Newfoundland.

Without any doubt great auks were taken from time to time by primitive people of those barren coasts until about the first century A.D., although bones have been found in only three localities in Scotland. They are in the Orkney Islands, in Caithness, on the far northern mainland, and on a small island in the Firth of Lorne off the west coast.[4] There have been great numbers of excavations, but in relatively few have great auks' bones been found. At only one site are they reported to have been common, and this in a chambered cairn near Hullion, Rousay, Orkney. These birds were eaten perhaps 4,000 years ago or later.[5] The second site on Orkney is a broch, once one of a system of small stone forts built close to each other, apparently for mutual protection between Ayr and St. Marys, near Kirkwall. These are thought to have been occupied as lately as the first century A.D. (Childe, 1947), but since the auk bones were found in a lower stratum beneath the floor, the bird no doubt was killed much earlier. Auks do not appear to have been common there at that time.[6]

Finds of bones in brochs near the villages of Elsay and Keiss, Caithness, on the northern coast of the mainland of Scotland, appear to have been of

---

[1] Macaulay, 1764 (not seen by me). Grieve cites p. 156.

[2] Sibbald, 1684, and "An Account of Hirta and Rona given to Sir Robert Sibbald by the Lord Register Sir George M'Kenzie, of Tarbat" (as printed by Pinkerton, **3**, 1809).

[3] Witherby *et al.*, **5**, p. 150: "Credible record for St. Kilda, 1840"; Harvie-Brown and Buckley (1888, p. 158).

[4] Apparently no effort has been made at St. Kilda.

[5] Platt (1936) says Neolithic. Four thousand years is perhaps vaguely approximate.

[6] A single bone, the lower half of a tibia, was found (Ticehurst, 1908). This was associated with bone implements such as combs and dice. Iron artifacts were found at a higher level.
Other bird bones found included gannet, cormorant, shag, great northern diver, gull, wild swan, and shearwater, and of mammals, ox, pig, red deer, sheep, and horse (Graeme, 1914).

a much later date.[1] Since remains of only two or three auks were found, we may conclude that they were not very common.

By far the oldest finds in Britain are of four shell mounds on the small island of Colonsay, near Oronsay, off the coast of Argyll in the Firth of Lorne. These Obanian middens were formed not less than 4,000 years ago.[2] Like the brochs of a much later age, they now appear as grass-grown mounds, called locally Casteal-nan-Gillean, Croch Sigach, Croch Sligach, and Croch Riach. There appears to be no accurate record of the exact number of great auks found in these, but the bones of "several" are recorded.

The record for the Farne Islands, off the east coast of Northumberland, is quite as good as other hearsay records. Hancock (1874) writes: "In Wallis's 'History of Northumberland', it is stated, under the heading 'Penguin', that 'a curious and uncommon bird' was taken alive a few years ago in the island of Farn, and presented to the late John William Bacon, of Etherstone, with whom it grew so tame and familiar, that it would follow him with its body erect to be fed."

An upper mandible was found in a cave near Whitburn Lizards in Durham,[3] thought to be prehistoric but ethnologically recent.

This completes the record for England and Scotland, except for two sight records, which are no better than the story of the last sighting at St. Kilda; they have been less believed. The first involves the account of a man who shot a bird "that he had never seen the like of before" off the coast of the island of Skye in 1844.[4] The second is a similar record for Lundy Island, off the coast of Devon. It is based on the story of an old resident who saw the "King and Queen of the Murres" about 1835.[5]

---

[1] Clark (1948) places them in the early Iron Age. The bones were found near the surface, however, and since the brochs are thought by Childe to have been occupied as late as the first century, it is possible that the birds were eaten there at an even later date. Furthermore, Grieve (1885, p. 46) points out that a fragment of reindeer horn was found, together with the bones, and that the reindeer was mentioned in the "Orkneyinga Saga" of the twelfth century A.D. Hinton (1926) says he believes that the reindeer survived until this late date. It appears to be stated in the saga that the Jarls came to the mainland from Orkney to kill both red deer and reindeer.

[2] Bishop (1914) says between Paleolithic and Neolithic. Garrod (1926) says Azilian. A harpoon of bone was found. The number of years involved is only an approximation. Movius (in verb.) calls this culture Obanian and considers it to be a local development of the Larnian.

[3] Howse, 1880. Although human bones were found, there were no implements in the deposits or any extinct animals.

[4] See Grieve (1885, p. 24), who thinks it doubtful. Grieve (pp. 4, 10) states that the birds may possibly have bred on the Shetland Islands. There is no evidence for this.

[5] Mathew, 1866. The old man said they were "like the Razorbill Murrs" and "stood up bold like." He claimed to have found an egg which later he lost. A dead bird is reported to have been found 'near Lundy' in 1829 (Moore, 1837).

In Ireland the last specimen to be taken in any of the British Isles was killed in May 1834 at Waterford, on the south coast.[1] This is the only great auk ever to be preserved in Ireland; except for two supposed sightings, one at Belfast Lough (bay) in September 1845,[2] and the third on the long strand of Castle Freke, Cork, on the west coast of Ireland,[3] there are no other records.

However much the accuracy of these sightings may be questioned, there can be no doubt that great auks happened upon these coasts frequently during the Stone Age of man. Remains of six individuals [4] were found in mounds among the sand hills of Traymore Bay, Waterford County, not far from the place where a bird was killed in 1834.

Likewise, in northern Ireland, bones have been found in kitchen middens of the men of the later Stone Age at Whitepark Bay, County Antrim.[5] On the west coast a few bones [6] have been found in middens behind the beaches in County Clare.

Perhaps great auks bred on the coasts of Scandinavia from a time soon after the retreat of the last glaciers, perhaps as early as 7000 B.C. until about 2000 B.C., and perhaps later. They were probably never as common as on Funk Island, but men of the Stone Age killed them on the shores of the Arctic Ocean and southward along the coasts of Norway, Sweden, and Denmark until about the year of the First Olympiad.

No remains have ever been found on the borders of the Baltic Sea. The subfossil record of Scandinavia is as complete as that of any other known bird anywhere in the world.

Finds of bones of young birds in Denmark [7] and, added to this, the

---

[1] Thompson (1849–56) says two specimens were shot; Ussher and Warren (1900) doubt this. At any rate, one bird is preserved in the Museum of Trinity College, Dublin.

[2] H. Bell, a wildfowl shooter, said he saw "two large birds, the size of Great Northern Divers . . . but with much smaller wings . . . their bills were more clumsy . . . they kept constantly diving." Grieve (App. p. 57) considers this doubtful.

[3] Ussher and Warren (1900, p. 360) seem to doubt the accuracy of this. Accounts of this kind, coming as they do after final extirpation and much publicity, often have the quality of the accounts of "marvelous female witnesses," who so often testify at murder trials even today.

[4] Ussher (1897), Ussher and Warren (1900). Found in a kitchen midden with remains of red deer, oxen, and domestic fowl.

[5] Knowles, 1895; Barrett-Hamilton, 1896. The former remarks that the birds "from the number of bones which have been found . . . must have been common." There is, however, no comparison between the numbers of great auk bones as against those of others found. Except perhaps at Orkney Island, they do not appear to have been common in Scotland, nor is the number of six individuals found at Waterford impressive, considering the number of years possible for Neolithic man to have thrown auks' bones in a trash heap.

[6] A tibia near Lahinch and a coracoid north of Doonbeg (Ussher, 1902), found on the surface together with horns of red deer.

[7] Meijlgaard (*fide* Winge), a Mesolithic or aeldre stenalder site.

theory that Neolithic man could catch but few great auks at sea, result in strong possibility that they nested in Scandinavia.

The supposed great auk egg, found at the Tillery of Hermanstorp, Parish of Reng, in southern Sweden, is now thought more likely to be

MAP 8.—Finds of great-auk bones in Scandinavia.
(Modified after Wiman and Hessland.)

that of a swan, according to Løppenthin (1952). Certainly it is not a locality where one would expect to find an auk's egg.

To be sure, this period, called Litorina, was slightly warmer than the preceding, elm and lime trees having been prominent in the flora; probably it was not so warm that the birds could not breed.

An estimation of populations on the basis of the numbers of bones found in these prehistoric sites must be wildly speculative. There are many completely unknown factors which must affect it. Bird bones are extremely fragile. Furthermore, although much archeological work has been done, more material might change such an estimate considerably. Finally, Neolithic peoples were not apparently particularly interested in catching birds. As compared to the numbers of sites where mammal bones have been found, and the numbers of individuals, but few birds were caught. No doubt birds were "small beef," as they are in parts of Africa today.

On the other hand, however, it is probable that great auk bones are no more or less liable to destruction by any agency, such as leaching elements in the soil, or to destruction by wild animals, than the bones of any other bird. If it is assumed, then, that the great auk could be caught rather more easily than other birds during the breeding season and less easily at other times, which assumption is justified by eyewitness accounts of great auks, it is likely not only that the birds were caught during the breeding season at one locality in Norway, and perhaps two in Denmark, but also that they were caught as often as most birds in those days. Certainly the total numbers of great auk bones found in Norwegian kitchen middens are as great as those of any other species, save one, and of the 21 sites at which bird bones were found, 14 contained those of the great auk. But these sites are of differing ages. The only one that contained bones of more than two great auks was Viste, Randeberg, an early Mesolithic site. Here were found remains of no less than six individuals, whereas no more than two were found of any of the other 27 species present, except for the cormorant, of which about 8 were found. It is probable that great auks and cormorants bred near Viste, but it cannot be said that there was any other breeding place in Norway during a period of thousands of years from the early Neolithic to the early Iron Age. Quite possibly the few birds taken at other localities were driven ashore by gales.

Of a list of 29 ducks and sea birds that were taken from Mesolithic (Aeldre Stenalder) through early Iron Age sites in Norway, all but five nest there today. These are the gannet (*Sula*), little auk (*Alle*), Brunnich's guillemot (*Uria lomvia*[1]), the king eider (*Somateria spectabilis*), and great auk. All but the last two are winter visitors. The king eider is rarely seen even in winter on the Danish coast.

In Denmark, as far as is known today, the history of the great auk is confined to Mesolithic time, roughly 3500–2500 B.C.[2] No doubt the birds

---

[1] Breeds on the Murman coast.

[2] Aeldre Stenalder or older Stone Age (Winge, 1903, 1910, 1912; Winge, *in* Schiøler). Ten localities (see map).

occurred on those coasts at a considerably earlier time, as the finds at Otterö Island in Sweden attest. The failure to find bones in earlier sites is to be explained in two ways. Between the approximate dates 7000–5000 B.C., the North Sea rose, gradually inundating the land from an original coast extending from Jutland via the Dogger Bank to the coast of Yorkshire (Hawkes, 1940). Thus the majority of sites where bones might have been found are now under water.

In the second place, a new people, late-comers, who lived nearer the sea shore and caught more sea birds, began to appear in Denmark sometime after 4500 B.C.[1]

It does appear that great auks were relatively common, though less so than in Norway. They were not abundant. Bones of at least five individuals have been found at Meijlgaard and seven at Klintesø. At these places no more than nine individuals of any other one species were found.[2]

It must be said, however, that at Sölager the numbers of individuals of several other species [3] were much higher than those of great auks, of which only one or two were found. This is the only kitchen midden where other species outnumber A. impennis to such an extent. Of the 32 Danish Mesolithic localities where bird bones were found, ten contained great auks.

Swedish finds of a later horizon are much smaller, and it is possible that all were strays. It is, of course, also possible that more bones will be found, but at present no localities in Denmark are known to be more recent than Mesolithic. There are two in Sweden.

The times of the five finds on Swedish coasts embrace the whole history of the birds in Scandinavia. The oldest contained remains of birds that died not much later than 7000 B.C.[4] This find, together with the two at the Soten Canal,[5] reveal only about a dozen individuals. In view of the fact that the early people of the Maglemose culture caught but few sea birds, this might indicate that the birds were not at all common at this

---

[1] Hawkes (1940), Clark (1936), Zeuner (1946). This was called the Ertebolle culture. The Cattegat flooded, and the Littorina Sea was at a maximum height by 5000 B.C. At this time the somewhat damper Atlantic period begins, and a forest of alder and oak mixed with hazel supersedes the older birch-pine complex.

[2] Winge (1903). The geographical position of these two places indicates nothing. Only three species: Cygnus musicus (8), Larus argentatus (9), and L. marinus (5) were found in numbers greater than two or three individuals.

[3] Somateria mollissima (68 coracoids), Anas platyrhynchos (at least 29 coracoids), Mergus serrator (29 coracoids). Winge (1903) so states when only fragments of bones were present.

[4] Wiman and Hessland (1942) describe this as a marine Quaternary deposit, or shell bed, and say that it dates from before the transition from Ancylus to Littorina times. It is on Otterö Island in the Skagerak.

[5] Jägerskiöld (1933, 1934, 1938); Wiman and Hessland (1942).

BIRD SPECIES FOUND IN MIDDENS

Species	Norway				Denmark			
	No. sites				No. sites			
	3 Aeldre	12 Yngre	8[a] Jern.	Present	31 Aeldre	11 Yngre	12 Jern.	Present
Fulmarus glacialis	—	—	x	B	—	—	—	AV
Sula bassana	x	x	—	—	—	x	x	AV
Pelecanus crispus	—	—	—	—	x	—	—	—
Phalacrocorax carbo[b]	x	x	x	B	x	x	x	R
P. aristotelis	x	x	x	R	—	—	—	AV
Cygnus olor	—	—	—	AV	—	—	—	B (R)
C. cygnus	x	x	x	B	x	x	x	WV (C)
C. bewickii	—	—	—	WV	x	x	—	WV (R)
Anser anser[b]	x	x	x	B	x	x	x	B
A. fabalis	x	x	—	B	—	—	—	M
A. brachyrhynchus	—	—	—	WV	x	—	—	WV
Anas acuta[b]	—	—	—	B	—	?	—	B
Anas platyrhynchos[b]	x	x	x	B	x	x	x	B
A. crecca	x	x	—	B	x	—	—	B
Tadorna tadorna	—	—	—	B	x	—	x	B
Mareca penelope	—	—	—	B	x	—	—	B
Spatula clypeata	—	—	—	B	x	—	—	B
Nyroca fuligula[b]	—	—	—	B	x	—	—	B (R)
N. marila	—	—	—	B	x	—	—	B
Bucephala clangula[b]	—	x	—	B	x	—	—	WV (C)
Clangula hyemalis[b]	—	—	x	B	—	—	—	WV
Somateria mollissima	x	x	x	B	x	x	x	B (R)
S. spectabilis	—	—	x	WV	—	—	—	WV (R)
Oidemia nigra	x	—	—	B	x	x	—	WV (C)
O. fusca	x	—	—	B	x	—	—	WV (C)
Mergus merganser[b]	—	—	x	B	x	—	—	B (R)
M. serrator	x	x	x	B	x	—	x	B
Larus canus	x	x	—	B	x	—	—	B (C)
L. argentatus	x	x	x	B	x	—	x	B (C)
L. marinus[b]	x	x	x	B	x	—	—	B
L. ridibundus[b]	—	—	—	AV	x	—	—	B (C)
Rissa tridactyla	—	x	x	B	—	—	—	WV (R)
Sterna paradisea	—	x	—	B	—	—	—	B
S. hirundo	—	—	—	B	—	x	—	B
Plautus alle	—	x	—	WV	—	—	—	WV
Alca impennis	x	x	x	—	x	—	—	—
A. torda	x	x	x	B[d]	x	—	—	WV
Uria aalge	x	x	x	B	x	—	—	WV (C)
U. lomvia[c]	x	—	—	B[d]	—	—	—	WV (R)
Cepphus grylle	—	x	x	B	—	—	—	B (R)
Fratercula arctica	x	x	x	B	—	—	—	WV

[a] Five sites are listed as "Yngre Stenalder og Jernalder" (Schaaning).
[b] Aeldste in Denmark.
[c] Aeldste in Norway.
[d] Breeds on Murman coast, Kola Peninsula.

AV—Accidental visitor. B—Breeds; (C) common, (R) rare. WV—Winter visitor. R—Resident (? occasional breeder).

time. Certainly if as large a colony as Funk Island's were imbedded in the corallina and marine shells 7,000 years ago, we should have expected to have made larger finds. A kitchen midden of the Mesolithic at Rotekärrslid (Henrici, 1935) revealed but two great auks together with one or two individuals of seven other species. Only a single gairfowl was found in the early Iron Age site near Greby.[1] There is no record of remains found with the egg in the parish of Reng mentioned above.

A selected list of ducks, geese, and sea birds taken in Scandinavian kitchen middens is given on page 288. The bones are of about the same size and durability as those of great auks

Authorities date the discoveries of bones in Denmark and Norway as follows:

Authority	Winge	Degerbøl	Movius	Hawkes, Clark
Aeldste Stenalder = Early Mesolithic	6000–5000 B.C.	Azilian— A. Tardenosian	8000–5000 B.C. Maglemosian	—
Aeldre Stenalder = Mesolithic	3500–2500 B.C.	Campignian	5000–2500 B.C Ertebolle	5000–2500 B.C.
Yngre Stenalder = Neolithic	2500–1500 B.C.	Robenhausian	2500–1500 B.C. Neolithic	2500–1500 B.C.
Aeldre Jernalder	400–1000 A.D.	—	—	—

Although it is possible that an occasional bird may have strayed to Scandinavian coasts in recent times, the records have not been taken seriously by naturalists there (Steenstrup, in Grieve, 1885, p. 24). The oldest is Sondmore, near Aalesund, at lat. 62° 30′ N.[2] The most northerly is Wardo (Vardö) at the mouth of Varanger Fiord near the Russian border.[3]

There are three sight records from the Skagerak and Cattegat. One in Norway near Frederikstadt in 1837, and two off the coast of Sweden, near Marstrand and Tistlarna. They have not been believed (Steenstrup, in Grieve, 1885, pp. 23, 24).

Very much the oldest finds of great auk bones are those made on the island of Jersey in the English Channel and at Gibraltar. These are probably between 70,000 and 90,000 years old. In a fissure called La Cotte de St. Brelade on Jersey the upper end of a right humerus was found.[4] In the

---

[1] Stolpe (1879). Also called Grebbstadt.

[2] Strom, H., 1762, Physical and Economical Description of Sondmore, Soroe (Denmark). Reference given by Grieve, 1885, p. 122, who doubts the validity of the record.

[3] See Grote (1914), who quotes Nilsson S., Skand. Fauna, Foglarna, Bd. 2, 1858, p. 562.

[4] Andrews, 1920. Apparently in the same locality were found woolly rhino, reindeer, cave bear. The culture is Mousterian, according to Movius (in verb.).

cave in a rock called "Devil's Tower," at Gibraltar, were found fragments of two humeri of great auks, together with bones of the alpine chough (*Pyrrhocorax graculus*), indicating a colder climate than presently prevails.[1]

On an islet called Er Yoh, off the coast of Brittany, a few bones have been found. These birds were probably killed about the same time that Neolithic man was eating great auks in Scandinavia.[2]

The most southerly locality in Europe is near the "toe of the boot" of Italy, south of lat. 40° N. in Apulia. The deposit is probably not much less than 60,000 years old.[3] The place is called Grotto Romanelli. The climate at that time is thought to have been colder than at present.

Habitat: Bred on rocky islands off the coasts of the North Atlantic Ocean. Usually to be found on soundings on or near the continental shelves. It was particularly common on the Grand Bank of Newfoundland in the later years preceding its extirpation.

Description: Largest of the auks, total length about 30 inches. Flightless, its wings shorter than those of the razorbill auk. Its sharply contrasting black and white plumage, high and compressed bill, and the white spot before the eye should have distinguished it easily from the loon, or great northern diver, with which it is said to have been confused. *Summer* (both sexes): Black above and white below. A large white oval spot before the eye. Sides of head, chin, and throat dark brown (these feathers have white bases). Wings dark brown, except for white tips of secondaries. Tail black. Bill and feet black. Eye chestnut (J. Fleming, 1824). *Winter* [4]: Like summer plumage except that chin and throat are white, not dark brown. There is no oval white spot before the eye, but a gray line runs through the eye to a point just below the ear, and there a broad band of white above the eye. The figure shows only four grooves

---

[1] Bate (1928) says the presence of these birds does not necessarily indicate a colder climate. It is not clear in which stratum the great auks' bones were found, but fragments of the skull of a Neanderthal boy were discovered in the cave (Buxton, 1928). The culture is Mousterian.

[2] Péquart (1926) says, "We think the utensils [found in the kitchen midden] belong to the time of the Dolmens."

[3] Blanc (1927, p. 79) indicates that the bones were found in layers C and D of the Terra Bruna, which are from one to three metres below the surface. Zeuner (1946, p. 225) says artifacts found were associated with upper Aurignacian culture, "and perhaps can be classified as a variety of the Gravettian of Garrod (1938), or the Grimaldian (Vaufrey)." Clark, 1948, p. 118, fig. 3, considers that drawings on the walls of El Pendo, a cave in northern Spain (of probably upper Paleolithic) are great auks, but I cannot so identify them from available drawings.

[4] No specimen examined. See Kjaerbolling, N., 1865, tab. lii, fig. 2; and Cat. Birds Brit. Mus., 26, 1898, p. 564. Only specimen in Copenhagen.

on the upper and three on the lower mandible, whereas there are 8 to 10 of these in mature specimens in summer plumage.

"Anser magellanicus seu Pinguini" of Olaus Wormius, the figure of which shows a great auk with a white ring around the neck, is to be explained by the fact that ornithologists of the day were confused as to the true identity of the great auk. Willughby (1678, p. 322, fig. 65) reproduced Wormius's figure with the white collar, even though he divided his account into two parts: "The Bird called Penguin by our Seamen which seems to be Hoiers Gorfugl" and "The Penguin of the Hollanders or Magellanic Goose of Clusius" (perhaps the jackass or Peruvian penguin, which has a white ring around the throat). In spite of the fact that he says he saw two specimens of the former, he is clearly confused. Edwards (1751, p. 147) points out the error and figures the great auk accurately without the collar (see also Mullens, 1921). Should anyone give a name to Wormius's figure, as has been done with travelers' accounts of parrots and other birds, he should suffer for it in this world and certainly will in the next.

Habits: Little has been recorded of life history and habits. The breeding season probably began in May and ended in mid-July. Martin (1698) records that they arrived at St. Kilda in May, upon reports of the inhabitants. Icelanders visited outlying islands to collect eggs late in June, according to traditions collected by Newton (1861). The last specimen ever taken was said to have been incubating on June 3. The incubation period has been estimated at 44 days.[1] A single egg was laid. Whether the birds laid a second is not known. Martin (1698) says, "Every Fowl (on St. Kilda) lays an egg three different times except the Gair-Fowl and the Fulmar, which lay but once." [2]

Location of specimens: I know of specimens in Aarau, Switzerland; Amsterdam; Andover, Mass.; Ann Arbor, Mich.; Autun, France; Berkeley, Calif.; Berlin; Bologna; Brussels; Buffalo; Cambridge, England; Cambridge, Mass.; Chicago; Graz; Lausanne; Leyden; Leipzig; Lisbon; London; Lund, Sweden; Metz, France; Munich; Nantes, France; Neuchâtel, Switzerland; New York; Newcastle-upon-Tyne; Oldenberg; Oslo; Paris; Philadelphia; Poitiers; Prague; San Francisco; Stockholm; Strasbourg; Stuttgart; Tokyo; Uppsala; and Rome.

---

[1] Worth, 1940, p. 50, on the basis of dimensions of egg, niche, habitat, and other factors.
[2] Macgillivray (1852) says that the fulmar lays a second egg if the first is removed. Fisher (1952, p. 129) says that this is not a proved fact.

Mauritius Blue Pigeon

Alectroenas nitidissima (Scopoli)

> *Columba nitidissima* SCOPOLI, Del. Flor. et Faun. Insubr., fasc. 2, 1786, p. 93 (Île de France=Mauritius). Plates in Sonnerat; Milne-Edwards and Oustalet, 1893; Rothschild, 1907.

Common names: Pigeon hollandais; colombe hérissée (Milne-Edwards, 1893).

Status: Extinct. Last specimen taken in 1826. (Milne-Edwards and Oustalet, 1893.) Extinct since 1830 (Meinertzhagen, 1912).

Only three specimens remain in museums. They are in the Museum of Science and Art in Edinburgh, in the Muséum National d'Histoire Naturelle in Paris, and in Port Louis, Mauritius. The last named was the last preserved. It was taken by E. Geoffroy in the Savane Forest in 1826. Sonnerat says that there was a bounty on "vermin" in 1775 and that even imported birds were shot. The Edinburgh specimen was probably purchased from the estate of Desjardins, who had traveled to Mauritius and was a dealer in natural-history specimens. The Paris specimen was collected during Sonnerat's voyage in 1774. It is in poor condition, owing to an attempt to fumigate cases in Paris with sulphuric-acid fumes (Milne-Edwards and Oustalet, 1893).

The group of obvious relatives called *Alectroenas* is to be found on widely separated islands in the western Indian Ocean: Madagascar, the Comoros, Aldabara, Seychelles, formerly Mauritius, and possibly long ago Rodriguez. Since they are more closely related to known forms of Oceania and Asia than to African birds, it is probable that they represent survivors of an ancient population movement, rather than remnants of a fauna of an older and once larger continent of Africa.

It is surprising that no evidence of the occurrence of the bird on Réunion Island has been discovered (Berlioz, 1946), but not without precedence. A sternum and other bones found on Rodriguez have been thought to be a form of this genus and have been called *Alectroenas ? rodericana* Milne-Edwards (Salvadori, 1893; Shelley, 1883; Rothschild, 1907).

Range: Formerly Mauritius, Indian Ocean.

Habitat: Desjardins, the only eyewitness whose impressions come to us, says that the birds lived near river banks. His other statements are open to question, however. In Madagascar their close relatives are birds of the tree tops in the forest. "It liked to perch on some tall dead stub in the forest, or on the edge of a clearing" (Rand, 1936). In the Seychelles Islands it is always to be found near *Ficus* trees when in fruit (Vesey-Fitzgerald, 1940).

Description: A medium-sized forest pigeon (total length about a foot). The long white feathers of head and neck contrasting with the red eye and blue back distinguish it.

Head, neck, and mantle white, the feathers extremely long and pointed. Bare skin around the eye and on forehead bright red. Breast and back dark blue. Upper tail coverts dark red, and tail dark red marked with black on outer webs of outer tail feathers and inner webs of certain inner tail feathers.

Fig. 40.—Mauritius blue pigeon (*Alectroenas nitidissima*).

Young birds of the Madagascar form are more greenish blue (Rand, 1936).

Alectroenas represents what was once a polytypic species. The island forms differ from each other by degree only, although these differences are striking. It is questionable as to whether, with the considerable passage of time since their separation, they have not developed characteristics which would prevent their breeding, were they brought together. It is perhaps better, therefore, to regard it as a superspecies. The relationships of the group to other genera are obscure, and it is not possible to abandon the generic concept of *Alectroenas*.

Habits: The only statement regarding the birds in life comes from Desjardins, whose observations may be questioned, according to Milne-

Edwards and Oustalet (1893). They reproduce a part of his manuscript, the translation of which follows: "This bird lives alone near the river banks. . . . It eats fruit and fresh-water mollusks."

Quite properly, these authors doubt that the bird ate mollusks. They say (p. 234): "The habits and food of *Alectroenas nitidissima* were probably the same as those of Funingus (*Alectroenas* of the Seychelles and Madagascar). Probably these pigeons formed large flocks; their food consisted of fruit, berries and seeds. Without doubt they nested in trees and their eggs were white. Finally, one may assume that the young, like those of Funingus, were to be distinguished from adults by the black shades being less well-marked and the feathers of the mantle much less decomposed."

Rand (1936) found that the *Alectroenas* of Madagascar bred during almost the entire year, from July to March. Its food was principally the small fruit of forest trees, and several stomachs contained nothing else. The birds were often found in flocks of 3 to 12 or more.

On the Seychelles Islands, where they are abundant, the habits of the birds are recorded by Vesey-Fitzgerald and Betts (1940), as follows:

Invariably to be found wherever a Ficus tree of native or introduced species is in fruit.

A habit of this bird is to take a short upward flight which is followed by a downward glide, the wings being depressed.This performance is repeated several times as the bird describes a wide circle in the air.

The nest is situated in a fork amongst the smaller branches of a tree. It is composed of fine twigs which are placed together to form a simple platform . . . nesting activity has been observed during October and November. A nest examined 19 November, 1938, contained one new laid egg. The egg was pure white and the two ends were quite symmetrical, the long axis was 40 mm. and the diameter across the widest part 26 mm.

Enemies: It is doubtful that this bird had any enemies other than man, except for introduced monkeys. Why it was extirpated, whereas *Nesoenas mayeri* remains, is explained by the theory that the latter was considered to be poisonous.[1] *Alectroenas* was a much more striking bird and therefore more likely to be killed.

Sonnerat bears witness to the killing of great numbers of birds in 1775. Even the thrushes, imported to kill locusts, were shot (3, p. 215). Rats were a great pest at that time, and it was told him that they were the reason that the Hollanders had abandoned the island in 1710, after 112 years of occupation.

Location of specimens: I know of specimens in Edinburgh; Paris; and Port Louis, Mauritius.

---

[1] Milne-Edwards and Oustalet (1893) quote the voyage of Bernardin de St. Pierre to this effect, with the remark that possibly *Nesoenas* ate something that made the flesh poisonous.

New Zealand Pigeon (*Hemiphaga*)

Hemiphaga novaeseelandiae spadicea (Latham)

> *Columba spadicea* LATHAM, Ind. Orn., suppl., 801, p. lx (Norfolk Island). Fig.: Mathews, 1928, pl. 1.

Status: Extinct.

Range: Norfolk Island, about 550 miles north-northwest of New Zealand.

Of the three subspecies of this polytypic species, one is extinct while two survive.

The form *Hemiphaga novaeseelandiae novaeseelandiae* (Gmelin) still lives in the original forest of North and South Islands, Little and Great Barrier Islands, Hen and Chicken, Mayor, Kapiti, and Stewart Islands, but in somewhat reduced numbers. The form does not appear to be in immediate danger.

The birds are recently reported to be not uncommon in forested hills above Tauranga, North Island. "Once or twice during the last ten years single birds have come down into the lowlands—even once into Tauranga itself—seeking puriri berries, etc., and stayed periods of a few months" (Hodgkins, 1949).

On Chatham Island, *H. n. chathamensis* (Rothschild) was reported to be quite common both among the Karaka groves of the northern sandhills and in the denser bush of the southern plateau by Archey and Lindsay (1924). It may be that here, as well as in New Zealand, they are adapting themselves to men and new environment, for they are reported to be seen quite often near houses. Residents believe that a noticeable increase in population is due to the Animals Protection Act.

That these two forms still exist is by far the more extraordinary fact. They are brightly colored and without fear of man; they are gregarious, flocking about trees in fruit and pools of water, and they lay only one egg in a clutch. Given this combination, together with the fact that they were much valued for food and ornament and heavily trapped by the Maoris, shot latterly by Europeans, and preyed upon by imported weasels and cats, their survival is astonishing. If they continue to exist, it will be only in places where the original forests have been preserved, and it is probable that reduction in the area of native forest is the primary reason for a considerable diminution of the populations of birds since the first settlements of Europeans. There is a clue to the reason for their survival and also a hope for the future in that they often been found nesting near human habitation (M'Lean, 1911; Buller, 1888, 1905), and they have, at least occasionally, fed upon introduced plants such as *Brassica napus,* or rape. If the off-season shooting can properly be controlled they may long survive because of their ability to adapt themselves. This is no doubt the most important requisite for continued existence of individuals and species.

By a curious coincidence many escaped death by the fact that their favorite food during at least a part of the breeding season made their flesh disgusting. But this was the fruit of the kahikatea (*Podocarpus dacrydioides*), a yew known to New Zealanders as "white pine," and because this is a workable, useful wood it had disappeared in parts of its range. The bitter, young roots of kohwai (*Sophora*) are said to make the birds' flesh bitter.

Other plants, the fruit of which they have fed upon, have gone as farming and grazing areas became larger and the forests were cut down and burned away to make room for these. The animals themselves, cattle, pigs, goats, and rabbits, help to destroy the native forests (Thomson, 1922).

On North Island, New Zealand, staple food is recorded by M'Lean (1911) as follows: Wineberry (*Aristotelia* sp.) and tawa (*Beilschmiedia tawa*) during January and February; in April, May, and early June, tawa and miro (*Podocarpus ferrugineus*)[1] as well as kahikatea (*Podocarpus dacrydioides*)[1]; in July the favorite food was the fruit of the hinau (*Elaeocarpus dentatus*) and karamu (*Caprosoma robusta* or *foetidissima*), supplemented by jack vine (*Rhipogonum scandens*) and porokaiwhiri, or pigeon wood (*Hedycarya arborea*); from August through October food is less apt to be plentiful. The leaves of rope vine (*Parsonsia capsularis*) and wineberry and young shoots of kowhai (*Sophora tetraptera*)[1] were eaten in these months, and in November and December *Fuschia* and *Hedycarya* and mahoe (*Melictus ramiflora*).

The nest is placed in the fork of a tree (recorded by M'Lean in the mahoe=*Melictus ramiflora*) and sometimes in low bushes. It is a loose collection of twigs. Only one pure white egg is laid. The period of incubation is not recorded.

Location of specimens: I know of specimens of *H. n. spadicea* in Bremen; Cambridge, Mass.; Frankfort; Leyden; Lisbon; Liverpool;

---

[1] Buller and Oliver record that the berries of the yew or "white pine" (*Podocarpus*) made the flesh of the pigeon taste of turpentine, but M'Lean says it was not unpalatable. These species came into fruit at 4-year intervals and were preferred where found. Because of extensive cutting, the trees have become rare in many localities. M'Lean records the pigeons as being poor and unpalatable in September and October when feeding on roots of kohwai.

The breeding season is recorded as November to April; its extreme length, and probable irregularity, may be due to the time of the ripening of certain fruits in particular localities. Studies of this problem would be of great interest (see Chapin, 1932, p. 301).

London; Milan; Naples; New York; Philadelphia; Rouen; Vienna; Wiesbaden; and Lyons; and of *H. n. chathamensis* in Cambridge, Mass.; Edinburgh; and New York.

Lord Howe Pigeon

Columba vitiensis godmanae (Mathews)

> *Raperia godmanae* MATHEWS, Austral Avian Rec., **3**, 1915, p. 24 (Lord Howe Island). Fig.: Emu, **40**, 1940, pl. 3, p. 2.

Range: Lord Howe Island, 300 miles east of Australia.

Status: Extinct. Known only by a painting and accounts of eighteenth-century travelers. The painting by Midshipman George Raper, of H.M.S. *Sirius,* about the year 1790 is now in the British Museum (Natural History). A second painting, by an unknown, is in the Alexander Turnbull Library in Wellington, New Zealand, and is considered to be a copy by Hindwood (1940), *q.v.* This author quotes Thomas Gilbert,[1] who visited the island in 1788: "On entering the woods I was surprised to see large fat pigeons . . . sitting on low bushes and so insensible to fear, as to be knocked down with little trouble." By the year 1853 the population had apparently been extirpated, for the naturalist MacGillivray was not able to find one.

Because of the picture and the accounts of these witnesses, it cannot be doubted that there was once a population closely resembling pigeons of New Guinea, the Solomons, the New Hebrides, Loyalties, New Caledonia, the Moluccas, and other East India Islands. How this form differed, if at all, from these subspecies cannot now be determined, for no specimens were preserved.

Puerto Rican Blue Pigeon

Columba inornata wetmorei Peters

> *Columba inornata wetmorei* PETERS, Check-List Birds World, **3**, 1937, p. 66— new name for *C. i. exsul* Ridgway, 1915 (Puerto Rico), not *C. i. exsul* Hartert, 1903.

Common names: Blue pigeon; plain pigeon; paloma ceniza; paloma sabanera; paloma boba; paloma de potrero; torcaza cenicienta; torcaza boba.

Status: Rediscovered by *N. Leopold (1963, p. 119) in numbers. Specimen taken. Last recorded in the hills near Añasco in November, 1926 (Danforth, 1931): last collected in 1912 (Worthington; specimen in

---

[1] Voyage from New South Wales to Canton in the year 1788. London, 1789.

Carnegie Museum, Pittsburgh)." It was reported as rare in 1927 by Wetmore and apparently had been a small population for many years previously.

No doubt men with guns are the principal enemies. This factor, together with the complete destruction of natural habitat, is probably the reason for the disappearance of the birds from Puerto Rico.

Range: Puerto Rico Island, West Indies. Last recorded from the hills near Añasco. It had been obtained previously near Lares in the vicinity of Cueva Pajita and at Utuado.

Other subspecies occur on Cuba and on the Isle of Pines, where it is rare and local, and on Hispaniola, where it is common in certain places (*inornata*). On Jamaica (*exigua*) it is also rare and local (Bond, 1950).

Habitat: There is no exact information about the Puerto Rican subspecies. On Cuba it was found nesting in Zapata Swamp, but usually to be found in arid savanna country of plains and scattered trees (Gundlach, 1874a). On Jamaica nests have been found in high trees near clearings or glades (March, 1863) but are also to be seen near "mangrove woods" (Gosse and Hill, 1847). "Frequents rather open country" (Bond, 1947).

Description: A large pigeon (total length 14–16 inches). Its size and the lack of conspicuous white or pale gray tips to the tail feathers should distinguish it.

Head, hindneck, and breast pinkish lavender (vinaceous). Chin gray. Upper back pale brown, tinged with gray. Rump and upper tail coverts gray, the latter darker. Flanks and under tail feathers gray. Wings grayish brown, a patch on the wing pinkish lavender and the coverts edged with white. The bill is black and the feet and legs red. The eye has two rings, the inner gray-blue and the outer orange.

"Female is duller than male and averages smaller."

"Immature is grayish brown becoming much paler on posterior underparts, wing coverts and chest with indistinct paler margins" (Bond, 1936).

Habits: Nothing is recorded of the habits of the Puerto Rican race and but little of others.

On Jamaica, Gosse records seeing two or three chasing each other around the tops of trees; "their movements were announced by a guttural *jug, jug* . . ." This took place in February and he "conjectured to be a symptom of pairing." He was told of a nest in April.

On Cuba it nests during May, for Gundlach (1856, 1874a) found a nest in that month on a horizontal branch of a tree in Zapata Swamp. It contained two white eggs that measured "0,038½ + 0,025½ mm." On Jamaica the nest is described as a platform, more massive than that of other native pigeons, often placed on some lofty tree in the vicinity of clearings and in open glades and hillsides. The egg is described as larger than that of *C. leucocephala* (March, 1863).

Gosse records the usual note to be like the barking of a cur, "bow-wow-wow" (*sic*).

Food is various fruits, berries, and grain, and the birds are to be found near the supply available, as is so with all pigeons.

"In January and February, in the early morning, it is seen in small companies of six or eight or as single pairs, passing from the hills to the fields of ripening Guinea corn, and again returning in the evening to the hills" (March, 1863). In Cuba Gundlach noted that they were fond of the berry of the "Ateye tree" (*Cordia callococca*).

The birds have earned the name *torcaza boba* (foolish) because of their lack of wariness. Gundlach syas, "If one shoots a pigeon sitting from a tree, the unwounded birds fly to the next tree and one can fire again and thus bag many."

Remarks: No record comes to us of the numbers of birds on Puerto Rico at any given time. Of course, overshooting and destruction of natural habitat furnish a ready answer to the question: Why are they rare? But March records the birds as more wary than the ringtail (*C. caribaea*), which is still common in the wilder parts of Jamaica. If its lack of shyness is the reason for its rarity on Jamaica, we may ask: Why is not *caribaea* rarer? It has been considered to be as good or better food than *inornata*.

Even though there may be another reason, there can be no doubt that *inornata* was once common over its range; Gundlach reported it so in 1874. Its extinction on Puerto Rico came concurrently with the almost total destruction of forests.

Nesoenas of the Mascarene Islands

Only because the genus *Nesoenas* has been from time to time thought to be extinct should it be mentioned here. *Nesoenas mayeri* ("Marchal," *in* Prévost and Knip) is still quite common on Mauritius, and it is not thought to be in any great danger. M. Georges Antelme, of Port Louis, writes that they are called locally "pigeon des mares," that they are found commonly enough, although localized, in the forests near Savan and Rivière Noire. He adds a curious detail: "It is easy to catch, at a certain season of the year, by putting a noose about its neck when it is intoxicated, having eaten seeds of a bush known commonly by the name tandramana."

This is probably *Terminalia arjuta,* introduced from India and known there as tandra. It is an intoxicant for men as well as pigeons.

There may once have been a form of this species on Réunion (Bourbon). It is known only from a description by Dubois, although it was formally described as a new species by Rothschild, who named it *Nesoenas duboisi.* Berlioz (1946) has presented the matter in unbiased terms which appear to be the final words possible on the subject. A translation follows:

"Wild Pigeons."
Identification very uncertain.

The two kinds of pigeons cited by Du Bois (perhaps, by the way, two sexes or two plumages of the same species) remain an enigma. It is without doubt those that are described as having the color of the plumage slate-gray, which Coquerel identifies as *Columba Schimperi* Bp., quite an imaginative identification, for this African bird appears never to have existed in the Malgache region.

Rothschild, on the other hand, shows himself better inspired when he considers those described as red tending toward orange (rouge roussâtre) as allied to, if not identical with, those wild pigeons that still exist in very small numbers on Mauritius, where they represent quite a distinct species: *Nesoenas mayeri* (Prév. and Knip). Because of the apparent relationship but also upon the underlying principle of the insular differentiation of faunas, Rothschild has named the reddish-orange pigeon of Réunion: *Nesoenas duboisi.*

Dove of Tanna

Gallicolumba ferruginea (Wagler)

Columba ferruginea WAGLER,[1] Isis von Oken, **22**, 1829, col. 738, *ex* Forster *ms.* (Island of Tanna.)

Status: Extinct. Known only from J. R. Forster's (1844, p. 265) descriptions; G. Forster's drawing, no. 142, in the British Museum. These fit no known species of pigeon, and it is necessary to assume that a population has been extirpated.

Some doubt has been cast on the provenance of the bird (see Peters's Check-List of Birds, **3**, 1937, p. 136, note), and of course it is possible that Forster was in error.

To the contrary, it must be said that although there were many mistakes in labels of specimens in those days, in this case G. R. Forster not only wrote the locality "Tanna, ♀, 17th August 1774" on the margin of his drawing, but also his father recorded a new species of dove as having been shot on Tanna, in his narrative, and therefore the record cannot simply be dismissed as an error. Stresemann (1950) points out that in the German edition of "Observations made during a voyage round the

---

[1] Wagler copies his description from Forster's Ms. See also Stresemann, 1950, p. 84, who has examined the Forster painting in the British Museum and is certain that it is a *Gallicolumba.*

world . . . ," Forster wrote under the date 17.8., 1774: "We came into a wood . . . where a dove of a new species was shot . . ." This note is contained neither in the English nor Italian editions available.

Range: Tanna Island, New Hebrides Islands, Western Pacific.

Description: A rather small pigeon (said by Forster to be only slightly larger than *Columba leucophrys=erythroptera*). The head was rusty brown (ferruginous), the back dark reddish purple, wing dark green, primaries brownish gray with very narrow pale edges, breast rusty brown (ferruginous), belly and crissum gray, bill black (with a somewhat swollen cere), eyes yellowish, feet red.

The yellowish eyes and dark green (not violaceous) wing coverts distinguish it from *Gallicolumba sanctaecrucis* Mayr.

Japanese Wood Pigeons

These large dark pigeons (*Columba janthina*) are distributed from Hondo, on the main islands of Japan, southward through the Seven Islands of Izu, the Bonin Islands (*nitens*), and the Riu Kius (Nansei Shoto). A related species (*C. vitiensis*) replaces them southward through the Philippines and the Solomon Islands to Samoa and Borneo. A distinct species (*versicolor*) once lived in the Bonin Islands. It is extinct. A third (*C. jouyi*), confined to the Okinawa Archipelago, has become very rare.

It seems probable that destruction of forests will put an end to all these small island populations. None of them are large, even on such islands as Luzon.

Bonin Wood Pigeon

Columba versicolor Kittlitz

> *Columba versicolor* KITTLITZ, Kupfertaf. Naturg. Vog., **1**, 1832, p. 5, pl. 5, fig. 2 (Bonin Islands: Peel Island or Chichi Jima).

Status: Extinct. Last specimen recorded as taken in 1889 on Nakondo Shima in the Parry Group (Seebohm, 1890).

Range: Recorded from the Bonin Islands only on Nakondo (or Nakodo) Shima in the Parry Group (Seebohm, 1891) and Peel Island (Chichi), where naturalists of H.M.S. *Blossom* found it in 1827 and Kittlitz in 1828.

Description: Differed from other wood pigeons of the group in being generally paler with metallic golden-purple back.

Top of the head metallic green, shading to purple and green on the hind neck, golden purple (not green) on the back; rump green and wing coverts with paler, more golden reflections. These colors change with the angle of the light. Light gray below.

Location of specimens: I know of specimens of *Columba versicolor* in Frankfurt, Leningrad, and London.

Fig. 41.—Bonin wood pigeon (*Columba versicolor*).

Riu Kiu Wood Pigeon

Columba jouyi (Stejneger)

> *Janthoenas jouyi* STEJNEGER, Amer. Nat., **21**, 1887, p. 583 ("Northern part of Liu Kiu," restricted to Okinawa by Stejneger).

Status: Rare. Possibly extinct on Okinawa; none were found there by

American collectors in 1945 (R. H. Baker, 1948). Probably still exists in small numbers on neighboring islets.

Range: Riu Kiu (Nansei Shoto) Islands: Recorded from Sheya Jima and Ijina (Izena) Shima, north of Okinawa; Yagaji Shima (or Yagachi) in the bight of northern Okinawa; Zamami Shima in the Kerama group southwest of Okinawa, and formerly on Okinawa. In the Borodino Islands, both on Minami and Kita Daito Jima (Kuroda, 1925).

Habitat: Heavy forest.

Description: A large black pigeon (length about 18 inches). The size along with the pale gray stripe across the shoulder distinguishes it.

Entirely slaty black with purple metallic reflections on the head and green on the lower back. A pale gray stripe across the upper back.

Madagascar Turtle Doves

Streptopelia picturata (Temminck)

> *Columba picturata* TEMMINCK, Pig. et Gall., **1**, 1813, pp. 315, 480 (Île de France =Mauritius).

The turtle doves of the Madagascar region are as a rule capable of occupying many types of habitat. They are to be found in deserts and mountain forests to 2,000 feet in Madagascar and almost all available islands in the western part of the Indian Ocean. They are quite able to maintain themselves in considerable numbers on the large islands, but they have disappeared from Rodriguez (Berlioz, 1946) and perhaps from the small island of Glorioso off the northwestern tip of Madagascar. There is no recent information from that island, however.[1]

Madagascar itself is apparently the center of distribution for these birds.

S. p. picturata has a pale gray head unlike all the other subspecies. The back and sides of the neck are violet, and the feathers of this region have central black spots. The upper back and wing coverts are purple, the lower back dark gray, merging in some specimens to green on the rump. The upper breast is violet shading on the belly to white. Wings and tail are brown.

They nest in low bushes in or near the forest and are often to be seen on the ground (Rand, 1936).

For many years it has been assumed that the birds were imported by man into the Mauritius and Réunion from Madagascar, perhaps because of the similarity of the two populations, as balanced against the very different forms on other islands. But this has been doubted because of

---

[1] Nicoll, 1908. Several men searched Glorioso in 1906 for a day, but did not find the birds.

the discovery of bones on Rodriguez (Berlioz, 1946, on the authority of Milne-Edwards). It seems possible, however, that the birds were extirpated and reintroduced later.

Somewhat darker populations, all with brown heads, have been described from Glorioso (*coppingeri*), Anjouan in the Comoros (*comorensis*), Aldabra (*aldabrana*), Assumption (*assumptionis*), Amirante Islands (*saturata*), and the Seychelles (*rostrata*).[1] The last is a very dark brown bird at once distinguishable from the rest. Birds have been reported to have been introduced from Madagascar into the Seychelles also.

Location of specimens: I know of specimens of *S. p. coppingeri* in London and in Washington.

Passenger Pigeon

Ectopistes migratorius (Linnaeus)

> *Columba migratoria* LINNAEUS, Syst. Nat., ed. 12, 1766, p. 285 ("in America septentrionali" = South Carolina).[2]

Common names: Passenger pigeon; wild pigeon; tourte (eastern Canada).

Status: Extinct. Last individual died in captivity in the Cincinnati (Ohio) Zoological Garden on September 1, 1914. Last wild specimen recorded certainly was taken at Babcock, Wis., between September 9 and 15, 1899.[3] Possible a few were to be found later. There are many sight records between 1900 and 1907. A reward of $1,500 was offered in 1909–1911 for information concerning a nesting pair. All claims were based on error, most of them proving to be mourning doves (*Zenaidura*).

Range: Before permanent settlement by man the breeding range included the deciduous forests of North America, from the Great Plains on the west extending eastward to the coast of the Atlantic Ocean in New Brunswick and Nova Scotia. The northern extremity was southern Canada in the provinces of Manitoba, Ontario, and Quebec; the southern, the Appalachian Mountains in northern Virginia southwestward to northern Mississippi.[4]

On the periphery of the range, where the forests straggle along river

---

[1] Vesey-Fitzgerald (1940) reports the two forms on Mahé. It would be extremely interesting to know the extent of hybridization and consequent morphological characters.

[2] It is desirable to retain the name *migratorius* in the spirit of the recommendations of the Zoological Congress of 1953. For an excellent complete account of the species, Schorger (1955) should be consulted.

[3] Schorger, 1938a. The specimen was in the possession of Emerson Hough.

[4] Wilson (1832) says he found abandoned nests in Choctaw country at lat. 32° N.

valleys, the birds did not form the vast breeding colonies so characteristic of them in more heavily forested areas.

At any given point in the range the more or less regular seasonal appearance of birds was marked. However, their appearance in great numbers was not regular. In 1688 they were seen in great flocks in the northeast only irregularly. ". . . They came in a flock thro' New England, New York and Virginia. Nothing of the like ever happened since, nor did I ever see past 10 in a flock I remember" (Clayton, *in* Wright, 1910).

The birds sometimes passed the winter as far north as Indiana, Pennsylvania, and Massachusetts, but usually farther south, in the region bounded by the Atlantic Ocean and the Gulf of Mexico.

Stragglers have been recorded in the extreme northwestern portion of North America (British Columbia, Washington) and eastward, singly and very occasionally, in northern Canada (Mackenzie District, Saskatchewan, northern Manitoba, and in the northeast at Baffin Bay). In winter such wanderers have been found on the islands of Bermuda, Cuba, in the Valley of Mexico (Sánchez and Villada, 1873), and in Europe.

Abundance and decline: That flocks of migrating passenger pigeons "darkened the sky," that the "weight of their numbers broke great branches from trees" in tracts of uncut forest of hundreds of square miles during the first half of the nineteenth century cannot be doubted.[1] However, even before Europeans came to harry them, their migrations were irregular; they did not necessarily appear every year in such numbers in any given place, either on migration or for nesting. Furthermore, such very large, crowded nesting areas were not to be found in every year (Schorger, *in* Scott, 1947). Nests were in twos, threes, and small "colonies" scattered throughout the birds' range. But in a year of plenty, Wilson (about 1810) estimated 2,230,272,000 birds in a flock, and Schorger calculated that there were probably as many as 136,000,000 birds in a concentrated Wisconsin nesting area, comprising 850 square miles, as late as 1871. But including the birds that nested in isolated pairs along the northern and western periphery of the range, which cannot even be estimated, and other smaller colonial nestings, it may be that half again as many should be added to reach a conservative total. If such a population existed in 1871, it is quite probable that Wilson's estimate is not completely inaccurate.[2] By 1870 birds had ceased to breed in large concentrated groups except in the northern States surrounding the Great Lakes.

---

[1] Alexander Wilson (1808–1814) and Kalm (1911) are good original accounts. Wright (1910, 1911, 1913) gives transcripts of many, and Forbush (1912) and Bent (1932) give bibliographies for others.

[2] Audubon probably never actually saw a great nesting. He divided Wilson's estimate by two.

FIG. 42.—Passenger pigeon (*Ectopistes migratorius*).

The following table illustrates the gradual decline in numbers during the years 1851–1899, when the last unquestionable record of a wild bird was made.

Region	Last large "colonial" nesting	Last nest	Last record
Northeastern States	1851	1880	1894
Mid-Atlantic States	1868	1889	1898
Central States	1855	1893	1897
Northern "Lake States"	1885	1894	1899

Although American Indians no doubt were a strong natural check upon the numbers of pigeons, for they lit fires in breeding areas and killed many birds, it is not probable that any large proportion of bird populations were extirpated in this way. It is known that great numbers occurred in the northeastern States bordering Atlantic coasts in certain years of the eighteenth and early nineteenth centuries. In 1851 the last nesting of considerable size was found near Lunenburg, Mass. This area was only about 5 acres in extent, not comparable in size to great nestings of that time farther west. Perhaps birds nested in New England in pairs and small parties until about 1880.[1] After 1884 they were extremely rare, and the last bird surely to have been seen was taken in Maine in 1894. It may be that a few old birds wandered about after that date; there are records for Connecticut in 1906 and Maine in 1904, but their accuracy is questioned.

In the more heavily populated regions of coastal New York and Pennsylvania, the pigeons never appeared in the very large flocks of former years after about 1860. However, on the forested hills of northern Pennsylvania, and in bordering northeastern New York State, birds were to be seen until about the year 1890. Certainly there was a great gathering and perhaps many nested in Potter County, Pa., in 1886 (French, 1919). Isolated pairs were found nesting at Lake Ross and Fairmont in Luzerne County in 1889 (French, in Todd, 1940). As elsewhere, there are records of the early twentieth century, but they have not been generally accepted.

Westward, in the remaining original forests of Ohio, the birds were still abundant during the decade 1850–60, but the last great nesting took place near Kirkersville, in Licking County, in 1885.[2] After 1860 they appeared in large flocks only infrequently; the last of these was noted about 1875. In 1884 the last bird was seen.

[1] Batchelder (1882) reported scattered pairs breeding near the border of Maine in New Brunswick in 1877. Forbush (1912, 1929) and Palmer (1949) are good summaries.

[2] Wheaton (1882) and Trautman (1940) are good histories of the pigeon in Ohio.

Because the original deciduous forests were confined almos. to the
river valleys in the State of Illinois, it is not surprising that there were no
great colonial nestings there. But Indiana, to the eastward, was densely
covered with oak and hickory woods before settlement by Europeans, and
although many roosts are recorded until about 1875, there is little evidence
that they nested in great numbers (Butler, 1899; Wright, 1911). There
were few persons inclined to report such events as early as 1832, and
perhaps the lack of records is due to the lack of observers.[1] Certainly near
Shelbyville, Ky., just across the river from Indiana, Alexander Wilson
found an extremely large nesting flock in the year 1810.

Nesting records south and west of that place are vague and unsatisfactory. Indeed, were it not for the fact that Alexander Wilson recorded
old, empty nests in "the country of the Choctaws in latitude 32°", now
called northern Mississippi, and were it not for the fact that "roosts" have
been recorded [2] in Missouri, there might be some question as to whether
the birds had ever nested in the States of Missouri, Arkansas, or Mississippi. It is recorded that in 1893 the last shipment of birds was received
by N. W. Judy, a game dealer of St. Louis, and this from "Sylvan
Springs," Ark.[3] The last specimen was taken in Missouri in 1896.

Perhaps we shall never be sure what the status of the birds was in these
forests, but from the few existing records we can be sure that great flocks
were seen on migration until about 1880. The strong probability is that
passenger pigeons ceased to breed near the rivers soon after 1810, when
Wilson saw the great flock nesting at Shelbyville, on the Ohio River.

The pigeons nested longer in the northern "Lake States" than elsewhere. From here almost all observations came. In Minnesota the last
large nesting is thought to have taken place about 1877 in Wabasha
County, in the southeastern corner of the State, on the Mississippi River.
Certainly there was a very large "roost" near Rochester in the same area
in 1879. Scattered single nests were found until 1895, and the last recorded
eggs were discovered then in what is now the center of the city of Minneapolis (Roberts, 1932).

There was a "colonial" nesting in 1886. About 1,200 birds are said to
have been seen in a swamp near Lake City, northwestern Michigan. In
1881 a much larger one, 8 miles long and containing thousands, is recorded (Brewster, 1889).

---

[1] Record of a nesting of about 40 birds in 1898 has been questioned.

[2] Wright (1910) quotes Mease to the effect that there was a large roost near the present
site of Vicksburg, Miss., in 1805.

[3] I cannot find this place. It is perhaps Sylvana, in the central part of the State, near the
great swamps.

Small by standards of earlier years in Wisconsin, a nesting area of about forty acres was found in the year 1885. This was apparently a section characterized by jack-pine woods of northeastern Wisconsin (Langlade County), surrounded, within feeding distance, by forests of birch, beech, and maple. This is the last known flock-nesting. Three years before there had been two nestings within 50 miles of one another in Adams County, central Wisconsin. One of these was said to have covered a square mile and the other to have been even larger. As elsewhere there were unconfirmed reports of birds after 1900, but the last sure record is the specimen shot at Babcock in 1899.

Description: A large pigeon, about half again as large as the mourning dove, with which it was often (and still is) confused. Total length about 12 inches. Head bluish gray, not buffy brown as in the mourning dove.

Male: Entire head bluish gray, except for irregular black blotches about the eyes and chin which is paler in some specimens. Upper back with metallic-violet reflections, lower back brownish gray, rump and upper tail coverts like the head. Wing coverts brownish gray with some irregular black markings. Primaries and secondaries browner, the latter with prominent white edges. Two middle tail feathers dark gray, remainder white. Upper breast pale cinnamon-rufous, becoming paler and more vinous on lower breast, white on vent and under tail coverts. Bill black, feet red, and eyes orange. Wing 196–214, tail 173–211 mm.

Female like the male but smaller and less brightly colored; above browner, less gray, and below grayer, less brown. Wing 175–210, tail 141–194 mm.

In the first-year plumage the sexes were alike. The metallic reflections of the upper back were lacking. Feathers of head, upper back, wing coverts, and breast with white tips. Primaries tipped and margined with rufous. Feet and irides paler than adult.

Down of the fledgling was pale brown.

Habitat: In the center of distribution, the original hardwood forests of central North America, the beech-oak-maple association of trees was chosen. Where spruce and hemlocks and pines occurred they were often chosen as roosting places, but not for nesting. According to Wilson (1832), beech (*Fagus*) forests were preferred above all. Dillin (1911) records them in southeastern Pennsylvania, a region characterized by *Pinus rigida,* with oaks in patches, during the autumn of 1864. Behr (1912) found them nesting in northern Pennsylvania, where spruce, hemlock, maple, and beech predominated—this in the spring of 1870. E. S. Wilson (1934) records them in a similar area of Michigan, although he adds white oaks

(*Quercus alba*) and cedars (*Thuya occidentalis*) to the association of trees.

Habits: As a rule the birds arrived in the forests of the North from early March through April and in almost incredible numbers. These great flock-nestings and generally gregarious dispositions were the important habitual attributes that distinguished them from all other pigeons, indeed predisposed the species to destruction by men, and prevented them from reproducing after the large nestings had been destroyed.

Craig's (1911) studies of prenesting behavior of caged birds (*q. v.*) demonstrated only slight differences in courtship from that of other pigeons, but nevertheless differences that impeded and even (sometimes) prevented copulation with members of other species even though the birds were caged. In his opinion, Audubon's descriptions of behavior are too imaginative to be accurate. He found that during courtship males and females sidled up to one another in manner characteristic of most pigeons but that males did not bow and strut in the same manner, and the performance of billing was almost never seen.

The nest itself was a simple thing of sticks and twigs piled loosely on the branches, often so carelessly that the whole would fall in a breeze.

In captivity no more than one egg was laid, according to Craig and Whitman, but from the second-hand accounts of professional hunters it may be inferred that occasionally two could be found in a nest.

The period of incubation was 14 days, according to Deane (1896), who had his information from observation of captive birds. Both male and female incubated and fed the young as well.

The favorite food of adults is thought to have been beech and oak mast, but the seeds of maples, elms, mulberries, wintergreen (*Gaultheria*), partridgeberry (*Mitchella*), elderberry (*Sambucus*), Juneberry (*Amelanchier*), wild cherry, and choke cherry (*Prunus*), as well as rye, wheat, and buckwheat, have been recorded as staples. No doubt there were many other foods.

Location of specimens: I know of specimens of *Ectopistes migratorius* in the following places: In the *United States:* University (Alabama); Berkeley, Los Angeles, Oakland, San Francisco (California); Denver (Colorado); New Haven (Connecticut); Chicago, Springfield (Illinois); Davenport, Grinnell (Iowa); Lawrence (Kansas); Cambridge, Pittsfield, Worcester (Massachusetts); Ann Arbor, Bloomfield Hills (Michigan); Minneapolis, Northfield, St. Paul (Minnesota); Hastings (Nebraska); Newark, Princeton, Trenton (New Jersey); Albany, Alfred, Buffalo; Granville, Syracuse, New York (New York); Bismarck (Nort!. Dakota); Cleveland, Dayton, Delaware (Ohio); Providence (Rhode Island); Charleston (South Carolina); St. Johnsbury (Vermont); Seattle (Washington); Fond du Lac, Green Bay, Madison, Milwaukee, New London (Wisconsin); and Washington, D. C. In *Canada:* Winnipeg (Manitoba); St. John (New Brunswick); Kingston, Ottawa, Toronto (Ontario); and Quebec (Quebec).

In *Europe:* Vienna (Austria); Brussels (Belgium); Exeter, Huddersfield, London, Newcastle-upon-Tyne, Nottingham, Sheffield, Tring, Wakefield (England); Lyons, Nancy, Paris, Poitiers, Rouen (France); Hamburg, Leipzig, Stettin (Germany); Amsterdam, Leyden, Nijmegen (Holland); Genoa (Italy); Bergen, Oslo, Gotenburg (Norway); Edinburgh (Scotland); Lund, Malmo, Stockholm (Sweden); Basle, Geneva, Neuchâtel (Switzerland); and Cardiff (Wales). And in *Asia:* Tokyo (Japan).

Crested Pigeon of the Solomons

Microgoura meeki Rothschild

> *Microgoura Meeki* ROTHSCHILD, Bull. Brit. Orn. Club, **14**, 1904, p. 78 (Choiseul, Solomon Islands). Fig.: Novit. Zool., **11**, 1904, pl. xxi.

Status: Most probably extinct. Five experienced bird collectors of the Whitney South Sea Expedition of the American Museum (New York) searched for a total of two months in 1927 and 1929 without success. Pictures were shown to the Melanesian natives who vaguely intimated that the birds had not been seen for many years. Those who had any opinions thought that perhaps introduced domestic cats were responsible (Mayr *in verb.*).

Only six specimens have been preserved. All were collected by A. S. Meek, an Australian who, cruising among the islands on a small vessel, found time to collect many birds. In 1904 the inhabitants of Choiseul were dangerous. Meek would probably not have been able to penetrate far inland. Whether he obtained the birds from natives who had trapped them at higher altitudes is not known. Certainly the mountains were searched carefully, but unsuccessfully, by members of the Whitney Expedition.

Range: Confined to Choiseul, Solomon Islands.

Habitat and habits: Unknown.

Description: A large (total length about a foot) ground pigeon with a bluish-gray crest of hairlike feathers. Forehead and face black. Top of the head bluish gray like crest; back darker with a gray wash; lower back and rump brownish. Upper throat almost bare. Belly bright cinnamon-brown. Wings brown. Upper bill blue with a black tip; lower bill reddish; iris dark lemon; legs purplish red (Mayr).

Location of specimens: I know of specimens in London and New York.

Fig. 43.—Crested pigeon of the Solomons (*Microgoura meeki*).

Norfolk Island Parrot

Nestor meridionalis productus (Gould)

> *Plyctolophus productus* Gould, Proc. Zool. Soc. London, 1836, p. 19 (no type locality=Philip Island off Norfolk Island).
> *Nestor norfolcenis* Pelzeln is based on an aberration in the Liverpool Museum.

Status: Extinct. Last specimen died in a cage in London sometime after 1851. It is probable that the birds were extirpated from their original habitat even before this, but there are no records, except that they were to be found on Phillip Island after having been extirpated from Norfolk.

Man was most probably the sole enemy of these birds, and he no doubt extirpated the birds for food.

Range: Formerly Norfolk and Philip Islands and perhaps Lord Howe Island, western Pacific north of New Zealand.

Habitat: Said to have been a forest bird like the kaka of New Zealand.

Description: A large (total length about 16 inches) parrot like the kaka of New Zealand, but with a yellow breast.

Head and neck brownish gray; chin, cheeks, and lores varying from yellow to orange; hind neck greenish yellow; back like the head but browner and darker; lower back and rump dark red; wing coverts greenish; wings and tail brown; upper breast grayish brown; lower breast bright yellow; belly and thighs reddish orange.

The species was variable. The amount of red and the relative brightness of this and other colors may be suffused with brown and sometimes areas normally red are yellow. A specimen in the Liverpool Museum has a very long bill with a strongly grooved culmen. It is very probably an aberration, which occurs sometimes in small populations on islands.[1]

Indeed individual variation in the kaka (*Nestor meridionalis*) of New Zealand is so great that if it were possible to bring together all the widely separated specimens of the Norfolk Island form, it might be shown that the differences between the two are bridged by variants.

Habits: Gould (1841) reports that a Mrs. Anderson, who had a captive bird in her possession, said that the birds bred among trees and rocks and laid four eggs in a hollow tree. This is quite probable because the kaka is known to prefer hollow trees as nesting sites.

Location of specimens: I know of specimens in Amsterdam; Besançon, France; Berlin; Dresden; Florence; Leyden; Liverpool; London; New York; Melbourne; Philadelphia; Prague; Vienna; and Washington.

New Caledonian Lorikeet

Vini diadema (Verreaux and Des Murs)[2]

    Psitteuteles diadema VERREAUX AND DES MURS, Rev. Mag. Zool. (2) **12**, 1860, p. 390 (New Caledonia). Fig.: Oiseau Rev. Franç. Orn., **15**, 1945, p. 7.

Status: Unknown. Apparently difficult to find, for it has not been taken since the types were collected, although search has been made. Sarasin (1913) says it was reported to exist in the northern forests near Oubatche. MacMillan, collecting for the American Museum, New York, did not find it. According to Berlioz (1945) the last specimen was taken before 1860. It is in the National Museum in Paris.

Range: Presumably northern New Caledonia.

---

[1] *"Dysmorodrepanis"* of Lanai Island and *"Oreomyza perkinsi"* of Hawaii are other examples.

[2] Following Mayr (1945a). Has been placed by authors in the genus *Trichoglossus, Glossopsittacus,* and *Charmosyna.*

Habitat: Upper levels in heavy forest. Other species of the genus are to be found in flowering trees.

Description: A small green parrot (total length about 7 inches). The female only is known. It is generally green, paler on the forehead, lores, ear coverts, and sides of the neck; cheeks and throat yellowish; top of the head blue; below paler green; vent red; tail feathers generally green, the four outer ones red at the base and with a black band, yellowish on the end of the outer webs.

Macaws of the West Indies

THE RED MACAWS

Ara tricolor BECHSTEIN, in Latham, Allgem. Uebers Vög., **4**, Th. 1, 1811, p. 64, pl. 1 ("South America" *recte* Cuba).

Range: Cuba.
Based on specimens.

Ara guadeloupensis CLARK, Auk, **22**, 1905, p. 272 (Guadeloupe).

Range: Guadeloupe, Martinique, ? Dominica.
Based on credible evidence.

Ara gossei ROTHSCHILD, Bull. Brit. Orn. Club, **16**, 1905, p. 14 (Jamaica: mountains of Hanover Parish, about 10 miles east of Lucea).

Range: Jamaica.
Based on credible evidence.

YELLOW AND BLUE MACAW

? *Ara martinica* ROTHSCHILD, Bull. Brit. Orn. Club, **16**, 1905, p. 14 (Martinique).
Ara erythrura ROTHSCHILD (1907) is a synonym.

Range: Martinique.
Based on poor hearsay evidence.

GREEN AND YELLOW MACAWS

? *Ara erythrocephala* ROTHSCHILD, Bull. Brit. Orn. Club, **16**, 1905, p. 14 (Jamaica: mountains of Trelawney and St. Ann Parishes).

Range: Jamaica.
Based on poor hearsay evidence.

? *Ara atwoodi* CLARK, Auk, **25**, 1908, p. 310 (Dominica).

Range: Dominica.
Based on a poor description.

MYTHICAL MACAW

> ? *Anodorhynchus* ? *purpurascens* ROTHSCHILD, Bull. Brit. Orn. Club, **16**, 1905, p. 13 (Guadeloupe).

Range: ? Guadeloupe.
Based on poor evidence.

These large, gay, noisy birds have been recorded by more or less credible witnesses from seven islands of the West Indies: Martinique, Guadeloupe, Dominica, Jamaica, Hispaniola, St. Croix, and Cuba. They are no longer to be found. There can be no possible doubt that they were once to be found on Cuba and St. Croix; we have specimens from the former and bones from the latter.[1]

The records of macaws on West Indian islands began with Ferdinand Columbus, Casas, and Oviedo, and there can be little doubt that the birds were endemic on many islands. On which islands they occurred and how the forms differed from one another are questions that cannot be answered accurately. It is most probable that the red macaw was a native on several.

The native endemic Cuban species (*tricolor*), of which there are 15 specimens in existence, was clearly a representative of *Ara macao* of Mexico and South America. Evidence that there were also other forms of this superspecies on Jamaica, Hispaniola, and Guadeloupe is fair, but for other islands the record is unsatisfactory. The questions whether the contemporary historians Du Tertre, Labat, Browne, Robinson, Hill, and Atwood described captive birds brought from Mexico or whether captive birds had gone wild in West Indian forests, and to what extent second-hand information was distorted, can probably never be answered. Since 1900 seven "scientific" names have been given to supposed species based on sixteenth- and seventeenth-century descriptions, but since no specimens were used and the descriptions are inadequate, modern authors have listed them as "hearsay species."

Sadly, little information comes to us even of the Cuban macaw.

The Cuban Red Macaw

Common name: Guacamáyo.

Status: Extinct. Last specimen recorded to have been shot at La Vega, near the Ciénaga de Zapata on the south coast, in 1864 (Bangs and Zappey, 1905). Probably there were a few birds still living until about 1885, for Cory (1886) says, "Dr. Gundlach writes me that he believes it is still to be found in the swamps of southern Cuba." Gundlach in 1876 had re-

---

[1] *Ara autochthones* Wetmore, Journ. Agr. Univ. Puerto Rico, **21**, 1937, p. 12.

Fig. 44.—Cuban macaw (*Ara tricolor*).

corded it as very rare, with the remark that in 1849 it could easily be found.

In spite of the fact that Gundlach found the flesh evil-smelling and bad to taste, it is recorded that the Cubans killed them for food. They captured them, usually in the nest, to be bred as pets also. There is no other known reason for their extirpation.

Range: Cuba. Actually reported by good witness (Gundlach) from the region of the Ciénaga de Zapata and the Encenada de Cochinos on the south coast, but by tradition from all of Cuba (except the Province of Oriente) and the Isle of Pines. Small populations bred in scattered localities.

Habitat: The Zapata Swamp is a forested area. Gundlach reported that the birds nested in holes in palm trees, according to reports of native Cubans.

Description: A relatively small (total length about 20 inches) macaw, predominantly red and yellow. Sexes alike, as in most macaws.

Forehead red, yellowish on the crown and yellow on the hind neck; upper back red, feathers with green edges, wing coverts darker red; rump and upper tail coverts pale blue; wings purplish blue; tail blue above with red bases to feathers, red below; underparts red; under tail coverts blue.

Habits: Gundlach (1876) records that they nested in holes in palm trees and lived in pairs and families. Their diet was fruit and seeds (especially of *Melia azdarach,* a large flowering tree, and palms), sprouts, and buds. The egg has not been preserved or described.

Remarks: Possibly there were once similar forms on other islands. Casas, although his descriptions are not satisfactory, indicates that there was a macaw on the island of Hispaniola also at the end of the fifteenth century (Casas, 1876, p. 298; Armas, 1888). He says there were three kinds of parrots on the island, a large, a smaller, and a very small one, by which it may be inferred that he meant the macaw, the parrot (*Amazona*), and the small conure (*Aratinga chloroptera*). Probably he meant the red macaw, for he says, "the largest . . . differ from those of the other islands in that they have [the region] over the bill or the forehead white, not green or red; those of this species that are in the island of Cuba have [the region] over the bill or the forehead red (colorado)." The Cuban bird (*tricolor*) has a red forehead.

A macaw is reported later (Buffon, 1779, **6**, p. 183) on the authority of a local resident, M. Deshayes, of the Parish of Jérémie, on the south coast.[1]

---

[1] I propose that a special committee be appointed at the next International Ornithological Congress to excoriate the man who attempts to give this a "scientific" name.

Two hundred years ago red macaws were reported from Guadeloupe, Martinique, and Jamaica. Ferdinand Columbus (see Oviedo), Du Tertre (1654), and Labat (1742) described [1] them from the French Islands. These have been called *Ara guadeloupensis* Clark. From descriptions it appears that they resembled *Ara macao,* but were much smaller and the tail was all red. From the Cuban *tricolor* they differed in having yellow on the wing but no blue and having a shorter tail.

Ferdinand Columbus records that red parrots "as big as chickens," called "Guacamayos" by the Caribs, were seen on Guadeloupe in April 1496. Since this was the same name used by the Cubans for *tricolor* in 1840, it appears possible that an endemic species was once on Guadeloupe.[2] To be sure, the Caribs did tell Columbus of the mainland and its direction, and it is possible that parrots were imported by them, but the name is indicative, especially in view of the evidence offered by the French priests a hundred years later.

Furthermore, there is a painting by Roeland Savery in the British Museum of a group of birds, including a macaw, which answers this description exactly. The picture was painted early in the seventeenth century. Since we know that parrots were constantly imported into Europe from all parts of the world—Christopher Columbus used them in his triumphal processions in Spain—it is even possible that Savery painted one of these.

There is disagreement about whether the red macaw (*macao*) was endemic on Jamaica. Such a bird was described by a Dr. Robinson, who saw a stuffed specimen. It is reported by Gosse (1847) to have been shot about 1765 some 10 miles east of Lucea, a locality not for from Montego Bay. Robinson described it as having the forehead, crown, and back of the neck bright yellow; sides of face around eyes, anterior and lateral part of the neck, and back a fine scarlet; wing coverts and breast deep sanguine red; winglet and primaries an elegant light blue; basal half of the upper mandible black, apical half ash colored; lower mandible black; tail and feet were missing. From these descriptions it has been inferred that the bird differed from *tricolor* in having a yellow, not red, forehead. It has been named *Ara gossei* Rothschild.

On the other hand, Dr. Browne, also a careful man, said in 1756, "red Mackaw of Edwards [this is *Ara macao*] . . . is not a native of Jamaica but they are frequently brought from neighboring parts of the main . . ."

---

[1] See Clark (1905). The account of Rochefort, Histoire naturelle et morale des Isles Antilles etc., 1658, p. 154, appears to be rewritten from Du Tertre, as the latter claimed.

[2] He unfortunately records the occurrence of honey and wax as well; Las Casas says there was none to be found on the islands or the mainland. The evidence points to the occurrence of bees on the mainland in pre-Columbian time, but not in the Antilles.

Perhaps Browne was right, but, on the other hand, perhaps he did not know of wild endemic birds living in the mountains.

In addition to the red macaw two other identifiable species, or perhaps their island representatives, have been recorded. These are the yellow and blue macaw (*araurana*) and the green-backed *militaris*.

Yellow and Blue Macaws

From Dr. Browne's evidence it is not clear whether the yellow and blue macaw was endemic. He says:

> Psittacus. The blue Mackaw of Edwards. This beautiful bird is a native of Jamaica, tho' seldom cached there; most of those that are generally seen about gentlemen's houses, being introduced there from the main, where they are more common. I have seen one or two of these birds wild in the woods of St. Ann's, and yet keep some of the feathers of one that was killed then by me; but they are very rare in the island, and keep generally in the most unfrequented inland parts.

Evidence that may certainly be called "hearsay" came early in the nineteenth century from the Reverend Mr. Comard of Jamaica, who saw two "large Parrots flying overhead" and ascertained from residents of the neighborhood that they were blue and yellow (Gosse, 1847, p. 261). Of course, it may be that these were wild birds.[1]

The only evidence we have that a blue and yellow macaw (*araurana*) ever occurred on any island other than Jamaica is based on Bouton (1635),[2] who says, "Les Aras sont deux ou trois fois gros comme les Perroquets et ont un plumage bien different en couleur. Ceux que j'ai vus avaient les plumes leleucs et orangées." It is assumed that he saw it on Martinique. Quite likely Bouton described a captive bird. This is the basis for the name *Ara martinica* Rothschild. No other historian has described any macaw, other than the red, from this island. However, in the painting by Savery mentioned above, there is a blue and yellow macaw that differs from the South American bird in having yellow, not blue, under tail coverts. It is not possible to determine from what island the original model came, however.

Green and Yellow Macaws

There is a possibility that a third species was once to be found on Jamaica. This is *Ara militaris* or the "Great Green Macaw of Mexico," as

---

[1] Rothschild (1907, p. 53) surmises that these must have been *A. erythrura* Rothschild, which is based on Rochefort. Since Rochefort never went to Jamaica, and indeed seems to have taken his account from Du Tertre, "*erythrura*" is not credible. If any disposition of the name is necessary, it is a synonym of *martinica* Rothschild.

[2] According to Rothschild (1907). I have not been able to examine this book.

Mr. Hill, a resident of Jamaica in the early nineteenth century, called it. Browne (1756) does not mention this, and the only evidence we have is from Mr. Hill. It is not apparent that he ever saw one.[1] He said, "The head *is spoken* of as red; the neck, shoulders, and underparts of a light and lively green; the greater wing coverts and quills blue; and the tail scarlet and blue on the upper surface, with the under plumage both of wings and tail a mass of intense orange yellow."

Mr. Hill said they were to be found "in a mountain district, very remote between Trelawney and St. Anne's" and the country of the "Accompong Maroons," which is the southern part of the "Cockpit." He assumed that they bred in Mexico and were to be found in Jamaica only in winter. Rothschild (1905) named this almost mythical bird *Ara erythrocephala.*

On the authority of one Thomas Atwood, who published a general account of the island of Dominica in 1791 (see Clark, 1908), the name *Ara atwoodi* Clark has been given to a supposed macaw. Atwood said, "The Mackaw is of the parrot kind, but larger than the common parrot, and makes a more disagreeable noise. They are in great plenty, as are also parrots on the island; have both of them a delightful green and yellow plumage, with a scarlet coloured fleshy substance from the ears to the root of the bill, of which colour is likewise the chief feathers of the wings and tail . . ." This is like no known Macaw or Parrot. The phrase "have both of them" indicates a factor of error in the description.

A Mythical Macaw

"Anodorhynchus (?) *purpurascens"* Rothschild is said to have inhabited Guadeloupe also. The description, according to Rothschild, is of a bird which was entirely deep violet and was called Oné Couli by the natives. It is attributed by Rothschild to "Don de Navaret, Rel. Voy. Christ. Colombe (or Rel. Quat. Voy. Christ.), 11, p. 425 (1838)."[2] Possibly this is a careless description of *Amazona violacea* Gmelin or possibly an *Ara leari* Bonaparte (=*Macrocerus hyacinthinus* Lear) imported by natives from South America.

Location of specimens: I know of specimens of *Ara tricolor* in Berlin; Cambridge, Mass.; Dresden; Havana; Leyden; Liverpool; London; New York; Paris; Stockholm; and Washington.

---

[1] Gosse, 1847, pp. 261–262. The implication in Rothschild (1907, p. 53) is that he did. Gosse is here misquoted.

[2] I have not been able to check this reference. Perhaps it is to Martín Fernández Navarrete, "Coleción de los viages y descubrimientos que hicerion por mar los Españoles," Madrid, 1825–1837. Armas (1888) did not find it in Casas's or Oviedo's accounts.

West Indian Conures

The small, green, sharp-tailed conures (*Aratinga*) have survived well compared to the macaws. Of the five forms recorded, three are still common on Hispaniola, Cuba, and Jamaica (Bond, 1950). A subspecies of the bird of Hispaniola has been exterminated on Puerto Rico quite recently.

Aratinga chloroptera maugei (Souancé)

Psittacara maugei SOUANCÉ, Rev. et Mag. Zool. (2), **8**, 1856, p. 59 (no type locality, but probably Puerto Rico).

Status: Extinct.
Range: Confined to Puerto Rico.

? Aratinga labati (Rothschild)

Conurus labati ROTHSCHILD, Bull. Brit. Orn. Club, **16**, 1905, p. 13 (Guadeloupe Island).

Status: Most probably there were conures on Guadeloupe 200 years ago, but the only evidence of them is an inadequate description of that time.
Range: Supposedly confined to Guadeloupe Island, West Indies.

It is possible that a very small number of the Puerto Rican bird still remain on Mona Island near Puerto Rico, but Bond (1950) thinks this form is probably extinct. The last specimen was taken on Mona Island by W. W. Brown, a collector, in 1892. The birds had disappeared from Puerto Rico at an earlier date, perhaps as early as 1860. Gundlach (1874b) had reports of it from the forests of the eastern part of the island at that time. According to Stahl (1887) it was known in 1883 only by reports of the parents of the oldest inhabitants.[1] Whether this population differed from that of Hispaniola is doubtful. Only three specimens exist,[2] and only one of these has been compared to *chloroptera* of Hispaniola.

According to Ridgway (1916) and Wetmore (1927) this specimen differs from *chloroptera* in having a slightly smaller and darker bill, in having lighter red under primary coverts, and in having the lesser primary coverts entirely green.

The bird of Hispaniola is green with the forehead sometimes narrowly marked with red. Bend of the wing and under wing coverts red. Under surface of wings and tail paler red with a yellowish tinge. Red feathers are often scattered through the body plumage. The sexes are alike. Birds of the year have the under wing coverts green.

---

[1] Wetmore (1927) believes later reports to be inaccurate.
[2] Paris, Leyden, Chicago.

Destruction of the forests is primarily responsible for the extinction of this population. It is possible that too many were captured for export as pets, but no direct evidence of this is available.

The best description of the bird of Guadeloupe is that of Labat, who said: "Those of Guadeloupe are about the size of a blackbird, entirely green, except a few small red feathers, which they have on their head." On this basis Rothschild postulated a distinct species, *Conurus labati*, although no specimens exist. But that such a bird once lived there is the more probable because Du Tertre, who also describes three sizes of parrots which he calls aras, perroquet, and perrique, says: "Those which we call Perriques are the small Perroquets, all green and big as magpies." [1]

Carolina Parakeet

Conuropsis carolinensis carolinensis (Linnaeus)

Psittacus carolinensis LINNAEUS, Syst. Nat., ed. 10, **1**, 1758, p. 97 (Carolina and Virginia).

Conuropsis carolinensis ludoviciana (Gmelin)

Psittacus ludovicianus GMELIN, Syst. Nat., **1**, 1788, p. 347 (Louisiana).

Two easily differentiated subspecies of the parakeet were once to be found in North America east of the Great Plains. A third form, known only from a humerus, is *C. fraterculus* Wetmore, Amer. Mus. Novit., no. 211, 1926, p. 3 [Snake Creek Quarry (Miocene), Nebraska].

Status: Extinct. Last bird died in the Cincinnati Zoological Garden in September 1914. Last specimen collected in the wild on Padget Creek, Brevard County (east coast), Fla., by. Dr. E. A. Mearns on April 18, 1901. An unquestionable sight record is by Chapman in April 1904; he saw two flocks and a total of 13 birds at Taylor's Creek on the northeastern side of Lake Okeechobee, Fla. There are sightings recorded as late as 1920, when a flock of about 30 was reported near Fort Drum Creek by a resident named Henry Redding.[2] Rumors of the existence of the bird persisted in Florida as late as 1938, but in spite of careful investigation by several agencies the birds could not be found.[3] However, Sprunt (1949) maintains stoutly that the birds inhabited the Santee Swamp near Charleston, S. C., in 1938. This region has since been destroyed by a power project.

---

[1] Bond points out that great care should be exercised in assuming that perroquet and perrique mean parrot today, for the inhabitants of the West Indies use these names for other birds. There can be no doubt that Labat and Du Tertre meant parrots, however.

[2] McCormick, 1915. Late sightings are less credible because of the possibility of confusion with imported foreign parrots. A specimen of *Aratinga h. holochlora* (Sclater), a form of eastern Mexico and Guatemala, was collected from a band west of Palm Beach in 1924 (Barbour, 1925).

[3] Auk, **56**, 1939, p. 215; **57**, 1940, p. 134.

A lone bird is described as "unaccountably straying" into the Courtney Bottoms of the Missouri River near Kansas City in 1912. It was observed for several days there (Harris, 1919).

During a period of about 90 years the range of the birds gradually contracted from east to west, toward the Mississippi River. Dates of their extirpation coincide well with the increase of human populations and destruction of forests in those regions. Proceeding down the Ohio River from the Pennsylvania border the dates for final sightings [1] are 1832 and 1856 in Ohio and Indiana; 1851 and 1878 in Kentucky. This is the last recorded sighting east of the Mississippi River, except for those in the Florida wilderness.[2]

Last records west of the Mississippi River range from 1857 on the Upper Missouri River to 1881 in Louisiana and 1891 in the less settled parts of southern Missouri (Stone County). It is probable that small numbers of birds existed in well-forested places far removed from human beings until later dates than are indicated. Sightings of single birds are reported in Missouri and Kansas as late as 1905 and 1912, and there are vague reports from Louisiana in 1910.[3]

Man appears to have been the direct cause of the extirpation of these birds. Too many were shot and too many taken captive. A comparison with the slower decrease in numbers of ivorybill woodpeckers shows that the habitat and distribution of the two species were almost identical. A hypothesis that the destruction of habitat was responsible is therefore untenable. Although the parakeets, except perhaps a few individuals, disappeared from the Mississippi Valley between 1808 on the upper reaches of the river and 1880 on the lower, timber cutting in the swamps of that region did not reach its zenith of destruction until about 1880 in the north and 1900 near the river in the south. A few such regions still exist in an untouched state and harbored woodpeckers.

Nor is it possible to assume that disease was an important factor, for the slow decline in numbers of birds coincides so well with an increase in human population. Epizootic disease would without doubt have caused a more rapid diminution in numbers. The habitat, so inaccessible in the great trees in trackless swamps, makes it unlikely that introduced animals

---

[1] Ridgway, 1916, pp. 147–149. These dates are almost all obtained at second-hand from local residents. Although this evidence has the quality of hearsay, there can be little doubt that they are accurate to within a few years, and none as to the general trend of destruction.

[2] The record (1879) for Coosada, Alabama (in Ridgway, 1916) is an error. See Brown, Bull. Nuttall Orn. Club, 4, 1879, p. 11.

[3] Widmann (1907); Oberholser (1938). The specimen said to have been shot in Kansas in 1904 was not preserved. McIlhenny (1943) says they were not seen at Avery Island, La., after 1881.

could have troubled the birds. However, the swamps were, of course, not impenetrable to practiced woodsmen, who captured numbers of birds in rookeries inside hollow trees. This, together with the habit of destroying fruit and consequent destruction by farmers, was probably primarily responsible for the extinction of the species. In a region of Louisiana little disturbed by man, the last were seen in 1881. "These birds . . . were eleven in number. My uncle . . took me with him and shot three of the birds" (McIlhenny, 1943) Had every man with a gun been so moderate the bird might have survived, but Wilson, Audubon, and other writers of the time describe the killing of the majority of the flocks. Because the living returned to hover and scream over the dead, this was easy. Year after year, as the guns and trappers increased, the birds decreased, until the death rate overtopped the birth rate over the entire range. Possibly the populations were not large. Perhaps there were less obvious reasons for the birds' disappearance, but they are not known and now never will be.

Range: Regularly and commonly found in the deciduous forests of North America, before the coming of Europeans, from southern Virginia (lat. 35° N.)[1] on the Atlantic coast, south through Florida and westward to the edge of the river-valley forests in eastern Texas, Oklahoma, Kansas, and Nebraska to about lat. 40° N. Birds sometimes wandered north as far as Maryland, eastern Pennsylvania, New York (Albany), and the Great Lakes in the northeast, Wisconsin, and Colorado (Denver) in the northwest. It is probable that the axis of the Appalachian Mountains through West Virginia, Kentucky, Tennessee, and Alabama was the area of intergradation between the eastern race (*carolinensis*) and the western (*ludoviciana*).

It is probable that the birds were usually to be found only in suitable localities over this wide range. Wilson (1808–14, **3**, p. 91) makes this plain: "For even in the states of Ohio, Kentucky, and the Mississippi territory [they were not found] unless in the neighborhood of such places as have been described [heavily forested river valleys]. The inhabitants of Lexington (Ky.), as many of them assured me, scarcely ever observe them in that quarter. In passing from that place to Nashville, a distance of two hundred miles, I neither heard nor saw any, but at a place called Madison's Lick. In passing on I next met with them on the banks and rich flats of the Tennessee River; after this I saw no more till I reached Bayo St. Pierre, a distance of several hundred miles."

Habitat: "They are partial to heavily timbered bottom lands bordering

[1] William Byrd in 1728 said they occurred only in winter here.

the larger streams and the extensive cypress swamps which are such a common feature of many of our southern states." [1]

". . . Low rich alluvial bottoms, along the borders of creeks, covered with a gigantic growth of sycamore trees or buttonwood—deep and almost

FIG. 45.—Carolina parakeet (*Conuropsis carolinensis carolinensis*).

impenetrable swamps, where the vast and towering cypress lift their still more majestic heads; and those singular salines, or, as they are usually called, *licks,* so generally interspersed over that country, and which are regularly and eagerly visited by Parakeets" (Wilson, 1808–14, **3**, p. 90).

---

[1] Bendire, 1895, p. 2, who saw the birds.

Description: A small parrot with a long pointed tail (total length about 12 inches); the only conure with a yellow and orange head. The immature bird, which has the forehead and region around the eyes orange but the rest of the head green, may be confused with other small parrots in Florida that have escaped from captivity. A Mexican conure (*Aratinga holochlora*), recorded from Florida, has no orange or yellow on head or wings.

Adult (sexes alike): Fore part of the head from bill to behind the eyes orange; hind part of the head and neck, the chin and throat yellow. Body plumage green, paler below. The flanks yellow. Wing generally green but with the outer edge red-orange, the coverts broadly tipped with yellow. Bases of primaries yellow. Tail green.

The western subspecies (*ludoviciana*) differed in being somewhat paler generally. The green of the upper back and rump more bluish and less yellowish than the back. Usually more yellow on wing coverts and primaries. Eyes light brown; bill yellow; feet gray.

For a few weeks of the first year the young differed from the adult in having the top, back, and sides of the head green, not yellow, and in being somewhat paler and more yellowish below and in having the edge of the wing green, not orange. The adult plumage was acquired gradually between summer and following March.

Habits: Accounts of the breeding habits in the wild are all at second hand. The majority of these, and the most probable, relate that two to five pure white eggs were laid in holes in large trees a considerable distance above the ground. Eggs have been taken in Florida on April 2 and 26 (Wilson, Audubon, Bendire).

Another version, collected by Brewster (1889) from local residents, is to the effect that the birds sometimes built nests of twigs in the forks of branches of small cypress trees in northern Florida. This has not been given credence. These eggs were reported as being greenish.

Both male and female fed the young in captivity, but they were careless and often were unsuccessful in rearing them (Childs, 1905b, 1906).

There can be no doubt that they roosted in hollow trees at night, since they were seen to do so in captivity. "They used their beaks for holding to the interior of the tree (Bent, 1940, p. 10). They are reported to have been found in large numbers in hollow trees.

That there was a period of hibernation does not seem probable, for the birds were seen constantly during the winter months. A flock was seen in a snowstorm by Alexander Wilson on the Mississippi, and they have been reported at the northern extremity of their range in winter.

In general their food was seeds. All observers speak of their fondness for thistles or cockleburs (*Xanthium* spp. in the Mississippi Valley and *Circeum lecontei* in Florida), which grow abundantly in marginal and waste lands. The seeds of several trees were eaten, particularly cypress (*Taxodium*), maple (*Acer*) and elm (*Ulmus*), which grow in damp places or near rivers. An examination of stomach contents of a single bird revealed "two rabbit hairs, two bits of the bird's own feathers, and two fragments of an indeterminable ant, which formed the only traces, (and) . . . the remains of no fewer than thirty-two seeds of loblolly pine (*Pinus taeda*)" (Cottam and Knappen, 1939).

They were said to be notoriously fond of the fruit and seeds of cultivated plants as well and were capable of destroying orchards and are said to have done so in a "wanton and mischievous manner". Wilson wrote in 1810, "I have known a flock of these birds to alight on an apple tree, and have myself seen them twist off the fruit, one by one, strewing it in every direction around the tree, without observing that any of the depredators descended to pick them up. To a Parakeet which I wounded and kept for some considerable time, I very often offered apples, which it uniformly rejected but burs, or beech nuts never." In Florida they cut off young green oranges and peaches before the fruit was formed (Scott, 1889). In feeding they used either the right or left foot, but were not usually ambidextrous.

As with other parrots, flight was rapid and undulating and voice was loud and harsh. And like most other parrots and many birds, they would return to hover over the body of a fallen bird and make it possible for shooters to kill many of the flock.

Location of specimens: I know of specimens of *Conuropsis c. carolinensis* in Albany, N.Y.; Amsterdam; Ann Arbor; Berkeley, Calif.; Berlin; Bloomfield Hills, Mich.; Brussels; Buffalo; Cambridge, Mass.; Charleston, S. C.; Chicago; Cleveland; Davenport, Iowa; Denver, Colo.; Edinburgh, Scotland; Exeter, England; Galesburg, Ill.; Geneva; Hamburg; Jupiter, Fla.; Lawrence, Kans.; Leyden; London; Los Angeles; Milwaukee, Wis.; Minneapolis, Minn.; New Haven, Conn.; New York; Oslo; Palo Alto, Calif.; Paris; Philadelphia; Pittsburgh; Princeton, N. J.; Providence, R. I.; Quebec; Raleigh, N. C.; Rouen, France; San Francisco; St. Johnsbury, Vt.; Santa Barbara, Calif.; Springfield, Ill.; Springfield, Mass.; Stockholm; Syracuse, N. Y.; Tokyo; Toronto; Tring, England; University, Ala.; and Worcester, Mass.

Amazonas of the West Indies

Extinct forms are:

Amazona violacea (Gmelin)

> *Psittacus violaceus* GMELIN, Syst. Nat., **1**, pt. 1, 1788, p. 337 ("Insulae aquarum Lupiarum").

Range: Guadeloupe.
Status: Extinct since early in the eighteenth century. Known by good descriptions of naturalists and travelers.

? Amazona martinicana Clark

> ? *Amazona martinicana* CLARK, Auk, **22**, 1905, p. 343 (Martinique).

Range: Martinique.
Status: Extinct since early in the eighteenth century. How this population may have differed from that of Guadeloupe is not clear.

Amazona vittata graciliceps Ridgway

> *Amazona vittata graciliceps* RIDGWAY, Proc. Biol. Soc. Washington, **28**, 1915, p. 106 (Culebra Island).

Status: Probably extinct.
Range: Culebra Island.

Amazona vittata vittata (Boddaert)

> *Psittacus vittatus* BODDAERT, Tabl. Pl. Enl., 1783, p. 49 ("Santo Domingo" in error: Puerto Rico).

Status: Rare and local.
Range: Puerto Rico Island, West Indies.

The astonishing Amazonas of the Lesser Antilles represent a very old group and a relict population, the relationships of which are now obscure. They are very large and strikingly colored.[1] The most surprising thing about them is that such specialized birds, being on small islands and surrounded by enemies, should still exist. There are not many of them.

On three islands, Dominica, St. Vincent, and St. Lucia, there are four forms. The large green and purple "sisserou" of Dominica (*A. imperialis*)

---

[1] The tail with three bands, which seems to me to be indicative of relationship, is shared with these by the species *vinacea* of southern Brazil. An indication of a third band occurs also in certain specimens of *farinosa*. I agree with Clark (1905) that the exaggerated erectile crest does not prove relationship with *Deroptyus,* since all Amazonas have this characteristic to a greater or lesser degree.

is the most specialized of these. It is dark green above with a large black ruff about the neck, and the purple feathers of the underparts have metallic reflections. There are no color bands on the tail. This, with the extinct forms of Guadeloupe and Martinique, probably represented a superspecies. Unfortunately there is no description of the nesting or the juvenal.

Fig. 46.—West Indian Amazona (*Amazona*).

According to Bond (1950) this is a rare bird, by which he means that but few individuals would be found during a search of several days. Even though the terrain is rugged and very difficult in the mountain forests of Dominica, it is quite probable that this population is small and scattered now and that it never was large. It is also probable, unless the human population of Dominica grows much larger, and much of the habitat of the birds is destroyed by cutting the forests, that the birds are in no great danger. This eventuality is unlikely because of the steep terrain of the island.

A related species, or perhaps a subspecies, once lived on Guadeloupe, but presumably because there is more flat country suitable for farming, and therefore more people than on Dominica, the birds are extinct. No specimens exist; they are known from Du Tertre's description [1] and that of Labat.

These differ principally in that the head, neck, and underparts were violet somewhat mixed with green and black, according to Du Tertre, whereas Labat says these regions were gray. This difficulty cannot be resolved, but it is quite probable that Labat confused this form with that of Martinique, for he apparently does not distinguish between the birds of the two islands.[2] This is *Amazona violacea* (Gmelin).

Du Tertre described a large, a medium-sized, and a small parrot on Guadeloupe, and says of the medium-sized bird: "Le Perroquet . . . is about the size of a hen. It has the beak and eyes bordered with red, all the feathers of the head, neck and stomach are violet somewhat mixed with green and black and changing like the throat of a pigeon. All the top of the back is green strongly [tinged] brown. Three or four of the primaries 'maistresses plumes' of its wings are black, all the others are yellow, green and red. It has two beautiful rosettes 'roses' composed of the same colors on the [? wing coverts or ? bend of wing: 'sur les deux gros des aisles']. When it ruffles the feathers of its neck it makes as though there were a frilled collar around the head . . ."

He adds that they were eaten by the French settlers, and no doubt the imported Africans ate them too. Destruction of habitat and the undue toll taken by a meat-hungry people together were responsible for the extinction of these birds, as well as those of Martinique. There can be no doubt that there were such birds in Martinique during the seventeenth century and that they disappeared early in the eighteenth. The only description is that of Labat, and there is no indication as to how it differed from the form of Guadeloupe. It has been named *A. martinicana* by Clark (1905).

The second superspecies includes *Amazona versicolor* (Müller) of St. Lucia, *A. arausiaca* [3] of Dominica, and *A. guildingi* (Vigors) of St. Vincent. Except for *arausiaca,* which is fairly common, these are rather rare birds and according to Bond are to be found scattered throughout the mountain forests.

The smaller green Amazonas are all fairly common, except for *vittata*

---

[1] 1667, **2**, p. 250; 1724, **2**, p. 214.

[2] Rothschild, 1907, quotes Labat but not Du Tertre, and his figure is from Labat's description.

[3] Bond (1950) remarks that this form might better be regarded as conspecific with *versicolor.*

on Puerto Rico, which was reported still to exist in reduced numbers in the Luquillo National Forest, where birds are protected (Bond, 1950). However, they must still be considered to be in great danger, for they are inclined to wander, and it is difficult to see how they can be kept in any restricted area. The human population, furthermore, has reached a peak of numbers far above the point of subsistence of its agrarian economy. Should some event upset the flow of goods between North America and the island, this small population would be eaten within a few days.

Populations of *Amazona vittata* have disappeared from Vieques and Culebra, small islands lying near the eastern coast of Puerto Rico. Birds of Culebra have been named *Amazona vittata graciliceps* by Ridgway and described as smaller than Puerto Rican birds. None have been found since 1899, although Wetmore (1927) searched for them in 1912. Only three specimens are preserved; they are in the U.S. National Museum. None were ever collected on Vieques Island.

Seychelles Island Parrot

Psittacula eupatria wardi (Newton)

Palaeornis wardi Newton, Proc. Zool. Soc. London, 1867, p. 346 (Seychelles). Also Ibis, 1867, p. 341; 1876, p. 382.

Range: Mahé and Silhouette, Seychelles Islands.

Status: Extinct. In 1866 Newton was told that because of their taste for ripe maize the birds had been practically exterminated. He added destruction of the original forest as another cause. No doubt shooting and trapping were the primary causes, for Mahé Island rises almost perpendicularly from the sea to 2,000 feet. It would be surprising if forest had not remained on those steep slopes.

The birds were still present in 1870, when a few skins were sent to Cambridge University, but apparently they had disappeared before 1906, for Nicoll could get no word of them. He obtained specimens of the rare owl (*Otus insularis*) and the black parrot (*Coracopsis barklyi*). Vesey-Fitzgerald, who spent considerable time on the islands between 1931 and 1936, did not find "Cateau Vert." In spite of its height, Mahé is not a large island, being only 17 miles long by 5 in width; it is not probable that birds will be found there. No record of their having been seen on the neighboring islands of Praslin and Silhouette exists. Vesey-Fitzgerald found a small population of the black parrot on Praslin, but by that time the green one, if ever it did live there, was gone.

It resembled the Mascarene parrots closely, although it belonged to the

Asiatic group that lacks the rosy collar. It was about 16 inches in length, with a long, pointed tail. Generally green, it had a red speculum on the wing, and the male had an incomplete black collar. In both sexes the back of the neck and tail were pale blue.

Location of specimens: I know of specimens in Cambridge, England; Cambridge; Mass.; London; New York; and Paris.

Fig. 47.—Seychelles Island parrot (*Psittacula eupatria wardi*).

Cynaoramphus Parrots

The extinct forms are:

Cyanoramphus novaezelandiae subflavescens Salvadori

> *Cyanoramphus novaezelandiae subflavescens* Salvadori, Ann. Mag. Nat. Hist. (6), **7**, 1891, p. 68 (Lord Howe Island). Fig.: Cat. Birds Brit. Mus., **20**, 1891, fig. 18.

Status: Extinct.

Range: Lord Howe Island, about 500 miles east of Australia.

Cyanoramphus novaezelandiae erythrotis (Wagler)

> *Psittacus erythrotis* WAGLER, Abh. K. Bayern Akad. Wiss., Math. Phys. Kl. **1**, 1832, p. 426 (Macquarie Island).

Status: Extinct.

Range: Macquarie Island, about 600 miles south of New Zealand.

Cyanoramphus zealandicus (Latham)   [See frontispiece]

> *Psittacus zealandicus* LATHAM, Index Orn., **1**, 1790, p. 102 (Tahiti. "New Zealand" is an error). Type (?) in Liverpool. Fig.: Des Murs, Icon. Orn., livr. **3**, 1849, pl. 16 (*"Platycercus phaëton"*).
> *C. erythronotus* KUHL (1820) is a synonym.

Status: Extinct.

Range: Tahiti, Society Islands, south Pacific.

Cyanoramphus ulietanus (Gmelin)   [See frontispiece]

> *Psittacus ulietanus* GMELIN, Syst. Nat., **1**, pt. 1, 1788, p. 328 *ex* Latham (Ulietea = Raiatea). Type in Vienna. Fig.: Zool. Journ., **1**, 1825, suppl. pl. 3.

Status: Extinct.

Range: Raiatea, Society Islands, south Pacific.

On New Zealand and on most of the surrounding islands, from New Caledonia, 800 miles north, to Macquarie Island, 600 miles south, live four species of parakeets. One of these species (*novaezelandiae*) comprises nine subspecies, of which two are extinct. Both of these were confined to small islands.

Almost 2,000 miles northeastward, in the Society Islands, there were once two species, both related to those of New Zealand and both extinct since the beginning of the nineteenth century.

All are medium-sized parrots with long, pointed tails. Generally green in color with blue wings, they may have crimson or yellow and red heads (see key, p. 334), and spots of the same color on each side of the rump. All have concealed yellow or white bases to feathers of the nape.

In Tahitian species crimson is replaced by brownish red (or maroon), and there is a greater extent of this, the entire rump being brownish red or chestnut and the forehead black or brownish black. The green of the head, back, and underparts is somewhat less bright a green, described sometimes as bluish and sometimes grayish green.

The species *ulietanus* has the entire head dark brown, the back dark olive-green suffused with brownish, and the rump brownish red (maroon). The underparts are greenish yellow. The flight feathers and tail are brown, with a suggestion of a pale blue wash on the outer webs of the upper sides.[1]

The following key identifies the species:

1. Forehead colored like crown..................................2
   Forehead and crown not the same color.......................3
   Forehead and crown green; plumage entirely green............*unicolor*
2. Forehead and crown dark brown; rump dark reddish brown...*ulietanus*
   Forehead and crown crimson; rump spotted with crimson.......*novaezelandiae*
3. Forehead red or orange (crown yellow).......................4
   Forehead black (crown green)...............................*zealandicus*
4. Forehead orange; size smaller (wing 92–99).................*malherbi*
   Forehead red; size larger (wing 109–114)...................*auriceps*

THE NEW ZEALAND SPECIES (*Cyanoramphus novaezelandiae* and subspecies; *C. unicolor, C. auriceps,* and *C. malherbi*).

Cyanoramphus novaezelandiae and subspecies: On the North Island of New Zealand the birds, formerly abundant, are to be seen only in government-owned forests. They now occur only on reserves such as Little Barrier and Kapiti Islands, however, although they are still recorded often from the South Island.[2]

According to recent reports [3] the New Caledonian form (*Cyanoramphus novaezelandiae saisetti*) is not rare but is shy, retiring, and difficult to find. Scarcely ever are they shot, and unless great changes are brought about in the environment the birds are probably in no great danger.

C. n. subflavescens of Lord Howe Island has been extinct for at least 50 years. The last pair was seen in 1869 by a Mr. Hill. The islanders are said to have shot and trapped the birds because of the damage they did to crops (Hindwood, 1940).

Curiously enough, the birds still exist on Norfolk Island (*C. n. cookii*),[4] from which so many other species have been extirpated.

Although it was reported at one time that the birds had been extirpated from the Kermadec Islands by collectors, this was later proved to be un-

---

[1] I am much obliged to Dr. R. W. Sims, scientific officer of the British Museum, for this information.

[2] Classified summarized notes in Notornis, formerly New Zealand Bird Notes.

[3] Anon. in Oiseau, 16, 1946, pp. 168–170. Notes.

[4] See Bull. Brit. Orn. Club., 73, no. 9, 1953, p. 104.

R. W. Sims, scientific officer of the British Museum (Natural History), has compared the types of *Cyanoramphus magnirostris* Forbes and Robinson, 1901, and *Platycercus rayneri* G. R. Gray, 1862, with that of *Cyanoramphus cookii* G. R. Gray, 1859, and finds that they belong to one subspecies. I agree with Peters (Check-List 3, 1937, p. 270) that *Psittacus verticalis* Latham is unrecognizable, although the individual variation in this subspecies is very great. I am much obliged to Dr. Sims and Dr. Wagstaffe, of the Liverpool Museum, who lent the type of *magnirostris*.

true. This race (*C. novaezelandiae cyanurus*) is in no danger, nor can *C. novaezelandiae hochstetteri* on the Antipodes Islands be so classified (Falla, *in litt.*), unless more enemies are brought to kill them.

Probably none of the small-island populations were large at any time. As long ago as 1840 Dieffenbach reported *chathamensis* to be rare. Fleming in 1949 thought there had been forces, such as settlement of the land, that caused the birds to be localized in smaller areas, but he found them on most of the islands of the Chatham group.[1]

C. n. erythrotis has been extirpated from Macquarie Island, perhaps by house cats abandoned by men who came to kill penguins for their oil (Porter, 1934). This was just before the war of 1914–18 or perhaps earlier. The subspecies was larger and paler than that of New Zealand. As there are no trees on Macquarie, it nested under bunches of tussock grass on the ground, as it does sometimes on other islands.

C. auriceps: The Maoris apparently did not distinguish between the large red-fronted and a distinct species, the medium-sized yellow-fronted parakeets (*auriceps*). Both were called kakariki. According to report (Buller; Oliver), they are usually to be found together, and there is no hint in the literature that their habits differ. Both frequent higher levels in the forest trees, both nest in holes, and both feed upon buds, shoots and crops as parrots usually do. Apparently the smaller yellow-fronted bird has been less common in the past, for it is found on none of the outlying island groups other than the Chathams, where the subspecies *forbesi* lives.

C. malherbi: Even more uncommon is this, the smallest of these three species. Apparently the Maoris took no notice of it, for it is not recorded that they had a name for it. It may always have been rare. The birds are reported [2] as having been seen on South Island from time to time, but there have been no recent records for the small island sanctuaries.

None of these species appear to be in immediate danger of extinction.

C. unicolor: C. n. hochstetteri and *C. unicolor* live together on Antipodes Island, nesting on the ground. Although there are no longer any official seal fisheries or other contact with the islands, these small specialized populations would disappear within a few years should rats come ashore from a wreck or by other means. The relationships of *unicolor* are obscure. We may only guess that it was an earlier arrival than its neighbor. Both forms are said to be quite common.

TAHITIAN SPECIES (EXTINCT): *C. zealandicus* (Latham) disappeared from the island of Tahiti, Society Islands, about the middle of the nine-

---

[1] Porter (1934) has reported these birds to be extinct, but the facts, as in other instances, proved to be otherwise.
[2] New Zealand Bird Notes.

teenth century or perhaps later; the last-known specimen, now in Paris, was taken in 1844 by Lieutenant Marolles. Anthony Curtiss, formerly a resident of Tahiti, has told me that there is no longer any tradition of the occurrence of the green and red parrot on the island, except the name á-á. Members of the Whitney South Sea Expedition were unable to find the birds, in spite of careful search.

There are two specimens in Liverpool, one of which is thought to have come from the collection of Sir Joseph Banks and may then be Latham's type of *zealandicus* (Salvadori, 1907). There is also a specimen in the British Museum, but the origin of this as well as the second Liverpool specimen is uncertain, although they probably were collected by naturalists of Captain Cook's voyages. A drawing in the British Museum by Sydney Parkinson, who accompanied Cook on his first voyage, completes the evidence for the former existence of this species. It is plate 8, entitled "no. 5, Green Peroquit Otahite, Aá." [1]

C. ulietanus (Gmelin) is also unquestionably extinct. The species is known by only two specimens thought to have been collected in 1773 or 1774, during Captain Cook's second voyage, at Ulieta (Raiatea), in the western or leeward group of the Society Islands.

A skin in Vienna has been accepted as the type of Latham's "Society Parrot" because it was purchased at the sale of the Leverian Museum in London in 1806 (Sassi, 1939), and because Latham (1781) said that he saw the bird in that museum. The origin of a specimen in the British Museum, received from the Massena collection, is not known. [2]

Latham's statement that his type came from Ulietea may perhaps be doubted because labels in the Leverian Museum were sometimes inaccurate. However, he may well have had the information from naturalists of Cook's voyages. Indeed G. R. Forster (1844, p. 238) refers to this species: "*Psittacus pacificus* . . . habitat in Otahaitee et Oriadea"—by which we may be almost certain that he saw and collected both *zealandicus* on Tahiti and *ulietanus* on Raiatea, even though he did not distinguish the two (see Stresemann, 1950).

Naturalists of Captain Cook's first voyage probably did no collecting on Raiatea or on any of the neighboring islands in September 1768 as a result of difficulty with the natives. It is likely that Forster obtained all the known specimens during Cook's second voyage. He must have had many

---

[1] Sharpe, 1906, **2**, p. 173. Dr. R. W. Sims, of the British Museum, writes me that this is identical with Kuhl's type (*erythronotus*).

[2] There has been considerable confusion about this specimen, owing to the locality on the label having been changed to Tanna, New Hebrides, by Gray. See Salvadori, Cat. Birds Brit. Mus., **20**, 1891, p. 579, footnote. This locality was suggested by Finsch (1868).

opportunities to collect on that island, as well as on Bora-Bora and others, during two visits in 1773 and 1774.

Neither Cook's nor Banks' published diaries mention parrots on Tahiti itself (save the name aá in a vocabulary), yet there can be no question that *C. zealandicus* existed there. Indeed Forster's remark in his "Observations" (p. 207) to the effect that the Tahitians so prized the crimson parrot feathers brought from the Friendly (Tonga) Islands that they gave their treasured pigs for them might lead one to believe that there was no green and red parrots on Tahiti; the remark has no such significance, nor apparently has his silence on the subject of parrots in the Society Islands. There is little reason to disbelieve Latham's locality. Furthermore there can be little doubt that these two species form together a superspecies that is distinct from their New Zealand relatives which are derived from the same stock.

Location of specimens: I know of specimens of the following species in the cities named: *C. ulietanus* in London and Vienna; *C. zealandicus* in London, Liverpool, and Paris.

Australian Grass Parakeets (Neophema)

Of the six species of little Australian grass parakeets (*Neophema*), three are extremely rare. Two were reported to be extinct in the early part of this century but have reappeared. No doubt their numbers have been reduced where there are large human populations, but there is no good explanation for the rarity of some of these species as against the others.

The following key illustrates the differences between them but probably not their actual biological relationships. All are small birds (total length 8–9 inches) with long pointed tails. Save one (*bourkii*), they are generally green in color and have blue wing coverts.

1. Black brown (pink breast) . . . .*bourkii*
   Back green 2
2. Tail feathers (from above) green 3
   Tail feathers (from above) blue 7
3. Cheeks blue; lesser wing coverts blue 4
   Cheeks greenish yellow; lesser wing coverts green . . .*chrysogaster* (orange-breast)
4. Breast red . . . .*splendida* ♂
   Breast yellow or greenish yellow 5
5. Red shoulder patch . . . .*pulchella* (turquoise) ♂
   No red shoulder patch 6
6. Lores blue . . . .*splendida* ♀
   Lores yellow . . . .*pulchella* ♀
7. Lores yellow or greenish 8
   Lores blue 9
8. Lesser wing coverts dark blue; tail 109–111 . . . .*chrysostoma* (blue-wing)
9. Lesser wing coverts pale blue; tail 118–120 . . .*elegans* (elegant) *petrophila* (rock)

If the relative length of wing and tail may be used as a clue to relationship, then *petrophila* and *chrysostoma* are closely related, since tails measure 108–112 mm. *N. chrysogaster* has a much shorter tail (104–105), and that of *elegans* is much longer (118–122). Wings of all vary from 105 to 112 mm., with complete overlap in measurement. However, on the basis of color characters *chrysostoma* and *elegans* agree in four of five characters, while no two others agree in more than two.

According to authorities all nest during August through December, and all lay 3–8 eggs. Some variation in these data is indicated by the fact that some breed in June in captivity at Adelaide.

One of the most interesting problems in nature is the wanderings, the fluctuation in numbers, and the competition for nesting places of these rare species. They must constantly compete with the larger and stronger parrots of the genus *Psephotus,* with the small but fecund budgerigars, with owls, and recently with starlings, imported from Europe for the holes in trees in which they must nest. The only species that nests on the ground (*petrophila*) usually lays its eggs on small islands; it is said to be by far the most numerous.

Perhaps two are relict of long ago, when the climate of at least a part of the land behind the ranges had a more regular rainfall, but this does not explain the relative abundance of the blue-wing as against the orange-breast, nor the rarity of the splendid compared with the relatively fertile elegant.

Details of their feeding habits might reveal that the more numerous are in some way more competitively effective.

The orange-breast (*N. chrysogaster* Latham) and the blue-wing (*N. chrysostoma* Kuhl) are so similar that they are impossible to distinguish in life except by experts, and it is probable that the accuracy of many records is to be doubted.[1] There can be no doubt that they are distinct sympatric species. The bends of the wings (lesser wing coverts) of the blue-wing are always dark blue; the tail, as seen from above, is blue, not green; the tail is longer as a rule (109–111 as against 94–105 mm.), and the fourth primary is not spotted with white. They breed together in the same region in central Tasmania; "each species appearing, however, to adhere to its particular locality on the estate in a general way" (Brent, *in* North, 1911). Observers on the breeding grounds (Giblin and Swindells, 1927) report that the blue-wings (*chrysostoma*) use the same nesting holes year after year when not molested, but that both swift parrots

[1] "Examination of large series of this and *chrysostoma* shows intergradation, making it difficult or impossible to separate the one from the other" (J. R. Kinghorn *in litt.*).

(*Lathamus discolor*) and, in increasing instances, European starlings (*Sturnus vulgaris*) take these places for themselves. The orange-breast (*chrysogaster*) is apparently so rare that it must be considered to be in danger of extinction. The blue-wing is more often seen, but the population is a small one.

Because of their similarity, their rarity, the extraordinary increases and decreases in numbers over periods of years, and their wandering habits, even the record of their breeding ranges is unsatisfactory.

The blue-wing (*chrysostoma*) is recorded to be breeding regularly in central Tasmania near Bothwell, and it is possible that they will be found to nest elsewhere. Formerly they bred near Hobart and also near Melbourne, Victoria, but apparently they do no longer. There is considerable migratory movement from Australia to Tasmania in August and September, with a return in January and February. Birds have been observed on the islands in Bass Strait and are seen in northern Tasmania not only during the season of migration but also in small numbers in June and July. They are considered now to be rare on the southeastern coasts of South Australia (see North, 1911; Cayley, 1938; Littler, 1910; Lord and Scott, 1924).

The orange-breast or orange-belly (*chrysogaster*) is known to breed in central Tasmania near Melton—Mowbray and Bothwell. Littler (1910) considered the southern part of the island to be its "stronghold," and a population may still exist unrecorded. North records eggs from the vicinity of Sydney taken in the latter part of the nineteenth century presumably. It may be that they will be found to breed there again. They are extremely rare visitors in restricted areas of extreme southeastern South Australia, where a flock was seen between Beachport and Robe between 1926 and 1930. A single female taken at Geelong, Victoria, is the only record for that state. Like the blue-wing (*chrysostoma*) a proportion of the populations migrate from Australia to Tasmania. Birds have been observed on those islands and in northern Tasmania in early September. It has been thought possible that they breed on the islands in Bass Strait, but because they are treeless, it is unlikely (see North, 1911; Cayley, 1938; Ashby, 1927; Dove, 1925).

The elegant parakeet (*N. elegans*) is the most numerous of the group. It has been recorded to have been found nesting in South Australia at Yultacowie Creek, 120 miles from Port Augusta, and in Western Australia from the extreme southwest, north to about lat. 31° S. (Moora and Merredin) (Serventy and Whittell, 1951). Since 1937, birds have been found in the forested coastal areas occupied by *Eucalyptus marginata* ("Jarrah") and in the Swan River Valley, where they had not before been recorded. Large flocks have been seen in the east at Chinchilla, south-western Queensland, near Lakes Alexandra and Albert, and on the coast between Adelaide and the Darwin River during the last century. In 1947 a few were seen in the Mallee country of Victoria. It is probable that they breed occasionally in unfrequented places in this wide area, unseen or unreported by man.[1]

The turquoise (*pulchella*) was common 100 years ago, when Gould visited New South Wales, and probably its range was greater, for it has been recorded from Port Denison in Queensland at lat. 20° S. (Ramsay, 1866).

Recorded breeding places include only Dubbo, Macquarie Fields, and an unnamed locality in the vicinity of Sydney, in New South Wales, where a few pairs were breeding in 1945 (Chaffer and Miller, 1946). It is considered to be localized but not uncommon in this area (Kinghorn *in litt.*). It has been suggested (Seth-Smith, 1931) that sheep and cattle grazing has been responsible for localization of this species. Certainly it was more common a century ago. Live-bird markets were readily supplied until the decade 1870–80, after which they were no longer. Probably destruction of habitat together with the lively market in caged birds will account for their great rarity at the beginning of this century, when they were reported to be extinct. Possibly the birds have become more numerous as trapping lessened.

Although, like the turquoise, the splendid or scarlet-chested (*splendida*) is a favorite cage bird and is bred in captivity, it has been brought to market by aborigines and the breeding range is apparently a mystery. The only record is from Pudnooka Station on the Murray River in South Australia (North, 1911), where eggs were taken. It was said to be extinct in 1917. However, birds were trapped near Yaminee on the west coast of the Eyre Peninsula, and there are recent (1948–49) records of sightings in dry country along the east-west railway in the Nullabor Plain at Kweda, between Pingelly and Corrigin and at Lake Cowan, 100 miles south of Kalgoorlie, in West Australia (Serventy and Whittell, 1951). No doubt the birds nest somewhere in this wide range.

Little can be said of the size of this population. It is probably extremely rare, but so few human beings live in its range that it must be that there are fewer to report them. However they have never been common in the cage-bird market, although they were thought to be worth £100 sterling a pair in 1931. Perhaps this species is one of the few known victims of

---

[1] North (pt. 6:448); Cayley (1938); Binns (1947). Mallee refers to areas of dense growth of various trees, usually eucalypts, 15–20 feet in height.

destruction of habitat by natural means, a large part of it having become a desert during the life of the species. No doubt fires, grazing, imported enemies, and general interference by mankind have been the cause of disappearance in the eastern part of the range during the past 200 years.

Like the two foregoing forms (*pulchella* and *splendida*), the Bourke parakeet (*bourkii*) of central and western Australia is seen seldom by Europeans and is reported as news when found. One hundred years ago the birds were to be found, although rarely, in western New South Wales, where they are now found in only very restricted areas.[1] Eggs have been found in the Moolah District. The type locality is on the Bogan River, New South Wales.

These are the only members of the group to be found in the deserts of northern South Australia and Central Australia. They are reported from the Musgrave Range on the border of those states [2] and from Alice Springs in middesert. In Western Australia they are to be found in the dry interior, on the upper Ashburton River in the northwest and Leonora (lat. 29° S., long. 121° 50′ E.) on the south, and they are thought to be increasing in numbers (Serventy and Whittell, 1951).

The rock parakeets (*petrophila*) are apparently quite common on the coasts of southwestern and southern Australia, east to Sharks Bay. They usually nest on islands near the coast, where they are relatively free from interference, and appear to be the most numerous population of the genus and in no danger.

Paradise Parrot

Psephotus pulcherrimus (Gould)

> *Platycercus pulcherrimus* GOULD, Ann. Mag. Nat. Hist., **15**, 1845, p. 115 (Darling Downs, Queensland).

Common names: Paradise parrot; Gould's beautiful parrot; anthill parrot; ground parrot.

Status: Probably never a common species. It is now rare, local, and, in the opinion of some Australian naturalists, in danger of being extirpated. Before the coming of man it is probable that the only enemies were snakes and "iguanas" (*Varanus*). Since that time it is obvious that the populations have been reduced to the danger point, even in places where there are few humans, and extirpated where there are many. The prin-

---

[1] MacGillivray (1927) says: "I know of only one restricted locality where they are still to be found. Where Sturt noted so many there is none today."

[2] Type locality of *pallida* Mathews.

cipal reason for this is thought to be extensive trapping for the cage-bird trade during the breeding season. Grass fires have been mentioned and are no doubt a great danger to the food supply. Rats are no doubt a very great danger.

Close relatives of this bird, *P. c. chrysopterygius,* the golden-shouldered, and *P. c. dissimilis,* the hooded, nest in termite mounds and are somewhat rare and localized even over a restricted range. The former is confined to Cape York and the latter to the Melville Peninsula of northern Australia. Two other species, *varius* (the mulga) and *haematonotus* (the red-backed), form with these a morphologically well-knit generic group. They are, however, both quite common, and the only ecological difference to be found in literature is that the small populations nest in termite mounds and the larger in holes in trees. If it were a certainty that fewer eggs were laid by the rarer birds, as there is some reason to think, this would be an important point, but it is not to be said with certainty. Nests in termite mounds are more accessible to predators, particularly rats, but there is no evidence that these are common in the range of the northern forms.

Range: Northeastern Australia. Recorded from widely separated localities in Queensland and northern New South Wales; in the north from the Archer River and Rockhampton, from 150 miles north of Brisbane, and from Goondiwindi and St. George in southwestern Queensland. In northern New South Wales it has been seen on the Severn River and Francis Creek, which flow into the Macintyre River, and at Narrango Casino (Chisholm, 1922; Irby, 1927).

Habitat: Nests in termite mounds and river banks. Usually to be found on the ground in open or sparsely wooded country. Natives have reported that they sometimes nest in holes in trees, but the accuracy of this has been questioned.

Habits: The nesting season begins in August and continues through December and sometimes April. Three to five eggs are laid in burrows made in termite nest hillocks. The entrance to this is in the side near the top. A passage about 9 inches long by 1½ inches leads to the nesting cavity, which itself is 15 to 18 inches in diameter and 5 inches in height. The female incubates the eggs.

The birds are usually to be found in pairs, not in flocks, and do not nest in communities as do their congeners *haematonotus.*

Their favored food is grass seeds. They strip them from the stem by drawing this sidewise through the bill (Jarrard, *in* Chisholm, 1922).

Description: A small parrot with a long, pointed tail. Total length about 12 inches. Its red forehead, bright green breast, and red patch on the wing distinguish it.

Male: Forehead red, head and nape black to dark brown. A ring of yellowish green feathers around the eyes. Cheeks, ear coverts, and sides of the head bright blue. Back grayish brown. Rump bright blue. Central tail feathers brown, outer blue, black at base, chin, throat, and breast bright green. Belly and under tail coverts rosy red.

Female: The general color pattern is similar to the male, but the red band on the forehead is reduced to a few feathers. Region of throat, ear coverts, and breast dull olive-green and the belly pale bluish with sometimes a few red feathers.

Location of specimens: I know of specimens in Brisbane; Brussels; Cambridge, Mass.; Dresden; Edinburgh; Exeter, England; Genoa; Leyden; New York; Oslo; Rouen; Stockholm; and Sydney.

Australian Ground and Night Parrots

GROUND PARROT—*Pezoporus wallicus* (KERR), Animal Kingdom, **1**, 1792, p. 581 (New South Wales).
P.w. flaviventris North, *P.w. leachi* Mathews, *P. formosus* Latham, and *P. terrestris* Ranzani are synonyms.

Status: Uncommon and localized.
Range: Southern coasts of Australia and west coast Tasmania (see below).

NIGHT PARROT—*Geopsittacus occidentalis* GOULD, Proc. Zool. Soc. London, 1861, p. 100 (near Lake Gairdner, South Australia).

Status: Rare and localized.
Range: Probably deserts of interior Western Australia (see below).

Both of these parrots differ from all others, save *Strigops* of New Zealand, which they resemble in many ways, in having spots of dark brown and yellow in the centers of the green feathers, presenting generally a wavy pattern on a green base. Both, like *Strigops,* are ground birds, and one is nocturnal. The ground parrot differs from the much rarer night parrot in being darker and less yellowish green and in having a red forehead and green, not yellow, checks and lores. The cere is not swollen, nor does it extend over the bill as it does in *Strigops,* and the tail is longer. Total length about 12 inches, as against about 8 inches for the night parrot.

Since both are vulnerable to attacks of mammals and both have disappeared rapidly, following the advance of human habitation, it is tempting to accept this as the cause of their rarity. No doubt in certain localities, where European settlers have burned the country for cattle ranges and have introduced rats and cats, a great diminution or even extirpation has occurred. This is clear in the case of the ground parrot near Sydney. The rarity of the night parrot all over its arid range where there are almost no men cannot be so explained.

One hundred years ago this species was easily to be found in open, swampy heath lands near the coasts, and on coastal islands from southern Queensland, New South Wales, Victoria, Flinders Island in Bass Strait, Tasmania, South Australia, and in the small area of greater rainfall on the southwestern tip of the continent. Now they are to be found only in restricted localities on the coasts of New South Wales, on Frazar Island and Marlo in Victoria (Cayley, 1938), and perhaps on certain islands in Bass Strait. On the west coast of Tasmania they are common (Hinsby, 1948), although rare elsewhere, as they are also in South Australia. They have disappeared from West Australia, where they were seen last in 1913 (Serventy and Whittell, 1951).

The nest is made on the ground under a protecting brush; it is a depression lined with leaves. Eggs have been found in September and November. Three to five pure-white eggs are laid.

Having survived difficulties and dangers for so many generations, it is probable that they will continue to survive in small numbers remote from man until their enemies become too numerous.

The night parakeet (*Geopsittacus occidentalis*) exists only in very small and scattered populations; perhaps in deserts of northern South Australia, Central Australia, and probably Western Australia (Serventy and Whittell, 1951). The birds were reported to be "not uncommon" in South Australia near Coopers Creek, northern Eyre Peninsula in 1875 (F. W. Andrews, 1883). Eleven of the 15 known specimens were collected in this region, nine from the Gawler Ranges, one from near Lake Eyre, and one from Lake Gairdner. That this is so is borne out by accounts of aborigines and resident Europeans for these and other localities to the westward in South Australia (Whitlock, 1924; McGilp, 1931; H. Wilson, 1937). Although the birds have been sought, none have been found for 70 years.

There are no specimens to prove that birds are or were to be found in Central Australia, although perhaps they were about 1922 when they were reported to have been seen by Europeans and natives. Possibly they still occur.[1]

Four specimens in the British Museum labeled "western Australia" are the only tangible evidence we have of their existence in that state. The first bird known to Europeans was found by the Austin Surveying Expedition

---

[1] Whitlock, 1923. Other Australian ornithologists believe that the birds will be found.

in 1854 about 8 miles southwest of Mount Farmer, in the Murchison District. They are said to have been numerous southeast of Oodnadatta between 1870 and 1890 (McGilp, 1931) and were seen at three localities in the neighborhood of Lake Nabberu or on a tributary of the Ashburton River between 1912 and 1935.[1] They were seen only occasionally and never in great numbers.

Reports of sightings in New South Wales and Victoria may be accurate, but it is not clear that there was no confusion with *Pezoporus wallicus* (Kershaw, 1943; Mathews, Birds Australia).

Fig. 48.—Australian night parrot (*Geopsittacus occidentalis*).

During daylight the birds live in the midst of thick clumps of porcupine grass, a spiky bush sometimes called spinifex (*Triodia*), the seeds of which they eat. They are to be found also on the margins of salt flats in samphire bushes (*Salicornia*). The nest is hollowed out in the center of such plants, and the eggs are laid on a platform of sticks or grasses. It is entered by a low tunnel. Four eggs have been found in a nest (H. Wilson, 1937).

The birds have been protected by law since 1937, but how this law can be enforced with respect to these birds is not clear. Even though this population is dangerously small, it will probably persist in the deserts so far from numbers of men.

Location of specimens: I know of specimens of *Pezoporus wallicus wallicus* in Brisbane, Australia; Cambridge, Mass.; Chicago; Launceston, Tasmania; London; New York; Perth, Australia; Rouen; and Sydney; and of *Geopsittacus occidentalis* in Adelaide; Basle; Cambridge, Mass.; Hornsey, England; London; Melbourne; New York; and Sydney.

[1] Bolgers Soak, Pinyerinya Pool, Windich Spring, and Nichol Spring; see Wilson, 1937.

Puerto Rican Short-eared Owl

Asio flammeus portoricensis Ridgway

> *Asio flammeus portoricensis* RIDGWAY, Proc. U.S. Nat. Mus., **4**, 1882, p. 366 (Puerto Rico).

Common name: Múcaro de sabana.

Status: Very rare and localized, a bird was last taken at Anegado Lagoon in 1942, according to Bond, 1950. Because the birds nest on the ground and are therefore vulnerable to predators, and because so many populations of birds have been extirpated from Puerto Rico, this one should be considered to be in danger of extinction.

Another local West Indian race is not uncommon in Haiti (Bond, 1950).

Range: Puerto Rico, West Indies.

Habitat: Open savanna country (Bond).

Description: A rather large (total length about 16 inches) buffy-brown owl heavily marked with black above and streaked with brownish black below. The mating call of the North American subspecies is a loud *toot* repeated, and when the bird is disturbed the cry is like that of an angry cat.

New Zealand Laughing Owls

Sceloglaux albifacies albifacies (G. R. Gray)

> *Athene albifacies* GRAY, Voyage *Erebus* and *Terror*, Birds, 1844, p. 2 (Waikouaiti, South Island).
>
> *Sceloglaux albifacies rufifacies* [1] BULLER, Ibis, 1904, p. 639 (Wairarapa, North Island). Fig.: Buller, 1905, pl. 7.

Common name: Whekau.

Status: Extinct on the North Island since 1889 (sight record). Last specimens taken in 1868. Rare and localized in the Southern Alps of the South Island. (See Marples, 1946; Check List of New Zealand Birds, 1953; *Williams, 1962, p. 18.)

Because they disappeared rapidly with the increase of the human population of New Zealand, and from the neighborhood of towns first, it may be assumed that predators introduced by man were primarily responsible. Myers (1923) was of the opinion that stoats and weasels preyed upon them, and no doubt the habit of nesting on the ground made them easy prey for introduced cats and rats. Buller (1905) suggested that scarcity of their staple, the Maori rat (*R. exulans* subsp.), which is said to have been killed off by introduced rats, might have been the cause.

---

[1] Perhaps a synonym. See Rothschild, 1907.

Range: New Zealand: North Island: Specimens taken at Wairarapa[1] District, about 50 miles from Wellington about 1868, at Mount Egmont, Taranaki District, rather vague sight records at Porirua Harbor of uncertain date, and Waikohu, Te Karaka, of 1889.

FIG. 49.—New Zealand laughing owl (*Sceloglaux albifacies*).

South Island: Formerly in open country in the vicinity of Canterbury and Otago, from which the birds disappeared about 1900, and the Southern Alps, where in localized areas they are still to be found.

---

[1] The type and only one preserved. See Buller, 1905.

Habitat: Nested in open or brushy country, usually in crevices or small caves in limestone cliffs.

Description: Largest of the New Zealand owls (total length about 16 inches), streaked above and below with buffy and dark brown.

Feathers of the head, back, and underparts with a wide brown streak surrounded by paler brown or buffy edges; buff is replaced by white on the cheeks. Wing coverts dark brown spotted with white. Wings and tail barred with dirty white. Tarsi covered with buff feathers, and toes with brown, hairlike feathers. Bill horn-colored at tip, black at base. Eyes brownish yellow.

Habits (W. W. Smith, 1884; Buller, 1888, 1905) : the breeding season is said to begin in September. Nests and eggs have been found in small caves and crevices in limestone rocks and cliffs. There are usually three eggs in a clutch and there may be a second laying. The male is said to incubate occasionally during a period of 25 days, but the female is more often on the nest. The young have been seen to be fed worms.

From examination of casts or pellets in caves, as well as observation in captivity, they are known to feed on rats, mice, lizards, and insects (elytra of beetles were found).

The note most commonly reported is a loud *coo-e-e-e.* In captivity they have been heard to utter a low chuckle like a turkey, to mew like a cat, yelp like a puppy, and whistle tunelessly.

Location of specimens: I know of specimens of *Sceloglaux albifacies albifacies* in Bremen; Cambridge, Mass.; Christchurch, New Zealand; Dunedin, New Zealand; Edinburgh; London; New York; Norwich, England; Tring, England; Wellington, New Zealand; and of *S. a. rufifacies* in Bremen, Germany.

Seychelles Island Owl

Otus insularis (Tristram)

> *Gymnoscops insularis* TRISTRAM, Ibis, 1880, p. 458 (Mahé, Seychelles).

Status: Rare. Last collected in 1940 (*Benson, 1960, p. 62).
Range: Mahé Island, Seychelles Islands, Indian Ocean.

Desmond Vesey-Fitzgerald wrote me in 1940: "Gymnoscops has always been a mystery to me. One would have thought a bird like this would not have been affected by the opening up of the country, but the fact remains, I never got a record of its occurrence."

Michael Nicoll, who spent two weeks on Mahé (the principal island of the group) in March 1906, says that he heard an owl but could not shoot it. The owner of the estate told him that the bird was frequently heard

FIG. 50.—Seychelles Island owl (*Otus insularis*).

and sometimes seen when driven from its hiding place amongst the rocks or hollow trees. No other owls were to be found there. Mr. Thaummesset, his informant, collected one of the three specimens there are in the British Museum. This appears to be the last record.

It is probable that it belongs to the Asiatic superspecies *bakkamoena* (see Peters, 1940, p. 97).

Location of specimens: I know of specimens of *Otus insularis* in Berlin, London, and New York.

Cormoro Scops Owl

Otus rutilus capnodes (Gurney)

Scops capnodes GURNEY, Ibis, 1889, p. 104 (Anjouan, Comoros).

*C. W. Benson (1960, p. 63) reports that the British Ornithologists' Union Centenary Expedition of 1959 was unable to find any of these birds. He writes also that the collector, Naidoo, could find none during his stay in 1906 and 1907. The subspecies is known by thirty or more specimens collected by Humbolt in the mid-eighties; he may have killed them all, according to Benson.

Specimens may be found in Cambridge, London, New York, Norwich, Paris, and a few other museums in France not known to me.

Closely related subspecies are to be found on neighboring islands of Comoros.

Burrowing Owls

Sixteen geographical forms of this widespread, though localized, American species (*Speotyto cunicularia*) still exist. Two West Indian forms are extinct:

Speotyto cunicularia guadeloupenis Ridgway

Speotyto cunicularia guadeloupensis RIDGWAY, *in* Baird, Brewer, and Ridgway, Hist. North Amer. Birds, **3**, 1874, p. 90 note.

Status: Extinct.
Range: Marie Galante Island, West Indies.

Speotyto cunicularia amaura Lawrence

    Speotyto amaura Lawrence, Proc. U.S. Nat. Mus., **1**, 1878, p. 234.

Status: Extinct.

Range: Antigua Island, Nevis Island, St. Christopher Island (St. Kitts), West Indies.

Fig. 51.—Burrowing owl (*Speotyto cunicularia*).

    Both disappeared at about the same time, at the end of the nineteenth century, shortly after the introduction of the mongoose, and they are thought to have been killed by these (Bond, 1950). There were probably never many of either.

    Only six specimens of the Marie Galante bird are known to exist, and all these were taken by L'Herminier in the early part of the nineteenth century. From these it would appear to have been a somewhat darker bird than those of other West Indian islands (*troglodytes* and *floridana*).

S. c. amaura is also darker and the five known specimens are smaller than *guadaloupensis* (wing 140–152 mm. as against 156–164 mm.).

It is not surprising that the mongoose made away with these colonies; their nests, made as they are at the end of long tunnels in the sand, are particularly vulnerable.

Both in California (*hypugaea*) and in Florida the colonies are smaller than they were 50 years ago (Howell, 1932; Grinnell and Miller, 1944). In general the encroachment of civilization, the increase of cultivated land, the numbers of semiwild house cats, and shooting are responsible. In California the United States Government has for years poisoned and sealed ground-nesting mammal burrows; many burrowing owl broods have inadvertently been destroyed.

Location of specimens: I know of specimens of *Speotyto cunicularia guadeloupensis* in Cambridge, Mass.; London; New York; Pointe-à-Pitre, Guadeloupe, West Indies; and Washington; and of *S. c. amaura* in Chicago, London, New York, and Washington.

West Indian Nighthawks

Caprimulgus vociferus noctitherus (Wetmore)

> Setochalcis noctitherus WETMORE, Proc. Biol. Soc. Washington, **32**, 1919, p. 235 (Puerto Rico).

Status: Localized.
Range: Puerto Rico Island.

Siphonorhis americanus americanus (Linnaeus)

> Caprimulgus americanus LINNAEUS, Syst. Nat., ed. 10, **1**, 1758, p. 193 ("America calidiore" = Jamaica).

Status: Probably extinct.
Range: Jamaica.

Interference by introduced mammals is said to have been responsible for the extermination of these two forms. Of the seven others that breed in the West Indies, two are rare and in danger of extinction. These are:

Siphonorhis brewsteri (Chapman)

> Microsiphonorhis brewsteri CHAPMAN, Bull. Amer. Mus. Nat. Hist., **37**, 1917, p. 329 (Túbano, Azua Prov., Dominican Republic).

Status: Rare and localized.
Range: Island of Hispaniola.

Caprimulgus rufus otiosus (Bangs)

Antrostomus otiosus BANGS, Proc. Biol. Soc. Washington, **24**, 1911, p. 188 (St. Lucia).

Status: Rare and localized.
Range: St. Lucia Island.

The Puerto Rican nighthawk was collected in 1961. Several were heard in the vicinity (*Reynard, 1962). The birds had probably not been seen since December 23, 1911 (Wetmore, 1927).

The form is known by only two specimens, a female in the Chicago Natural History Museum, and a male in Washington. Bones [1] have been found in caves; they indicate a shorter wing than other forms.

Probably one of the earliest immigrants in the West Indies, the longest isolated, and certainly the most strongly differentiated nighthawks are related to the South and Central American genus *Nyctidromus*. They are called pauraqué because of their cries. There are two species. One, *Siphonorhis americanus,* is now presumably extinct (Bond, 1950). The other, *S. brewsteri,* is still to be found in restricted localities on Hispaniola. They have been reported by Bond only from Túbano in the Dominican Republic, the type locality, and from Magasin Caries between l'Arcahaie and Mont Rouis in Haiti. They must be considered to be in danger. The mongoose is a comparatively recent arrival on that island. If the populations of that animal follow the usual pattern they can be expected still to be increasing rapidly. It is unlikely that the small bird populations can endure any considerabe predation.

The Jamaican species must always have been localized. Neither Gosse nor any of his ornithological acquaintances in 1830–40 knew them, although the last known specimen was sent to London in 1859. Since the mongoose was not introduced into Jamaica until 1872, it has been assumed that the birds were extirpated by rats before that date. It is even possible that a small number may still be found, although authorities in Jamaica believe that they are extinct (Stuart Panton, *in* Bond, 1950).

Only three specimens are known to exist; these were taken in the western part of the island in Westmoreland and at Freeman's Hall in Trelawny.

These are smaller species than any other nighthawk in the West Indies, the Haitian bird being only 7–8 inches long and the Jamaican 9–10. The nostrils are very prominent and the tarsi long (20–22 mm.). A broad

---

[1] Humeri and metacarpal in American Museum, New York.

white band between the upper breast and throat also distinguishes them.

Above brown, mottled and streaked with gray and darker brown. A rufous collar at the back of the neck spotted with black and white. Primaries barred irregularly with light and dark brown. The tail feathers are mottled and flecked with gray and brown and have darker brown spots in the centers and edges the two central feathers sometimes narrowly tipped with white, all others with a conspicuous white tip. Males of *brewsteri* are darker than females.

Location of specimens: I know of specimens of *Caprimulgus vociferus noctitherus* in Chicago and Washington and of *Siphonorhis americanus americanus* in London and New York.

Fig. 52.—West Indian nighthawk (*Caprimulgus vociferus*).

In addition to these, two species of potoos, representatives of the essentially Central and South American genus *Nyctibius,* are to be found on Jamaica and Hispaniola. The status of both of these subspecies is uncertain. Thirty years ago the Jamaican form (*N. g. griseus* Gmelin) was said to be "not uncommon" (Bangs and Kennard, 1920), and according to Bond (1950) the Hispaniolan bird (*N. g. abbotti* Richmond) is "probably not uncommon."

Extremely shy and retiring during daylight, feeding only at night, they are difficult to find. In South America they are said to nest on stumps of trees. Although no nests have been found in the West Indies, it is most probable that habits do not differ from those of their relatives in South America. If this proves to be so, it may be that rats and mongooses have found the eggs less readily than they would those of the species that nest on the ground.

There is no such explanation for the survival of the birds called killy-kadick, or, in Spanish, querequété (*Chordeiles minor gundlachii*). Like other species they lay only one or two eggs and these on the ground. As far as is known, their habits differ from other nighthawks only in their being less markedly nocturnal and in their nesting in open country. Either they have some ability to protect eggs and young about which we know nothing or the mongoose is not entirely responsible for the extirpation of other ground-nesting birds.

It has been suggested by Danforth (1935, p. 52) that the venomous snake called fer-de-lance (*Bothrops atrox*) protects the rare and local whippoorwill of St. Lucia (*Caprimulgus rufus otiosus*) by its presence. The West Indian authority, James Bond, writes: "From what I am told, the mongoose readily captures the young of the fer-de-lance, although it usually avoids the larger ones." Whatever the reason, the birds appear to occupy the same restricted range as the snake.

The range of *Caprimulgus cubanensis cubanensis* (Lawrence) (confined to Cuba) is said to have become somewhat restricted, but the population appears to be in no great danger. On Hispaniola *Caprimulgus ekmani* (Lönnberg) is nowhere a common bird, although widely distributed.

Riu Kiu Island Kingfisher

Halcyon miyakoensis Kuroda

> *Halcyon miyakoensis* KURODA, Dobuts. Zasshi (Tokyo), **31**, 1919, p. 231 (Miyakoshima, Riu Kiu Islands). Fig.: Kuroda, 1925, pl. 6, no. 6.

Status: Extinct (Yamashina *in litt.*, 1938). Collectors (particularly Orii for Kuroda) have searched these islands carefully. No bird has been found since 1887, when the type and only existing specimen was collected by Y. Tashiro. It was in the Zoological Museum, Science College, Tokyo.

Range: Supposedly confined to Miyako Island, about 140 miles southwest of Okinawa and 180 from the northern tip of Formosa, in the south Riu Kiu (Nansei Shoto) Islands.

Description: Resembled *cinnamomina* of Guam (other well-marked subspecies occur in the Carolines and Palaus), but the primaries are longer in proportion to the secondaries and it lacks the black nape band (Orn. Soc. Japan, 1942, p. 85 note; Kuroda, *l.c.*). Top of the head cinnamon; upper back, wings, and tail dark greenish blue, lower back and upper tail coverts bright blue. Underparts cinnamon. Wing 105 mm. Feet red.

The female of *cinnamomina* has white underparts.

Remarks: Except for the red feet, no differences of importance have been pointed out. The relationship between length of primaries and length of secondaries is often distorted when the ulna is stripped during preparation. Because the subspecies of *cinnamomina* are well marked, we might expect a readily recognizable geographical representative in the Riu Kius, had a population ever existed there. When evaluating single puzzling specimens it must be remembered that there is always a possibility that there has been a mistake in the label.

Guadalupe Island Flicker

Colaptes cafer rufipileus Ridgway

Colaptes mexicanus rufipileus RIDGWAY, Bull. U.S. Geol. Geogr. Surv. Terr., **2**, no. 2, 1876, p. 191 (Guadalupe Island, off the west coast of Lower California).

Status: Most probably extinct. Last specimen collected and last recorded sighting June 1906 (Thayer and Bangs, 1908), when W. W. Brown, a collector, estimated that there were about 40 birds. Whether this was before or after he took 12 is not recorded. In 1885 the birds were not rare in the restricted grove of cypress at the top of the ridge and occasionally were seen in other places where there were stands of pine or palm (Bryant, 1887).

In the spring of 1897 they were "rarely seen" (Kaeding, 1905). Nothing is recorded of them between 1906 and 1922, when none could be found by naturalists from the California Academy of Sciences. None have been reported since then, although they have been sought.

The destruction of habitat principally by goats is thought to have been the reason for their disappearance. House cats roamed the island freely during those years. Brown saw bodies of birds obviously dismembered by cats. No other such predator has ever been seen on the island. Obviously these contributed to the disaster.

Range: Confined to Guadalupe Island, in the Pacific Ocean off the coast of Baja (Lower) California.

Habitat: In the small grove of cypress (*Cupressus guadalupensis*) at the top of the ridge and sometimes among pines (*Pinus radiata*) on the slopes.

Description: A large gray and white woodpecker with red cheeks, wings, and tail. Impossible to confuse with any other bird on the island. Similar to *C. c. collaris* of California but has the top of the head browner, less grayish. This region is a rusty brown in *C. c. cafer* of Mexico.

Location of specimens: I know of specimens in Berkeley, Calif.; Chicago; Cambridge, Mass.; New York; Providence, R. I.; and Washington.

Ivory-billed Woodpeckers

Campephilus principalis principalis (Linnaeus)

Picus principalis LINNAEUS, Syst. Nat., ed. 10, **1**, 1758, p. 113 (North America = South Carolina *ex* Catesby).

Campephilus principalis bairdii Cassin

Campephilus principalis bairdii CASSIN, Proc. Acad. Nat. Sci. Philadelphia, 1863, p. 322 (Monte Verde, Cuba).

Common names: Ivory-billed woodpecker; carpintero real (Cuba).

Status: The species is in immediate danger of extinction. Extremely small populations exist in northern Florida, where two were seen in 1950[1] and perhaps in Louisiana. Rediscovered in eastern Texas (Dennis, 1967).

An estimate of the populations as of 1939[2] follows. Since then the cypress forests of the Singer Tract have been cut down and the small population dispersed.

Locality	Number of individuals	Basis for estimate of numbers
Big Cypress area, Florida	6	Estimate based on carrying capacity of region and locations of reports.
West-central Florida	2	Known to be there late in 1937 by O. E. Baynard.
Gulf Hammock-Suwannee region, Florida	4	Estimate based on carrying capacity of region.
Apalachicola region, Florida	4	Estimate based on carrying capacity of region and reports of natives.
Singer Tract, Louisiana	6	Five individuals observed and sign of at least one other individual.
Approximate total	22	

The disappearance of this species has been contemporaneous with the destruction of habitat and consequent deprivation of food. Neither those birds that compete with the ivorybill for food nor predators of any kind are to be held responsible, although shooting has been increasingly dangerous to the existence of the species as man encroaches. James T. Tanner, in his excellent account of the few remaining ivorybills, points out that if the birds are to be preserved their remaining habitat must be set aside as a refuge. This is true both in North American and in Cuba. In the moun-

---

[1] Auk, **67**, 1950, p. 320.
[2] From Tanner, 1942. Reprinted by Dover Publications in 1966. Reproduced by permission.

tains of Cuba there is a belief that the birds hold the key to good fortune
and they are sometimes killed to gain possession of this prize, Dr. A.
Moreno has informed me.

FIG. 53.—Ivory-billed woodpecker (*Campephilus principalis*).

In Cuba a few are still to be found in northern Oriente Province from
the pine ridges of Mayari eastward (Dennis, 1948) and possibly near
Artemisa, Pinar del Río (Barbour, 1943).

Range: Within historical times in river-valley forests of the United
States from the Carolinas and western Florida westward, along rivers

flowing into the Gulf of Mexico, to the Brazos River in Texas, and in the Mississippi Valley, north to southern Illinois, Indiana and Ohio.[1] In Cuba, formerly in suitable localities everywhere, in all probability. They occurred in the Organ Mountains north of San Diego de los Banos and near Artemisa[2] in Pinar del Río Province. In Santa Clara Province in forests near Bahía de Cochinos, as well as near Guantánamo in southeastern Oriente Province. Traditions of occurrence in Matanzas Province lingered at the villages of Banaguises and Calimete during the latter part of the nineteenth century (Gundlach, 1856).

Habitat[3]: In heavily forested and usually flooded alluvial land bordering rivers. In the eastern part of the range oaks (*Quercus nigra* and *laurifolia*) and sweetgum (*Liquidambar*) are the common trees, while in Florida birds are usually found in regions where cypress (*Taxodium*) predominates, although they are not confined to this habitat. In the Mississippi River Delta the birds were more apt to be found in drier country lying near swamps where sweetgum (*Liquidambar*), oaks (*Q. rubra* subsp.), and green ash (*Fraxinus pennsylvanica* subsp.) are most common. Their food requirements of wood-boring insects postulates a habitat where recently dead and dying trees are numerous. Only in great stands of virgin timber can this necessity be supplied.

Recently in Cuba, birds have been found only in the mountains covered with pines and scattered hardwoods.

Habits (Tanner, 1942): The breeding season may begin at any time during the months January through April. The factors governing this are not known. Birds have been found to be incubating eggs as early as January 20 (Scott, 1888) and as late as April 13 (Allen and Kellogg, 1937). They are recorded as having nested twice, the second as late as May 19.[4]

The only courtship performance recorded involves the clasping of bills (Allen and Kellogg; Tanner).

Nests have been found in almost every species of tree occurring in the birds' habitat. In Cuba pines[5] are preferred. Nests are in holes 15 to 70 feet from the ground, although usually above 30 feet. Dead or partially dead trees are used. As a rule one to three eggs are laid, although five

---

[1] Wetmore (1943) records bones found in an old Indian encampment at Clay Township, Scioto County, Ohio. The bird has not been seen in that region in modern times.

[2] An unconfirmed report of the latter part of the decade 1930–1940 (Barbour, 1943).

[3] See J. T. Tanner, 1942, for an excellent study.

[4] In Louisiana, *fide* Bendire.

[5] This is *P. cubensis* (*fide* Dennis, 1948). Formerly it is probable that other trees were used, for there are but few pines in the vicinity of the Zapata Peninsula. *Pinus tropicalis* occurs in Pinar del Río.

have been found. The exact period of incubation is not known; Tanner thinks 20 days probable. The sexes alternate in the incubation of eggs, the female during most of the day and the male by night. Periods vary from 15 minutes to 3½ hours. The nest territory is vigorously defended from other birds by both parents, but the birds soon become accustomed to the presence of humans.

Both parents feed the young, sharing the work equally during the entire day. How long the young remain in the nest is not known.

The staple is larvae of beetles of the families Buprestidae, Cerambycidae, and Elateridae, which are to be found beneath the bark of moribund trees.

Location of specimens: I know of specimens of *Campephilus p. principalis* in Cambridge, Mass.; Chicago; Hamburg; New York; Paris; Princeton, N. J.; Rouen; Springfield, Ill.; Springfield, Mass.; San Francisco; Syracuse, N. Y.; and of *C. p. bairdii* in Berlin; Cambridge, Mass.; Chicago; Guantánamo and Havana, Cuba; New York; Philadelphia; and Washington.

Tristram's Korean Woodpecker

Dryocopus javensis richardsi Tristram

> *Dryocopus javensis richardsi* TRISTRAM, Proc. Zool. Soc. London, 1879, p. 386 (Tsushima Island, Japan Sea). Fig.: *l.c.,* pl. 31.

Status: Rare. Once said to have been extirpated from Tsushima Island, it has recently been found breeding there again (T. Udagawa, *in litt.*).

Range: Korea and Tsushima.

Description: Tristram's woodpecker is one of the largest in the world (total length about 14 inches); it has a crimson crest. Confusion with the great black woodpecker (*Dryocopus martius*) is possible, but the pure-white belly and black bill of Tristram's woodpecker distinguishes the bird, for the plumage of *martius* is all black and the base of the bill is yellowish. Furthermore, it is possible that the two do not occur together in exactly the same habitat [1], at least at present.

Habitat: In 1948 (Wolfe, 1950), Tristram's woodpecker was to be found breeding in isolated valleys of giant spruce trees, surrounding tombs, apparently not far north of Seoul. Although nests of the great black woodpecker have not been found recently, birds are thought to breed farther to the northward in heavily forested areas in northeastern Kyonggi-Do Province. Large trees in this region are being cut down, however, and the population must be in danger there.

---

[1] Microhabitat is more exact.

FIG. 54.—Tristram's Korean woodpecker (*Dryocopus javensis richardsi*).
(After J. Smit.)

Both these subspecies are representative of widespread polytypic species. *Dryocopus javensis,* apparently having a center of distribution in Burma, is to be found in India, Sumatra, Java, the Philippine Islands, and in China as far north as Koko Nor in Kansu (now Htsinghai in northwestern China).

The habitat on Tsushima (Island) is said to be similar to that of Korea.

The following is a quotation from Ijima (1892): "It finds its abode in dense, more or less extensive forests of tall pines, firs, Cryptomaria, oaks, camphor trees, etc. Such forests usually exists in valleys between hills, and is known to the natives as kuromi (a dark place)."

Ijima has this to say of the habits: "The bird is never found in any numbers together; perhaps a pair is the utmost that a kuromi might harbor . . . Dead trees or branches naturally attract it when in search of food, and it is said that the bird goes regular rounds for its favorite trees every day. Should one of its trees be recognized, a collector would do well therefore to lie in ambush awaiting its arrival. Its manner of flight is similar to that of other woodpeckers . . . on more than one occasion they [collectors] discovered it again on the ground, whence it climbed up a tree-trunk in the usual manner."

Noguchi's Okinawa Woodpecker

Saphaeopipo noguchii (Seebohm)

> *Picus noguchii* SEEBOHM, Ibis, 1887, p. 178, pl. 7 ("Great Loo-Choo Island" = Okinawa Island).

Status: Rare and localized.

Range: Okinawa Island, Nansei Shoto, western Pacific.

Saphaeopipo noguchii has apparently been isolated on the island of Okinawa for a long time. Relationships are not clear. The bird cannot be mistaken for any other on Okinawa or indeed any in the world. It is a medium-sized woodpecker (total length about 10 inches). Top of head red (male; brown in female); sides of head pale brown fading to paler on the throat. Back dark red. Breast dark brown. Belly pink. Wings and tail darker brown, the wings with three prominent white bands or spots.

A small population is confined apparently to the dense forests near Hedo in northernmost Okinawa. Here two specimens were taken in 1945 (Baker, 1948). It is reported to be extremely secretive and possibly less rare than has been supposed. Although the birds may in the future be exposed to dangers because of the military occupation, these will probably be confined to destruction of habitat. Whether this will be brought about in the near future cannot be determined.

Location of specimens: I know of specimens in Cambridge, Mass.; New York; and Tokyo.

Fig. 55.—Noguchi's Okinawa woodpecker (*Saphaeopipo noguchii*).

Delalande's Madagascar Coucal

Coua [*Cochlothraustes*] *delalandei* (Temminck)

    Coccycus Delalandei TEMMINCK, Pl. Col., livr. **74**, 1827, pl. 440 (Madagascar).

    Status: It is possible, but not probable, that a very small population of this bird still lives in the great humid forests of northeastern Madagascar. According to Lavauden (1932), birds were occasionally trapped by natives in the forest near Fito and Maroantsetra, presumably during the years 1920–1930, and that the natives prized the feathers. Rand (1936) is of the opinion that the birds must be extinct for the reason that none have been taken by native hunters in recent years. Delacour (1932) says the forests in that area were destroyed. Certainly large prizes were offered in 1932 to dealers in Tananarive, whose hunters should have been able to supply them. None have been collected or recorded by Europeans since 1834, when Bernier obtained one for the Paris Museum.

Range: Formerly forests of the northeastern coast of Madagascar, perhaps from Antongil Bay south to Tamatave, and Île Sainte Marie. There

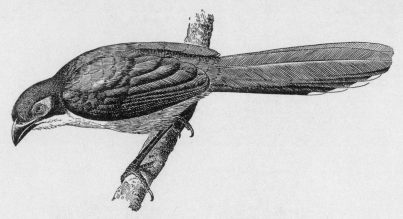

Fig. 56.—Delalande's coucal (*Coua delalandei*).

are no exact records of the provenance of mainland specimens and Tamatave is probably the address of a dealer. Populations have probably been small and localized for a long time.

Habitat: Lived on the ground in humid forests.

Description: Largest of the couas (total length about 22 inches). Dark blue above and white and chestnut below. Head blue with a blue area of bare skin around the eye, this surrounded by black feathers. Back blue. Tail with a greenish tinge, outer feathers with white tips. Throat and upper breast white, belly chestnut. Bill, legs, and feet black. Eyes brown.

Location of specimens: I know of specimens in Cambridge, Mass.; Leyden; London; New York; Paris; Philadelphia; and Tananarive, Madagascar.

Ani of Baja California

Crotophaga sulcirostris pallidula Bangs and Penard

Crotophaga sulcirostris pallidula BANGS AND PENARD, Bull. Mus. Comp. Zool., **64**, 1921, p. 365 (San José del Cabo, Lower California).

It has been intimated by many authors during the past 25 years [1] that this population is extinct. Its migratory tendencies were suggested by Brewster in 1902, Nelson in 1921, and confirmed by van Rossem in 1945. It seems probable that it is still present in localized areas in certain seasons.

It is also probable that this is not a valid subspecies. Many skins prepared by Frazar are faded, and the type of this supposed subspecies is no exception.

Bush and Rock Wrens (New Zealand)

Xenicus longipes stokesii Gray

Xenicus longipes stokesii G. R. GRAY, Ibis, 1862, p. 219 (Rima-taka Hills, New Zealand). Fig.: Ibis, 1905, p. 392.

Status: Very rare and local, if not quite extinct.
Range: North Island, New Zealand.

Xenicus longipes longipes (Gmelin)

Motacilla longipes GMELIN, Syst. Nat., **1**, 1789, p. 979 ("in nova Seelandia").

Status: Not uncommon in restricted localities.
Range: South Island, and probably Stewart and nearby islets, New Zealand.

Xenicus gilviventris Pelzeln

Xenicus gilviventris PELZELN, Verh. zool.-bot. Ges. Wien, **17**, 1867, p. 316 (New Zealand).

Status: Common in restricted localities.
Range: South Island, New Zealand.

Xenicus lyalli (Rothschild)

Traversia lyalli ROTHSCHILD, Bull. Brit. Orn. Club, **4**, no. 22, 1894, p. 10 (Stephen Island, New Zealand).

Status: Extinct.
Range: Stephen Island, New Zealand.

Only three species represent an entire family which is confined to New Zealand. One of these is certainly extinct, the other two, although shy and retiring, are reported from time to time. However, a subspecies of the

---

[1] E.g., Grinnell, 1928; Bancroft, *in* Bent, 1940; Friedmann, Griscom, and Moore, 1950.

North Island (*Xenicus longipes stokesii*), although perhaps it still exists, has actually not been recorded for many years and the population may have disappeared altogether. Nesting as they do in logs near the ground, they are particularly vulnerable to rats and other imported predators.

On the South Island *Xenicus longipes longipes* has been seen near the Waiau River, and near Lake Waikareiti in recent years.[1] A sighting "almost certainly" of this species is recorded at Stewart Island in January or February 1951 between Deep Bay and Ringaringa,[2] and the note implies its continued existence on Kotiwhenu, off Stewart Island.

FIG. 57.—New Zealand bush wren (*Xenicus longipes*).

These are very small birds (total length about 4 inches) with short wings and almost no visible tail, usually to be seen running and hiding among bushes and rocks. The top of the head is mouse brown with a white streak above the eye; the back and wings are olive-green, the underparts gray. There is a conspicuous white spot at the bend of the wing.

The North Island subspecies (*stokesii*) is described as having the sides of the neck shining slate blue and in having a patch of bright yellow feathers on the flanks, which is concealed when the wings are folded.

[1] New Zealand Bird Notes, 3, no. 6, 1949, pp. 164, 172.
[2] Notornis (formerly New Zealand Bird Notes), 4, no. 6, 1951, p. 148.

The rock wren, *Xenicus gilviventris,* of the mountains of South Island, is apparently quite common. Recent sightings (between 1948 and 1950) have been reported at Harris Saddle, Rantebourn, and Lake Adelaide in western Otago. A nest with five eggs was found at McKinnon Pass.[1]

This species resembles the young of *longipes,* being all brown except for an olive tinge on the lower back and tail, but differs in having no white patch at the bend of the wing, and being somewhat duller colored, without the purplish wash on the back.

Fig. 58.—Stephen Island wren (*Xenicus lyalli*).

The Stephen Island wren, *Xenicus* [*Traversia*] *lyalli* is famous for being the little bird that was discovered and extirpated by the lighthouse keeper's cat during the year 1894. If this story is true, the population must have been a very small one. Other than that it was a skulker among rocks and appeared to be seminocturnal, nothing is known of the bird.

Like so many island representatives, this was somewhat larger. It was brown above and greenish yellow below, and had a distinctly mottled appearance due to dark brown edges of the feathers.

[1] New Zealand Bird Notes, **3**, no. 8, 1950, p. 216.

Location of specimens: I know of specimens of *Xenicus lyalli* in Cambridge, Mass.; Christchurch, New Zealand; London; New York; and Pittsburgh.

Australian Scrub-birds ("The Feathered Mouse")

Atrichornis clamosus (Gould). "The Noisy."

> *Atrichia clamosa* GOULD, Birds Australia, **3**, pl. 34 [pt. 14, pl. 11], 1844 ("Between Perth and Augusta, western Australia" = Drake's Brook, Waroona, W.A.).

Atrichornis rufescens (Ramsay). "The Rufous."

> *Atrichia rufescens* RAMSAY, "Grafton Examiner," June 12, 1866 ("Cowlong" now Eltham near Lismore, New South Wales). (See Chisholm, 1951.)

A. c. tweedi Mathews, 1917 (Tweed River), and *A. c. jacksoni* White, 1920 (MacPherson Range), are thought to be synonyms.

Status: Localized near Albany, western Australia, in 1962 (*Webster, 1962, no. 3, p. 57, and no. 4, p. 81; *Serventy, 1962, p. 213). None had been found since 1889, when A. J. Campbell collected a specimen at Torbay, near Albany, in extreme southwestern Australia (Serventy and Whittel, 1951).

The eastern species (*rufescens*) is rare and apparently confined to the MacPherson Range on the border of Queensland and New South Wales, although it is possible that birds may still be found elsewhere.

Range: A. clamosus was once to be found on a considerable stretch of the coast of southwestern Australia, although the birds were restricted to a special habitat. Specimens have been taken in the past at Drake's Brook in 1842, at the Margaret River, and at King Georges Sound, not far from the type locality, in 1866.

The eastern species (*rufescens*) once occurred southward, not far from the coast, from the MacPherson Range for about 75 miles in the forests of the Richmond River and the vicinity of Dorrigo and the Bellinger (Bellenger or Bellingen) River (Jackson, 1911, 1921). Large tracts of forests have been destroyed in these lower altitudes during the past century. Fortunately the present range of the birds lies in the Lamington National Forest, where they are not likely to be disturbed. They are said to be most common on a spur north of the border in Queensland called the Tweed Range.[1]

---

[1] Emu, **47**, 1948, p. 381.

Habitat: Usually in dense, tangled jungle, often called "scrub," under fallen limbs and trees, but often in small tracts of grass land within the forest.

Description: So adept at hiding themselves are these birds, and so difficult to see, that they are best found by following their notes, described as loud and piercing, with a sharp *whip crack*. Both are small (about 8½ inches long) brown birds with long, rounded tails and short wings. They almost never fly but scuttle with great speed.

Both are brown above, the feathers giving an effect of thin, irregular black bands. The chin is white. The throat and upper breast black, washed sometimes with whitish.

Fig. 59.—Australian scrub-bird (*Atrichornis*).

A. clamosus is larger than *rufescens* (wing more than 10 mm.), the tail is not regularly barred with black but rather vermiculated, and the sides of the breast and belly are less rufescent.

The nestling is said to be naked, except for a line of down on the mid-back and head.

No complete account of the nesting of the western spicies exists. The reliable naturalist William Webb wrote in 1895 [1]: "I have seen but one nest of Atrichia and that was at Torbay Inlet. It was on the ground under a low spread bush and was composed mostly of thin strips of paper bark, wire grass and a few feathers. It was of very loose and slovenly construction and contained only one egg." Descriptions of the eastern

---

[1] In "The Albany Mail." See Whittell, 1943.

bird's nest indicate a somewhat more elaborate structure (Jackson, 1911, 1920, 1921). It is made in tufts of *Carex* grass, domed and lined with "wood pulp."[1] Grass (xerotes) and bits of tree fern leaves were used. One or two eggs are laid during September and perhaps October. As far as is known only the female incubates. The male is said to defend a territory of about 50 square yards. Although the birds are able to fly they are loath to do so; only under great stress do they, and then only for a few yards (Chisholm, 1921). They can and do run with great speed and agility and are extremely secretive and difficult to find, as are all birds with such habits.

Their notes have been described as "chits" and "pops" varied with mimicking songs, having the effect of ventriloquism.

Remarks: Long before the advent of man, and the cats and rats he brought with him, it is probable that a larger part of the continent of Australia had a heavier rainfall. The gradual desiccation of the country has led to the partition and diminution of many populations. The same factors—poor reproductive power and small numbers—that have led to the extinction of many island populations have operated to reduce the numbers of these. We may hope that the rufous scrub-bird will continue long to survive in Lamington National Park.

Because of the possession of only two pairs of vocal muscles, as against the seven usual in "song birds," *Atrichornis* is said to be primitive, or nearer the base of the phylogenetic tree than other small "perching birds." However, if we may believe an eyewitness, its nesting habits are rather special. Peculiarities, such as its rudimentary clavicles and much reduced sternum, may probably be due to the partial loss of flight.

Location of specimens: I know of specimens of *Atrichornis clamosus* in Cambridge, Mass.; London; Melbourne; New York; Perth, Australia; Philadelphia; Sydney; and Washington; and of *A. rufescens* in Brisbane, Australia; Brussels; Cambridge, Mass.; Edinburgh; Melbourne; Munich; New York; Perth, Australia; Stockholm; Sydney; and Washington.

Guadalupe Island Wren

Thryomanes bewickii brevicauda Ridgway

Thryomanes brevicauda RIDGWAY, Bull. U.S. Geol. Geogr. Surv. Terr., **2**, no. 2, 1876, p. 186 (Guadalupe Island).

Status: Most probably extinct. The last specimens were collected by A. W. Anthony in 1892 (Anthony, 1901). Except for a sight record by

---

[1] Jackson, 1920, p. 266, implies that decayed wood is somehow processed by the bird itself. It would be of great interest to know that this is the case.

Kaeding (1905) in 1897, and in spite of intensive search by parties of trained naturalists, not one bird has been seen since.

This was probably never a large population. In 1875 the species was said "not to be numerous" (Palmer, *in* Ridgway, 1876).

Range: Formerly confined to Guadalupe Island, in the Pacific Ocean off the coast of Lower (Baja) California.

Habitat: Only three eyewitness accounts of the birds exist. In 1875 Palmer (*l.c.*) wrote: "They live in the brush, being rarely seen on trees. . . . Stomach contained remnants of some small black insect which fed upon blossoms of the white sage." This indicates that in 1875 the habitat was, at least sometimes, in the thick, white-leaved sagebrush (*Senecio palmeri*), which grew to a height of about 4 feet and covered the slopes of the mountain. This plant was almost extirpated by introduced goats during the next few years, as were many others.

In 1885 there must still have been some of this left, for Bryant (1887) observed that the yellow blossoms were a favorite food of the goats, during dry years especially. In that year he found the birds in an area only 60 x 300 feet, among the dying pines (*Pinus radiata binata*) at the northern edge of the 3,000-foot ridge. Here they took refuge among the dry branches of the many fallen trees. Seven years later Anthony (1901) found a lone bird in this grove and two were taken on the side of the hill below, apparently in open country.

Description: A very small wren with a long bill, much smaller and darker than the common rock wren (*Salpinctes*), the only other recorded from Guadalupe Island.

A representative subspecies of *T. bewickii,* which is to be found from British Columbia and Alberta south to southern Mexico, it is brown above and grayish white below. A distinct white stripe runs through and behind the eye. Apparently closest to *correctus* of southern California, it differs in having narrower black bars on the tail feathers and in being much smaller.

Wing 48–49 mm.; tail 42–45 mm.; bill 17 mm.

Habits: All we know is contained in the few observations of two naturalists. In 1875 Palmer wrote: ". . . Their motions are very quick; their general habits restless, impatient and shy. Their almost incessant activity, together with their shyness, renders them difficult to secure."

Bryant (1887) said of them: "The birds were timid rather than shy, being alarmed by the crushing of dry branches as I worked my way amidst the dense windfalls of pines, where they were found, they fled into the thickest parts. When all was quiet they would cautiously approach until within a few feet of me, seemingly prompted by curiosity. . . . A frightened female uttered a few "twit twits" of alarm, but with this exception they were utterly silent."

The breeding habits were never observed, nor were any nests or eggs found.

Remarks: Destruction of habitat was probably the primary cause for this sad event, although no doubt the feral house cats contributed. When Palmer referred to the "brush," in which he thought the birds were restricted, he meant the species of low shrubs which covered the open country, not only on the slopes of the mountain but also on the table land at the edges of the cypress groves there (Watson, 1876; Greene, 1885). This brush habitat had been entirely destroyed by 1906 (Thayer and Bangs, 1908), and no doubt it had been much reduced in extent even in 1885, for Bryant said "there grows in places a small white sagebrush." By this he means *Senecio,* and he speaks also of the common *Alfileri.* At that time there was no cover for the birds except in the dry brush of the sparse pine grove.

There is no direct evidence that the food supply of the birds was curtailed or altered in any way, but it is not improbable. No less unlikely is it that the cats destroyed them, for although there is actually no evidence that they killed wrens, they did kill the flickers and other birds, according to all witnesses. However, since the rock wren, which all have found to be so confiding that it would perch on a man's hat, is still quite common, it is not conceivable that cats were primarily responsible.

Location of specimens: I know of specimens in Cambridge, Mass.; Chicago; and Washington.

West Indian Wrens

Troglodytes musculus guadeloupensis (Cory)

> *Thryothorus guadeloupensis* Cory, Auk, **3**, 1886, p. 381 (Grande Terre, Guadeloupe).

Status: Probably extinct.
Range: Guadeloupe Island, West Indies.

Troglodytes musculus martinicensis (Sclater)

> *Thryothorus martinicensis* Sclater, Proc. Zool. Soc. London, 1866, p. 321 (Martinique).

Status: Probably extinct.
Range: Martinique Island, West Indies.

Troglodytes musculus guadeloupensis was last seen when G. K. Noble collected a female in the forest ("which had been more or less cut over") near Sainte Rose on July 13, 1914 (Noble, 1916; see also Bond, 1950). He did not find the birds elsewhere. Neither J. L. Peters in 1924, nor J. Bond in 1930, nor S. T. Danforth in 1937, nor the resident ornithologist Biaggi saw or heard them since. Perhaps the populations were not so large as sometimes has been indicated, for in 1878 Ober said,[1] "I found this bird only in the second growth of the hills, and in a wood in the flat portion of the island."

Exactly when the Martinique bird (*T. m. martinicensis*) disappeared is not known; the last specimen was taken in 1886, at which time the birds are said to have been fairly common (Cory, 1886; Bond, 1950). However, in 1879 Ober found that birds in general were rare on Martinique, and presumably the wren was no exception. He said of it: "An inhabitant of the woods; I have not seen it near houses or sugar mills, only in forests of the hills, and along the borders of streams where the bushes are thick" (Lawrence, 1879).

Four subspecies of this widespread and generally common South American species still occur in the West Indies. On St. Vincent and Dominica, Ober had found the wrens near houses and quite common generally. At present they are rare and localized on St. Vincent (*musicus*) and St. Lucia (*mesoleucus*), but on Dominica *rufescens* is abundant and on Grenada *grenadensis* is not uncommon (Bond, 1950).

No doubt interference by man and introduced mammals, cats, rats, and mongooses, are responsible for these sad events. Ober (*l.c.*) observed in 1879 that, generally speaking, birds were relatively uncommon on Martinique. He thought this was due to dense populations of human beings, many of whom persecuted birds constantly, as well as to destruction of birds' habitat by extensive cultivation.

Neither this theory nor the more recent one that the mongoose is entirely responsible answers this question: Why are the birds common on Grenada? At the turn of the century there were almost as many people on that island (459 per square mile) as there were on Martinique (534), and the mongoose was present to prey upon birds, as it is now.

It has been said (Bond, 1936)[2] that on the island of St. Lucia the range of the large, fierce native boa (*Constrictor orophias*) and that of the wren and other ground-nesting birds coincide because the snake presumably controls the mongoose. *Constrictor orophias* is not be found on Grenada, and there is no evidence that the two tree boas (*Boa cooki* and *B. hortulana*) have the same powers. The problem of the ecological rela-

---

[1] *In* Lawrence, 1879. Phillips (1926) says "not rare in 1902" but gives no authority.

[2] Danforth (1935) suggests that the poisonous *Bothrops atrox*, or fer-de-lance, protects the birds, but Bond disagrees.

tions of the snakes, the mongoose, and the birds on these islands is most interesting.

No nests or eggs have been described. If, as is probable, their habits are similar to their relatives on the mainland, they nest on the ground, in stumps and fallen logs, and in crevices in walls. Here they would undoubtedly be vulnerable to predators.

Only one species [1] is to be found on the Windward Islands of the West Indies. These, like all related wrens, are small (total length 4–5 inches), nervous birds with long curved bills. They are dark brown above; wings and tail barred with black. Some of the subspecies have the underparts white, others brown.

It is probable that they have been long established. Each island has its well-marked form.

The following key will help to distinguish these:

I. *Underparts white*

    Larger (wing 59–60)..................................*musicus* (St. Vincent)
    Smaller (wing 52–54)................................*mesoleucus* (St. Lucia)

II. *Underparts brown*

    1. Larger as a rule (wing 53–59).....................3
       Smaller as a rule (wing 48–54).....................2
    2. Under tail coverts with white tips..................*guadeloupensis*
       Under tail coverts without white tips...............*rufescens* (Dominica)
    3. Lores brown like head............................*martinicensis*
       Lores white ....*grenadensis*

Location of specimens: I know of specimens of *Troglodytes musculus guadeloupensis* in Cambridge, Mass.; Chicago; London; and New York; of *T. m. martinicensis* in Cambridge, Mass.; Chicago; London; and New York; of *T. m. musicus* in Cambridge, Mass.; Chicago; London; and New York; and of *T. m. mesoleucus* in Cambridge, Mass.; London; Mayagüez, Puerto Rico; Chicago; and New York.

Some Pacific Island Thrushes

Turdus poliocephalus vinitinctus (Gould)

> *Merula vinitincta* GOULD, Proc. Zool. Soc. London, 1855, p. 165 (Lord Howe Island). Fig.: Mathews, 1928, p. 46.

Status: Extinct.

---

[1] Confined to the Zapata Swamp on the south coast of Cuba is *Ferminia* (or *Thryomanes*) *cerverai* Barbour. It is said to be common there.

Range: Lord Howe Island, about 500 miles off the east coast of Australia.

Three populations of the island thrush (*Turdus poliocephalus*) have been extirpated or are very nearly extinct. Closely related to the black

Fig. 60.—Island thrush (*Turdus poliocephalus*).

bird of Europe (*Turdus merula*), the species occurs in many slightly differing color forms on islands from Formosa in the north, through the Philippines, southward to Sumatra, Java, eastward to Samoa, but never on the continents. The southernmost representative is extinct. The people of Lord Howe Island called it doctorbird. Before 1909 the birds were quite numerous but perhaps had been slowly diminishing in numbers, for by

1913 they were said not to be as common as formerly. This would not be surprising, for they nested on the ground and would have been vulnerable to the introduced domestic cats, dogs, and hogs. Soon after 1918, when rats came ashore from a wreck, the birds disappeared entirely (Hindwood, 1940).

An attempt was made without success to introduce the closely related *Turdus poliocephalus* from Norfolk Island nearby. This form would, of course, be in great danger if it were exposed to the same enemies.

Description: About the size of the English blackbird and the American robin (total length about 9 inches), the head was olive-brown, the back chestnut, wings and tail dark brown. Chin and throat were pale brown tinged with olive, and the underparts chestnut-brown tinged with lavender (vinous).

The bird of Norfolk has the head grayish white, and is otherwise blackish brown.

Certainly the representative subspecies *T. p. pritzbueri*[1] of Lifu, in the Loyalty Islands near New Caledonia, is extremely rare, if not extinct, for the experienced naturalist Macmillan searched for them without success (see Macmillan, 1939, p. 32; Mayr, 1944). *T. p. mareensis*[2] is most probably extinct.

Location of specimens: I know of specimens of *T. p. vinitinctus* in Berlin, London, New York, Sydney, and Washington, and of *T. p. pritzbueri* in Berlin and New York.

Bonin Island Thrush

Zoothera [Aegithocichla] terrestris Kittlitz

Zoothera [*Aegithocichla*] *terrestris* KITTLITZ, Mém. Acad. Imp. Sci. St. Pétersbourg, 1, 1830 (1831), p. 245 (Boninsima = Peel Island, Beechey Group). This series is usually cited under "Akademia Nauk S.S.S.R., Leningrad," in libraries. Fig.: Seebohm and Sharpe, Monograph Turdidae, pt. 3, pl. 33.

Range: Peel Island, Beechey Group, Bonin Islands, western Pacific.

Status: Extinct. None have been found since 1828, when four were collected. Neither the capable collector P. A. Holst, who spent three months in the islands in 1889, nor Allan Owston's Japanese collectors 10 years later, nor the Japanese naturalists Momiyama and Hachisuka, between 1920 and 1930, were able to find one. Perhaps the species was confined to Peel Island, largest of the group. Nothing save their appearance is known of them, but it is probable that they were extirpated by cats and rats that

---

[1] *Turdus pritzbueri* Layard, Ann. Mag. Nat. Hist., 1878, p. 374 (Lifu Island).

[2] *Turdus mareensis* Layard and Tristram, Ibis, 1879, p. 472 (Maré Island).

escaped from American and British whaling ships careened on the beaches for repair.

A medium-sized ground thrush, the head was brown faintly streaked with black, back streaked with paler brown and black, back and tail rusty, reddish brown; throat white, streaked on the sides with brown; flanks brown; underparts generally white.

Location of specimens: I know of specimens in Frankfurt, Leningrad, Leyden, and Vienna.

FIG. 61.—Bonin Island thrush (*Zoothera terrestris*).

Mysterious Bird of Ulieta

Turdus ulietensis Gmelin

> *Turdus ulietensis* GMELIN, Syst. Nat., **1**, 1789, p. 815 (Ulieta, now Raiatea, Society Islands); *ex* Latham, "The Bay Thrush," Gen. Syn., Birds, **2**, p. 35, no. 31.
>
> *Turdus badius* FORSTER, Descript. Animal., 1844, p. 239 ("Oriadea insula," now Raiatea, Society Islands).

Status: Extinct. No specimen now exists, although both Latham and Forster saw one without question. Known from Johann Georg Adam Forster's drawing, pl. 146, in the British Museum marked "Raiatea ♀, June 1, 1774" (*fide* Sharpe in History Collections . . . Brit. Mus., **2**, 1906,

p. 194). Johann Reinhold Forster (*l.c.*) says his son drew the bird accurately. We must therefore assume that a population has been extirpated.

Late in May 1774 a specimen of this bird was collected by a member of Captain Cook's second expedition, probably by one of the Forsters. It is not surprising that none were recorded in 1777, when Cook's ships anchored there again, for the natives gave trouble. No doubt the species was extinct before 1850, however, for Andrew Garrett, an indefatigable collector, was unable to find one. None have been discovered since then. Forster was a good naturalist; it is extremely unlikely that his description is distorted.

Range: Known only from Raiatea (Ulieta, Oreadea), in the western group of the Society Islands, about 200 miles from Tahiti.

Description: Following is Latham's description: Size of the song thrush: length eight inches and a half. Bill an inch and a quarter, notched at the tip, and of a reddish pearl-colour: general colour of the plumage rufous brown: quills edged with dusky: tail rounded in shape and dusky, legs dusky black.

"Inhabits Ulietea. In collection of Sir Joseph Banks."

An abridged (and free) translation of Forster's (*l.c.*) Latin description is as follows: "Head dusky marked with brown. Above dusky, all the feathers edged with reddish brown; wings dusky, the primaries edged with brown, as are the wing coverts and the tail feathers. Below ochraceous. Iris dark yellow. Twelve tail feathers. Tibiae compressed and with seven scutes. Tongue bifid at the tip and ciliated."

The scutellated tarsi and bifid tongue do not indicate a thrush. Whether this bird was a peculiarly adapted thrush or belonged to some other group cannot now be determined.

West Indian Mockers and Thrushes

Ramphocinclus brachyurus brachyurus (Vieillot)

> *Turdus brachyurus* VIEILLOT, Nouv. Dict. Hist. Nat., nouv. éd., **20**, 1818, p. 255 (Martinique).

Status: Very rare and localized.

Range: Confined to Presqu'île de la Caravelle, Martinique Island, West Indies.

Ramphocinclus brachyurus sanctae-luciae Cory

> *Ramphocinclus brachyurus sanctae-luciae* CORY, Auk, **4**, 1886, p. 94 (St. Lucia).

Status: Rare and localized.

---

[1] See Bond 1936, 1948, 1950, 1951.

Range: Confined to a small area east of Morne la Sorcière, St. Lucia Island, West Indies.

Cinclocerthia ruficauda gutturalis (Lafresnaye)

    Ramphocinclus gutturalis LAFRESNAYE, Rev. Zool., **6**, 1843, p. 67 (Martinique).

Status: Very rare.
Range: Martinique Island, West Indies.

FIG. 62.—West Indian mocker (*Ramphocinclus brachyurus*).

On the islands of the West Indies there are two genera of thrushes and four of mockingbirds that are to be found nowhere else in the world. So markedly different from other members of their families are they that their relationships cannot now be traced. It is likely that they came originally to the islands from North America over water.

Of these six genera there are two rare mockingbirds, as well as a thrush, of which certain populations are in danger of extinction.

Rarest of these, and for that reason in greatest peril, are the white-breasted thrashers (*Ramphocinclus*), which occur only on Martinique (*R. b. brachyurus*) and St. Lucia (*R. b. sanctae-luciae*). They are medium-sized (total length about 8 inches) with rather long, thin bills, chocolate-brown above and pure white below They have a characteristic habit of cocking the tail and chattering.

From time to time in recent years both species have been thought to be extinct, but later were found to persist in small numbers. On St. Lucia they were to be seen in the bush north of Grande Anse and east of Morne la Sorcière in 1932 (Bond, 1950), and no doubt they still are. On

FIG. 63.—Trembler of Dominica (*Cinclocerthia ruficauda*).

Martinique, where they had been thought almost certainly to be extinct because they had not been reported since 1886, they were found at Presqu'île de la Caravelle in 1950 (Bond, 1951, p. 14).

Next of the mockers in order of rarity are the tremblers (*Cinclocerthia*), large (total length about 10 inches) and with very long thin bills, varying in color among the islands from dark brown to gray on the back and paler brown to very pale gray below. Their tendency to shiver or tremble violently has never been explained. On St. Lucia, the only island on which they occur together with the rare thrasher, this characteristic, as well as the pure white underparts of the smaller thrasher, should distinguish them.

The genus contains but a single polymorphic species. Of the six well-marked subspecies of the single species (*Cinclocerthia ruficauda*), four are said to be fairly common, although the small populations are probably hard to find in the thick forests. These are *ruficauda* of Dominica, *tremula* of Guadeloupe, *macrorhyncha* of St. Lucia, and *tenebrosa* of St. Vincent. On Martinique *gutturalis* is now extremely rare. Even in 1879 Ober (*in* Lawrence), said of it: "This Trembleur, known as the 'Grive Trembleuse,' is not found so easily as on Dominica. This I attribute wholly to the fact that it is pursued here with greater vigor than on the other island. So dense is the population of Martinique that nearly every bird is considered as fit for food."

Of the race *pavida,* a few are still found in mountain forest on the small islands of Saba and St. Christopher (St. Kitts), and on Montserrat there are rather more of them. They are thought to have been extirpated from St. Eustacius and Nevis. Little forest habitat exists—or perhaps ever existed—on these islands; and they are now confined to the peaks of extinct volcanoes on both. Bird populations were probably never large. Only two specimens have been taken on Nevis, and those about 1850.[1] Danforth (1936), who spent several days on the island, could not find a trembler there. There exists only a single specimen from St. Eustacius. It is in the United States National Museum.

It is doubtful that the birds ever were to be found on Barbuda. A specimen collected by Ober in the United States National Museum is said to have come from there (Lawrence, 1879), but it is a low island without forest and quite possibly an error was made, for the consignment of specimens was mislaid for a period during which Ober was subject to attacks of fever.

It is possible that before the advent of Europeans and an agricultural economy caused the destruction of large trees, tremblers were to be found at sea level. They are to be found in forests only 800 feet above the tide line on Montserrat today (Danforth, 1939b). However, evidence for their former occurrence on the low-lying Barbados is not very convincing, for the authors of the seventeenth and eighteenth century natural histories (Ligon, 1673; Hughes, 1750) [2] make some strange misstatements. They do say that thrushes called "grive" or "quaking thrush" were to be found, and no other bird is known to shiver as these do.

---

[1] Sclater and Salvin, 1866–1869. Two specimens in British Museum.
[2] See also Clark, 1905.

Forest thrushes (*Cichlerminia l'herminieri*) are found only on a few of the Lesser Antilles. On Dominica (*dominicensis*) and Montserrat (*lawrencei*) they are not rare, but on Guadeloupe (*l'herminieri*) and St. Lucia (*sanctae-luciae*) they have become so in recent times. Possibly they once occurred on Martinique and St. Vincent, but if they did they have disappeared without leaving a trace.[1] The obvious reasons for their rarity are not only that they are inclined to be terrestrial and therefore are prey for the mongoose, but also that men consider them to be a delicacy (Bond, 1936).

They occur together with tremblers on Dominica and St. Lucia, but there is little difficulty in distinguishing them, for they have short thick yellow bills; the feet and legs are yellow, not brown, and the underparts are spotted with white.

Two mockers (*Allenia* and *Margarops*) and a thrush (*Mimocichla*) are the three remaining endemic genera not to be found elsewhere. With the exception of the more heavily settled islands of Martinique and St. Lucia, *Margarops* (pearly-eyed thrashers) are common. Even on these islands *Allenia* (scaly-breasted thrashers) are not difficult to find, and only on Grenada are they said to be rare.

The red-legged thrush (*Mimocichla*) is more or less common on the northern Bahamas, Cuba, Hispaniola, Puerto Rico, and Dominica. It is rare on Cayman Brac and has perhaps been extirpated from tiny Swan Island, off the western end of Cuba, as well as from some of the southern Bahamas, as the bones found on Great Exuma indicate (Wetmore, 1937).

The relative abundance of these three genera is to be explained, at least partly, by the fact that they are much less often found on or near the ground and therefore less vulnerable to predators.

Although there are four genera nesting in the West Indies and also in North America, one in particular should be mentioned. Nesting on the ground or in stumps as they do, the solitaires (*Myadestes elizabethae* and *M. genibarbis*) are extremely vulnerable to rat and mongoose, and it is remarkable that they should still occur commonly on the Greater as well as many of the Lesser Antilles. However, they may once have been found on Guadaloupe, although there is no record. Races of the Isle of Pines (*M. e. retrusus*) and St. Vincent are inexplicably rare. That they are not extinct is indicated by sightings on the Isle of Pines in 1934 and on St. Vincent in 1950 (Bond, 1950, 1951).

Presence or absence of genera on certain islands requires more explanation even if it be speculative, for it has been suggested that these genera are

---

[1] Present records are believed by Bond (1951, p. 15) to be erroneous.

relicts, formerly more widespread in the West Indies (Bond, 1948), imply-
ing that many populations have become extinct without leaving a trace.
The range of *Margarops* has been used as an example. With an apparent
center of distribution in Dominica-Guadeloupe-Martinique, it is found
sporadically as far north as the Bahamas, occurring on Puerto Rico and
small islands, but not the larger islands of Hispaniola, Cuba, or Jamaica.
It may be that this indeed was the case; subfossil remains of *Margarops*
and *Mimocichla* found on Exuma, Bahamas, where neither now is to be
found, appear to support such a theory. Certainly there has been insufficient investigation to prove or disprove an hypothesis that there once
were such forms on the Greater Antilles.

It is quite as possible—even likely—that the peculiar, discontinuous
distribution of many endemic West Indian birds is due principally to
their having migrated to the islands under stress of wind and weather
and their having established themselves where lack of competition permitted. When this occurred, and what the ecological situation in which
they originally found themselves was, it is not impossible to determine.
The fortuitous manner of their arrival appears to explain most of the facts
of their distribution and the remainder may quite well be due to ecological
requirements now imperfectly understood or to competitors now unknown.
For example *Mimus gundlachii* is to be found only in limestone formation.

Whether or not species have disappeared from Cuba or Jamaica, it is

DISTRIBUTION OF MOCKERS (MIMIDAE) AND THRUSHES (TURDIDAE)

Locality	Allenia	Mar-garops	Cinclo-certhia	Rampho-cinclus	Mimo-cichla	Cichler-minia
Bahamas	—	C	—	—	LC	—
Cuba	—	—	—	—	C	—
Hispaniola	—	a	—	—	C	—
Puerto Rico	—	C	—	—	C	—
Jamaica	—	—	—	—	—	—
Virgin Islands	—	C	—	—	—	—
St. Kitts	C	C	R	—	—	—
Antigua	C	C	—	—	—	—
Montserrat	C	C	C	—	—	LC
Guadeloupe	C	C	LC	—	—	LR
Dominica	C	C	C	—	C	C
Martinique	C	R	R	LR	—	—
St. Lucia	C	R	C	LR	—	LR
St. Vincent	C	—	C	—	—	—
Barbados	R	—	—	—	—	—
Grenada	R	—	—	—	—	—

a Confined to Beata Island, off the coast. C—Common. R—Rare. L—Localized.

certain that neither the ancestors nor the relations of these genera are to be found. Can it be that they were extirpated from the mainland, or has the mode of evolution been so different that we can no longer know them?

Location of specimens: I know of specimens of *Cinclocerthia r. gutturalis* in Cambridge, Mass.; Chicago; Dresden; London; New York; and Philadelphia; of *C. r. tremula* in Cambridge, Mass.; Chicago; London; and New York; of *C. r. pavida* in Cambridge, Mass.; Chicago; and New York; of *Ramphocinclus b. brachyurus* in Cambridge, Mass.; Chicago; Dresden; London; Philadelphia; and Stockholm; of *R. b. sanctae-luciae* in Cambridge, Mass.; Chicago; Mayagüez, Puerto Rico; and Philadelphia; of *Cichlerminia l'herminieri l'herminieri* in Cambridge, Mass.; Chicago; London; Philadelphia; and New York; and of *C. l. sanctae-luciae* in Cambridge, Mass.; Chicago; and New York.

Hawaiian Thrushes

Confined to the Hawaiian Islands and apparently long isolated there, the relationships of the genus *Phaeornis* are with *Myadestes,* an American thrush distributed through California and Mexico, south to the highlands of Nicaragua and in the West Indies (Stejneger, 1889; Amadon, 1942b).

Phaeornis palmeri Rothschild

> *Phaeornis palmeri* ROTHSCHILD, Avifauna Laysan Island, etc., pt. 2, 1893, p. 67 (Halemanu, Kauai). Fig.: Wilson and Evans, Aves Hawaiiensis, pt. 6.

Status: Rare and localized. Sighted by *Richardson (1961).
Range: Kauai Island.

Phaeornis obscurus myadestina Stejneger

> *Phaeornis obscurus myadestina* STEJNEGER, Proc. U.S. Nat. Mus., **10**, 1887, p. 90 (Kauai). Fig.: Rothschild, Avifauna Laysan, p. 61.

Status: No recent information. Localized and probably rare.
Range: Kauai Island.

Phaeornis obscurus oahensis Wilson and Evans

> *Phaeornis obscurus oahensis* WILSON AND EVANS, Aves Hawaiiensis, 1899, p. xiii (Oahu).

Status: Extinct.
Range: Oahu Island.

Phaeornis obscurus rutha Bryan

> *Phaeornis obscurus rutha* BRYAN, Occ. Pap. B. P. Bishop Mus., **4**, no. 2, 1908, p. 81 (Molokai).

Status: Probably extinct.

Range: Molokai Island.

Phaeornis obscurus lanaiensis Wilson

> *Phaeornis obscurus lanaiensis* WILSON, Ann. Mag. Nat. Hist. (6), **7**, 1891, p. 460 (Lanai). Fig.: Rothschild, Avifauna Laysan, p. 61.

Status: Probably extinct.

Range: Lanai Island.

Phaeornis obscurus obscurus (Gmelin)

> *Muscicapa obscura* GMELIN, Syst. Nat., **1**, 1789, p. 945 ("in insulis Sandwich" = Hawaii). Fig.: Rothschild, Avifauna Laysan, p. 61.

Status: Not uncommon.

Range: Island of Hawaii.

Like their American relatives they are all of medium size (total length 6 to 8 inches) and plain colors—adults brown above, gray below, and young spotted with pale brown. They can scarcely be confused with any other birds in the Hawaiian Islands.

A curious habit of trembling is shared by the genus *Chasiempis,* which is apparently of Asiatic origin. Their song is said to be the most pleasing of any of the Hawaiian birds.

The following key distinguishes the forms:

1. Outer tail feathers tipped with white or buff.....................2
   Outer tail feathers not tipped with white or buff.................6
2. Feet and legs pink (yellow when dry)...........................4
3. Feet and legs blackish gray.......................................5
4. Smaller; wing (87–90 mm.)....................................*palmeri*
5. Larger; wing (99–107 mm.)...................................*myadestina*
6. Darker gray below; larger (wing 101–104 mm.).................*obscurus*
7. Paler gray below; smaller (wing 89–96 mm.)..................*lanaiensis*

P. o. rutha was described as slightly larger, throat and breast grayer, and back darker than *lanaiensis,* by which it may be inferred that the birds are paler below than *obscurus* (wing 90–97 mm.).

Phaeornis palmeri, the puaiohi of the Hawaiians, may once have been common and may have inhabited all of the forests of the island of Kauai, but there is no evidence that it did. When first made known to science in 1893, it was said to be rare indeed. Only one was taken after months of work. Even the redoubtable naturalist R. C. L. Perkins saw few; he considered it one of the rarest of the islands. The last specimen was collected by him in 1892. Only a sight record of 1940 [1] permits us to hope that some still live. Amadon (1950) says the species is probably extinct.

Naturalists who saw the birds alive believed that they were chiefly insectivorous, feeding for the most part on beetles to be found in the great koa trees (*Acacia koa*), and Palmer (Rothschild, 1893–1900), collector of the type specimen, found them only in forests largely composed

FIG. 64.—Thrush of Hawaii (*Phaeornis obscurus*).

of koa at Halemanu. This diet they found to be in strong contrast with that of the physically larger birds, *Phaeornis obscurus myadestina,* called kamau or amaui. The two occupied the same general habitat, and the latter, as do other Hawaiian subspecies of the same species, ate fruit and berries.

No doubt these observations should not be taken too seriously, for they are few, and negative evidence is always unsatisfactory. Indeed there is positive evidence from Henshaw (1902) and Munro (1944) that the representative subspecies of the island of Hawaii and Lanai eat insects as well as fruit.

At any rate, the situation on Kauai in 1892 was that there were two species, one small in size and numbers, the other quite the contrary.

[1] By Walter Donagho (Munro, 1944).

Theorizing about the simultaneous occurrence of these two species, Amadon (1947) says that in all likelihood the smaller *palmeri* colonized the island of Kauai before the larger *myadestina*. Perhaps we may theorize further that those individuals of *palmeri* which ate insects would have been under selective pressure and their progeny would have been more likely to survive, for in competition the smaller fruit-eating individuals would have been at a disadvantage with newcomers in obtaining favored berries in season. In time the weaker species as a whole suffered numerically because of specialized diet. At present *myadestina* is apparently confined to the high mountain regions. Munro (1944) found the birds in 1936 only at Kaholuamanu, in a forest reserve 3,700 feet above the sea. In 1891 they had been common in forests from sea level to the tops of the mountains. Partial destruction of forest, or perhaps a predator such as the roof rat, was responsible for the reduction in numbers, but the true reasons remain obscure.

Little can be said of the form that no doubt lived once on Oahu (*oahensis*), the next island to the eastward. In 1825, when Bloxam, naturalist in H.M.S. *Blonde,* collected at least one bird, he found the form common on the sides of the hills. Exactly when it was extirpated is not known. There are none now, nor are there any specimens in museums, for although Bloxam collected a few they were lost.

The Molokai thrush (*rutha*) disappeared sometime between 1907 and 1936 (Munro, 1944). The most recent search attained to the summit of Mount Olokui, where the forests have never been disturbed. No thrushes were seen or heard, but an important clue to the destruction of these populations is the discovery of *Rattus rattus* in the trees there (Richardson, 1949).

According to the authority on Hawaiian birds George Munro, the decline in population of the Lanai thrush began in 1923, when Lanai City was built and the human populations increased. In his opinion bird diseases, endemic in imported poultry and transmitted by mosquitoes, were the cause of this. There can be little hope that any still live. Munro (1944) states that he has not seen or heard one since 1931.

Only on Hawaii have the birds been reported to be common in recent years. They were numerous in Ohia Lehua (*Metrosideros polymorpha*) forests north of Makaopuhi crater in 1940 and less commonly on the sides of Kilauea Volcano, 4,000 feet, and Mauna Loa from 8,000 to 8,500 feet in 1940 (Baldwin, 1941).

Location of specimens: I know of specimens of *Phaeornis palmeri* in Cambridge, England; Cambridge, Mass.; Honolulu; and London; and of *P. obscurus lanaiensis* in Cambridge, England; Cambridge, Mass.; Edinburgh; London; New York; and Washington.

New Zealand "Thrushes"

Turnagra capensis tanagra (Schlegel)

Otagon capensis tanagra SCHLEGEL, Nederlandsch Tijd. Dierk. (Amsterdam), **3**, 1866, p. 190 (New Zealand = North Island). Fig.: Buller, 1888, p. 26.

Common names: North Island thrush [1]; piopio.

FIG. 65.—North Island thrush (*Turnagra capensis tanagra*).

Status: Extremely rare and localized, if not extinct. The last specimen was taken in 1900 at Waitotara.[2] In 1949, however, W. P. Mead reported having seen and heard birds similar to the native thrush on the Wanganui River near Te Auroa. This report has never been confirmed, but there are many sight records.[3]

---

[1] According to Oliver (1945, p. 148), these birds are closer to the Muscicapidae or Old World flycatchers than to the thrushes. See also Mayr and Amadon (1951) and Ripley (1951), who think it related to *Pachycephala*.

[2] In Dominion Museum, Wellington. Oliver, 1930.

[3] Notornis, **4**, no. 1, 1950, pp. 3–5. There are many unpublished reports as well (R. A. Falla *in litt.*).

The species was quite common in the southern part of North Island when Europeans arrived, but apparently it began to disappear quickly as civilization advanced. In 1887 they were still to be found in wooded regions of the west coast, but were then rare for they were said to have been common fifteen years before (Buller, 1888, 1, p. 237). Probably men as well as imported house cats and rats were responsible for a great diminution in numbers of birds, and destruction of habitat was a later factor accounting for their almost total extinction.

Range: North Island, New Zealand.

Description: A large perching bird (total length about 11 inches) having the general appearance of a thrush, with a distinctive whistle *pio-pio*. Head and back are olive-brown, wings and tail rufous, throat white, breast olive-gray, and belly yellowish white. The sexes are alike. The immature is spotted.

Turnagra capensis capensis (Sparrman)

Tanagra capensis SPARRMAN, Mus. Carlsonianum, 2, 1787, no. 45 (Dusky Sound, South Island, New Zealand).

Common name: South Island thrush.

Status: Extremely rare and localized in southwestern South Island. A sighting of 1947 near Lake Hauroko appears to be the most recent (Dunckley and Todd, 1949).

Naturalists who wrote in the latter part of the nineteenth century all speak of rats as a reason for the disappearance of bird populations, and it does appear possible that they were directly responsible. Large populations of birds disappeared between 1880 and 1890 from the west coast of South Island, where in 1863 they had been common (Buller, 1888, 1, p. 32). There can be little doubt that imported rats are extremely destructive to birds' nests and eggs (Bull, 1946). Introduced Norway rats were found in large colonies on the west coasts during those years. It must be admitted that Norway rats are perhaps not such a great danger to tree-nesting birds as are black rats and Alexandrine rats, but the records of those species in the South Island are not clear. However, in other parts of the world black rats usually preceded the Norways and were themselves displaced by the larger species.

Range: South Island, New Zealand.

Description: Like the North Island thrush but with throat and underparts heavily streaked with olive-brown. Young differ only in having more rusty brown on wing coverts and cheeks (Oliver, 1930).

Location of specimens: I know of specimens of *Turnagra capensis tanagra* in Chicago, London, Philadelphia, and Wellington; and of *T. c.*

capensis in Basle; Bremen; Brussels; Cambridge, Mass.; Chicago; Dresden; Exeter; Gotenburg; London; and Stockholm.

Lord Howe Island Flycatcher

Gerygone igata insularis Ramsay

> *Gerygone igata insularis* RAMSAY, Proc. Linn. Soc. New South Wales, **3**, 1878, p. 117 (Lord Howe Island).[1] Fig.: Mathews, 1928, p. 35.

Status: Extinct.

Apparently the little rainbird of Lord Howe Island disappeared shortly after rats came ashore from the wreck of the *Mokambo* on Ned's Beach

FIG. 66.—Riroriro of New Zealand (*Gerygone igata igata*).

in 1918 (Hindwood, 1940). No doubt these were roof rats, for they are known to destroy nests and eggs even in trees. One hundred owls were imported later to kill the rats, and undoubtedly these predators reduced the song bird population to some extent. Whatever the immediate cause

[1] For reasons for considering this a subspecies of *igata,* see Meise, Nov. Zool., **36**, 1931, p. 351.

was, the birds ceased to breed in numbers after 1919 and disappeared, even though they had formerly been common.

Range: Lord Howe Island, about 500 miles off the east coast of Australia.

This was a clearly differentiated representative of the polytypic species called riroriro, the gray warbler, in New Zealand, where it is apparently still quite common in the forests. Another relative still lives on nearby Norfolk Island.

These were small (total length about 5 inches) active birds of the forests. The head, back, wings, and tail were brown with an olive tinge; throat and upper breast gray and belly yellow. The outer tail feathers were spotted with white; this character, as well as a white ring around the eye, was a prominent distinguishing mark.

Lord Howe Island Fantail

Rhipidura fuliginosa cervina Ramsay

> *Rhipidura cervina* RAMSAY, Proc. Linn. Soc. New South Wales, **3**, 1879, p. 340 (Lord Howe Island), Fig.; Mathews, 1928, pl. 28, p. 42.

Status: Extinct.

Range: Lord Howe Island, about 500 miles off the east coast of Australia.

These birds were most closely related to a group which is still quite common in the forests of Australia and New Zealand. They disappeared from Lord Howe Island between 1924 and 1928. Perhaps this was a result of the accidental introduction of rats from a wreck of 1918.

These were small birds (total length about 5 inches) with long tails rounded at the end, which they characteristically spread like a fan. They were most often to be seen darting about in the trees after insects.

The entire top of the head and wings (except for two whitish bars were dark brown, underparts buff, outer tail feathers tipped and edged with white.

Pomarea of the South Seas

Only because they have often been spoken of as "rare birds" are these species mentioned here. *Pomarea dimidiata* (Hartlaub and Finsch), confined to Rarotonga in the Cook Group, is said still to be not uncommon, having been seen by members of expeditions from the Bishop Museum, Honolulu, and the American Museum of Natural History, New York during the past 25 years.

Pomarea nigra (Sparrman) was collected on Tahiti by members of the Whitney Expedition of the American Museum, in spite of the fact that residents had reported that it was to be seen no more. *Pomarea mendozae* subsp. are still to be found not uncommonly on the Marquesas Islands.

Location of specimens: I know of specimens of *Pomarea dimidiata* in Bremen; Cambridge, England; Hamburg; Honolulu; London; New York; and Philadelphia.

Laysan Millerbird

Acrocephalus familiaris familiaris (Rothschild)

> *Tatare familiaris* ROTHSCHILD, Ann. Mag. Nat. Hist. (6), **10**, 1892, p. 109 (Laysan Island).

Status: Extinct.

Range: Laysan Island, 790 miles northwest of Honolulu, Hawaiian Islands.

FIG. 67.—Laysan millerbird (*Acrocephalus familiaris familiaris*).

This subspecies disappeared between 1904, when rabbits were introduced on Laysan Island, and 1923, when a party of naturalists from the Smithsonian Institution landed and found that almost all vegetation had been destroyed by them and no birds remained (Wetmore, 1925a).

The birds were island representatives of a large and reproductively powerful species called reed warblers, with representatives extending from the coasts of eastern Asia through the Marianas, eastern Carolines, reappearing in the Society Islands and Tuamotus in eastern Polynesia. Of all this vast number of forms only this one is known to have become extinct, although perhaps some, of which nothing now is known, may have lived once in the Marshall Islands, where they do not now occur. Nothing is known of *Acrocephalus luscinia rehsei* (Finsch) [1] of Nauru. That small island was occupied by troops, and the birds may be extinct.

These were small (total length about 4 inches) flycatcherlike birds, grayish brown above and brownish white below, usually to be seen chasing miller moths.

Location of specimens: I know of specimens in Bremen; Cambridge, Mass.; Chicago; Denver; Honolulu; New York; and Pittsburgh.

Australian Bristlebirds

None of the species of the peculiar Australian genus *Dasyornis* are extinct, but some are rare, localized, and presumably in danger of extinction because they are few, often menaced by bush fires and, nesting near the ground as they do, by rats and other predators. Although they are somewhat shy and retiring, both the species have been found within the past 10 years in widely separated populations not far from the coasts of New South Wales, South Australia, and West Australia. The forms with some details of their status follow:

Dasyornis brachypterus brachypterus (Latham)

Turdus brachypterus LATHAM, Index Orn., Suppl., 1801, p. 43 (Sydney).

Status and range: Eastern coast of Australia. Sightings are recorded from the Cordeaux River, the Dorrigo Tableland, the Jambero District in New South Wales, and Lamington National Forest in northern Queensland during the past 10 years.[2]

---

[1] *Calamoherpe rehsei* Finsch, Ibis, 1883, p. 143 (Nawado = Nauru Island, Gilbert Islands, South Pacific). Now placed in Sylviinae.

[2] McNamara, 1946; Robertson, 1946; Goddard, 1948. The New York Times reported a sighting at Kiama, south of Sydney, in January 1939.

? *Dasyornis brachypterus victoriae* (Mathews)

Sphenura brachyptera victoriae MATHEWS, Austral Av. Rec., **3**, 1916, p. 61 (Muddy Creek, Victoria).

Status and range: Perhaps a sighting (Bryant, 1936) at Womboyne Inlet in New South Wales, near the border of Victoria, in 1936 was of this supposedly darker form. We may doubt that the race can be distinguished. Gippsland, a locality recorded by Mathews,[1] is not far from there. In view of the general interest shown in these rare birds during the past 30 years, it is strange that there are no records other than Mathews's from Gilburn River.

Dasyornis brachypterus longirostris Gould

Dasyornis brachypterus longirostris GOULD, Proc. Zool. Soc. London, 1840, p. 170 (Swan River, West Australia).

Status and range: A specimen was collected in 1945 at Two People Bay, east of Albany. The birds formerly were to be found along the coasts of extreme southwestern Australia, from Perth to the vicinity of Albany, but latterly had become so rare that many believed them to have disappeared (Serventy and Whittell, 1951).

Dasyornis broadbenti broadbenti (McCoy)

Sphenura broadbenti McCOY, Ann. Mag. Nat. Hist. (3), **19**, 1867, p. 185 (Portland Bay, Victoria).

Status and range: The rufous bristlebird is apparently common all along the south coasts, from Marlo in eastern Victoria to the mouth of the Murray River and Otway Range in South Australia.[2]

Dasyornis broadbenti littoralis (Milligan)

Sphenura littoralis MILLIGAN, Emu, **1**, 1902, p. 69 (Ellensbrook, South West Australia).

Status and range: Rare and confined to about 30 miles of coastline between Cape Naturaliste and Cape Mantelle, in extreme southwestern Australia.

These birds are found only with difficulty skulking among ferns and grasses. They are often spoken of as frequenting tea-tree thickets behind the beaches.

Location of specimens: I know of specimens of *Dasyornis b. brachypterus* in Brisbane, Australia; Cambridge, Mass.; London; and New York;

---

[1] Birds Australia, **10**, 1922–23, p. 153.
[2] Lang, 1946; Serventy and Whittell, 1951.

of *D. b. longirostris* in London; Melbourne, Australia; New York; Perth, Australia; Philadelphia; and Sydney, Australia; of *D. b. broadbenti* in Sydney and New York; and of *D. b. litoralis* in Adelaide, Australia; London; Melbourne; New York; and Sydney.

Mata, the Fernbird

Bowdleria rufescens (Buller)

> *Sphenoeacus rufescens* BULLER, Ibis, 1869, p. 38 ("small rock isle, a satellite of Chatham Island").

Status: Extinct.

Range: Recorded only from Pitt Island and Mangare Island, near Chatham Island, off the coast of New Zealand. Both of these islands were settled during the last part of the nineteenth century, and much habitat was destroyed by burning. Introduced rabbits and goats destroyed more. Even before that cats had accidentally been introduced, and they are thought to have been very destructive [1] to birdlife. No doubt rats had come with the men also, and they would naturally be a great danger to birds nesting near the ground, as these did.

There is still some heavy scrubby bush on the steep cliffs of Mangare Island, and perhaps some birds may still be found there (E. G. Turbott, *in litt.*) or on one of the smaller islets nearby, but it is not probable.

On the Snares (Island), where there are no men and no introduced mammals, fernbirds (*caudata*) [2] are to be found commonly in all available habitat, including the steep bluffs (Stead, 1948). Like the birds of the larger islands they nest in grass tussocks or clumps of fern or in hollow logs and would therefore be in extreme danger should rats come to the island. Escapes from a wreck would breed thousands within a few months, and the birds would disappear even though it has been declared a sanctuary by the government.

Fernbirds are all rather small, brownish, heavily streaked with black, and with long pointed tails. They are usually to be seen only with difficulty, skulking among ferns and grasses near the ground.

B. punctata (and subspecies) is the smallest (total length about 6½ inches; wing 58–65 mm.). The head is rusty brown, spotted with darker brown, sometimes with a white "eyebrow"; back and flanks are pale brown, streaked with darker, underparts white (or pale brownish white), spotted with dark brown. Wing and tail are dark brown, the latter very decomposed and "spiny."

---

[1] See Fleming (1939) for a good account.

[2] *Sphenoeacus caudatus* Buller, Trans. New Zealand Inst., **27**, 1895, p. 127 (Snares).

B. caudata of Snares Island is longer (total length about 7½ inches; wing 70–74 mm.) and much paler; the belly and flanks are a uniform fawn color, not streaked or spotted.

B. punctata and subspecies [1] is still seen regularly although in restricted areas on North Island, Barrier Island, and South Island, as well as Stewart and the smaller islands—Big South Cape, Solomon, Pukeawa, Cundy, Jaques Lee, Codfish, and probably others. On the larger islands, where

FIG. 68.—Chatham Island fernbird (*Bowdleria rufescens*).

rats and other predators are common, the birds are confined to swamps but are seen often [2] by those interested enough to search for such shy elusive birds.

---

[1] *Bowdleria punctata punctata* (QUOY AND GAIMARD). *Synallaxis punctata* Quoy and Gaimard, Voyage *Astrolabe,* Zool., **1**, 1830, p. 255 ("baie de Tasman," South Island).

B. p. vealeae Kemp, Austral Av. Rec., **1**, 1912, p. 124 (North Island).

B. p. stewartiana Oliver, New Zealand Birds, 1930, p. 451 (Stewart Island).

B. p. insularis Stead, Trans. and Proc. Roy. Soc. New Zealand (Wellington), **66**, 1936, p. 312 ("Stewart Island and surrounding islets"). Unless it be a distinct species, this name is inevitably a synonym of *B. p. stewartiana.* No doubt this population is extremely variable individually and long series are necessary to distinguish between geographical variation and those due to other causes. I have no material on which to base a decision.

B. p. wilsoni Stead, *l.c.* ("confined to Codfish Island").

[2] Potter, 1949; Classified notes in Notornis (New Zealand Bird Notes), **4**, 1952, p. 191.

B. *rufescens* (the extinct form of the Chatham Islands) is larger than *punctata* but smaller than *caudata* (total length about 7 inches; wing 59–67 mm.). It differs sharply from the other two species in having unspotted white underparts and in its maroon crown, distinct white loral spot, and darker brown back.

They are best included in the subfamily Sylviinae, close to *Megalurus* (see Mayr and Amadon, 1951).

Location of specimens: I know of specimens of *Bowdleria rufescens* in Auckland, New Zealand; Cambridge, Mass.; Berlin; Chicago; Christchurch, New Zealand; London; Liverpool; New York; Paris; Pittsburgh; Stockholm; and of *B. punctata punctata* in Auckland; Cambridge, Mass.; Chicago; Christchurch; New York; and Pittsburgh.

Eyrean Grass-wren

Amytornis goyderi (Gould)

> *Amytis goyderi* GOULD, Ann. Mag. Nat. Hist. (4), **16**, 1875, p. 286 (Lake Eyre, South Australia). Fig.: Mathews, Birds Australia, **9**, 1921–22, p. 206 ("Eyramytis"); Emu, **23**, 1924, pl. 14.

Status: Rediscovered "in small numbers" at Christmas Water Hole, South Australia (lat. 27°30′ S., long. 136°46′ E.) on September 3, 1961 (*Morgan, Robinson and Ashton, 1961, and *H. T. C. [Condon], 1962, p. 216). Not one had been found since the two cotypes were collected in 1875. These specimens are in the British Museum.

Range: Found near the Macumba River, which flows into the northern end of Lake Eyre, Australia. Presumably it is dry except in unusually rainy weather. Whitlock (1924, p. 273) describes the region: "A remarkably cold wind was blowing half a gale in our faces and the air was full of particles of flying dust . . . the slope of a low hill clothed with salt brush, and isolated masses of low dead scrub . . . . There was no camel feed and they were in need of water."

Description: A small (total length about 147 mm.) brown warbler (Sylviidae) streaked above with white and with a long tail. It differs from similar forms by its thick sparrowlike bill.

Head and back cinnamon-brown with short white streaks, rump unstreaked, breast white, belly and under tail coverts pale buff. Wings and tail darker brown, with a grayish cast to the center of the feathers.

Semper's Warbler

Leucopeza semperi Sclater

> *Leucopeza semperi* SCLATER, Proc. Zool. Soc. London, 1876, p. 14, pl. 2 (St. Lucia).

Common name: Semper's warbler.

Status: Extremely rare. According to Bond (1950) this bird was last collected on the summit of Piton Flore in 1934 by Stanley John and last sighted by the same man between Piton Lacombe and Piton Canaries in March 1947. Because the population has been decreasing during the past 50 years, for causes unknown, the species may well be in danger of extinction.

Nothing is known of its habits.

Range: St. Lucia, West Indies.

Habitat: Confined to undergrowth of mountain forests.

Description: A rather large wood warbler (total length about 6 inches), dark gray above and grayish white below.

Green Solitaire

Viridonia sagittirostris Rothschild

> *Viridonia* [1] *sagittirostris* ROTHSCHILD, Ann. Mag. Nat. Hist. (6), **10**, 1892, p. 112 (Lower Hilo slopes of Mauna Kea, Hawaii).

Common names: The Hawaiians had no name for this bird.

Status: Certainly extremely rare, if not more probably extinct, for not one has been reported since 1900.[2] Its discoverers found it a small and very localized population, difficult to find even in 1892 (see Rothschild, 1893–1900; Perkins, 1903). Amadon (1950) says of the habitat: "This forest gave way to sugar cane, and *sagittirostris* is extinct."

Range: Mountain rain forest on the windward (northeastern)side of the island of Hawaii, near the Wailuku River, between 1,200 and 4,000 feet.

Description: A small green bird usually to be seen creeping on the branches of trees in search of insects. Similar to *Chlorodrepanis virens* (the amakihi) but almost twice as large, with a straighter bill. It is more olive green both above and below, which together with its larger size, should distinguish it from the olive-green creeper (*Paroreomyza mana*)[3] also.

---

[1] Placed in the genus *Loxops* by Amadon, 1950.

[2] Specimens in Museum of Comparative Zoology collected by H. W. Henshaw. See also Baldwin, 1941.

[3] Variously named *Himatione, Oreomyza* by authors, and recently "lumped" with *Loxops* by Amadon, 1950.

The Creepers (*Paroreomyza*)[1]

Two subspecies of this group have been sought in vain and are extremely rare, if not quite extinct; they are:

Loxops (*Paroreomyza*) *maculata flammea* Wilson

Loxops (*Paroreomyza*) *maculata flammea* WILSON, Proc. Zool. Soc. London, 1889 (1890), p. 445 (Kalae, Molokai).

Status: Probably extinct.
Range: Molokai.

Loxops (*Paroreomyza*) *maculata montana* (Wilson)

Himatione montana WILSON, Proc. Zool. Soc. London, 1889 (1890), p. 446 (Lanai).

Status: Probably extinct.
Range: Lanai.

On Molokai and Lanai much habitat has been destroyed. Both subspecies were quite common 50 years ago, and both have been sought during recent years in vain. An expedition of 1949 to Mount Olokui, Molokai, where original forest still exists, failed to find any (Richardson, 1949). None have been sighted on Lanai since 1937 (Munro, 1944). No reason can be given for the extinction of these forms other than the very considerable destruction of habitat. Why the presumably closely related amakihi (*Chlorodrepanis*) should survive in numbers cannot be explained.

On the island of Kauai *bairdi,* on Oahu *maculata,* on Maui *newtoni,* and on Hawaii *mana*[2] are still to be found, although much reduced in numbers.

These are all small grayish-green, yellow, or scarlet birds, usually to be seen crawling on the trunks and branches of the koa trees (*Acacia koa*) in search of insects. By this habit they may be distinguished from the very similar amakihi ("*Chlorodrepanis*"), still quite a common bird, which feeds among the leaves and twigs as a rule. The straight bill of the creeper and the song also set them apart.[3]

---

[1] Regarded as a subgenus by Amadon (1950), who perhaps rightly includes all the small creepers in the genus *Loxops.*

[2] See "Elepaio," **7,** no. 7, pp. 39, 41 (Kauai); **11,** pp. 29, 56, 58 (Oahu); and **12,** no. 5, p. 32 (Maui).

[3] The plates in Northwood, 1940 (frontispiece), and Munro, 1944 (pl. 12, fig. 3), show white wing bars, but these are characteristic of females and immatures, not mature males.

Akepa (Loxops)

Loxops coccinea (Bloxam)

Fringilla rufa BLOXAM, Voyage of the *Blonde,* 1826, app., p. 250 (Oahu).
Loxops wolstenholmei ROTHSCHILD, Avif. Laysan, p. 177, pl., is a synonym.

Common name: Oahu akepeuie.

Range: Oahu, Hawaiian Islands.

Status: Most probably extinct, the last specimen was obtained on May 20, 1893, by Palmer for the Rothschild collection. Perkins (*in* Munro, 1944) reports having seen one in the forests of the Waialua district about 1900, but in spite of search not even the most enthusiastic observer has claimed to have seen one since. Like all Hawaiian honeycreepers of Oahu, save the two small green ones, *chloris* and *maculata,* these have probably been extirpated by the destruction of the forests and the sources of food. Disease cannot be discarded as a possible reason.

Three close relatives (or geographical races) are still to be found on Hawaii and Maui, where they are very rare and localized, and on Kauai, where they are apparently more common.

Lanai and Molokai, so close to Maui, may once have had forms of this species, but if so they must have become extinct before the arrival of Europeans, for they have never been reported.

Habitat: Mountain forests from about 3,000 to 6,000 feet, where now they are most commonly to be found on Hawaii (Munro, 1944). They prefer *Acacia koa* forests but are also to be found in *Metrosideros* and other native trees.

Description: This genus (or subgenus) of small red, green, or brown birds is characterized by a distortion of the lower jaw, the lower mandible being twisted to the right or left across the upper. Their small size and relatively long forked tail distinguished them.

Males of the species *Loxops coccinea* may be distinguished by the following key (after Rothschild):

1. Above green, below greenish gray.....................*caeruleirostris* (Kauai)
2. Orange-scarlet ....*coccinea* (Hawaii)
3. Rufous-orange (sometimes with a brownish wash).......*rufa* (Oahu)
4. Ochraceous or brownish orange (sometimes yellow)......*ochracea* (Maui)

Females are all greenish gray, as are the juveniles. So rare are they that the plumages, characterized apparently by much individual variation, are not well understood.

Habits: Nesting begins in March, for a female with an egg in the ovary was taken on Maui on April 4 (Perkins, 1903), and perhaps earlier, for Munro (1944) says that they seemed to be mating in February on Hawaii. He noted young in first plumage in September.

The staple is thought to be caterpillars and spiders although, to a much lesser extent, they suck nectar from the flowers of ohia (*Metrosideros*). Perkins (1903) was of the opinion that the long tubular tongue, which appears to be specialized for nectar feeding, is also useful in procuring larvae which live in the terminal buds of trees as well as between leaves that have been fastened together. The crossed bill is useful also in this type of feeding, according to him. Munro (1944) thought the bill useful also for scraping scale insects from leaves.

FIG. 69.—Akepa (*Loxops coccinea*).

Location of specimens: I know of specimens of *Loxops coccinea rufa* in Berlin; Bremen; Cambridge, England; Cambridge, Mass.; Kassel, Germany; Frankfurt, Germany; Liverpool; London; New York; Philadelphia; and Vienna.

Akialoas and Nukupuus (Hemignathus)

The genus *Hemignathus* is divided sharply into two sections, both with extraordinarily long curved bills and tubular tongues for extracting nectar from flowers. One group, however, is set apart sharply from the other by having a short adzedged lower mandible, and these birds prefer insects, using the "adz" efficiently to extract them from the bark of koa trees. Only the Hawaiian representative, *H. wilsoni,* can surely be said still to exist, and that species is the best adapted to procure insect food, for the lower mandible is shorter and stronger and the birds are said to chop away soft bark with them (Perkins, 1903; Munro, 1944). Polynesians called these "nukupuu," translated "moundbeak," and by some ornithologists in the past they were called *"Heterorhynchus."* [1]

The group named *Hemignathus,* or "akialoa" (curvebill) by Polynesians, had upper and lower mandibles about equal in length, and these, although they ate insects as well, were nectar-eaters. Perkins (1903, p. 423), a great naturalist, said of them: "I am satisfied that the different kinds of Akialoa are all still largely nectar-eaters, although possibly on the way to becoming entirely insectivorous. In the . . . genus Heterorhynchus [Nukupuu], we see a still further advance in the latter direction, indeed one of the species (*H. wilsoni*) already feeds on insects and spiders alone."

Probably there was considerable competition among these forms for, as the following table shows, species which were primarily nectar-eaters did not live together on the smaller islands; only on the large island of Hawaii was this possible. Here the size of the island itself permitted more species to coexist. That no form of *"Heterorhynchus"* occurred on Molokai indicates that perhaps a subspecies became extinct before it could be known to science, although the presence of the bellicose mamo (*Drepanis*) may have prevented colonization.

The true reason for the disappearance of these curious birds is not known. Most of them were few in numbers even before cattle began to roam the forests and destroy the undergrowth. Disease has been mentioned as a possible factor (Munro, 1944; Amadon, 1950). It is also possible that insects and flowers upon which the birds fed at certain times of year have disappeared without our knowledge. Certainly it is to be deplored that they have gone without our having enough intimate detail of their mode of life to be able to assign a cause.

FOOD OF SELECTED HONEYCREEPERS

Form	Kauai	Oahu	Molokai	Maui	Lanai	Hawaii
"Heterorhynchus" lucidus Nukupuu ...	I	I	—	I	—	I
	Ex.?	Ex.?		Ex.?		
Hemignathus obscurus Akialoa	N	N			N	N
		Ex.?	—	—	Ex.	Ex.?
				I		
Pseudonestor	—	—	—		—	—
			N			N
Drepanis	—	—	Ex.	—	—	Ex.

I—Considerable nectar (insects also) N—Rarely nectar (insects as a rule). Ex.—Extinct.

[1] As constituted by Lafresnaye. See also Rothschild, 1893–1900; Perkins, 1903; Bryan and Greenway, 1944; Amadon, 1950. They are best designated as superspecies.

These are all small perching birds (total length about 4–5 inches). Their long curved bills should distinguish them. Males have yellow heads (with a slight greenish tinge), grayish-green backs and yellow (or grayish-yellow) underparts. Females, as well as immature birds, are grayish green above and pale gray below.

Fig. 70.—Nukupuu (*Hemignathus lucidus*).

The following key identifies the forms:

1. Lower mandible about half length of upper 4
   Upper and lower mandibles about equal 2
2. Wing shorter (76–85 mm.) 3
   Wing longer (85–94 mm.) . . . . *procerus* (Kauai)
3. Bill longer (52–53 mm.) . . . . *lanaiensis* (Lanai)
   ellisianus (Oahu)
   Bill shorter (41–47 mm.) . . . . *obscurus* (Hawaii)
4. Lower mandible curved 5
   Lower mandible straight . . . . *wilsoni* (Hawaii)
5. Top of head yellow or greenish 6
   Top of head grayish green (a pronounced yellow stripe above
      the eye) . . . . *lucidus* (Oahu)
6. Wing longer (♂ 79–87, ♀ 76–82 mm.) . . . . *hanapepe* (Kauai)
   Wing shorter (♂ 72–79, ♀ 69–71 mm.) . . . . *affinis* (Maui)

Hemignathus obscurus procerus Cabanis

Hemignathus obscurus procerus CABANIS, Journ. für Orn., 1889, p. 331 (Kauai).

Range: Mountain forests of Kauai Island.

Status: Probably still exists in the mountains of Kauai, but the population is very small and in danger of extinction. Munro (1944) reports on his observations of 1891 as follows (see also Perkins, 1903, p. 425): "Although these birds were quite numerous, it was evident that they were susceptible to disease. Their habit of coming to the forest's edge and to low elevations exposed them to introduced diseases. One was so disabled with lumps on legs and bill that it could scarcely fly. Another had a tumor a quarter of an inch thick in its throat full of small worms, and a tumor on its ovaries containing a brown paste. Perkins found them with tapeworms. Its decline was likely starting at that time. It may survive at the higher elevations."

Hemignathus obscurus lichtensteinii Wilson

Hemignathus obscurus lichtensteinii WILSON, Ann. Mag. Nat. Hist. (6), **4**, 1889, p. 401 (Oahu).

Drepanis (Hemignathus) ellisianus GRAY, Catalogue Birds Tropical Islands Pacific, 1860, p. 9 is a synonym of obscurus (fide Stresemann, Mitt. Zool. Mus. Berlin **30** (1), 1954, p. 44).

Range: Mountain forests of Oahu.

Status: Although two sightings have been recorded (in 1937 and 1940),[1] none have been reported since, even though careful searches have been made, and it is probable that this form is extinct; it was last seen surely in 1837.

Hemignathus obscurus lanaiensis Rothschild

Hemignathus obscurus lanaiensis ROTHSCHILD, Bull. British Orn. Club, **1**, 1893, p. xxiv (Lanai).

Range: Mountain forests of Lanai.

Status: Probably extinct. Last sure sighting in January 1894 by R.C.L. Perkins. Munro (1944) says: "I watched for it for 20 years and on only one occasion saw a bird that might reasonably be supposed to be this species. It was across a small valley from my position, moving up the steep hillside, flying from tree to tree. It was more yellow in color than any other Lanai bird, about the size and with the action of an akialoa, but I could not say for certain that it was this bird. I made two collec-

---

[1] Northwood, 1940. The record of 1937 was by a young amateur of birds named H. Craddock; see Munro, 1944, p. 115. Both these records are of questionable accuracy.

tions of the forest landshells of Lanai, two botanical collections, and hunted goats all through the forest; I explored for water in its valleys and rode its bridle trails scores of times, but I never again saw a bird I even remotely thought might be an akialoa."

Destruction of habitat is thought to be the primary reason for the extirpation of this form, as of other native land birds of Lanai.

FIG. 71.—Akialoa (*Hemignathus obscurus*).

Hemignathus obscurus obscurus (Gmelin).

> *Certhia obscura* GMELIN, Syst. Nat., **1**, pt. 1, 1788, p. 470 ("Sandwich Islands" = Island of Hawaii).

Range: Mountain forests of Hawaii. From 1,500 feet above sea level to tree line at close to 7,000 feet, distributed through the continuous forest in 1894 and 1895 and not uncommon, although less common than *H. wilsoni* (Perkins, 1903).

Status: Probably extinct. Although a bird is said to have been seen high on the windward side of the mountain in 1940[1], there have been no

---

[1] Baldwin 1941, p. 19, who says, "Mr. Chester E. Blacow has seen it" (Blacow to Baldwin in conversation).

records since then, in spite of search by ornithologists who patrol the Hawaiian National Park, and special searches.

Hemignathus lucidus hanapepe Wilson

Hemignathus lucidus hanapepe WILSON, Ann. Mag. Nat. Hist. (6), **4**, 1889, p. 401 (Kauai).

Range: Mountain forests of Kauai.

Status: Possibly extinct, although perhaps a small population may still exist in the mountain forests of Kauai. Apparently even in 1898 the birds were found to be rare and localized at the headwaters of the Hanapepe River (Munro, 1944; Henshaw, 1902; Perkins, 1903).

Hemignathus lucidus lucidus Lichtenstein

Hemignathus lucidus lucidus LICHTENSTEIN, Abh. Akad. Wiss. Berlin, 1839, p. 451, pl. 5, figs. 2–3 (Oahu).

Range: Mountain forests of Oahu.

Status: Extinct. According to Perkins (1903) the bird was common in the lower forest belt, largely composed of large koa trees, until about 1860. Perhaps a few survived the destruction of this forest for a little while, but Perkins found none during the years 1893–1902, nor did any professional collectors or other experienced naturalists.

Hemignathus lucidus affinis Rothschild

Hemignathus lucidus affinis ROTHSCHILD, Ibis, 1893, p. 112 (Maui).

Range: Mountain forests of Maui.

Status: Considerable native forest remains on the eastern and southern slopes of Haleakala; perhaps a few birds may still be found. However, none have been seen since 1896, when Perkins obtained specimens.[1] He found birds on the northeastern side of Haleakala between 4,000 and 4,500 feet.

Hemignathus wilsoni (Rothschild)

Heterorhynchus wilsoni ROTHSCHILD, Avifauna Laysan, etc., 1893, pt. 2, p. 97 (Hawaii).[2]

Common names: Akiapolaau; nukupuu.

Range: Mountain forests of the island of Hawaii.

---

[1] Munro, 1944. Mr. Munro writes me that he has searched the forests of the east but not the south side of Haleakala Mountain.

[2] Amadon (1950) says, "Bill is sufficiently different to indicate specific distinctness. They form a superspecies."

Status: Uncommon. Birds probably may be seen occasionally [1] above 4,500 feet on the mountains of Mauna Loa and Mauna Kea. Cattle and hogs have destroyed the undergrowth in much of these forests, which accounts for the disappearance of the birds at lower altitudes.

Location of specimens: I know of specimens of *Hemignathus obscurus lichtensteinii* in Berlin, Germany; Leyden, Holland; and New York; *H. o. lanaiensis* in New York and Tring, England.

Maui Parrotbill

Pseudonestor xanthophrys Rothschild

> *Pseudonestor xanthophrys* ROTHSCHILD, Bull. Brit. Orn. Club, 1, 1893, p. 36 (Maui).

Status: Extremely rare; a small population still exists on the wooded slopes of the extinct volcano, Haleakala. It was found in 1950, and the record is of unquestionable accuracy (P. H. Baldwin, *in litt.*).

Even in 1894 this was a rare species, localized on the northwest slope of Haleakala between 4,000 and 5,000 feet (Perkins, 1903; Henshaw, 1902). About that time, or perhaps a little later, Henshaw noticed that the koa trees were "suffering from the ravages of insect pests," and many were dead and dying. Certainly there are few trees of any sort now in that part of the mountain. If the birds were truly so restricted in range, this destruction of habitat would account for their plight.

Range: Confined to the mountain forests of Haleakala, Maui, Hawaiian Islands.

Habitat: Like the long-billed *Hemignathus,* with which Perkins found the birds associated, these were to be seen searching the branches of koa trees (*Acacia koa*) for insects.

Description: A small bird (total length about 5 inches), its proportionately much larger bill and the distinct yellow stripe above the eye should distinguish it from *Psittirostra.*

Above greenish gray with a bright yellow streak from the lores above the eye; below grayish washed with yellow. The female is slightly smaller.

Habits: Perkins (1903, p. 430) observed a nest that he thought belonged to a pair of these birds, but which might have been that of *Hemignathus,* presumably in April or May 1894. He described it as a simple cuplike form, resembling the nests of other drepanids. It was covered on the outside by gray lichens like those growing on the koa tree in

---

[1] See Donagho, 1951, pp. 58, 62 (for sightings in 1937), and Baldwin, 1941.

which he found it. He saw young birds as well, and Henshaw (1902, p. 63) records seeing them in June.

Both men were impressed with their parrotlike actions as they hunted longicorn beetles and larvae such as looper caterpillars, as well as their extreme tameness.

Location of specimens: I know of specimens in Cambridge, England; Cambridge, Mass.; Honolulu; and New York.

Laysan Honeyeater

Himatione sanguinea freethii Rothschild

> *Himatione sanguinea freethii* [1] ROTHSCHILD, Ann. Mag. Nat. Hist. (6), **10**, 1892, p. 109 (Laysan Island).

Common names: Redbird; Laysan apapane.

Status: Extinct. Only three individuals were seen by naturalists of the *Tanager* Expedition [2] in 1923, and no doubt those last few died soon afterward, for their habitat had been destroyed by rabbits, imported in 1903 as a commercial venture. The year before, A. K. Fisher (1904) found the "Redbirds" less common than other land birds (rail and "finch") but "by no means rare, for in a short walk we could always see plenty of them." In 1911 a visiting party of naturalists estimated a population of 300 individuals (Dill and Bryan, 1912). Although many searches have been made (the last by the Vanderbilt Pacific Equatorial Expedition in 1951), no "Redbirds" have been found.

Range: Confined to Laysan Island, Pacific Ocean (lat. 25°42′ N., long. 171°44′ W., 790 miles northwest of Honolulu.)

Habitat: Laysan is a raised coral reef about 2 miles long by 1 mile in width. Fisher (1904) describes the birds' haunts: "Found all over the island, but is most abundant in the interior among the tall grass and low bushes, bordering the open stretches near the lagoon, where all the land birds seem fond of congregating. Its favorite nesting place is in this same area, and the proximity of broad patches, acres in fact, of a prostrate succulent portulaca with yellow and a sesuvium with pink flowers has many attractions for the Honey-eaters."

Description: A small (total length about 5 inches) red bird, differing from the apapane (*Himatione*) of the large islands in being slightly paler. Wings and tail dark brown. The immature plumage is brown, paler below, the wing coverts edged with green.

---

[1] Originally spelled "Fraithii," which is misspelling the name of Capt. Freeth, later corrected by Rothschild.

[2] Sponsored by the U.S. Biological Survey, the U.S. Navy, and the B. P. Bishop Museum, Honolulu. See Wetmore, 1925.

Habits: The nest was placed in the middle of a tuft of grass. It was made of fine grass and rootlets and was lined with brown down of young albatrosses. Nests with one and four eggs were found between May 16 and May 23.[1] Their staple, like that of *sanguinea,* was both insects and nectar.

Location of specimens: I know of specimens in Ann Arbor, Mich.; Berlin; Bremen; Cambridge, Mass.; Chicago; Denver; Honolulu; London; New York; Palo Alto, Calif.; Pittsburgh; and Stockholm.

FIG. 72.—Laysan honeyeater (*Himatione sanguinea freethii*).

Hawaiian Crested Honeycreeper

Palmeria dolei (Wilson)

Himatione dolei WILSON, Proc. Zool. Soc. London, 1891, p. 166 (Maui).

Common name: Akohekohe.

Status: Probably extinct on Molokai, where it has not been taken since the early part of this century. A few are still to be found on Maui. Dr. Paul H. Baldwin (*in litt.*) saw them on the northeastern slopes of the extinct volcano Haleakala in 1944. Reported again in 1964 (*Bu. Sp. Fish. Wildl.).

In 1895 the birds were "locally abundant," according to Perkins, who said (1903): "In habits it closely resembles the Apapane (*Himatione*

---

[1] Fisher, 1904. Wilson and Evans record a female of *sanguinea* taken with an egg in the ovary in May. Munro (1944) found nests containing three eggs on Lanai. There are records of more than this number of eggs, but these require confirmation, according to Baldwin, 1941.

sanguinea), which is often to be found in company with it. Why the range of the larger bird should be now so restricted on the islands on which it is found, as compared with that of *Himatione,* is not manifest, but we know that on both the islands named it had a far wider range than is now the case, and it is possibly more susceptible to those changes that have in recent years taken place in the forests than is the other.

Fig. 73.—Hawaiian crested honeycreeper (*Palmeria dolei*).

Woods opened by the invasion of cattle, it sooner or later, as I have myself noticed, deserts, retiring to deep gulches, where they cannot penetrate, and from these depths it can be called, while in the surrounding and more open forest not one can be seen or heard."

Range: Formerly mountain forests of Molokai and Maui, now reduced to a small population on Maui, or perhaps extinct.

Habitat: Apparently confined to rain forests.

Description: A medium-sized perching bird (about 4 inches long). The plumage is generally black, the feathers both above and below tipped with orange. Its prominent gray crest and the ring of reddish-orange feathers at the back of the neck distinguish it from any other bird in the world.

Habits (Perkins, 1903; Henshaw, 1902): Nothing of the breeding biology which might have thrown some light on the reasons for the virtual extinction of this species has been recorded. We know that it ate (and fed its young) on caterpillars and that it was particularly fond of the nectar of the ohia (*Metrosideros*) blossoms. It was aggressive and was seen to drive away the smaller apapane (*Himatione*), but it was dominated by the o-o (*Moho nobilis bishopi*). Its loud clear whistle was unlike other Hawaiian birds.

Location of specimens: I know of specimens in Basle; Bremen; Cambridge, England; Cambridge, Mass.; Dresden; Honolulu; London; Oslo; Paris; Philadelphia; and Stockholm.

Ula-Ai-Hawane

Ciridops anna (Dole)

Fringilla anna DOLE, Hawaiian Almanac, 1879, p. 49 (Hawaii).

Status: Extinct.[1] Not one has been found since Palmer obtained a specimen on February 20, 1892, near the headwaters of the Awini River on Mount Kohala for Rothschild.

Perkins (1903) believed that the species was once widely distributed, having been found on both the windward side, in the Hilo District and the Kohala Mountains, and the leeward in the Kona District. Apparently the population was extremely small in 1892, however.

Range: Confined to the island of Hawaii (see above).

Habitat: Mountain forests, usually to be found in the vicinity of the small groves of *Pritchardia* palms, called hawane or loulu by Hawaiians, and the native name implies this.

Description: A small finchlike bird (total length about 4½ inches); black and gray are the predominant colors and the lower-back is red.

Forehead and lores black, top of the head and upper back pale gray, wing coverts and rump red. Throat and sides of the head black, finely streaked with gray, breast black; belly red. Wings are black with the outer webs of the secondaries pink, and the tail is black.

---

[1] Munro (1944) says "This characteristic color made me almost sure I saw one in 1937 . . . this was on the Kahua ditch trail. . . . It seems unlikely . . . but it is possible."

Habits: Other than that the birds were wild, shy, and pugnacious (Perkins, 1903), no information has been brought to light.

Location of specimens: I know of specimens in Cambridge, Mass.; Honolulu; London; and New York.

Ou

Psittirostra psittacea (Gmelin)

> *Loxia psittacea* GMELIN, Syst. Nat., **1**, pt. 2, 1788, p. 844 ("in insula Sandwich" = Hawaii).
> *P.p. deppei* Rothschild and *P.p. oppidana* Bangs are synonyms. (See Bryan and Greenway, 1944, Amadon, 1950.)

Status: Extinct on Oahu since the beginning of this century and on Lanai since about 1932 (Munro, 1945). Recent searches on Molokai have been fruitless (Richardson, 1949), and presumably the birds have become very rare on Kauai and Maui as they have on Hawaii, where they are only occasionally seen (Baldwin, 1941). Several were found with difficulty in the Upper Olaa Forest Reserve in 1939 and two in the Napau Crater in 1940.

All observers reported them to be quite common 50 years ago and even when other honeycreepers were becoming rare the populations of ou apparently maintained themselves. One reason for this was their power of flight, which carried them from island to island, especially during storms. Interchange of population probably permitted survival of the species for longer than other more isolated birds and, of course, smaller colonies were re-enforced from time to time. Survival of the species on the island of Hawaii is probably due to the fact that the island is larger and that food and habitat more available.

Because of our lack of knowledge of the intimate details of the life history of this species, all the reasons for the extirpation of populations cannot be given. However, Perkins (1903) noticed that where cattle ranged through the original forests the birds became scarce and wary. On Oahu, where it was almost extinct in 1894, he found that a favorite fruit, the red ieie (*Freycinetia*), had been eaten and befouled by foreign rats. On Lanai he found remains of birds killed by cats and "twice I shot these vermin while in the act of devouring this very bird." Munro (1944) is of the opinion that the birds were much exposed to mosquito-born disease, owing to their habit of descending to low altitude to find fruit of the imported Guava, and that this was the probable cause.

Range: Hawaiian Islands: Kauai, Oahu (formerly), Molokai (formerly), Lanai (formerly), Maui, and Hawaii.

Habitat: Except that the birds inhabited forests little has been recorded. They were to be found from sea level to 6,000 feet and perhaps occasionally higher as they wandered about in search of food.

Description: Male: The head is yellow, the back olive-green, as are wings and tail; underparts greenish gray. The female has a green head and the plumage is duller.

Habits: Flowers and fruit are the staples of this species. According to Perkins (1903) the female flower of *Freycinetia* was a favorite; he mentions also the berries of Lobeliaceae and the large yellow ones of *Clermontia*, flowers of ohia and unidentified young leaves, but he felt that they had not entirely lost the nectar-eating habit. The young are fed on insects and caterpillars.

Their tameness and sweet song were noted by all naturalists of 50 years ago. Nothing of the breeding biology is known, nor has a nest ever been found.

Psittirostra palmeri Rothschild

The Koa "Finches"

Psittirostra [*Rhodacanthis*] [1] *palmeri* ROTHSCHILD, Ann. Mag. Nat. Hist. (6), **10**, 1892, p. 111 (Kona, Hawaii).

Status and range: Probably extinct. Competent naturalists of the United States Park Service have searched for years without success; no bird has been found since Perkins collected a series in 1896.[2]

Found only in the koa forests at about 4,000 feet and above in the Kona and Kau Districts on the western and southern slopes of the volcano Mauna Loa. Perhaps almost the whole population was seen when the koa was in fruit.

Description: Largest of the perching birds of the islands (total length about 8½ inches), it has a very large finchlike bill resembling that of the hawfinch (*Coccothraustes*) of Europe or the evening grosbeak (*Hesperiphona*) of America. The song is distinctive—"several whistled flutelike notes, the last ones prolonged" (Munro).

The male has a bright reddish-orange forehead, and the throat is washed with orange. The back is dark olive, the rump tinged with orange, breast and belly paler, more yellowish.

Females have yellow foreheads and are greener than males.

Young males are like females but have darker underparts.

---

[1] Placed in the genus *Psittirostra* by Amadon, 1950.
[2] Reported to have been heard in 1937 by Walter Donagho (Munro, 1944).

Psittirostra flaviceps Rothschild

Psittirostra [Rhodacanthis] [1] *flaviceps* ROTHSCHILD, Ann. Mag. Nat. Hist. (6) **10**, 1892, p. 111 (Kona, Hawaii).

Status and range: Probably extinct. No one has seen a bird since October 1891, when Palmer collected a few for Rothschild in the same region that *palmeri* was found. It was a very rare bird then, for neither Perkins in 1893 nor Henshaw in the years following found one, although they searched the koa forests carefully.

Habitat and habits: Other than that the birds were found in large koa trees together with *palmeri* and that stomachs contained seeds of that tree, nothing is known.

Description: Resembles *palmeri* but differed in having the whole head and neck yellow and in being smaller.

Location of specimens: I know of specimens of *Psittirostra palmeri* in Cambridge, England; Cambridge, Mass.; London; New York; and Philadelphia; and of *P. flaviceps* in London and New York.

Kona "Finch"

Psittirostra kona Wilson

Psittirostra [Chloridops] [2] *kona* WILSON, Proc. Zool. Soc. London, 1888, p. 218 (Kona, Hawaii).

Common names: Neither Europeans nor Hawaiians have ever invented one.

Status and range: Probably extinct, although Munro (1944) says there is a remote possibility that some may still be found at higher altitudes in the lava flows. Last seen in 1894 by Perkins (1903), who found the birds rare and localized in an area of about 4 square miles on the southwest slopes of the volcano Mauna Loa (Kona District), Hawaii, at about 3,500 to 5,500 feet.

Habitat and habits: All naturalists saw the birds most often in bastard sandalwood trees (*Myoporum*), where they fed upon the hard, dry fruit. These trees often grow in old lava flows and, according to Munro (1944), the birds seldom strayed from the more recent flows, where the trees were of medium size and there was no undergrowth. As might be expected they sometimes fed upon caterpillars and were sometimes seen among true sandalwood trees (*Santalum*). Other than the observations

---

[1] Placed in the genus *Psittirostra* by Amadon, 1950, who, after examining all six known specimens, feels sure that this is a valid species.

[2] Placed in the genus *Psittirostra* by Amadon, 1950.

of Perkins that they were sluggish birds and that they had a short sweet song but were most often heard to squeak, nothing comes to us.

Description: A large (total length about 7 inches) olive-green bird with a huge, rounded nut-cracking bill. The sexes are alike.

Location of specimens: I know of specimens in Berlin; Cambridge, England; Cambridge, Mass.; Edinburgh; Honolulu; Dresden; London; New York; Stockholm; and Tring.

FIG. 74.—Kona "finch" (*Psittirostra kona*).

Palila

Psittirostra [Loxioides] bailleui Oustalet

> *Psittirostra [Loxioides] bailleui* [1] OUSTALET, Bull. Soc. Philom. Paris (7), **1**, 1877, p. 100 (Hawaii).

Status: A small population is still to be found in the upper forests on the sides of the volcano Mauna Kea, but this is an exceedingly rare bird; in danger because of the small size of the population if for no other reason. The only reports of sighting in the past 50 years are those of 1937 and 1940, when two birds were seen singly on Mauna Kea (Donagho, 1940, 1952). Munro (1944) says that in 1891 it was common in the Kona District at 4,000 feet, but apparently the population wandered and perhaps may have been common only locally.

Range: Now confined to the upper edge of the forests on Mauna Kea

[1] This very distinct species is included in the genus *Psittirostra* by Amadon, 1950.

above 6,000 feet, it was known 50 years ago from the upper forest zone of Kona and Hamekua Districts, from 4,000 to 7,000 feet.

Habitat: Upper mountain forests on the leeward side. According to

D. M. HENRY

FIG. 75.—*Upper left:* Palila (*Psittirostra bailleui*); *upper right:* Maui parrotbill (*Pseudonestor*); *lower left:* Ou (*Psittirostra psittacea*); *lower right:* Koa "finch" (*Psittirostra palmeri*).

Perkins (1903) it was most numerous among large trees called "mamane" (*Sophora caryphylla*) and in a "naio" (*Myoporum santalinum*).

Description: A small (total length about 4 inches) bird. The yellow breast, pale belly, and thick bill should distinguish it from the ou (*Psit-*

tirostra psittacea). Head, neck, upper breast yellow, back gray, lower breast and belly white, wings and tail black, edged with green. The female duller than the male, with the head washed with green.

Habits: Nests were found in early summer by both Wilson and by Perkins, but none contained eggs. According to all observers the favorite food was the seeds of *Sophora chrysophylla,* which is still common at higher altitudes; the birds cracked the pods with their bills. Their musical whistle was said by the natives to be a sign of coming rain.

Mamo

Drepanis pacifica (Gmelin)

Certhia pacifica GMELIN, Syst. Nat., **1**, 1788, p. 470 (*"in insulis amicis,"* in error for Hawaii).

FIG. 76.—Mamo (*Drepanis pacificus*).

Status: Extinct, last sighted in forests of Kaumana, about 1,000 feet above the town of Hilo, in 1898 (Henshaw, 1902). A specimen was taken in July 1892 by R. H. Palmer, collector for Rothschild.

Although the Hawaiians trapped this species by various methods [1] for their decorative yellow plumes, and presumably for food, it is more probable that destruction of the forest (and perhaps disease) were primarily responsible for its extinction.

Range: Confined to the island of Hawaii.[2]

Habitat: Mountain forests.

Description: A medium-sized perching bird (total length about 8 inches) with a very long, curved bill. Except for yellow feathers at the

[1] See Perkins, 1903, pp. 400–401, for an account.
[2] Peale's (1848) record for Kauai is thought to be an error.

bend of the wing, on the rump and under tail coverts (these feathers have black bases) the entire plumage is black.

Habits: Except for the fact that they fed upon nectar of flowers (particularly lobelias and *Pritchardia* palms) nothing is recorded of the mamo's habits.

Location of specimens: I know of specimens in Cambridge, England; Cambridge, Mass.; Honolulu; London; New York; Paris; Tring; Vienna; and Leyden.

Black Mamo

Drepanis funerea Newton

> *Drepanis funerea* NEWTON, Proc. Zool. Soc. London, 1893, p. 690 (Molokai).

Common names: Oo-nuku-umu; hoa; mamo; Perkins's mamo.

Status: Most probably extinct. Last recorded specimen taken 1907 (W. A. Bryan, 1908). Interested residents as well as naturalists have searched the last remaining native forests, even to the top of Mount Olokui, without success (Richardson, 1949). They did find imported brown rats (*R. norvegicus*) as well as mongooses, and these are thought to be at least partly responsible. The population was probably never a large one. Perkins (1903) wrote: "The Oo-nuku-umu would appear to be the rarest of all the birds of which I obtained specimens . . . I have spent a whole day or several days in its haunts, and in exclusive search for it, without seeing a single one, and this at a time when I had become perfectly familar with its habits."

It is said (Perkins, 1903) to have been a bird of the underbrush of the forests, and further that cattle and deer had destroyed much of this habitat. No doubt this was also a factor in the extinction of the species.

Range: Confined to the island of Molokai.

Description: Similar to the mamo (*D. pacifica*) but plumage all black, except for some gray on the outer edges of the primaries. It is without yellow feathers. The nostril and operculum are more elongated than in *pacifica.*

Habits: We know only that this species inhabited the lower and middle levels of the forests, was tame and curious, fed upon nectar of flowering plants (particularly *Lobelia*) and had a call similar to the mamo's.

Location of specimens: I know of specimens in Bremen; Cambridge, England; Cambridge, Mass.; Honolulu; London; Molokai, Hawaiian Islands; and New York.

New Zealand Stitchbird

Notiomystis cincta (Du Bus)

Meliphaga cincta Du Bus, Bull. Acad. Sci. Bruxelles, **6**, 1839, p. 295 (North Island). Fig.: Buller, 1888, pl. 11.

Status and range: Small population confined to Little Barrier Island, near the coast of North Island, New Zealand.

Fig. 77.—New Zealand stitchbird (*Notiomystis cincta*).

Within the memory of Europeans this population has had a somewhat restricted range. Although the birds occupied all forested North Island at one time, according to Maori tradition, naturalists found the birds only south of the latitude of Auckland and then only until the decade 1870–80, after which they began to disappear. After the year 1900 they were no longer to be found anywhere except on Little Barrier Island in Hauraki Gulf, which had been declared a government sanctuary (Buller, 1888, 1905) and where they are still to be found. It was during this period (1875–90) that cutting and burning of native forests by European settlers in southern North Island was at its height, and no doubt destruction of habitat is the reason for the final extinction of populations on North Island itself (Cumberland, 1941). It must be said, however, that not all the native forest has been destroyed, and probably an unknown factor of disturbance such as was noticed by naturalists in the Hawaiian Islands played a part. Although not enough is known of the food habits to warrant generalization perhaps the birds would have suffered from a failure of some favorite tree to flower during certain years, since Guthrie-Smith (1925) indicated that in his opinion the birds (and even nestlings) fed exclusively on nectar. Since the tuis (*Prosthemadera*) and bellbirds, which belong to the same family, feed also on insects, perhaps their more catholic tastes might explain their relatively greater power of survival.

These are medium-sized honeysuckers (total length about 7 inches). Their note, a high *tee-tee-tee,* is said to be distinctive. The head, neck, upper back, throat, and upper breast are velvety black. Against this black background white tufts show prominently on the sides of the head. A band of yellow extends across the breast and shows brightly at the bend of the wing. Lower back and rump are brown or greenish brown, the feathers having paler edges. Lower breast and belly are paler, some of the feathers having a yellowish tinge. Wings and tail dark brown.

Location of specimens: I know of specimens in Auckland, Bremen, Brussels, Christchurch, Dresden, London, New York, and Stockholm.

New Zealand Bellbirds

Anthornis melanocephalus Gray

> *Anthornis melanocephalus* GRAY, *in* Dieffenbach, Travels in New Zealand, **2**, 1843, p. 188. Fig.: Gray, Zool. Voyage *Erebus* and *Terror,* **4**, 1845.

Status and range: Extinct. Chatham Islands, New Zealand.

This species disappeared from Chatham Island during the early years of this century, perhaps as a result of the destruction of the native trees that were the birds' habitat, as well as depredations of cats and rats. This could be true both of Chatham and the smaller Mangare Island nearby. On Little Mangare Island, which is free of predators, a few pairs were reported in 1906 (Fleming, 1939), but no birds were seen after that.

Their complete disappearance cannot be explained with certainty, for after they had been extirpated on the larger islands they persisted in small numbers on Little Mangare, where there were no men or other predators, according to Fleming (1939). Even assuming that the final sight record was in error, or that the birds were stragglers, not breeding residents on that steep islet, no explanation is adequate, for the birds' habitat was in the forests. Even on Chatham Island much forest was found by Cockayne in 1900. He reported that the trees resisted fire, although the low bush of the bogs was burned constantly. At that time but little forest had been actually cut away for cultivation. It can only be said that these, like other members of the family Meliphagidae on the Hawaiian Islands, disappeared quite suddenly following relatively little disturbance.

On the New Zealand mainland, although more localized and not as common as they were when Captain Cook landed, the birds are to be seen and heard in favored localities, where native forests remain, on both North and South Islands, as well as nearby Stewart Island and Three Kings Islands.[1] They have been recorded as plentiful recently in forest reserves near Rotorua and Kaingaroa and in the Tararua Ranges of the

FIG. 78.—New Zealand bellbird (*Anthornis melanocephalus*).

North Island and are to be found on Little Barrier and Kapiti Islands. On the South Island, as well as the Stewart Islands, and probably other small islets near the coast, they are common.[2]

There is less recent information on the population of the Auckland Islands (*melanura*),[3] but apparently birds are still to be found.

The voices of these birds serve best to locate them. In 1770 Sir Joseph Banks, anchored at last after so many weeks at sea, said of them, ". . . the most melodious wild music I have ever heard, almost imitating small

---

[1] *Certhia melanura* Sparrman, Mus. Carlson., fasc. 1, no. 5, 1786 (Queen Charlotte Sound, South Island, New Zealand). *A. m. obscura* Falla (1948), of Three Kings, is recognized by the Committee of the New Zealand Ornithological Society (1953).

[2] See "New Zealand Bird Notes" ("Notornis").

[3] ?*Anthornis incoronata* Bangs, Proc. Biol. Soc. Washington, 24, 1911, p. 23 (Auckland Islands). Marples (1946) records it as breeding still on the Aucklands. The 1953 edition of the Checklist of New Zealand Birds does not recognize this subspecies.

bells, but with the most tunable silver sound imaginable" (Banks, 1896).

Medium-sized (total length about 9 inches), dark-green honeysuckers (Meliphagidae), they are usually to be seen sucking nectar from flowers. The Chatham Island birds were larger and had dark glossy purple heads, whereas those of the mainland have only a faint wash of purple.

The O-o of Hawaii

Moho braccatus (Cassin)

> Mohoa braccata CASSIN, Proc. Acad. Nat. Sci. Philadelphia, **7**, 1855, p. 440 (Kauai).

Status: A small population, probably in danger of extinction (1964).
Range: Mountain forests of Kauai Island, Hawaiian Islands.

Moho apicalis Gould

> Moho apicalis GOULD, Proc. Zool. Soc. London, 1860, p. 380 ("Owhyie" = Oahu).

Status: Extinct.
Range: Mountain forests of Oahu Island, Hawaiian Islands.

Moho bishopi (Rothschild)

> Acrulocercus bishopi ROTHSCHILD, Bull. Brit. Orn. Club, **1**, 1893, p. xli (Molokai).

Status: Probably extinct.
Range: Mountain forests of Molokai Island, Hawaiian Islands.

Moho nobilis (Merrem)

> Gracula nobilis MERREM, Avium Rar. Icon. et Descript., fasc. **1**, 1786, p. 7 (Hawaii).

Status: Probably extinct.
Range: Mountain forests of the island of Hawaii.

Five, or perhaps six, forms of the essentially Australian family of honeyeaters (Meliphagidae) were once to be found on as many of the Hawaiian Islands. Of these, all save perhaps one are quite probably extinct. The reasons for this, as in the case of the Hawaiian honeycreepers (Drepaniidae), are not well understood, but it is probable that destruction of the native forests is the primary cause. Because the natives captured them for their yellow plumes it has often been said that they were extirpated by the Hawaiians, but it is not probable that this was the sole cause for their disappearance.

The o-o of Kauai (*Moho braccatus*)[1] probably still exists. Munro (1944, p. 84) wrote of the status of the population:

Endemic to Kauai and now in danger of extinction. In 1891 it was a common bird over all the Kauai forests. Its notes could be heard from near sea-level in the valleys on the north side to near the top of Mount Waialeale at over 4,000 feet elevation. It was still not uncommon when I left Kauai in 1899, but on four visits to the Kauai forests between 1928 and 1936 I failed to hear or see it. In 1936 I thought I heard one sing but could not be sure. Had it given its call note I could easily have identified it. Donagho penetrated deep into the Kauai forest in 1940 and is sure that he heard it. He had no previous experience with the Hawaiian oos so may have been mistaken. However, if it still exists no effort should be spared to save what would be the last of the famous Hawaiian oos.

FIG. 79.—O-o (*Moho apicalis*).

There can be no doubt that the birds of Oahu (*Moho apicalis*) will never be seen again. Indeed none have been recorded since 1837, when Herr Deppe collected a series in the hills behind Honolulu. No doubt they were to be seen on the island for a few years after that, but none of the naturalists who searched for them during the decade 1890–1900 could find them.

Deppe collected probably three specimens of which there is one in Berlin, one in Vienna (see Sassi, 1939) and one in New York. There is a second specimen in New York, acquired from Lord Rothschild's museum at Tring, which was collected in 1824 by Bloxam, naturalist on H. M. S. *Blonde*. The British Museum possesses a skin which was apparently purchased from John Gould, but who the collector was is not

---

[1] This genus is called *Acrulocercus* by many writers. Amadon (1950) says there are recent sight records.

known. The origin of a specimen in the Muséum National in Paris is also obscure. J. K. Townsend, who collected for three months with Deppe, secured at least one, which is in Cambridge (Massachusetts), and perhaps more.

In all probability *Moho bishopi* of Molokai Island is now also extinct. Munro (1944) records having seen about half a dozen in 1904, and that is the last definite record. He was told in 1915 that the birds were still then to be found, but later searches have been in vain. The last of these (in 1949) gained the top of Mount Olokui where there is considerable native forest left. Of native birds only *"Himatione"* and *"Chlorodrepanis"* were found, but black rats were seen in the trees in daylight, which means they are a terrible danger to the birds (Richardson, 1949).

Not since 1934, when one was heard on the slope of the volcano Mauna Loa, has there been any evidence that *Moho nobilis* [1] still lives on Hawaii. The species is probably extinct.

Perhaps there was once a form representing this group on Maui, for Henshaw (1902, p. 74), who knew the group well, saw and heard one in the forest northeast of Olinda at about 4,500 feet elevation.

These were all strikingly beautiful, either brown or jet black, with metallic reflections and with bright yellow ornamental plumes. Their voices described as a deep *took-took* carried for great distances. They were usually to be seen in the tops of the trees, where they fed on both nectar and insects.

The following key will distinguish the species:

1. Color generally black or gray (thighs not yellow)...........2
   Color generally brown (thighs yellow)....................*braccatus* (Kauai)
2. Shafts of feathers of underparts not white.................3
   Shafts of feathers of underparts white.....................*bishopi* (Molokai)
3. Outer tail feathers tipped with white......................*apicalis* (Oahu)
   Outer tail feathers black ....*nobilis* (Hawaii)

Location of specimens: I know of specimens of *Moho braccatus* in Ann Arbor; Berlin; Bremen; Cambridge, Mass.; Cambridge, England; Dresden; Edinburgh; Honolulu; London; Oslo; Paris; Philadelphia; Pittsburgh; New York; and Stockholm; of *M. apicalis* in Berlin; Cambridge, Mass.; London; New York; Paris; Tring; and Vienna; of *M. bishopi* in Bremen; Cambridge, Mass.; Honolulu; London; Molokai, Hawaiian Islands; New York; and Stockholm; and of *M. nobilis* in Berlin; Bremen; Brussels; Cambridge, Mass.; Dresden; Honolulu; Hamburg; London; Leyden; Los Angeles; New York; Paris; and Stockholm.

---

[1] Information on status from Munro, 1944.

FIG. 80.—O-o (*Moho nobilis*).

The Kioea of the Island of Hawaii

Chaetoptila angustipluma (Peale)

Entomiza angustipluma PEALE, U.S. Exploring Exped., **8**, 1848, p. 147 (Hawaii).

Status and range: Extinct. Island of Hawaii.

This species has not been found since about 1859, and there can be no doubt that the population is extinct, for many naturalists have searched diligently for the birds. We cannot be absolutely certain exactly what the range of the species was. According to Perkins (1903) it was confined to the high plateau between the mountains and the upper edges of the forest bordering this, where it was observed [and a specimen was collected] [1] by Pickering and Peale in 1840. It is not certain where the other three birds were collected. According to Palmer,[2] who collected for Rothschild in 1891–92, it was between the lower Volcano House and the crater of Kilauea.

The Hawaiian name means "standing high on long legs" and was applied also to the bristle-thighed curlew, a migrant. Other than this we

FIG. 81.—Kioea (*Chaetoptila angustipluma*).

know nothing of this large (its total length was about a foot) greenish bird that was lost to us so soon after its discovery.

Location of specimens: Cambridge, England; Honolulu; New York; and Washington.

Chestnut-flanked White-eye

(*Seychelles Islands*)

Zosterops semiflava Newton

> *Zosterops semiflava* E. NEWTON, Ibis, 1867, p. 354 (Marianne Island, Seychelles).
> Fig.: Shelley, Birds Africa, **2**, 1900, pl. 6 (p. 172).

Common name: Serin.

---

[1] *This is the type;* it is in the United States National Museum.
[2] As quoted by Rothschild, *in* Birds of Laysan etc., p. 216.

Status: Probably extinct. Vesey-Fitzgerald (1940) was unable to find the birds or obtain any report of them during long residence on the islands. He states "apparently extinct" with proper caution. Apparently destruction of the forests is one cause for the extinction or extreme rarity of this species. A little forest still remains in the hills of Mahé Island, where another species of white-eye is still to be found.

Range: Definitely reported only from Marianne Island, Seychelles, Indian Ocean. Somewhat indefinite records for Praslin, Ladigue, Silhouette, and Mahé Islands have never been substantiated.

Description: A small (total length about 4 inches) greenish-yellow bird with chestnut flanks and prominent white ring around the eye. The only other *Zosterops* on the islands is recorded from Mahé Island and is all grayish brown.

The forehead and a line above the eye are yellow; top of the head and back yellowish olive; wings and tail black. Underparts are bright yellow and flanks chestnut.

Now listed as a distinct species only because of lack of knowledge.

Location of specimens: I know only of a specimen in London.

Lord Howe Island White-eye

Zosterops strenua Gould

> *Zosterops strenuus* GOULD, Proc. Zool. Soc. London, 1855, p. 166 (Lord Howe Island). Fig.: Mathews, 1928, p. 50.

Status: Extinct. According to Hindwood (1940), "the extinction of this species occurred after 1918 when the rats reached the island. Sharland was unable to record the presence of this species on the island in 1928 . . . and in 1936 I made every effort, both by enquiry and observation, to ascertain if the bird still existed, without result.

"Bell records (*ms.* October 31, 1913) that the 'Big Grinnells', as they were known locally, were in thousands and highly destructive not only to fruit and other crops but also to other birds, sucking their eggs."

Range: Confined to Lord Howe Island, about 400 miles off the east coast of Australia.

Description: A small (total length about 3 inches) perching bird, with a green head, wings and tail, a yellow throat and white belly. It was considerably larger than the other resident white-eye on Lord Howe (*Zosterops lateralis tephropleura*), and its bright yellow (not greenish) chin, throat, and upper breast should distinguish it.

From a closely related species of Norfolk Island it may be distinguished by its white (not yellow) belly and paler flanks.

Location of specimens: I know of specimens in London, New York, and Sydney.

FIG. 82.—Lord Howe Island white-eye (*Zosterops strenua*).

Lord Howe Island Starling

Aplonis fuscus hullianus Mathews

> *Aplonis fuscus hullianus* MATHEWS, Novit. Zool., **18**, 1912, p. 451 (Lord Howe Island).

Status and range: Extinct. Lord Howe Island, western Pacific.

The population disappeared soon after the accidental introduction of rats from a wreck in 1918 (Hindwood, 1940). In the years before 1918 the birds had been extremely common and were thought to have eaten the eggs of other birds and destroyed fruit. The local inhabitants called them "cudgimaruk," in imitation of their calls, or "red eyes." They nested in holes in trees, often in tree ferns near the ground.

Description: About the size (total length about 7 inches) and general appearance of a common starling, the whole head, neck, upper back, chin, and throat glossy metallic green. The back dark gray overlaid with faint greenish tinge, rump and underparts gray; tail gray tipped with brown; wings brown and eyes dark red.

Location of specimens: I know of specimens in New York and Sydney.

Kusaie Island Starling

Aplonis [*Kittlitzia*] *corvina* (Kittlitz)

Lamprotornis corvina KITTLITZ, Kupfertaf. Vög., **2**, 1833, p. 12. pl. 15 (Kusaie Island).

Status and range: Extinct. Kusaie Island, Caroline Islands, southwestern Pacific.

FIG. 83.—Kusaie Island starling (*Aplonis corvina*).

Once inhabited the mountain forests of Kusaie Island in the eastern Caroline group, southwestern Pacific. Finsch (1881) was unable to obtain a specimen in 1880, and he quotes Kittlitz to the effect that it was a rare and solitary species even 50 years before. Japanese collectors, as well as W. F. Coultas for the Whitney Expedition of the American Museum of Natural History, searched for months without success during the 1930's. There can be little doubt that the species is extinct. Coultas (MS.) speaks of enormous populations of rats on Kusaie. Perhaps these were responsible. Certainly the whalers used these beaches to careen their ships during the years, and it is more than probable that rats escaped from them into the forests and there bred up to and above the point of subsistence, as they did on other islands such as the Bonins. Nothing is known of the nesting habits; perhaps such knowledge would explain why the smaller species (*opacus*), which presumably nests in holes in trees, is quite common even though much preyed upon by rats (R. H. Baker, 1951, p. 288, 296).

Description: This was a very large starling (total length about 10 inches), all glossy black and with a long curved bill and a long tail.

Location of specimens: I know of specimens in Leningrad. A skin formerly in Frankfurt is said to have disappeared.

Mysterious Starling

Aplonis mavornata Buller   [See frontispiece]

> *Aplonis mavornata* [1] BULLER, Birds New Zealand (2d ed.), **1**, 1887, p. 25 (type in Brit. Mus.; locality unknown).
> *Aplonis inornata* SHARPE, 1890, is preoccupied by *Calornis inornata* Salvadori 1880.

Status: Extinct.

Range: Unknown, but probably one of the Pacific Islands. Sharpe (1890; 1906) claims that the type was probably collected on one of Cook's voyages.

This unique specimen was exhibited in the collection of mounted birds in the British Museum for many years. Sharpe suggested in 1890 that the specimen may be identified with a drawing by Forster, naturalist on Captain Cook's second voyage, which is the type of *Turdus ulietensis* Gmelin (*T. badius* Forster). However, Stresemann (1949) has compared the two and finds that this is not the case. He says: "The skin, as Sharpe describes, is entirely brown above and below, with a slight browny gloss

---

[1] In spite of the fact that this name is a *lapsus,* I prefer to retain it in the interest of stability.

on the head, while the drawing shows a reddish-brown bird with a distinctly dark, almost black, tail and wings." Stresemann is in agreement with Wiglesworth (1891) who points out that the original descriptions indicate this to be a smaller bird than *Turdus ulietensis,* the bill not notched (see Seebohm, Cat. Birds Brit. Mus. **5**, 1881, p. 276, note).

Ernst Mayr kindly informs me that he examined the type in 1951 and that it belongs to no known species. He believes it to be related to *tabuensis.*

Huia

Heteralocha acutirostris Gould

> *Heteralocha acutirostris* GOULD, Synopsis Birds Australia, **1**, 1837, pl. 2, Fig.: Buller, 1888, p. 9.

Status: Almost certainly extinct according to *Williams (1962, p. 18) and to *Fleming (1964, p. 877). There has been no unquestioned report since December 28, 1907 (Myers, 1923).

Probably the most important reason for this loss is the destruction of forests by fire and the ax. Buller (1905) quotes the Maoris as saying, "You have permitted the killing of the Huia under heavy penalty, and yet you allow the forests, whence it gets its subsistence, to be destroyed." Killing of birds for decoration by the Maoris and museum specimens by Europeans has also been brought forward as a cause. However, this would not appear to have been important, since even if we double the possible number of skins and skeletons in museums (I have traced only 65) there could be no more than 150. Furthermore, the fact that the Maoris hunted the birds for generations before the final debacle indicates that this cannot be important. The third cause that has been suggested is disease. Ticks (*Haemaphysalis leachi* and *Hyalomma aegyptium*) of African and Asiatic origin and presumably introduced with mynahs (*Acridotheres tristis*) have been found on huias and may have carried diseases (Myers, 1923), but there is no proof of this.

Range: Forests of North Island, New Zealand.

Habitat: Usually to be seen on the ground or in the lower reaches of mixed beech (*Nothofagus*) and podocarp forests. Nested in hollow trees.[1]

Description: A rather large black bird (total length 18–19 inches) with a broad white band at the tip of the tail and orange wattles at the base of the white bill. Male and female differ sharply in the shape of the

---

[1] Buller, 1888; Potts, 1885.

bill, which is comparatively short and strong (like a starling) in the male but long and curved (like a nectar feeder) in the female.

Habits: The breeding season is said to have begun in October and been completed by the last of November (Potts, 1885; Buller, 1888). The nest was built in hollow trees. The staple was insects. Buller and others describe the birds' feeding, the male using his stout bill like an ax or adz to chop rotten bark and wood, the female probing into crevices he could not reach.

Location of specimens: I know of specimens in Amsterdam; Ann Arbor, Mich.; Auckland; Basle; Berlin; Bremen; Brussels; Cambridge, Mass.; Chicago; Christchurch; New Zealand; Dresden; Edinburgh; Exeter, England; Geneva; London; Melbourne; Munich; Newcastle-upon-Tyne; New York; Oslo; Philadelphia; Pittsburgh; Princeton, N. J.; Stockholm; and Vancouver.

Guadalupe Island Towhee

Pipilo erythrophthalmus consobrinus Ridgway

> *Pipilo maculatus* [1] *consobrinus* RIDGWAY, Bull. U.S. Geol. Geogr. Surv. Terr., 2, no. 2, 1876, p. 189 (Guadalupe Island).

Status: Most probably extinct. Last sighted in June 1897.[2] A single bird had been seen in September 1896 by Gaylord (Gaylord, 1897a). The last specimens were collected by Bryant in 1885 (Bryant, 1887). Like the Guadalupe wren, this was probably never an abundant species, for even in 1875 Palmer so recorded it (Palmer, *in* Ridgway, 1876).

Many parties of naturalists have looked carefully but unsuccessfully for the birds since then.

Range: Confined to Guadalupe Island, off the coast of Lower (Baja) California, Pacific Ocean.

Habitat: Formerly confined to groves of cypress on the plateau below the highest point, Mount Augusta, at an altitude of about 4,000 feet, according to Bryant (*l.c.*). Ten years earlier, however, Palmer (*l.c.*) found them "chiefly in the woods" and recorded them as "frequenting brushwood, fallen logs, fences, etc., rather than trees," indicating that they were perhaps not altogether confined to the groves.

Description: A large finch (total length about 7 inches), black above,

---

[1] Although it can never be proved that this form would have bred readily with those of the mainland had they had the opportunity, it is probable that they would have.

[2] Thorburn, 1899. Record probably by Professor Greene, a botanist, since Thorburn says he confined his own investigations to the coast.

with a black throat, white belly and brown flanks. Usually to be seen in low brush or on the ground.

Head black, back and wing coverts blackish brown, the wing coverts with outer margins of some of the feathers white, giving a speckled effect. Throat and upper breast blackish brown, sides and flanks the color of dead leaves; belly white. Tail brown, the outer feathers tipped with white.

Resembles the spotted towhees of western North America, particularly *oregonus,* but having the back browner and having a shorter wing (76–80 mm.) and tail (79–86 mm.).

Habits: No observation of the breeding biology was ever made, nor were any nests or eggs described. Palmer (*l.c.*) observed only that stomachs contained seeds as well as insects, whereas Bryant observed that they ate only insects. The song included a trill resembling that of a bluebird (*Sialia*), never heard in that of the mainland form. Nothing more has come to us.

Remarks: We know that the undergrowth not only in the cypress grove but also in open country was destroyed by introduced goats during the time that the birds were disappearing. Feral house cats, recorded as often seen by all the naturalists who visited the island after the birds had disappeared, would naturally be a great danger to any ground-nesting bird. We may assume that these two factors were primarily responsible.

Bryant (*l.c.*), noted that there seemed to be an imbalance in the sex ratio, but the observation has in it so many possibilities for error that it cannot be put forward as a cause, but only suggested.

Location of specimens: I know of specimens in Cambridge, Mass.; New York; and Washington.

Bonin Island Finch

Chaunoproctus ferreirostris (Vigors)

> *Coccothraustes ferreirostris* Vigors, Zool. Journ. (London), **4**, 1828, p. 354 (Bonin-Sima = Peel Island).

Status and range: Extinct. Peel Island, Bonin Islands, western Pacific.

It is possible that a small population of these beautiful little grosbeaks lingered on the Bailley Group of the Bonin Islands until about 1890, for the naturalist A. P. Holst was so informed by local residents at Port Lloyd on Peel Island in the Beechey Group (Seebohm, 1890). As far as is known, however, the only specimens ever collected came from Peel Island, where the types, now in the British Museum, were certainly found in 1827. A year later Kittlitz collected a few. Neither Stimpson in 1854 nor anyone since then has been able to find a bird. All we know of them comes from the following quotation from Kittlitz (1832): "This bird lives on Bonin-sima, alone or in pairs, in the forest near the coast. It is not common but likes to hide, although of a phlegmatic nature and not shy. Usually it is seen running on the ground, only seldom high in the trees. Its call is a single soft, very pure, high piping note, given sometimes shorter, some-

Fig. 84.—Bonin Island finch (*Chaunoproctus ferreirostris*).

times longer, sometimes singly or sometimes repeated. In its muscular crop and spacious gullet I found only small fruits and buds."

These were rather large finches (total length about 7 inches) with very large bills. The male had the forehead, a wide stripe above the eye, cheeks, throat and upper breast reddish orange the lower breast was paler shading to white on the belly. The rest of the plumage was olive-brown, striped with darker brown on the back. The female differed in being all brown, the forehead yellowish and the feathers of the flanks with dark brown tips giving a mottled appearance.

Location of specimens: I know of specimens in Berlin, Frankfurt, Leningrad, Leyden, London, New York, and Tokyo.

West Indian Bullfinches

Loxigilla portoricensis portoricensis (Daudin)

   Loxia portoricensis DAUDIN, Traité d'Orn., **2**, 1800, p. 411 (Puerto Rico).

Status: Rare and localized.
Range: Island of Puerto Rico, West Indies.

Loxigilla portoricensis grandis Lawrence

   Loxigilla portoricensis grandis LAWRENCE, Proc. U.S. Nat. Mus., **4**, 1881, p. 204
   (St. Christopher).

FIG. 85.—St. Kitts bullfinch (*Loxigilla portoricensis*).

Status: Extinct.
Range: St. Christopher (St. Kitts), West Indies.

   Only two forms of this peculiar West Indian species are known. Both
were once common, but now *portoricensis* is confined to the mountain of
El Yunque in Puerto Rico, and that of St. Christopher is extinct. According to Bond (1950), the birds of St. Christopher were extirpated by
the introduced African monkey (presumably *Cercopethicus aethiops
sabaeus*). All the other bullfinches of the West Indies (*violacea, noctis,
barbadensis*) have adapted themselves to men and introduced animals,

including this same monkey on Barbados. This, as well as the fact that a species as sensitive to human interference as the trembler (*Cinclocerthia*) still exists in small numbers in the small forest at St. Christopher, leaves the disappearance of the bullfinch somewhat mysterious.

Local residents of St. Christopher (St. Kitts) called the birds mountain blacksmith, and the Puerto Rican birds have various names—gallito, churri, capitan. The Puerto Rican birds are medium-sized black finches (total length about 4 inches) with a rusty-brown crown, throat, and under tail coverts. The birds of St. Christopher were similar but considerably larger (total length about 6 inches).

Darwin's Ground Finch
Geospiza magnirostris magnirostris Gould

> *Geospiza magnirostris magnirostris* GOULD, Proc. Zool. Soc. London, 1837, p. 5 ("Galapagos Islands").

Status: Localized. The population is known by seven specimens collected by Charles Darwin and others of the complement of H. M. S. *Beagle* in September 1835, and now in the British Museum of Natural History, and a single female collected on Charles Island, Galápagos in September 1957 (*Bowman, 1961, pp. 25, 271). Bowman saw a number of individuals, however.

Range: Galápagos Islands: Chatham and probably formerly Charles Islands.

Habitat: Closely related forms on other islands are to be found on the dry coasts and in a Transition Zone between that and the forests. They fed on the ground as a rule. Darwin found them in clearings near the houses in the upper forested zone.

Description: A rather large finch (total length about 6 inches) with a very large bill. The male is black except for the under tail coverts which in some specimens are white and others with white tips. The female is generally dark brown, the olive edges of the feathers giving a mottled appearance. Throat and breast like the back except that the edges of the feathers are whitish buff; lower breast and belly whitish buff spotted with dark brown.

This population differs from those that breed on the other islands of the Galápagos in being slightly larger and particularly in the size of the bill. There are no intermediates bridging the size differences, as Lack (1947, p. 174) and others (see Swarth, 1931) have shown.

Remarks: Historical evidence is not quite satisfactory, for Darwin's remarks (1839, p. 474) make it apparent that he did not indicate specific islands on the labels of his specimens. Salvin in an essay on the fauna of the Galápagos (1876, p. 466) remarks: "Mr. Darwin's specimens [of *Geospiza*] have no locality record," and again, "*G. magnirostris* of which Mr. Darwin is the only naturalist who has obtained specimens." Yet he seems to have believed that these large-billed birds came from Chatham

Fig. 86.—Darwin's ground finch (*Geospiza magnirostris*).

and Charles for he records *G. magnirostris* from those Islands only.

Perhaps this was because Darwin himself (*in* Gould, 1841, p. 100) records these birds from Chatham and Charles, remarking also, "I have strong reasons for believing this species is not found on James's Island," and further, "I seldom saw these birds in the upper and damp regions . . . excepting . . . near the houses on Charles Island." Perhaps it may be urged that his recollection after five years may have been at fault, but it was certainly not necessarily so when Darwin's intellectual attainments are taken into account. Lack (1947) took this view. For reasons to believe that the birds came from James Island, see Swarth (1931). In my opinion this theory, which involves a mutation of the species within a period of 60 years, is not as acceptable.

The extirpation of birds from Charles Island could well have been brought about by men and introduced animals. The birds were extremely tame and confiding. Darwin saw a small boy killing them with a stick near a well on Charles. This island had been a resort of pirates and whalers for 200 years, but the first settlers came only six years before Darwin's arrival. He noted pigs and goats. No doubt rats came in numbers with the settlers.

Bibliography

ABBOTT, C. G.
   1933. History of the Guadalupe caracara. Condor, **35**, pp. 10–14.
AHRONI, J.
   1927. Syrische Strausseier [*Struthio c. syriacus* eggs]. Beitr. Fortpfl.-biol. Vög. Berlin, **3**, pp. 186–188.
ALDRICH, J. W.
   1946a. The United States races of the bob-white. Auk, **63**, pp. 493–508.
   1946b. Speciation in the white-cheeked geese. Wilson Bull., **58**, pp. 94–103.
ALEXANDER, W. B.
   1954. Notes on *Pterodroma aterrima* Bp. Ibis, 1954, pp. 489–491.
ALLBEE, E. A.
   1911. Passenger pigeons in eastern Iowa 1856–1860. Auk, **28**, p. 261.
ALLEN, A. A., and KELLOGG, P. P.
   1937. Recent observations on the ivory-billed woodpecker. Auk, **54**, pp. 164–184.
ALLEN, J. A.
   1876. Extinction of the great auk on Funk Island. Amer. Nat., **10**, p. 48.
ALLEN, R. P.
   1952. The whooping crane. Nat. Audubon Soc. Res. Rep. no. 3, 220 pp. New York.
AMADON, D.
   1924a. [Notes on *Megapodius*.] Amer. Mus. Novit., no. 1175, pp. 9–10.
   1942b. Relationships of the Hawaiian avifauna. Condor, **44**, pp. 280–281.
   1947. Ecology and the evolution of some Hawaiian birds. Evolution, **1** (1, 2), pp. 63–68.
   1950. The Hawaiian honeycreepers (Drepaniidae). Bull. Amer. Mus. Nat. Hist., **95**, art. 4, pp. 157–257.
   1953. Avian systematics and evolution in the Gulf of Guinea. Bull. Amer. Mus. Nat. Hist., **100**, no. 3, pp. 397–451.
AMERICAN ORNITHOLOGISTS' UNION.
   1939. Report of Committee on Bird Protection. Auk, **56**, p. 212.
ANDERSON, C.
   1937. Sternum of a ratite bird from Wellington Caves, N.S.W. Rec. Australian Mus., **20**, pp. 76–77.
ANDERSON, H. T.
   1935. Condors in northern Los Angeles County. Condor, **37**, p. 170.
ANDERSON, J.
   1747. Nachrichten von Island, Gronland und der Strasse Davis etc. Frankfurt und Leipzig. (Reference to great auk, p. 52.)
ANDERSON, R. M.
   1907. Birds of Iowa. Proc. Davenport Acad. Sci., **11**, pp. 135–417. (*Ectopistes,* p. 239.)
ANDERSSON, J. G.
   1923. Fossil remains of Struthionidae in China. Mem. Geol. Surv. China, ser. A, no. 3, pp. 53–78.
ANDREWS, C. W.
   1910. *Psammornis,* a fossil ostrich from southern Algeria. In Rothschild, Berichte, 5th Int. Orn. Congr., Berlin.
   1920. Remains of great auk in Channel Islands [Jersey]. Ann. Mag. Nat. Hist., **9** (6), p. 166.

ANDREWS, F. W.
  1883. Notes on the night parrot (*Geopsittacus occidentalis*). Trans. Roy. Soc.
    South Australia, **6**, pp. 29–30.
ANONYMOUS.
  1906. Amer. Ornith. (Worcester), **6**, pp. 122–124. [Title varies.]
ANTHONY, A. W.
  1893. [Record of *Gymnogyps* in Lower California.] Zoë, **4**, p. 233.
  1898. Petrels of southern California. Auk, **15**, pp. 140–144.
  1901. The Guadalupe wren. Condor, **3**, p. 73.
  1925. [Birds and mammals of Guadalupe Island]. Proc. California Acad. Sci.,
    4th ser., **14**, pp. 277–320.
ARCHEY, G.
  1941. The moa, a study of the Dinornithiformes. Bull. Auckland Inst. and Mus.,
    no. 1, pp. 5–101.
ARCHEY, G., and LINDSAY, C.
  1924. Notes on the birds of the Chatham Islands. Rec. Canterbury Mus. [Christ-
    church, N.Z.], **2**, pp. 187–201.
ARMAS, J. I. DE.
  1888. La zoología de Colón etc., 183 pp. Habana.
ARRIGONI, E.
  1914. Specimen of *Alca impennis* sold to a Russian at Moscow. Riv. Ital. Orn.
    Bologna, **3**, pp. 1–3.
ASHBY, E.
  1921. Notes on the supposed "extinct" birds of the southwest corner of Western
    Australia. Emu, **20**, pp. 123–124.
  1927. Two *Neophema* parrots. Emu, **27**, pp. 1–2.
ATKINSON, G.
  1922. The extraordinary voyage in French literature from 1700–1720. Paris.
AUDUBON, J. J.
  1832–1839. Ornithological biography, vols. 1–5. Edinburgh.
AUSTIN, O. L., JR.
  1932. Birds of Newfoundland and Labrador. Mem. Nuttall Orn. Club, no. 7.
  1948. The birds of Korea. Bull. Mus. Comp. Zool., **101** (1), 301 pp.
  1949. Status of Steller's albatross. Pacific Sci. (Honolulu), **3** (4), pp. 283–295.
AUSTIN, O. L., JR., and KURODA, N.
  1953. The birds of Japan: their status and distribution. Bull. Mus. Comp. Zool.,
    109, no. 4, pp. 280–612.
BAHR, P. H.
  1911. [On a journey to the Fiji Islands etc.] Ibis, 1911, pp. 282–314 (292).
BAILEY, A. M.
  1948. Birds of Arctic Alaska. Colorado Mus. Nat. Hist. Pop. Ser. no. 8, 304 pp.
BAILEY, F. M.
  1928. Birds of New Mexico, 807 pp. New Mexico Department of Fish and
    Game, Santa Fe.
BAILEY, H. B.
  1881. Forest and stream bird notes, an index and summary of all ornithological
    matter contained in Forest and Stream, vols. I–XII (1873–79).
BAILEY, H. H.
  1913. The birds of Virginia, 362 pp. Lynchburg.
BAIN, F.
  1891. Birds of Prince Edward Island, Charlottetown.
BAIRD, S. F., BREWER, T. M., and RIDGWAY, R.
  1884. The water birds of North America, 2 vols. Mem. Mus. Comp. Zool.,
    12–13.

BAKER, E. C. S.
  1908. The Indian ducks and their allies, 292 pp. London.
  1929. Fauna of British India etc., Birds. 8 vols., 1922–1930. London.
  1931. Game birds of the Indian Empire. Journ. Bombay Nat. Hist. Soc., **34** (1), pp. 1–12. (*Rhinoptilus*, p. 5.)
BAKER, R. H.
  1948. Report on collections made by United States Naval Medical Research Unit no. 2 in the Pacific War Area. [New Hebrides, Solomons, Admiralties, Philippines, Marianas, Carolinas, Palau, Volcanos, and Riu Kius.] Smithsonian Misc. Coll., **107** (15), 74 pp.
  1951. The avifauna of Micronesia. Univ. Kansas Publ. Mus. Nat. Hist., **3** (1), 359 pp.
BALDWIN, P. H.
  1941. Checklist of the birds of the Hawaii National Park, Kilauea-Mauna Loa section, with remarks on the present status and a field key for their identification. (Mimeographed.)
  1945. The Hawaiian goose: Its distribution and reduction in numbers. Condor, **47** (1), pp. 27–38.
  1947. Foods of the Hawaiian goose. Condor, **49** (1), pp. 108–121.
  1949. The life history of the Laysan rail. Condor, **51**, pp. 14–21.
BANGS, O., and KENNARD, F. H.
  1920. A list of the birds of Jamaica, 18 pp. The Handbook for Jamaica, Kingston.
BANGS, O., and ZAPPEY, W. R.
  1905. Birds of the Isle of Pines. Amer. Nat., **39**, pp. 179–215.
BANNERMAN, D. A.
  1930–1951. The birds of Tropical West Africa. 8 vols. London.
  1915. [Report on birds collected by Boyd Alexander in Africa.] Ibis, 1915, pp. 89, 227, 473, 643.
BANKS, J.
  1896. Journal of the Rt. Hon. Sir Joseph Banks during Capt. Cook's first voyage, etc. Sir Joseph Hooker, *ed.*, 466 pp. London.
BARBOUR, R.
  1906. *Numenius borealis* at sea at 49° N. x 27° W. Auk, **23**, p. 459.
BARBOUR, T.
  1923. The birds of Cuba. Mem. Nuttall Orn. Club, no. 4, pp. 3–141.
  1925. [*Aratinga holochlora* collected in Florida.] Auk, **42**, p. 132.
  1943. Cuban ornithology. Mem. Nuttall Orn. Club, **9**, 129 pp.
BAROARSON, G. G.
  1911. [Bones of *Alca impennis* at Hunafloi, North Iceland.] Vid. Medd. Naturh. For. Kjøbenhavn, 1910, p. 73.
BARRETT-HAMILTON, G. E.
  1896. The great auk as an Irish bird. Irish Nat., **5**, p. 121.
BARROWS, W. B.
  1884. Birds of Lower Uruguay. Auk, **1**, pp. 313–319.
  1912. Michigan bird life, 822 pp. Lansing, Mich. [Curlew, p. 316.]
BARTSCH, P.
  1922. A visit to Midway [list of birds]. Auk, **39**, pp. 481–488.
BATCHELDER, C. F.
  1882. [*Ectopistes* breeding in New Brunswick.] Bull. Nuttall Orn. Club, **7**, p. 151.
BATE, D. M. A.
  1928. [Remains of great auk at Gibraltar, p. 107.] Journ. Roy. Anthrop. Inst. Great Britain and Ireland, **58**, pp. 92–110.

BECKHAM, C. W.
1885. Birds of Nelson County (Kentucky), 58 pp. Kentucky Geological Survey.
BEEBE, W.
1935. Re-discovery of the Bermuda "cahow." Bull. New York Zool. Soc., **38**, pp. 187–190.
BEECHEY, F. W.
1839. The zoology of Capt. Beechey's voyage to the Pacific . . . in the *Blossom* . . . 1825–1828; pp. 12–186. London.
BEHR, H.
1912. Recollections of the passenger pigeon. Cassinia, **15**, pp. 24–27.
BELKNAP, J.
1784–1792. The history of New Hampshire, etc., 3 vols. Philadelphia and Boston.
BENDIRE, C. E.
1892–1895. Life histories of North American birds. U.S. Nat. Mus. Spec. Bull. 1, 2 vols.
BENHAM, W. B.
1899. [Description of a specimen of *Notornis* taken in 1898.] Trans. and Proc. New Zealand Inst., **31**, pp. 146–156.
BENNETT, F. D.
1940. Narrative of a whaling voyage around the world from the year 1833 to 1836, 2 vols. London.
BENNETT, G.
1863. Notes on the kagu (two specimens in captivity). Proc. Zool. Soc. London, 1863, pp. 385–386, 439–440.
BENNITT, R., and NAGEL, W. O.
1937. A survey of the resident game and furbearers of Missouri. Univ. Missouri Stud., **12** (2), 215 pp.
BENT, A. C. (The Bent life histories have been reprinted by Dover Publications)
1912. Notes on birds observed during a brief visit to the Aleutian Islands and Bering Sea in 1911. Smithsonian Misc. Coll., **56**, no. 32, pp. 1–29.
1919. Life histories of North American diving birds. U.S. Nat. Mus. Bull. 107, 245 pp.
1921. Life histories of the North American gulls and terns. Order Longipennis. U.S. Nat. Mus. Bull. 113, 345 pp.
1922. Life histories of North American petrels and pelicans and their allies. Order Tubinares and Order Steganopodes. U.S. Nat. Mus. Bull. 121, 343 pp.
1923. Life histories of North American wild fowl. Order Anseres (part). U.S. Nat. Mus. Bull. 126, 250 pp.
1925. Life histories of North American wild fowl. Order Anseres (part). U.S. Nat. Mus. Bull. 130, 376 pp.
1926. Life histories of North American marsh birds. Orders Odontoglossae, Herodiones, and Paludicolae. U.S. Nat. Mus. Bull. 135, 490 pp.
1927. Life histories of North American shore birds. Order Limicolae (part 1). U.S. Nat. Mus. Bull. 142, 420 pp.
1929. Life histories of North American shore birds. Order Limicolae (part 2). U.S. Nat. Mus. Bull. 146, 412 pp.
1932. Life histories of North American gallinaceous birds. U.S. Nat. Mus. Bull. 162, 490 pp.
1937. Life histories of North American birds of prey. Order Falconiformes (part 1). U.S. Nat. Mus. Bull. 167, 409 pp.
1938. Life histories of North American birds of prey. Order Falconiformes (part) and Strigiformes. U.S. Nat. Mus. Bull. 170, 482 pp.

1939. Life histories of North American woodpeckers. Order Piciformes. U.S. Nat. Mus. Bull. 174, 334 pp.

1940. Life histories of North American cuckoos, goatsuckers, hummingbirds, and their allies. Order Psittaciformes, Cuculiformes, Trogoniformes, Coraciiformes, Caprimulgiformes, and Micropodiiformes. U.S. Nat. Mus. Bull. 176, 506 pp.

1942. Life histories of North American flycatchers, larks, swallows, and their allies. Order Passeriformes (pt.). U.S. Nat. Mus. Bull. 179, 555 pp.

1946. Life histories of North American jays, crows, and titmice. Order Passeriformes (pt.). U.S. Nat. Mus. Bull. 191, 495 pp.

1948. Life histories of North American nuthatches, wrens, thrashers, and their allies. Order Passeriformes (pt.). U.S. Nat. Mus. Bull. 195, 475 pp.

1949. Life histories of North American thrushes, kinglets, and their allies. Order Passeriformes (pt.). U.S. Nat. Mus. Bull. 196, 454 pp.

1950. Life histories of North American wagtails, shrikes, vireos, and their allies. Order Passeriformes (pt.). U.S. Nat. Mus. Bull. 197, 411 pp.

BERLIOZ, J.

1929. Catalogue systematique des types de la collection d'oiseaux du Muséum (1. Ratites- 2. Palmipedes). Bull. Mus. Nat. Hist. Nat. [Paris] (2), **1**, pp. 58–69.

1935. Notice sur les specimens naturalisés d'oiseaux éteint existant dans les collections du Muséum. Arch. Mus. Hist. Nat. [Paris] (6) **12**, pp. 485–495.

1940. Note critique sur . . . *Lophopsittacus mauritianus* (Owen). Bull. Mus. [Paris] (2), **12**, pp. 143–148.

1944. La vie de colibris, 198 pp. Paris.

1945. Les psittacidés de la Nouvelle Caledonie et des Establissements Français d'Oceanie. Oiseau et Rev. Franç. Orn., **15**, pp. 1–9.

1946. Oiseaux de la Réunion, *in* Faune de l'Empire Franç., LV, pp. 1–81. Paris.

1950. Évolution actuelle des oiseaux espèce recemment éteintes, pp. 836–844 in Traité de Zoology, etc. (P. Grassé ed.), **15**, Oiseaux, 1,120 pp., Paris.

BERTONI, A. DE W.

1901. Aves neuvas del Paraguay, etc., 216 pp. Asunçion.

BEYER, G. E., ALLISON, A., and KOPMAN, H. H.

1908. Passenger pigeon in Louisiana. Auk, **25**, p. 440.

BIGGAR, H. P.

1924. Voyage of Jaques Cartier, published from original, etc. Public Arch. Canada, no. 11 (French and English texts). Ottawa.

BINNS, G.

1947. Appearance of the elegant parrot at Ouyen, Victorian mallee [*Neophema elegans*]. Emu, **47**, pp. 7–10.

BISHOP, A. H.

1914. [An Oronsay shell mound.] Proc. Soc. Antiq. Scotland, **48**, pp. 52–108.

BLACKMAN, T. M.

1944. Concerning the Hawaiian goose or nene. Nat. Hist., **53** (9), p. 407.

1945. War casualties among the birds. Nat. Hist., **54** (7), pp. 298–299.

BLANC, G. A.

1927. Sulla presenea di *Alca impennis* Linn. nella formazione superiori di Grotta Romanelli in Terra d'Otranto. Arch. Antrop. Ethnol. (Firenze), **58**, pp. 155–186.

BLANFORD, W. T.

1898. *Rhinoptilus bitorquatus*. Fauna of British India (Birds), **4**, p. 213.

BLASIUS, W.

1883. Über die letzten Vorkommnisse des Riesen-Alks *Alca impennis* und die in Braunschweig und an anderen Orten befindlichen Exemplare dieser Art. III. Jahresb. Ver. Naturw. Braunschweig, 1883, pp. 89–115.

1884. Zur Geschichte der Uberreste von *Alca impennis* L. Journ. für Orn., **4**, (3), pp. 58–176.
1900. Der Riesen-Alk in der ornithologischen Literatur der letzten funfzehn Jahre. Orn. Monatsschr., **25**, pp. 434–446.

BOLAU, H.
1898. Die Typen der Vogelsammlung des Naturhistorischen Museums zu Hamburg [*Pareudiastes pacificus, Porphyrio melanotus*]. Mitt. Naturh. Mus. Hamburg, **15**, pp. 45–71.

BOND, F.
1921. The later flights of the passenger pigeon. Auk, **38**, pp. 523–527.

BOND, J.
1928a. A remarkable West Indian goatsucker [*Siphonorhis brewsteri*]. Auk, **45**, pp. 471–474.
1928b. Distribution and habits of the birds of Haiti. On the birds of Dominica, St. Lucia, St. Vincent and Barbados, B.W.I. Proc. Acad. Nat. Sci. Philadelphia, **70**, pp. 483–545.
1936. Birds of the West Indies, xxiv + 434 pp. Academy of Natural Sciences of Philadelphia.
1942. Additional notes on West Indian Birds. Proc. Acad. Nat. Sci. Philadelphia, **94**, pp. 89–106.
1947. A plea for conservation in the West Indies. Audubon Mag., **49** (6), pp. 348–354.
1948. Origin of the bird fauna of the West Indies. Wilson Bull., **60** (4), pp. 207–229.
1950. Check list of the birds of the West Indies, 184 pp. Academy of Natural Sciences of Philadelphia.
1951. First supplement to the check list of birds of the West Indies (1950), 22 pp. Academy of Natural Sciences of Philadelphia.

BORRI, C.
1927. Great auk in Pisa Mus. Atti Soc. Tosc. Sci. Nat., **36** (1), pp. 6–8.

BOWES, ARTHUR.
[1788]. A journal of a voyage from Portsmouth to New South Wales and China in the *Lady Penrhyn* 1787–89. Ms. in Mitchell Library, Sydney. See Hindwood, 1932.

BRADLEE, T. S., MOWBRAY, L., ET AL.
1931. A list of birds recorded from the Bermudas. Proc. Boston Soc. Nat. Hist., **39** (8), pp. 279–382.

BRASIL, L.
1914. The emu of King Island. Emu, **14**, pp. 88–97.

BRENCHLEY, J. L.
1873. Jottings during the cruise of H. M. S. *Curaçoa* among the South Sea Islands in 1865, 487 pp. London. [Birds by G. R. Gray.]

BREWSTER, W.
1889. Present status of the wild pigeon (*Ectopistes migratorius*), etc. Auk, **6**, pp. 285–291.
1902. Birds of the Cape region of Lower California. Bull. Mus. Comp. Zool., **41** (1), 241 pp.

BROEKHUYSEN, G. J.
1948. Observations on the great shearwater in the breeding-season. British Birds, **41** (11), pp. 338–341.

BROEKHUYSEN, G. J., and MacNAE, W.
1949. Observations on the birds of Tristan da Cunha Islands, etc. Ardea, **37**, pp. 97–122.

BRØGGER, A. W.
   1908. Vistefundet etc. Stavanger Mus. Aarshefte, **18** (2), pp. 1–102. [Remains of great auk, p. 10.]
BROOKLING, A. M.
   1943. [Reduction in numbers of whooping cranes apparent in last decade.] Nebraska Bird Rev., **11**, pp. 5–8.
BROWN, H.
   1904. Masked bob-white (*Colinus ridgwayi*). Auk, **16**, pp. 209–213.
BROWNE, PATRICK.
   1756. The civil and natural history of Jamaica, etc., London.
BRYAN, E. H., JR., and GREENWAY, J. C., JR.
   1944. Contribution to the ornithology of the Hawaiian Islands. Bull. Mus. Comp. Zool., **94** (2), pp. 80–142.
BRYAN, W. A.
   1908. Some birds of Molokai. Occ. Pap. B. P. Bishop Mus. (Honolulu), **4** (2), pp. 43–86.
BRYANT, J. J.
   1936. Bristle bird near N.S.W.-Victoria border [*Dasyornis* near Womboyne Inlet]. Emu, **36**, p. 56.
BRYANT, W. E.
   1887. Additions to the ornithology of Guadalupe Island [Pacific]. Bull. California Acad. Sci., **6**, pp. 269–318.
   1891. [California vulture.] Zoe, **2**, p. 52–53.
BUCKLEY, T. E., and HARVIE-BROWN, J. A.
   1388. A vertebrate fauna of the Outer Hebrides, 279 pp. Edinburgh. (2d ed., 1891.)
BUCKNILL, J. A.
   1924. Disappearance of the pink headed duck. Ibis, 1924, pp. 146–151.
BUFFON, G. L., DE
   1770–1786. Histoire naturelle des oiseaux, 10 vols., Paris. (**6**, p. 183.)
BULL, P. C.
   1946. Notes on the breeding cycle of thrush and blackbird in New Zealand. Emu, **46**, pp. 198–208.
BULLEN, R. P.
   1949. Excavations in northeastern Massachusetts. Pap. R. S. Peabody Foundation, Andover, Mass., **1** (3), 166 pp.
BULLER, K. G.
   1945. A new record of the western bristle bird. Emu, **45**, pp. 78–80.
BULLER, W. L.
   1881. *Notornis hochstetteri* of South Island. Trans. and Proc. New Zealand Inst., **14**, p. 243.
   1887–1888. A history of the birds of New Zealand, 2 vols. (2d ed.). London.
   1905–1906. Supplement to 'Birds of New Zealand', 2 vols. London.
BUNKER, C. D.
   1913. The birds of Kansas. Kansas Univ. Sci. Bull., **7** (5), pp. 137–158.
BUREAU, L.
   1911. Sur le capture en France d'un pigeon migrateur d'Amérique. Bull. Soc. Acclimat. France, **58**, pp. 353–356.
BURNS, F. L.
   1910. [Status of the passenger pigeon.] Wilson Bull., **22**, pp. 47–48.
BUTLER, A. W.
   1892. [Carolina parakeet.] Auk, **9**, pp. 49–56.
   1899. [*Ectopistes* in Indiana.] Proc. Indiana Acad. Sci., 1899, p. 150; id., 1902, p. 98.

1903. Notes on some rare Indiana birds. Proc. Indiana Acad. Sci., 1902, pp. 95–99.
1913. Further notes on Indiana birds. Proc. Indiana Acad. Sci., 1912, pp. 59–63.
BUXTON, L. H. D.
1928. Human remains in Devil's Tower, Gibraltar. Journ. Roy. Anthrop. Inst. etc., **58**, pp. 57–85.
BYERS, D. S., and JOHNSON, F.
1940. Two sites on Martha's Vineyard. Papers R. S. Peabody Foundation, Andover, Mass., **1** (1), 104 pp.
CALDWELL, J.
1875. Notes on the zoology of Rodriguez. Proc. Zool. Soc. London, 1875, pp. 644–647.
CALDWELL, H. R., and CALDWELL, J. C.
1931. South China birds, 447 pp. Shanghai.
CAMPANA, D.
1925. Gli uccelli della Grotta di Parignana (Monte Pisano). Atti Soc. Tosc. Sci. Nat. Pisa Mem., **36**, pp. 206–225.
CAMPBELL, A. J.
1905a. Notes on the kagu in captivity. Emu, **4**, pp. 166–168.
1905b. [Notes from H. E. Finckh with description of downy young of *Rhynochetos jubatus*.] Emu, **5**, p. 32.
1915. Missing birds [*Psephotus pulcherrimus, Neophema pulchella, Geopsittacus*]. Emu, **14**, pp. 167–168.
1923. The long-lost Eyrean grass wren. Emu, **23**, p. 81.
CARIÉ, P.
1916. Bull. Soc. Nat. Acclimat. France: No. 1, pp. 10–19; no. 2, pp. 37–46; no. 3, pp. 72–79; no. 4, pp. 107–111; no. 5, pp. 152–160; no. 6, pp. 191–199; no. 7, pp. 245–251; no. 9, pp. 355–364; no. 10, pp. 401–405.
1930. Le *Leguatia gigantea* Schlegel, a-t-il existé? Bull. Mus. Hist. Nat. Paris (2), **2**, pp. 204–213.
CARMICHAEL, D.
1818. Some account of the island of Tristan da Cunha. Trans. Linn. Soc. London, **12**, pp. 483–513.
CARRUTHERS, D.
1910. A journey to north-western Arabia. Geogr. Journ., **35**, pp. 225–245.
1922. The Arabian ostrich. Ibis, 1922, pp. 471–474.
CARTIER, J.
——. Voyages to . . . Newfoundland and Canada in . . . 1534 and 1535. [Seen in Pinkerton (vol. 12), Hakluyt (vol. 3), and in Biggar, H. P., 1924, q.v.]
CARTWRIGHT, GEORGE.
1911. Captain Cartwright and his Labrador journal. Edited by C. W. Townsend, 33+385 pp. Boston.
CASAS, B. DE LAS.
1875–1876. Historia de las Indias, 5 vols. Madrid.
CASE, W. L.
1930. Passing of the passenger pigeon. Michigan Hist. Mag., **14**, pp. 262–267.
CASSIN, J.
1862. Birds collected by U.S. North Pacific surveying expedition. Proc. Acad. Nat. Sci. Philadelphia, 1862, pp. 312–327.
CATESBY, M.
1731–1743. Natural history of Carolina, Florida and the Bahama Islands, etc., 2 vols. London.

CAUCHE, F.
1651. Relation du voyage. Relations Veritables et Surieuses, Paris.
CAYLEY, N. W.
1938. Australian parrots [etc.], 324 pp. London and Sydney.
CHAFFER, N., and MILLER, G.
1946. Turquoise parrot near Sydney [Neophema pulchella]. Emu, 46, pp. 161–167.
CHAMBERLAIN, M.
1882. [Record for Ectopistes.] Bull. Nat. Hist. Soc. New Brunswick, 1, p. 50.
CHAMPLAIN, S. DE.
1878–1882. Voyages of Samuel de Champlain. Translated by C. P. Otis. Prince Society, Boston.
1906. The voyages and explorations of Samuel de Champlain, 1604–1616, 2 vols., Translated by A. N. Bourne. (First edition, 1613.)
CHAPIN, J. P.
1932. Birds of the Belgian Congo, vol. 1. Bull. Amer. Mus. Nat. Hist., 65, 723 pp.
CHAPMAN, F. M.
1903. A letter from G. H. Van Note regarding the heath hen. Bird Lore, 5, pp. 50–51.
1915. The Caroline paroquet in Florida. Ibis, 17, p. 453.
CHEESEMAN, R. E.
1923. Recent notes on Arabian ostrich. Ibis, 1923, pp. 208–211, 359, and pl. 4.
CHILDE, V. G.
1935. The prehistory of Scotland, 285 pp. London.
1947. The dawn of European civilization, 350 pp. London.
CHILDS, J. L.
1905a. Personal recollections of the passenger pigeon. Warbler (2), 1, pp. 71–73.
1905b. Eggs of the Carolina paroquet. Warbler (2), 1, pp 97–98. (J. L. Childs edit.)
1906. Eggs of Aratinga carolinensis taken in Florida. Warbler (2), 2, p. 65. (Childs edit.)
CHISHOLM, A. H.
1921. Adventures among Atrichornis. Emu, 21, pp. 2–8.
1922. The "lost" paradise parrot [Psephotus pulcherrimus]. Emu, 22, pp. 4–17.
1951. Story of the scrub-birds [Atrichornis]. Emu, 51, pp. 89–112.
CHRISTIANI, A.
1917. [Bones of Alca impennis found near Vardö, Norway.] Dansk Orn. Foren. Tidsskr., 11, pp. 1–4.
1929. [Alca impennis bones in Pleistocene beds in N. Norway (Vardo).] Dansk Orn. Foren. Tidsskr., 23, pp. 79–83.
CHRISTOPHERSEN, E.
1940. Tristan da Cunha, the lonely isle. London.
CLARK, A. H.
1905. Lesser Antillean macaws. Auk, 22, pp. 266–273. (Conurus in the West Indies, pp. 310–312. West Indian parrots, pp. 337–344. Great macaws, pp. 345–348.)
1908. The macaw of Dominica. Auk, 25, pp. 309–311.
CLARK, G.
1866. Account of the late discovery of dodos' remains in the island of Mauritius. Ibis, 1866, pp. 141–146.
CLARK, H. L.
1918. Pterolysis of the wild pigeon [Ectopistes]. Auk, 35, pp. 416–420.

CLARK, [J.] G. [D.]
1932. The Mesolithic Age in Britain, 115 pp. Cambridge.
1936. The Mesolithic settlement of northern Europe, etc.
1948. Fowling in prehistoric Europe. Antiquity, **22** (87), pp. 116–130.
CLARKE, J. M.
1918. Alleged re-discovery of the passenger pigeon. Science, **48**, pp. 445–556.
CLARKE, W. E.
1905. On the birds of Gough Island, South Atlantic Ocean. Ibis, 1905, pp. 247–268.
CLELAND, J.
1940. [Behaviour of emus.] South Australian Orn., **15**, p. 109.
CLYMAN, J.
1928. [Account of *Gymnogyps* in 1845.] California Hist. Soc. Spec. Publ., **3**, p. 182.
COCKAYNE, L.
1904. A botanical excursion during winter to the southern islands of New Zealand. Trans. New Zealand Inst., **36**, pp. 225–333.
COHEN, N. W.
1951. [California condors in Madera County, California.] Condor, **53**, p. 158.
COLLETT, R.
1866. [Brief note on *Alca impennis* in Norway.] Journ. für Orn., **14**, p. 70.
CONDON, H. T.
1942. The ground parrot (*Pezoporus wallicus*) in South Australia. South Australian Orn., **16**, p. 47.
COOK, J.
1777. A voyage toward the South Pole and around the world in the years 1772, 3, 4, 5, 2 vols., 2d ed.
1784. A voyage to the Pacific Ocean . . . in the years 1776, 7, 8, 9, 80, 3 vols. London.
COOKE, W. W.
1912. Passenger pigeon in Alberta. Auk, **24**, p. 539.
1914. Winter birds of Oklahoma. Auk, **31**, pp. 473–493.
COOPER, J. G.
1860. [Record of *Ectopistes* in Nebraska.] U.S. War Dept. Rep. Expl. Railroad Route Mississippi R. to Pacific, **12** (2), p. 218. Variously cited "Pacific railroad report" etc.
1871. [Notes on California condor.] Amer. Nat., **4**, pp. 756–758.
CORY, C. B.
1886. [Note on *Ara tricolor*.] Auk, **3**, p. 454.
COTTAM, C., and KNAPPEN, P.
1939. Food of some uncommon North American birds. Auk, **56**, pp. 138–169. [Carolina parakeet, p. 158.]
COUES, E.
1861. Notes on the ornithology of Labrador. Proc. Acad. Nat. Sci. Philadelphia, **13**, pp. 215–257.
1866. List of the birds of Fort Whipple, Arizona, etc. Proc. Acad. Nat. Sci. Philadelphia, **18**, pp. 39–100.
1874. Birds of the Northwest; a handbook of the ornithology of the region drained by the Missouri River and its tributaries. U.S. Geol. Surv. Terr. Misc. Publ. no. 3, 791 pp.
COVERT, A. B.
1898. [*Ectopistes* sight record.] Bull. Michigan Orn. Club, **2**, p. 37.

CRAIG, W.
1911. Expression of emotions in pigeons: III, The passenger pigeon (*Ectopistes migratorius*). Auk, **28**, pp. 408–427.
1913. Recollections of the passenger pigeon in captivity. Bird Lore, **15**, pp. 93–99.

CRANDALL, L. S.
1912. [Carolina parakeet in zoological gardens.] Bull. New York Zool. Soc., **49**, pp. 834–836.

CUMBERLAND, K. B.
1941. A century's change: Natural to cultural vegetation in New Zealand. Geogr. Rev., **31**, 529–554.

DALGLEISH, J. J.
1880. Occurrence of North American birds in Europe [*Numenius*]. Bull. Nuttall Orn. Club, **5**, p. 210.

DANFORTH, S. T.
1931. [Puerto Rican ornithological records. *Columba inornata*, p. 68, and *Amazona vittata*, p. 70.]. Journ. Dept. Agr. Puerto Rico, **15** (1), 106 pp.
1935. Birds of St. Lucia. Mongr. Univ. Puerto Rico, Ser. B, no. 3.
1936. Birds of St. Kitts and Nevis. Trop. Agriculture (Port of Spain, Trinidad), **13** (8), pp. 213–217.
1939a. Birds of Guadaloupe and adjacent islands. Journ. Agr. Univ. Puerto Rico, **23**, pp. 9–66.
1939b. Birds of Montserrat. Journ. Agr. Univ. Puerto Rico, **23**, pp. 47–66.

DARWIN, C.
1839. Journal of researches into the geology and natural history of the various countries visited during the voyage of H. M. S. *Beagle,* etc. (Titles of later editions vary.)

DAVID, A., and OUSTALET, M. E.
1877. Les oiseaux de la Chine, 2 vols., text 573 pp., atlas 124 col. pls. Paris.

DAVIDSON, M. E. M.
1928. [*Oceanodroma macrodactyla*.] Condor, **30**, pp. 355–356.

DAVIS, M.
1946. Morning display of California condor. Auk, **63**, p. 85.

DAVIS, S. M.
1929. Notes on the passenger or wild pigeon. Maine Nat., **9**, pp. 97–98.

DAVISON, V. E.
1940. An 8-year census of lesser prairie chickens [in Oklahoma]. Journ. Wildlife Management, **4** (1), pp. 55–62.

DAWSON, E.
1949. Sub-fossil bird remains from Lake Grassmere (Marlborough, N. Island, New Zealand). New Zealand Bird Notes, **3** (5), p. 132.

DEANE, R.
1895. Additional records of passenger pigeon in Illinois and Indiana. Auk, **12**, pp. 298–299.
1896. Additional records of the passenger pigeon in Illinois and Wisconsin; habits in confinement. Auk, **13**, pp. 81, 234.
1897. [Records of passenger pigeon in Missouri.] Auk, **14**, p. 316.
1908. The passenger pigeon (*Ectopistes migratorius*) in confinement. Auk, **25**, pp. 181–183.
1909. The passenger pigeon—only one pair left. Auk, **26**, p. 429.
1911. Passenger pigeon—only one bird left (in Cincinnati Zoo). Auk, **28**, p. 262.
1913. Last survivor of *Ectopistes migratorius* still alive in Cincinnati Zoo. Auk, **30**, p. 433.

DEEVEY, E. S., JR.
1954. The end of the moas. Sci. Amer., **190** (2), pp. 84–90; **190** (5), pp. 5, 6.
DEGERBØL, M.
1926. [Bibliography of H. Winge—*Alca impennis*, p. 38.] Vid. Medd. Dansk Naturh. For. København, **82**, pp. 1–42.
1933. Danmarks Patteydyr i. Fortiden, etc. [An account of mammals found in kitchen middens, etc.] Vid. Medd. Dansk Naturh. For. København, **96**, Festskrift, etc., pp. 357–641.
1934a. [Animal bones from the Eskimo settlement in Dødemandsbugten, Clavering Island.] Medd. Grønland, **102** (1) (Zool. App.), p. 173.
1934b. [Animal bones from the Norse ruins at Brattahlid.] Medd. Grønland, **88** (1) (Zool. App.), pp. 149–156.
1935. [Animal bones from King Oscar Fjord region in East Greenland.] Medd. Grønland, **102** (2) (Zool. App.), p. 94.
1936. [Animal remains from the West Settlement in Greenland.] Medd. Grønland, **88** (3), p. 52.
DEGERBØL, M., and HØRRING, R.
1936. Bones from Julianhaab District [Greenland]. Medd. Grønland, **118** (1), p. 131.
DEHÉRIAN, H.
1926. Le voyage de François Leguat dans l'océan Indien (1690–1698) est-il imaginaire? Ministère de l'instruction publique et des beaux arts. Comité des travaux historiques et scientifiques. Bull. section geographie, **41**. (Not sighted; see Swaen, 1940.)
DEKAY, J. E.
1844. Zoology of New York, etc. (pt. 2, Birds), 380 pp. Albany, N. Y.
DELACOUR, J.
1932. Les oiseaux de la mission zoologique . . . à Madagascar. Oiseau Rev. Franç. Orn. (n.s.), **2** (2), pp. 1–96 (84).
1951a. The pheasants of the world, 347 pp.
1951b. Preliminary note on the taxonomy of Canada geese, *Branta canadensis*. Amer. Mus. Novit., no. 1537, pp. 1–10.
DELACOUR, J., and MAYR, E.
1945. The family Anatidae. Wilson Bull., **57**, pp. 3–55 (9).
DEMENT'EV, G.
1951. Ptitsy sovietskovo soyuza, etc. [Birds of U.S.S.R.], 3 vols. Moscow.
DENNIS, J. V.
1948. Last remnant of ivory-billed woodpeckers in Cuba. Auk, **65**, pp. 497–506.
DERSCHEID, J. M.
1939. An unknown species—the Tahitian goose (?). Ibis, 1939, pp. 756–760.
DESJARDINS, J.
1837(?). Huitième rapport sur les travaux de la Sociéte de Histoire Naturelle de l'île de Meurice, lu à la séance du 24 Aout 1837. (Not sighted, see Oustalet, 1896b, pp. 35–38.)
DE VIS, C. W.
1888. A glimpse of the post-Tertiary avifauna of Queensland. Proc. Linn. Soc. New South Wales (2), **3**, pp. 1290–1292.
1884. The moa (*Dinornis*) in New Zealand. Proc. Roy. Soc. Queensland, **1**, pp. 23–28.
1892. Residue of the extinct birds of Queensland as yet detected. Proc. Linn. Soc. New South Wales (2), **6**, pp. 437–456.
DIDIER, R.
1934. Le grand pingoiun, Plautus impennis l. La Terre et la Vie (Paris, **4**, pp. 13–20.

DIEFFENBACH, E.
  1843. Travels in New Zealand, etc., 2 vols. London.
DILL, H. R., and BRYAN, W. A.
  1912. Report of an expedition to Laysan Island in 1911. U.S. Biol. Surv. Bull.
       42, pp. 1–30.
DILLIN, J. G.
  1911. Recollections of wild pigeons in southeastern Pennsylvania. Cassinia, 14,
       1910, pp. 33–36.
DILLON, P.
  1829. Narrative . . . of a voyage in the South Seas to ascertain the actual fate
       of La Perouse's expedition . . . 2 vols., 302, 486 pp. London.
DIONNE, C. E.
  1906. Oiseaux de la Province de Québec, 403 pp. Quebec.
DIXON, D.
  1885. Ornithology of St. Kilda. Ibis, 1885, pp. 69–97.
DONAGHO, W. R.
  1940. A list of birds mated on the island of Hawaii. Elepaio, 1 (4), pp. 3–5.
  1951. Journal of ornithological work during the summer of 1937. Elepaio, 11,
       pp. 56–58, 62–65.
  1952. Journal of ornithological work during the summer of 1937. Elepaio, 12
       (7), pp. 46–48.
DOUGLAS, D.
  1904. Sketch of a journey to the northwestern parts of the continent of North
       America during the years 1824, '25, '26, '27. Journal and Letters of D.
       Douglas. Reprinted (at least in part) from The Companion to the Botani-
       cal Magazine, London, 2, 1836, in Quarterly Oregon Hist. Soc., 5, 1904,
       pp. 230–271, 325–369; 6, 1905, pp. 76–97, 206–227, 288-309, 417–449.
DOVE, H. S.
  1925. Some rare birds in Tasmania. Emu, 24, pp. 280–282.
DOWNS, A.
  1888. Catalogue of birds of Nova Scotia [record for Ectopistes]. Proc. and
       Trans. Nova Scotia Inst. Nat. Sci., 7, p. 157.
DOWNS, T.
  1946. Birds on Tinian in the Marianas. Trans. Kansas Acad. Sci., 49 (1), pp.
       87–106.
DRUMMOND, J.
  1908. The Little Barrier Bird Sanctuary. Trans. and Proc. New Zealand Inst.,
       40, pp. 501–506.
DRUMMOND-HAY, H. M.
  1877–1878. "On Migration." Account of birds seen on Bermuda. Scottish Nat.,
       4, pp. 229–241.
DRURY, C.
  1910. The passenger pigeon, a reminiscence. Cincinnati Journ. Soc. Nat. Hist.,
       21, pp. 52–56.
DU BATY, R. R.
  1912. Notes on Tristan da Cunha [p. 95]: 15,000 miles in a ketch, 348 pp.
       London.
DUBOIS, PÈRE.
  1674. Les voyages faits par le Sieur D. B. aux Isles Dauphine ou Madagascar et
       Bourbon ou Mascarene, années 1669, 1670, 1671, 1672.
DUFF, R. S.
  1949. Moas and man. Antiquity, 23, pp. 172–179.
DUGAND, A.
  1947. Aves del Departamento Atlántico, Colombia. Caldasia, 4 (20), pp. 499–
       648 (637).

Du Mont, P. A.
1933. The passenger pigeon as a former Iowa bird. Proc. Iowa Acad. Sci., **40**, pp. 205–211.

Duncker, H.
1953. Mitteilungen aus der Bremer Vogelsammlung. Erganzungen zum Beitrag von Gustav Hartlaub von April 1896. Abh. naturw. Ver. Bremen, **33**, pp. 216–246.

Dunckley, J. V., and Todd, C. M.
1949. Birds of the Waiau River. Notornis, **3** (6), pp. 163–164.

Dutcher, W.
1891. The Labrador duck, a revised list of extant specimens in N. America. Auk, **8**, pp. 301–216.
1894. Another specimen of Labrador duck, etc. Auk, **11**, pp. 4–12.

Du Tertre.
1667. Histoire générale des Antilles habitées par les François. Paris.

Eastman, W. H.
1949. Hunting for ivory bills in the big cypress. Florida Nat., **22**, pp. 79–80,

Eastwood, A.
1929. Studies of the flora of Lower California and adjacent islands. Proc. California Acad. Sci. (4), **18**, pp. 393–484.

Eaton, E. H.
1910, 1914. Birds of New York. New York State Mus. Mem. 12, 2 vols. Albany, N. Y.

Edwards, G.
1743–1751. A natural history of uncommon birds, pts. I–VII. London.

Elliot, D. G.
1898. The wild fowl of the United States and British Possessions, etc., 316 pp. New York.

Esten, S. R.
1933. Notes on the prairie chicken in Indiana. Auk, **50**, p. 357.

Evermann, B. W.
1889. List of birds seen in Ventura Co., Calif. Pacific Sci. Monthly, **1**, pp. 77–78. Reprinted from Auk, **3**, 1886, pp. 86–94.

Ezra, A.
1942. [Note on *Rhodonessa*.] Avicult. Mag. (5), **6**, p. 141.

Faber, F.
1822. Prodromus der isländischen Ornithologie oder Geschichte der Vögel Islands. Kopenhagen.

Fabricius, O.
1780. Fauna Groenlandica. Hafniae et Lipsiae.

Falla, R. A.
1937. Birds. B(ritish) A(ustralian) N(ew) Z(ealand) Antarct. Res. Exp. Rep., ser. B., **2**, pp. 18–24.
1949. *Notornis* rediscovered. Emu, **48**, pp. 316–322.
1951. Nesting season of *Notornis*. Notornis (continuation of New Zealand Bird Notes), **4** (5), pp. 97–100.

Falla, R. A., and Stead, E. F.
1938. Plumages of *Nesonetta aucklandica*. Trans. Proc. Soc. New Zealand, **68**, pp. 37–39.

Fanning, E.
1833. Voyages round the world, etc., 499 pp. New York. Reprinted by the Marine Research Soc., Salem, Mass., 1924, "Voyages and Discoveries in the South Seas 1792–1832."

FARLEY, F. L.
1944. [Saskatchewan records for *Grus americana*.] Can. Nat., **58**, p. 142.
FEILDEN, H. W.
1872. Birds of the Faeroe Islands. Zoologist (2), **7**, pp. 3210–3225, 3245–3257, 3277–3299. (Reference to *Alca impennis*, pp. 3280–3285.)
1878. On the reported occurrence of a gare-fowl in the Faeroes. Zoologist (3), **2**, pp. 199–201.
1889. On the birds of Barbados. Ibis, 1889, pp. 477–503. [*Numenius*, p. 498.]
FINCKH, H. E.
1915. Notes on kagus (*Rhynochetos jubatus*)—in captivity. Emu, **14**, pp. 168–170.
FINLEY, W. L.
1906. Life history of California condor. Condor, **8**, pp. 135–142.
1908. Life history of California condor. Condor, **10**, pp. 5–10, 59–65.
1910. Life history of California condor, pt. IV. Condor, **12**, pp. 5–11.
FINN, F.
1915. Indian sporting birds, 280 pp. London.
1919. Bird behaviour, psychical and physiological, 363 pp. London.
FINSCH, O.
1867–1868. Die Papagaien. 2 vols. in 3, Leiden. (Vol. 2, p. 271.)
1875. Zur Ornithologie der Sudsee-Inseln, I. Journ. Mus. Godeffroy, **5** (8), pp. 1–51 (133–183).
1876. Zur Ornithologie der Sudsee-Inseln, II. Journ. Mus. Godeffroy, **5** (12), pp. 1–42.
1881. Ornithological letters from the Pacific. [Notes on rare birds of the Caroline Islands.] Ibis, 1881, pp. 102–114.
FINSCH, O., and HARTLAUB, G.
1867. Beitrage zur Fauna Centralpolynesiens, Ornithologie der Viti-, Samoa und Tonga-Inseln, 290 pp. Halle.
FISCHER, M.
1913. A vanished race [passenger pigeon]. Bird Lore, **15**, pp. 77–84.
FISHER, A. K., and WETMORE, A.
1931. [Notes on *Nannopterum harrisi* in "Birds of Pinchot Expedition of 1929," etc.] Proc. U.S. Nat. Mus., **79**, pp. 1–66 (p. 32).
FISHER, H. I.
1947. The skeletons of recent and fossil *Gymnogyps*. Pacific Sci., **1**, pp. 227–236.
1949. Populations of birds on Midway and the man-made factors affecting them. Pacific Sci., **3**, pp. 103–110.
1951. The avifauna of Niihau. Condor, **53**, pp. 31–42.
FISHER, H. I., and BALDWIN, P. H.
1945. A recent trip to Midway Islands, Pacific Ocean. The Elepaio, Journ. Honolulu Audubon Soc., **6** (2), pp. 11–16 (mimeographed).
1946. War and the birds of Midway Atoll. Condor, **48**, pp. 3–15. [Had previously appeared (in substance) in the Elepaio, Journ. Honolulu Audubon Soc., **6** (2), pp. 11–16 (mimeographed).]
FISHER, J.
1952. The fulmar, 489 pp. London.
FISHER, W. K.
1906. Birds of Laysan, etc. Bull. U.S. Fish Comm., **23** (for 1903), pt. 3, pp. 767–807.
FLEMING, C. A.
1951. *Notornis* in February, 1950. Notornis (continuation of New Zealand Bird Notes), **4** (5), pp. 101–106.

FLEMING, J.
1824. Gleanings of natural history during a voyage along the coast of Scotland in 1821. Edinburgh Philos. Journ., **10**, art. 13, pp. 95–101.
FLEMING, J. H.
1901. [Passenger pigeon in Ontario.] Auk, **18**, p. 37.
1903. Recent records of the passenger pigeon. Auk, **20**, p. 66.
1907. The disappearance of the passenger pigeon. Ottawa Nat., **20**, pp. 236–237.
1935. A new genus and species of flightless duck. Occ. Pap. Roy. Ontario Mus., no. 1, pp. 1–3.
1938. A new genus and species of flightless duck from Campbell Island. Ibis, 1938, pp. 590, 591.
1939. Birds of the Chatham Islands [New Zealand]. Emu, **38**, pp. 381–413; **38**, pp. 492–509; **39**, pp. 1–15 [*Thinornis*].
FORBES, H. O.
1892. [Extinct avifauna of New Zealand.] Trans. and Proc. New Zealand Inst., **24**, pp. 185–189.
1893. [Birds of Chatham Islands.] Ibis, 1893, pp. 521–545.
1898. On . . . a species . . . from the Mascarene Islands, provisionally referred to *Necropsar*. Bull. Liverpool Mus., **1** (2), pp. 29–35.
1901. [An account of the type of *Notornis stanleyi*.] Bull. Liverpool Museum, **3** (2), pp. 62–68.
1923. [Note on *Notornis mantelli* and *N. hochstetteri*.] Nature, **112**, p. 762.
FORBUSH, E. H.
1912. Game birds, wild fowl and shore birds of Massachusetts and adjacent states. Pt. II—Species extinct or extirpated, pp. 397–494.
1913. The last passenger pigeon. Bird Lore, **15**, pp. 99–103.
1925–1929. Birds of Massachusetts and other New England States. 3 vols. Boston.
FORSTER, J. R.
1778. Observations made during a voyage round the world . . . , 649 pp. London.
1844. Descriptiones animalium quae in itinere ad Maris Australis Terras per annos 1772, 1773 et 1774 suscepto. (Lichtenstein ed.) Berlin.
FRENCH, J. C., ET AL.
1919. The passenger pigeon in Pennsylvania, its remarkable history, habits and extinction, 257 pp. Altoona, Pa.
FRIEDMANN, H.
1929. The cowbirds. Springfield, Ill., and Baltimore, Md.
1927. Classification of cowbirds. Auk, **44**, pp. 495–508 (506).
1941. Birds of North and Middle America, part IX, by Robert Ridgway and Herbert Friedmann. U.S. Nat. Mus. Bull. 50.
FRIEDMANN, H., GRISCOM, L. and MOORE, R. T.
1950. Distributional checklist of the birds of Mexico, pt. 1. Cooper Orn. Club, Pacific Coast Avif., no. 29, 202 pp. (134).
GALSTOFF, P. S.
1933. Birds on Pearl and Hermes Reef, etc. B. P. Bishop Mus. Bull. 107, 49 pp.
GAMBLE, W.
1847. Remarks on the birds observed in Upper California, etc. (condor). Journ. Acad. Nat. Sci. Philadelphia (2), **1**, pp. 25–57.
GANIER, A. F.
1933. A distributional list of the birds of Tennessee. Tennessee Ornithological Society, Nashville, Tenn.
GARROD, D. A. E.
1926. The Upper Paleolithic Age in Britain, 194 pp. Oxford. [Great auk, p. 183.]
GAULT, B. T.
1895. [Records of *Ectopistes* in Minnesota and Illinois.] Auk, **12**, p. 80.

GAYLORD, H. A.
   1897a. Notes from Guadalupe Island. Nidologist, **4**, pp. 41–43.
   1897b. Remarkable confidence in Guadalupe junco. Osprey, **1**, p. 98.
GIAI, A. G.
   1950. [*Mergus octocetaceus* in Misiones.] Hornero, **9** (2), p. 164.
GIBLIN, W. W., and SWINDELLS, A. W.
   1927. Field notes on the blue winged parrot [*Neophema chrysostoma*]. Emu, **27**, pp. 5–15.
GIFFORD, E. W.
   1919. Field notes on the birds of the Galapagos. Proc. California Acad. Sci., **2**, pp. 189–258.
GIRAUD, J. P.
   1844. The birds of Long Island, 397 pp., New York.
GLEIZE, D. A.
   1945. Birds of Tinian [Marianas Islands]. Bull. Massachusetts Audubon Soc., **29** (7), p. 220.
GODDARD, M. T.
   1948. Another record of the eastern bristle-bird, and other notes. Emu, **47**, pp. 311–312.
GOSS, N. S.
   1891. History of the birds of Kansas. 692 pp. Topeka.
GOSSE, P. H.
   1840. The Canadian naturalist, 372 pp. London. [*Ectopistes*, pp. 199, 293.]
GOSSE, P. H., and HILL, R.
   1847. The birds of Jamaica, 447 pp. London.
GOULD, J.
   1840–1848. Birds of Australia, 7 vols. London.
   1841. *In* Darwin, Zoology of voyage of the *Beagle*, pt. 3, Birds.
GRABA, C. J.
   1930. Tagebuch, geführt auf einer Reise nach Färö im Jahre 1928, 24 pp. Hamburg.
GRAEME, H. S.
   1914. Excavation of the Broch of Ayre, St. Mary's Holm, Orkney. Proc. Soc. Antiq. Scotland, **48**, pp. 31–52.
GRAVES, A.
   1949. Wild pigeons in Indiana. Year Book Indiana Audubon Soc., **27**, pp. 34–35.
GRAY, G. R.
   1843. *In* Dieffenbach's Travels in New Zealand, p. 197.
GRAY, R.
   1871. The birds of the west of Scotland, including the Outer Hebrides etc., 520 pp. Glasgow.
GREENE, E. B., and HARRINGTON, V. D.
   1932. American population before the Federal census of 1790, 206 pp. New York.
GREENE, E. L.
   1885. [Notes on Guadalupe Island (flora).] Bull. California Acad. Sci., **1**, pp. 214–228.
GREENWAY, J. C., JR.
   1952. *Tricholimnas conditicius* is probably a synonym of *T. sylvestris* (Aves, Rallidae). Breviora Mus. Comp. Zool., no. 5, pp. 1–4.
GREGG, W. H.
   1879. *Camptolaemus labradorius*. Amer. Nat., **13**, pp. 128–129.
GREGORY, W. K.
   1935. Remarks on the origins of the ratites and penguins. Proc. Linn. Soc. New York, **45**, **46**, pp. 11–18.

456   EXTINCT AND VANISHING BIRDS OF THE WORLD

GRIEVE, S.
1885. The great auk, or garefowl (*Alca impennis* Linn.), its history, archaeology, and remains, 540 pp. London.
1897. Supplementary note on the great auk. Trans. Edinburgh Field Nat. and Microscop. Soc., **3** (6), pp. 237–272.
1898. Additional notes on the great auk. Edinburgh Field Nat. and Microscop. Soc., **3** (7), pp. 327–340.
GRINNELL, G. B.
1888. Restocking with foreign game. Forest and Stream, **31**, no. 23, p. 453.
GRINNELL, J.
1928. A distributional summation of the ornithology of Lower California. Univ. California Publ. Zool., **32** (1), pp. 1–300.
GRINNELL, J., DIXON, J. S., and LINSDALE, J. M.
1937. Fur bearing mammals of California, 2 vols. Berkeley, Calif.
GRINNELL, J., and MILLER, A. H.
1944. The distribution of the birds of California. Pacific Coast Avif. no. 27, 576 pp.
GROSS, A. O.
1928. The heath hen. Mem. Boston Soc. Nat. Hist., **6** (4), pp. 491–588.
1930. Progress report on the Wisconsin prairie chicken investigation, 112 pp. Wisconsin Conservation Commission, Madison.
1931. The last heath hen. Bull. Massachusetts Audubon Soc., **15** (5), p. 12.
1944. The present status of the American eider on the Maine coast. Wilson Bull., **56**, no. 1, pp. 15–26.
GROTE, H.
1914. *Alca impennis* im Jahre 1848 im Norwegen erbeutet? Orn. Monatsb., **22** (1), pp. 5–6.
GUNDLACH, J.
1856. [*Campephilus principalis*, p. 102; *Ara tricolor*, p. 105; *Columba inornata*, p. 106.] Journ. für Orn., **4**, pp. 1–16, 97–112, 337–352, 417–432.
1872. Neue Beiträge zur Ornithologie Cubas. Journ. für Orn., **20**, pp. 401–432.
1874a. Neue Beiträge zur Ornithologie Cubas. Journ. für Orn., **22**, 113–166, 286–303.
1874b. Beitrage zur Ornithologie der Insel Puerto Rico. Journ. für Orn., **22**, pp. 304–315.
1876. Contribución a la ornitología Cubana, 360 pp. Havana.
GUNTHER, A., and NEWTON, E.
1879. Extinct birds of Rodriguez. Philos. Trans. Roy. Soc. London, **168** (extra), pp. 423–437.
GURNEY, J. H.
1921. Early annals of ornithology, 240 pp. London.
GUTHRIE-SMITH, H.
1925. Bird life on island and shore, 195 pp. London.
HACHISUKA, M.
1953. The dodo and kindred birds, 250 pp. London.
HADLEY, P.
1930. The passenger pigeon. Science, 71, p. 187.
HADLOCK, W. S.
1939. The Taft's Point shell mound at West Gouldsboro, Maine. (Great auk, p. 17.) Robert Abbe Mus. (Bar Harbor, Maine) Bull., **5**, pp. 1–29.
HAGEN, YNGVAR.
1952. Birds of Tristan da Cunha. Norwegian Scientific Expedition Tristan da Cunha 1937–38, no. 20, 248 pp. Oslo.

HAKLUYT, R.
1552–1616. The principal navigations, voyages, traffiques and discoveries of the English nation, etc., George Bishop, Ralph Newberrie and Robert Barker, 3 vols. London.

HALL, H. M., and GRINNELL, J.
1919. [Life zones in California.] Proc. California Acad. Sci. (4), 9, pp. 37–67.

HALL, W. L.
1904. The forests of the Hawaiian Islands. U.S. Dept. Agr. Bur. Forestry Bull. 48.

HAMERSTROM, F. N., and HAMERSTROM, F.
1949. Daily and seasonal movements of Wisconsin prairie chickens. Auk, 66, pp. 313–337.

HAMILTON, A.
1893. On the fissures and caves at the Castle Rocks, Southland: etc. Trans. Proc. Roy. Soc. New Zealand, 25, pp. 88–106.
1895. Notes on a visit to Macquarie Island. Trans. New Zealand Inst., 27, p. 559.
1904. Note on the remains of some extinct birds of New Zealand. Trans. Proc. Roy. Soc. New Zealand, 36, 1903, pp. 474–477.

HAMLIN, H.
1931. In Mayr and Hamlin. Birds collected during the Whitney South Sea Expedition, xv. Amer. Mus. Novit., no. 488, pp. 1–11.

HANCOCK, J.
1874. Catalogue of the birds of Northumberland and Durham. (Great auk, p. 165.) Nat. Hist. Trans. Northumberland and Durham, 6, pp. 1–174.

HANNA, G. D.
1925. Expedition to Guadalupe Island. Proc. California Acad. Sci. (4), 14, 217–275.

HARDY, F. P.
1888. Testimony of some early voyagers on the great auk. Auk, 5, pp. 380–384.

HARRIS, H.
1919. Birds of the Kansas City region. Trans. Acad. Sci. St. Louis, 23 (8), pp. 213–371.
1941. The annals of Gymnogyps to 1900. Condor, 43 (1), pp. 3–55.

HARTERT, E.
1898. [Notes on Megapodius laperouse.] Novit. Zool., 5, pp. 61, 69.
1924. [No title.] Bull. Brit. Orn. Club, 45, p. 48.
1927. Types of birds in the Tring Museum. (Hypotaenidia wakensis and H. owstoni, p. 22.) Novit. Zool., 34, pp. 1–38.

HARTING, J. E.
1883. The last great auk. Zoologist, 7 (83), p. 470.
1884. The last great auk. Zoologist, 8 (88), pp. 141–142. [Record found in Labrador by Indians in 1870 and sold in France probably rests on error.]

HARTLAUB, G.
1893. Vier seltene Rallen [notes on rare insular rails]. Abh. naturw. Ver. Bremen, 12 (3), pp. 389–402.
1896. Ein Beitrag zur Geschichte der ausgestorbenen Vögel der Neuzeit, etc. Abh. naturw. Ver. Bremen, 14 (1), pp. 1–43 (in second edition only).

HARTT, C. E., and NEAL, M. C.
1940. The plant ecology of Mauna Kea, Hawaii. Ecology, 21 (2), pp. 237–266.

HARVIE-BROWN, J. A., and BUCKLEY, T. E.
1888. A vertebrate fauna of the Outer Hebrides. 279 pp. Edinburgh. (Reference to Alca impennis, p. 158.)

HAWKES, C. F. C.
1940. The prehistoric foundations of Europe to the Mycenean Age, 384 pp. London.
HAY, O. P.
1902. On the finding of the great auk in Florida. Auk, 19, pp. 255–258.
HAYDEN, F. V.
1863. On geology and natural history of the Upper Missouri [River]. Trans. Amer. Philos. Soc., 12, art. 1, pt. 3, pp. 151–176.
HELLIESEN, T.
1901. [Alca impennis remains at Kvernevig near Stavanger.] Stavanger Mus. Aarsheft, 11, p. 59.
HELLMAYR, C. E.
1932. Birds of Chile. Publ. Field Mus. Nat. Hist., zool. ser., 19 (308), 472 pp.
HELMS, O.
1926. Birds of Angmagsalik [east coast of Greenland]. Medd. Grønland, 58, pp. 207–274.
HENRICI, P.
1935. En Bohuslänsk Kokkenmödding på Rotekärrslid, Dragsmark. Göteborg och Bohusläno Fornmim. Tidskr., 4 (3), pp. 38–42.
HENRY, R.
1899. [Record of Porphyrio melanotus at sea.] Trans. and Proc. New Zealand Inst., 32, pp. 53–54.
HENSHAW, H. W.
1874. An annotated list of the birds of Utah. Ann. New York Lyc. Nat. Hist., 11, 1876, pp. 1–14.
1902. Birds of the Hawaiian Islands, being a complete list of the birds of the Hawaiian possessions, with notes on their habits. 135 pp. Honolulu.
HERBERT, T.
1638. Some yeares travels into Africa and Asia the Great, etc., 364 pp. London. [Picture of "Dodo," "a Hen," "a Cocato" of Mauritius, p. 212.]
HERRICK, H.
1873. Partial catalogue of birds of Grand Manan. [Bones of Alca impennis found in kitchen midden at Nantucket Island, Grand Manan Group, New Brunswick, p. 41.] Bull. Essex Inst., 5, pp. 28–41.
HINDWOOD, K. A.
1932. An historic diary. Voyage of the Lady Penrhyn, 1787–1789, etc. Emu, 32, pp. 17–29.
1938. The extinct birds of Lord Howe Island. Austr. Mag., 6, pp. 319–324.
1940. Birds of Lord Howe Island. Emu, 40, pp. 1–86.
HINDWOOD, K. A., and CUNNINGHAM, J. M.
1950. Notes on the birds of Lord Howe Island. Emu, 50, pp. 23–35.
HINSBY, K. B.
1948. [Pezopoporus in Tasmania.] Emu, 47, p. 313.
HINTON, M. A. C.
1926. The Pleistocene Mammalia of the British Isles and their bearing upon the date of the Glacial Period. Proc. Yorkshire Geol. Soc., 20, pp. 325–483.
[HOBLEY], C. W.
1932. Dominica diablotin (Pterodroma hasitata, reported rediscovery of). Journ. Soc. Pres. Faun. Emp. (N.S.), pt. 17, pp. 17–20.
HODGE, C. F.
1912. A last word on the passenger pigeon. Auk, 29, pp. 169–175.
HODGES, J.
1950a. Record of the passenger pigeon. Auk, 29, pp. 169–175.
1950b. Record of the passenger pigeon for Scott County (Iowa). Iowa Bird Life, 20 (1), p. 26.

HODGKINS, M.
  1949. Bird notes in Tauranga District. New Zealand Bird Notes, **3** (5), pp. 116–125.
HOLM, G.
  1918. Gunbjørns-Skaer og Korsøer. Medd. Grønland, **56** (8), pp. 289–308.
HORREBOW, N.
  1752. Tilforladige Efterretninger om Island, etc. Kjøbenhavn. (Ref. great auks, pp. 175–176.)
HØRRING, R., in Mathiassen.
  1931. [Bird bones in Eskimo archeological site.] Medd. Grønland, **91** (1), Zool. App., pp. 137–138.
HØRRING, R.
  1934. Danmarks Fauna, etc., Fugle III, 309 pp. Copenhagen.
HOWARD, H.
  1930. A census of Pleistocene birds of Rancho La Brea, California. Condor, **32**, pp. 81–88.
  1937. A Pleistocene record of the passenger pigeon in California. Condor, **39**, pp. 12–14.
  1938. [*Polyborus prelutosus,* sp. nov.] Publ. Carnegie Inst., no. 487, p. 226.
  1940. [*Polyborus pulutotus grinnelli,* fossil described from Nuevo León, Mexico.] Condor, **42**, p. 41.
HOWE, F. E., and ROSS, J. A.
  1933. On the occurrence of *Psophodes nigrogularis* in Victoria. Emu, **32**, pp. 133–148.
HOWE, R. H., and ALLEN, G. M.
  1901. The birds of Massachusetts, 154 pp. Cambridge, Mass.
HOWELL, A. H.
  1932. Florida bird life, 479 pp. Florida Department of Game and Fresh Water Fish, Tallahassee, Fla.
HOWELL, T. R., and CADE, T. J.
  1954. The birds of Guadalupe Island. Condor, **56**, pp. 283–294 (289).
HOWLAND, L.,
  1955. Howland Island, its birds and rats, as observed by a certain Mr. Stetson in 1854. Pacific Sci., **9** (2), pp. 96–106.
HOWSE, R.
  1880. Preliminary note on the discovery of old sea caves . . . at Whitburn Lizards. Nat. Hist. Trans. Northumberland, Durham, and Newcastle-on-Tyne, **7**, pt. 2, pp. 361–365.
HUBER, W.
  1938. Two specimens of the heath hen from New Jersey, Auk, **55**, pp. 527–528.
HUEY, L. M.
  1924. [A trip to Guadalupe Island (Pacific).] Nat. Hist., **24**, pp. 578–588.
HUGHES, G.
  1750. The natural history of Barbados, 314 pp. London.
HULL, A. F. B.
  1909. [Notes on the birds of Norfolk and Lord Howe Islands.] Proc. Linn. Soc. New South Wales, **34**, pp. 636–693.
  1910. Further notes on the birds of Norfolk and Lord Howe Islands. Proc. Linn. Soc. New South Wales, **35**, pp. 783–787.
HUME, A. O., and MARSHALL, C. H. T.
  1879–1881. The game birds of India, Burmah and Ceylon, 3 vols. Calcutta.
HUNT, H.
  1948. Various bird notes [*Dromiceius*]. Victorian Nat., **64** (11), pp. 211–212.

HUTCHINSON, G. E.
1950. The biogeochemistry of vertebrate excretion. Bull. Amer. Mus. Nat. Hist., 96, xviii + 554 pp.
HUTTON, F. W.
1871. Catalogue of the birds of New Zealand, with diagnoses of the species, 85 pp. Wellington.
1874. On a new genus of Rallidae [*Cabalus*]. Trans. Proc. New Zealand Inst., 6, p. 108, pl. 20.
1894. On *Dinornis* (?) *queenslandiae*. Proc. Linn. Soc. New South Wales (1), 8, pp. 7–10.
HUXLEY, J. S.
1925. Studies on the courtship and sexual life of birds—V. The oystercatcher. Ibis, 1925, p. 868.
IJIMA, I.
1892. Notes on a collection of birds of Tsushima. Journ. Coll. Soc. Japan, 5, pp. 105–128 (not seen. See Austin, 1953).
IRBY, F. M.
1927. Further notes on rare parrots. Emu, 27, pp. 13–16.
IREDALE, T.
1910. [Note on the birds of Norfolk and Lord Howe Islands.] Proc. Linn. Soc. New South Wales, 35, pp. 773–782.
1913. On some interesting birds in the Vienna Museum. Austral Avian Rec., 2 (11), pp. 14–32.
1930. Avian sea-toll. Australian Zool., 6 (2), pp. 112–116.
JACKSON, S. W.
1911. Haunts of the rufous scrub bird (*Atrichornis rufescens* Ramsay). Emu, 10, pp. 327–336.
1920. [Haunts of the rufous scrub bird (*Atrichornis rufescens*).] Emu, 19, pp. 258–272.
1921. Second trip to MacPherson Range, S.W. Queensland (*Atrichornis*). Emu, 20, pp. 195–209.
JACOBSEN, W. C.
1918. History of ground squirrel control in California. Monthly Bull. California State Comm. Horticulture, 7, p. 725.
JÄGERSKIOLD, L. A.
1933. En utdod svensk fågel, Garfageln, *Alca impennis*. Göteborgs Mus., arstryck 1933, pp. 13–20.
1934. [No title.] Göteborgs Mus., arstryck 1935, arsredogorelse, 1934, zoologiska avelningen pp. 6–21. [*Alca impennis* pp. 8–9.]
1938. [No title.] Göteborgs Mus., arstryck 1938, Naturhistoriska Mus., berattelse för år 1937, pp. 8–33. [*Alca impennis* p. 10.]
JAMES, H. E. M.
1887. The long white mountain. London.
JARDINE, W.
1849, 1850. Contributions to ornithology for 1848–52, 5 vols. Birds of Bermuda 1849, pp. 76–87; 1850, pp. 5–14, 35–38, 67.
JERDON, T. C.
1864. Birds of India, 3 vols. Calcutta.
JOHANSSON, I.
1927. Studies on inheritance in pigeons, VI. Number of tail feathers and uropygial gland. Genetics, 12, pp. 93–107.
JOHNSON, H. T.
1945. [Sight record of *Gymnogyps*.] Condor, 47, p. 38.

JONES, J. M.
    1859. The naturalist in Bermuda, etc. xii + 200 pp. London. [Migrations of
        plover and curlew, pp. 36–37, 71–77.]
JORDAN, D. S., ET AL.
    1898–99. The fur seal and fur-seal islands of the North Pacific, 4 vols. U.S.
        Treasury Department, Commission of Fur Seal Investigation, Washington.
JOSSELYN, J.
    1672. New England's rarities discovered; in birds, beasts, fishes, serpents and
        plants of that country, 4 + 114 pp. London. (Reprint 1678, 1865.)
    1675. Account of two voyages to New England, 2d ed. London. (Reprint 1865,
        279 pp., Boston, Mass.)
KAEDING, H. B.
    1905. Birds from the west coast of Lower California and adjacent islands. Con-
        dor, 7, pp. 105–111, 134–138.
KALM, P.
    1759. A description of wild pigeons which visit the southern English colonies in
        North America during certain years in incredible multitudes. Kongl. Vet.-
        Akad. Handl., 20. (Translation by Gronberger, Auk, 28, 1911, pp. 53–66.)
KERR, R.
    1824. A general history . . . collection of voyages and travels, vol. 6.
KERSHAW, J. A.
    1943. Concerning a rare parrot [Geopsittacus]. Victorian Nat., 59, p. 196.
KILLERMAN, S.
    1915. Die ausgestorbenen Maskarenenvögel. Naturw. Wochenschr., 30, pp. 353–
        378.
KINNEAR, N.
    1935. Zoological notes from the voyage of Peter Mundy 1655–56, (a) Birds.
        Proc. Linn. Soc. London, 1935, pp. 32–37.
KITTLITZ, F. H. VON.
    1832–1833. Kupfertafeln zur Naturgeschichte der Vögel, Heft 3. Frankfurt am
        Main.
    1834. Nachricht von der Brüteplätzen einiger tropischen Seevogel etc. Mus.
        Senckenberg. Abh. Gebiete beschreib. Naturg., 1, pp. 115–126. (Sometimes
        cited under Senckenbergische Naturforschende Gesellschaft, Frankfurt am
        Main.)
    1858. Denkwürdigkeiten einer Reise nach dem russischen Amerika, nach Mik-
        ronesien und durch Kamtschatka, xiii + 459 pp. Gotha.
KJAERBOLLING.
    1851–1856. Danmarks Fugle, 422 pp. Copenhagen.
KNOWLES, W. J.
    1895. Remains of great auk in County Antrim, Ireland. Proc. Roy. Irish. Acad.
        (3), 3, pp. 650–663.
KOBAYASHI, K.
    1930. Collection of bird skins and eggs from the Riu Kiu Islands. Tori, 6, pp.
        341–387.
KOFORD, C. B.
    1953. The California condor. Nat. Audubon Soc. Res. Rep. no. 4, 154 pp. Re-
        printed by Dover Publications in 1966.
KUBARY (see Finsch, 1875).
KUHL, H.
    1820. Conspectus psittacorum, 104 pp. Bonn.
KUMLIEN, L.
    1891. A list of the birds known to nest within the boundaries of Wisconsin, etc.
        [Ectopistes, p. 148]. Wisconsin Nat., 1, pp. 103–105; 125–127; 146–148.

KURODA, N.
1925. A contribution to the knowledge of the avifauna of the Riu Kiu Islands, 234 pp. Tokyo.
1940. An old record for a pair of *Pseudotadorna cristata* obtained near Hakodate. Tori, **10**, 739–741.
1941–42. A study of *Anas oustaleti* [in Japanese]. Tori, **11**, pp. 99–119, 443–448.
1953. On the Japanese taken *Branta canadensis*. Tori, **13**, 1952, pp. 4–9.
LABAT, J. B.
1722. Nouveau voyage aux isles de l'Amérique, etc., 6 vols. Paris. (Also 1742, 8 vols.)
LACK, D.
1947. Darwin's finches, 208 pp. Cambridge.
LAFRESNAYE, M. F. DE.
1844. Description de quelques oiseaux de la Guadaloupe. Rev. Zool., 1844, pp. 167–168.
LAING, H. M.
1925. Birds collected and observed during the cruise of the *Thiepval* in the North Pacific, 1924. Canada Dept. Mines, Victoria Mem. Mus., Biol. Ser., **9**, pp. 1–43.
LAMBEYE, J.
1850. Aves de la isla de Cuba, 136 pp. Havana.
LAMBRECHT, K.
1921. Fossilium catalogus. (Ed. C. Diener, part 12, Aves.) 104 pp., Berlin.
1929. *Stromeria fajumensis* n. g., n. sp., die kontinentale Stammform de Aepyornithidae, mit einer Übersicht über die fossilen Vögel Madagaskars und Afrikas. Abh. Bayer. Akad. Wiss., math. natur. Abt., N. F., **4**, 18 pp.
1933. Handbuch der Palaeornithologie, xix + 1024 pp. Berlin.
LANG, C. L.
1946. Notes on the rufous bristle-bird. Emu, **45**, pp. 257-259.
LATHAM, J.
1781–1787. A general synopsis of birds, 3 vols. London.
LAROCQUE, A.
1930. The passenger pigeon in folklore. Can. Field Nat., **44**, pp. 49–50.
LA TOUCHE, J. D.
1925–1934. A handbook of the birds of eastern China, 2 vols. London.
LAVAUDEN, L.
1932. Étude d'une petite collection d'oiseaux de Madagascar. Bull. Mus. Nat. Hist. (Paris) (2), **4**, pp. 629–640.
LAWRENCE, G. N.
1879. Catalogue of birds collected in Martinique and Guadaloupe by Mr. Fred A. Ober. Proc. U.S. Nat. Mus., **1**, pp. 349–360, 449–462.
LAYARD, E. L.
1875. Notes on Fijian birds. Proc. Zool. Soc. London, 1875, pp. 423–442.
1876. Notes on some little known birds of the Fiji Islands. Ibis, 1876, pp. 137–156.
LAYARD, E. L., and LAYARD, E. L. C.
1882. Avifauna of New Caledonia. Ibis, 1882, pp. 493–546.
LEACH, J. A.
1928. Notes made on a holiday trip to New Caledonia. Emu, **28**, pp. 20–42.
1929. [Australian birds etc.] Emu, **29**, pp. 16–38.
LEGUAT, F.
1708. Voyage et avantures en deux isles desertes des Indes Orientales. London. (Also 1891, Voyage to Rodriguez, etc., Hakluyt Society, London, translation.)

LEY, W.
1935. The great auk [general account]. Nat. Hist., **36**, pp. 351–357.

LIGON, J. S.
1952. The vanishing masked bob white. Condor, **54**, pp. 48–51.

LIGON, R.
1673. A true and exact history of the island of Barbados, 144 pp. London [1st ed. 1657].

LINSLEY, J. H.
1843. A catalogue of the birds of Connecticut, etc. Amer. Journ. Sci. and Arts, **44** (2), art. 3, pp. 249–274.

LISTER, J. J.
1911. *Megapodius* in the Pacific. Proc. Zool. Soc. London, 1911, p. 749.

LITTLER, F. M.
1910. A handbook of the birds of Tasmania, 228 pp. Launceston, Tasmania.

LOFBERG, L. M.
1936. [*Gymnogyps* seen.] Condor, **38**, p. 177.

LOOMIS, F. B., and YOUNG, D. B.
1912. Shell heaps of Maine [great auk, p. 24]. Amer. Journ. Sci. (4), **34**, pp. 17–42.

LOOMIS, L. M.
1918. A review of the albatrosses, petrels and diving petrels. Proc. California Acad. Sci., **2**, pt. 2, no. 12, pp. 1–187.

LØPPENTHIN, B.
1952. The egg from Hermanstorp. A late-glácial find from Scania. Dansk Orn. foren. Tidssk., **46** (2), pp. 12–30.

LORD, C. E., and SCOTT, H. H.
1924. A synopsis of the vertebrate animals of Tasmania, 340 pp. Hobart. (Birds, pp. 118–231.)

LOW, G.
1813. Fauna Orcadensis; or the natural history of . . . Orkney and Shetland. 230 pp. Edinburgh. [Great auk, p. 170.]

LOWE, P. R.
1922. Reminiscence of the last great flight of the passenger pigeon in Canada (1870). Ibis, 1922, pp. 137–141.

1923. [Notes on birds of Tristan da Cunha Group, S. Atlantic.] Ibis, 1923, pp. 514–529.

1927. Anatomy and systematic position of *Aechmorhynchus cancellatus* (Gm.), etc. Ibis, 1927, pp. 114–132.

1928. A description of *Atlantisea rogersi* [many notes on birds of Tristan and Gough Islands]. Ibis, 1928, pp. 99–131.

1929. [*Hypselornis sivalensis* probably not a struthionid]. Ibis, 1929, pp. 571–576.

1931a. Some further notes on *Aechmorhynchus cancellatus.* Ibis, 1931, pp. 241–243.

1931b. Struthious remains from northern China and Mongolia. Palaeont. Sinica, ser. C., **6**, fasc. 4, pp. 5–47.

1933. [Egg-shell fragments of *Struthio* found in southern Arabia.] Ibis, 1933, pp. 652–658.

LUCAS, F. A.
1888. On the bird rocks of the Gulf of St. Lawrence in 1887. Auk, **5**, pp. 129–135.

1890. The expedition to Funk Island, with observations upon the history and anatomy of the great auk. Rep. U.S. Nat. Mus. for 1888, pp. 493–529.

1891. Some skeletons from Guadalupe Island. Auk, **8**, pp. 218–222.

1901. A flightless auk, *Mancalla californiensis,* from the Miocene of California. Proc. U.S. Nat. Mus., **24**, pp. 133–134.

MACFARLANE, R.
1891. Notes on and list of birds collected in Arctic America, 1861–1866 [*Numenius,* p. 429]. Proc. U.S. Nat. Mus., **14**, pp. 413–446.

MACGILLIVRAY, W.
1837–1852. A history of British birds, 5 vols. London.

MACGILLIVRAY, W.
1927. The charming Bourke parrot. Emu, **27**, pp. 65–67.

MACKAY, G. H.
1893. [*Numenius borealis.*] Auk, **10**, p. 79.
1894. [*Numenius borealis.*] Auk, **11**, pp. 75–76.
1897. [*Numenius borealis.*] Auk, **14**, pp. 212–214.
1911. [Records for Nantucket Island, Mass.] Auk, **28**, pp. 119–120, 261–262.

MACMILLAN, L.
1939. Études Melanésiens, **1** (2), p. 193.

MARCH, W. T.
1864. Notes on the birds of Jamaica [*Columba inornata,* p. 301]. Proc. Acad. Nat. Sci. Philadelphia, 1863, pp. 150–154, 283–305.

MARIÉ, E.
1870. Mélanges ornithologiques sur la faune de la Nouvelle Caledonie [notes on the kagu]. Actes Soc. Linné Bordeaux (3), **27**, 7, pt. 1, pp. 323–328.

MARPLES, B. J.
1946. List of the birds of New Zealand. Bird Notes, **1** (Suppl.), pp. i–vii.

MARSHALL, J. T., JR.
1949. The endemic avifauna of Saipan, Tinian, Guam and Palau. Condor, **51**, pp. 200–222.
1951. Vertebrate ecology of Arno Atoll, Marshall Islands. Atoll Res. Bull., nos. 3, 4, pp. 1–38.

MARTIN, M.
1698. A late voyage to St. Kilda, 158 pp. London. [Great auk p. 48.]
1716. A description of the western islands of Scotland (ed. 2, very much corrected), 392 pp. London.

MASON, G. E.
1936. On a hitherto unrecorded egg of the lesser emu of Kangaroo Island. Emu, **35**, pp. 331–332.

MATHEW, M. A.
1866. [Great auks on Lundy Island.] Zoologist (2), **1**, pp. 100–101.

MATHEWS, G. M.
1910–1928. Birds of Australia, 12 vols.
1928. The birds of Norfolk and Lord Howe Islands, 139 pp. London.
1932. Birds of Tristan da Cunha. Novit. Zool., **38**, pp. 13–48.
1934. Check list of the Procellariiformes. Novit. Zool., **39**, pp. 151–206.
1936. A supplement to the birds of Norfolk and Lord Lowe Islands, 177 pp. London.
1948. Eggs of the *Spinifex* or night parrot [*Geopsittacus occidentalis*]. Emu, **48**, p. 84.

MATHEWS, G. M., and IREDALE, T.
1913. A reference list of the birds of New Zealand. Ibis, 1913, pp. 201–263 (209).
1926. [New generic names.] Bull. Brit. Orn. Club, **46**, p. 76.

MAUCAULAY, K.
1764. The history of St. Kilda, 278 pp. London.

MAYFIELD, H.
1953. A census of Kirtland's warbler. Auk, **70**, pp. 17–20.

MAYNARD, C. J.
  1870. The naturalist's guide in collecting and preserving objects of natural his-
        tory, etc., 170 pp. Boston. [Reference to great-auk bones found with In-
        dian relics at Ipswich, Mass., p. 159.]
MAYR, E.
  1931. Birds collected during the Whitney South Sea Expedition, XIII. Amer.
        Mus. Novit., no. 486.
  1933. Die Vogelwelt Polynesians. Mitt. Zool. Mus. Berlin, **19**, pp. 306–323.
  1938a. [Taxonomy of *Megapodius*.] Amer. Mus. Novit., no. 1006.
  1938b. Birds collected during the Whitney South Sea Expedition, XL. Amer.
        Mus. Novit., no. 1007.
  1941a. Taxonomic notes on the birds of Lord Howe Island. Emu, **40**, pp. 321–
        322.
  1941b. The origin and history of the bird fauna of Polynesia. Proc. 6th Pacific
        Sci. Congr., **4**, pp. 197–216.
  1941c. Borders and subdivision of the Polynesian Region as based on our knowl-
        edge of the distribution of birds. Proc. 6th Pacific Sci. Congr., **4**, pp. 191–
        195.
  1942. Systematics and the origin of species, 334 pp. New York.
  1945a. Birds of the Southwest Pacific, xx + 316 pp. New York.
  1945b. Bird conservation problems in the Southwest Pacific. Audubon Mag., **47**
        (5), pp. 279–282.
  1949. Notes on the birds of Northern Melanesia, 2. Amer. Mus. Novit., no. 1417.
MAYR, E., and AMADON, D.
  1951. A classification of recent birds. Amer. Mus. Novit., no. 1496.
McCORMICK, W. F. J.
  1915. Sighting of *Conurus carolinensis* in southern Florida in 1915. Bird Lore,
        17, p. 453.
McCULLOCH, A. R.
  1921. Lord Howe Island: A naturalist's paradise. Austral. Mus. Mag., **1** (2),
        pp. 30–47.
McGEE, W J
  1910. Notes on the passenger pigeon. Science, **32**, pp. 958–964.
McGILP, J. N.
  1931. [Habits of *Geopsittacus*.] South Australian Orn., **9**, p. 68.
McGILP, J. N., and PARSONS, F. E.
  1937. Mallee whipbird, *Psophodes nigrogularis* (? *leucogaster*). South Australian
        Orn., **14**, pt. 1, pp. 3–5.
McILHENNY, E. A.
  1943. Major changes in the bird life of southern Louisiana during sixty years.
        Auk, **60**, pp. 541–549.
M'LEAN, J. C.
  1911. Bush-birds of New Zealand. Emu, **11**, pp. 1–17, 65–79, 171–187, 223–236.
McNAMARA, E.
  1946. Field notes on the eastern bristle-bird. Emu, **45**, pp. 260–265.
MEAD, J. C.
  1910. The last of the passenger pigeons breeding in North Bridgeton, Maine.
        Journ. Maine Orn. Soc., **12**, pp. 1–3.
MEADOWS, D.
  1933. California condor in San Diego County. Condor, **35**, p. 234.
MEARNS, E. A.
  1879. Notes on some of the less hardy winter residents in the Hudson River
        Valley. Bull. Nuttall Orn. Club, **4**, pp. 33–37.

MEINERTZHAGEN, R.
1912. On the birds of Mauritius. Ibis, 1912, pp. 82–108.
MERRIAM, C. H.
1882. [Birds of Point de Monts, Quebec.] Bull. Nuttall Orn. Club, **7**, pp. 233–242.
MERRIAM, J. C.
1911–1912. Fauna of Rancho La Brea. Mem. Univ. California, **1** (1–2), pp. 199–262.
MERSHON, W. B.
1947. The passenger pigeon, 217 pp. New York.
MEYER, A. B.
1879–1897. Abbildungen von Vogel-Skeletten, 2 vols.: Vol. 1, 71 pp; vol. 2, 120 pp. Berlin. [*Notornis hochstetteri,* vol. 1, pp. 28–31, figs. 34–39.]
1885. [Fig. *Notornis hochstetteri.*] Zeitschr. Ges. Orn., **2**, p. 45.
MILLER, A. H.
1937. Structural modifications in the Hawaiian goose (*Nesochen sandvicensis*). Univ. California Publ. Zool., **42**, 80 pp.
MILLER, A. H., and FISHER, H. I.
1938. Pterylosis of the California condor. Condor, **40**, pp. 248–256.
MILLER, L.
1925. Birds of Rancho La Brea. Carnegie Inst. Washington Publ. no. 349, p. 103.
1933. The Lucas auk of California. Condor, **35**, p. 34.
1942. Succession in the cathartine dynasty. Condor, **44**, p. 212.
1943. The Pleistocene birds of San Josecito Cavern, Mexico. Univ. California Publ. Zool., **47**, pp. 143–168.
MILNE-EDWARDS, A.
1874. Recherches sur la faune ancienne des isles Mascareignes. Bibliotèque École Hautes Études, Sect. Sci. Nat., **9**, art. 3, pp. 1–31. [Apparently printed also in Ann. Sci. Nat. Zool. (Paris) (5), **19**, 1874 (Dec. 1873), art. 3, pp. 1–31.]
1875. Observations on the period of extinction of the fauna of Rodriguez. Ann. Mag. Nat. Hist. (4), **15**, pp. 436–439.
MILNE-EDWARDS, A., and OUSTALET, E.
1893. Notice sur quelques espèces d'oiseaux actuellement éteintes, etc. Pp. 190–252 *in* Centenaire de la fondation du Muséum d'Histoire Naturelle (Paris).
MITCHELL, M. H.
1929. A brief outline of the history of the passenger pigeon. Bull. Roy. Ontario Mus. Zool. 3, pp. 1–12.
MOMIYAMA, T. T.
1930. On the birds of Bonin and Iwo Island. Reprint of Bull. Bio-Geogr. Soc. Jap., **1**, 1930, no. 3. Annotationes Ornithological Orientalis, **2** (Supplementi numerus), pp. 89–186.
MONTAGU, G.
1931. Ornithological dictionary, 2d ed., pp. 40–592. London.
MONTBEILLARD, P. G.
1778. *In* Buffon's "Histoire Naturelle des Oiseaux," 9 vols. (16–24, 1770–1783) [vol. 19 of series, vol. 4 of oiseaux, p. 366].
MOORE, E.
1837. [Body of great auk on Lundy Island.] Mag. Nat. Hist., **1**, p. 361.
MORAN, J. V.
1946. Birds [sighted] on Saipan and Tinian. Bull. Massachusetts Audubon Soc., **33** (8), pp. 261–262.
MORENO, A.
1953. Considerations about the systematic value of *Laterallus j. jamaicensis* and *L. j. pygmaeus.* Torreia, Publ. Occ. Mus. Poey, Havana, no. 20, pp. 1–8.

MORGAN, A. M., and SUTTON, J.
  1928. Critical description of some recently discovered bones of the extinct Kangaroo Island emu [D. diemenianus]. Emu, **28**, p. 1–19.
MORRELL, B.
  1832. A narrative of four voyages etc., pp. 29–492. New York.
MORRISON, C. P. (and others).
  1942. The "majestic" emu (a symposium). Victorian Nat., **59**, pp. 57–79.
MORTENSEN, TH.
  1932. Om "Solitairen" paa Rodriguez. Naturens Verden, 1932, pp. 446–448.
  1934. On François Leguat and his voyage . . . and on Leguatia gigantea Schl. Ardea, **23**, pp. 67–77.
MOSLEY, H. N.
  1892. Notes by a naturalist, an account of observations made during the voyage of H.M.S. Challenger round the world during the years 1872–1876. London. [Notes on Tristan da Cunha Island, pp. 93–118.]
MOUSLEY, H.
  1919. Further notes on the birds of Hatley, Stanstead Co., Quebec. Auk, **36**, pp. 472–487. [Ectopistes, p. 486—Gosse records.]
MUGGERIDGE, J., and COTTIER, W.
  1931. [Food habits of Porphyrio melanotus.] New Zealand Journ. Sci. Tech., **13**, pp. 36–38.
MULLENS, W. H.
  1921. Notes on the great auk. British Birds, **15** (5), pp. 98–108.
MUNDY, P.
  1914. Travels of Peter Mundy in Europe and Asia, 1608–1667. Hakluyt Soc., Works, no. 46, vol. 3, pt. 2, pp. 317–577 (352).
MUNNS, E. N.
  1938. The distribution of important forest trees in the United States. U.S. Dept. Agr. Misc. Publ. no. 287.
MUNRO, G. C.
  1944. Birds of Hawaii, 189 pp. Honolulu.
  1945. Endangered bird species of Hawaii. Elepaio, Journ. Honolulu Audubon Soc., **5** (8), pp. 50–51; (12), pp. 76–79; **6** (1), pp. 1–6.
  1946a. The war and Pacific birds. Nature Mag., **39** (3), pp. 125–127.
  1946b. Laysan Island in 1891 [notes on birds]. Elepaio, **6** (10), pp. 51, 60, 66.
  1947. Notes on the Laysan rail. Elepaio, **8** (5), pp. 24–25.
MUNTHE, H.
  1914. Finds of late-glacial bird remains hitherto made in Sweden. Sv. Geol. Unders. Arsbök, **8** (4) (Afhandlingar o. uppsatser, Ser. C, no. 263) (English summary, pp. 24–26).
MURDOCH, J.
  1885. Bird migration at Point Barrow, Arctic Alaska. Auk, **2**, p. 63.
MURPHY, R. C.
  1933. Probable records of the Eskimo curlew (Numenius borealis) at Montauk Point, New York. Auk, **50**, pp. 101–102.
  1936. Oceanic birds of South America, 2 vols. American Museum of Natural History, New York.
  1952. Larger petrels of the genus Pterodroma. Amer. Mus. Novit., no. 1580.
MURPHY, R. C., and MOWBRAY, L. S.
  1951. New light on the cahow, Pterodroma cahow. Auk, **68** (3), pp. 266–280.
MURPHY, R. C., and PENNOYER, J. M.
  1952. Larger petrels of the genus Pterodroma. Amer. Mus. Novit., no. 1580.
MURPHY, R. C., and VOGT, W.
  1933. The dovekie influx of 1932. Auk, **50**, pp. 325–349.

Musgrave, T.
  1865. Castaway on the Auckland Isles: A narrative of the *Grafton* and of the escape of the crew after twenty months suffering. Edited by John Shillinglaw, Melbourne. (London edition, 1866.)
Myers, J. G.
  1923. The present position of the endemic birds of New Zealand. New Zealand Journ. Sci. Tech., **6** (2), pp. 65–99.
Nauman, E. D.
  1924. Birds of early Iowa. Palimpsest, **5**, pp. 133–138.
Nelson, E. W.
  1887. Report upon natural history collections made in Alaska between the years 1877 and 1881. Arctic series no. 3, Signal Service, U.S. Army, Part I, Birds. [Note on *Numenius borealis* p. 121.]
  1921. [Notes on Guadalupe Island pp. 93–95]. Mem. Nat. Acad. Sci., **16** (1st mem.).
Newberry, J. S.
  1855–1860. Reports of explorations and surveys to ascertain the most practicable and economical route for a railroad from Mississippi River to the Pacific Ocean made under the direction of the Secretary of War, in 1854–5, 12 vols. "U.S. Pacific Railroad Survey" (variously cited), **6** (2) (Report upon the Zoology), p. 73, 1857.
Newton, A.
  1861. Abstract of J. Wolley's researches in Iceland respecting the gare fowl or great auk. Ibis, 1861, pp. 374–399.
  1865. The gare-fowl and its historians. Nat. Hist. Rev., **5**, pp. 467–488.
  1867. On a picture supposed to represent the didine bird of the Island of Bourbon (Réunion). Trans. Zool. Soc. London, **6**, pp. 373–376.
  1875a. Additional evidence as to the original fauna of Rodriguez. Proc. Zool. Soc. London, 1875, pp. 39–40.
  1875b. Note on *Palaeornis exsul*. Ibis, 1875, pp. 342–343.
  1897. A specimen of *Alectroenas nitidissima* in Edinburgh. Proc. Zool. Soc. London, 1897, pp. 2–4.
  1898. On the Orcadian home of the gairfowl (*Alca impennis*). Ibis, 1898, pp. 587–592.
  1905. Ootheca Wolleyana: An illustrated catalogue of the collection of birds eggs formed by the late John Wolley, Jun., M.A., F.Z.S. Edited from the original by Alfred Newton. Part III, Columbae-Alcae. London.
Newton, E.
  1861. Ornithological notes from Mauritius. Ibis, 1861, pp. 270–277.
Newton, E., and Gadow, H.
  1893. On additional bones of the dodo and other extinct birds of Mauritius. Trans. Zool. Soc. London, **13**, pp. 281–302.
Nicholson, D. J.
  1948. Escaped paroquets breeding in Florida. Auk, **65**, p. 139.
Nicoll, M. J.
  1908. Three voyages of a naturalist, being an account of many little-known islands in three oceans visited by the *Valhalla*, 246 pp. London.
Noble, G. K.
  1916. Resident birds of Guadaloupe [island, West Indies]. Bull. Mus. Comp. Zool., **40** (10), pp. 359–396.
Nordhoff, C.
  1943. Notes on birds of Tahiti. Avicult. Mag. (5), **8**, pp. 119–121.
North, A. J.
  1901–1914. Australian Museum, Sydney; special catalogue no. 1, nests and eggs of birds found breeding in Australia and Tasmania, 2d ed., 4 vols. Sydney.

NORTHWOOD, J. DE A.
1940. Familiar Hawaiian birds, 69 pp. Honolulu.
NORWEGIAN SCIENTIFIC EXPEDITION.
1946. Results of the Norwegian scientific expedition to Tristan da Cunha, 1937–1938, 2 vols., 330 pp. Oslo. (In English.)
NUMMEDAL, A.
1923. On Flintpladsene [on Norwegian archeological sites]. Norsk. Geol. Tidssk., 7 (2), pp. 89–141.
NUTTALL, T.
1832. A manual of the ornithology of the United States and Canada: Land Birds, 683 pp.
1840. A manual of the ornithology of the United States and Canada, 2d ed.: Land Birds. 832 pp.
OBERHOLSER, H. C.
1918. The scientific name of the passenger pigeon. Science, 48, p. 445.
1938. The bird life of Louisiana. Dept. Conservation State of Louisiana Bull. 28, 760 pp. New Orleans.
OBERHOLSER, H. C., and CHAPMAN, F. M.
1932. Migrations of North American birds [range and plumages of Conuropsis]. Bird Lore, 34, pp. 328–332.
OLAFSEN, E.
1774–1775. Des . . . E. Olafsens und . . . des B. Povelsens Reise durch Island . . . aus dem danischen Ubersetz, 2 vols. Kopenhagen and Leipsig. (First edit., 1772.) Reise igienem Island (not sighted). Variously cited in libraries "Olafson, Olafssen, Olafsson," etc. Also K. Dansk Videns. Selskab.
OLIVER, W. R. B.
1930. New Zealand birds, 523 pp. Wellington.
1945. Avian evolution in Australia and New Zealand. Emu, 45, pp. 55–77 119–152.
1949. The moas of New Zealand and Australia. Dominion Mus. Bull. no. 15, 195 pp.
ORNITHOLOGICAL SOCIETY OF JAPAN.
1942. A hand-list of the Japanese birds, 3d and rev. ed., 238 pp. Tokyo.
ORNITHOLOGICAL SOCIETY OF NEW ZEALAND.
1953. Checklist of New Zealand birds, 80 pp.
OTTAWA FIELD NATURALIST CLUB.
1891. Birds of Ottawa [Ectopistes bred in Ottawa though rare there]. Ottawa Nat., 5, p. 38.
OTTOW, B.
1950. [Bone disease in Pezophaps solitaria of Rodriguez.] Kungl. Svenska Vet. Handl. (1), 9, pp. 1–37. (English summary.)
OUDEMANS, A. C.
1917. Dodo-Studien, etc. Verh. Akad. Wetensch. Amsterdam, 19 (4), 140 pp.
OUSTALET, E.
1878. Étude sur la faune ornithologique des Seychelles. Bull. Soc. Philom. (Paris) (7), 2, pp. 161–206.
1896a. [Notes on Megapodius laperouse.] Nouv. Arch. Mus. (Paris) (3), 8, pp. 26–32.
1896b. [Faune des Îles Mascareignes.] Ann. Sci. Nat. Zool. (Paris) (8), 3, 128 pp.
OVIEDO Y VALDES, G. F. DE.
1535. Historia generale y natural de las Indias., 193 pp.
OWEN, R.
1848. [Description of Notornis mantelli.] Trans. Zool. Soc. London, 3, pp. 347, 366.

1864. [Description of the skeleton of the great auk.] Trans. Zool. Soc. London, **5** (4), pp. 317–335.
1879. Memoirs on the extinct wingless birds of New Zealand, etc., 2 vols., text, 465 pp. London.
PALMER, R. S.
1949. Maine birds. Bull. Mus. Comp. Zool., **102**, 656 pp.
PALMGREN, P.
1935. Ein Exemplar von *Phalacrocorax perspicillatus* Pall. in den Sammlungen des Zoologischen Museums in Helsingfors. Ornis Fennica, **12**, pp. 78–80.
1944. [On the bird species extirpated by man and the reason for their disappearance.] Ornis Fennica, **21** (1), pp. 15–25.
PARKER, T. J.
1882. On a skeleton of *Notornis*. Trans. and Proc. New Zealand Inst., **14**, pp. 245–258.
1886. On a skeleton of *Notornis*. Trans. and Proc. New Zealand Inst., **18**, p. 78.
PAYNTER, R. A., JR.
1955. The ornithogeography of the Yucatán Peninsula. Peabody Mus. Nat. Hist., Yale Univ., Bull. 9, 347 pp.
PEALE, T. R.
1848. United States Exploring Expedition, during the years 1838, 1839, 1840, 1841, 1842, etc. Mammalia and ornithology. Philadelphia.
PENARD, F. P., and PENARD, A. F.
1908. De Vogels van Guyana (Suriname, Cayenne en Demerara), 2 vols., xliii+ 587 pp. Paramaribo.
PENNOCK, C. J.
1908. [Passenger pigeon in Delaware.] Auk, **25**, p. 285.
1912. Nesting of passenger pigeon in New York. Auk, **29**, pp. 238–239.
PÉQUART, SR. J.
1926. Un kjokkenmodding Morbihannais: Er Yoj. Rev. Anthrop., Paris, **36**, (4–6), pp. 206–211. [Great auk, p. 207.]
PERKINS, R. C. L.
1903. Vertebrata. *In* Fauna Hawaiiensis, or The zoology of the Sandwich (Hawaiian) Isles, vol. 1, pt. 4, pp. 365–466. Cambridge, England.
PERRY, M. C.
1856. Narrative of the expedition of an American squadron to China Seas and Japan performed in the years 1852, 1853, 1854, compiled by F. L. Hawkes, 3 vols. Washington.
PETERS, J. L.
1931–1951. Check-list of birds of the world, 7 vols. Cambridge, Mass.
PETERS, J. L., and GRISCOM, L.
1929. A new rail and a new dove from Micronesia. Proc. New England Zool. Club, **10**, pp. 99–106.
PETTINGILL, O. S., JR.
1939. Bird life of Grand Manan Archipel. Proc. Nova Scotia Inst. Sci., 1937–38, **19** (4), pp. 293–372.
PHILLIP, A.
1789. The voyage of Governor Phillip to Botany Bay, etc., 398 pp. London.
PHILLIPS, J. C.
1922–1926. A natural history of the ducks, 4 vols. Boston.
1926. An attempt to list extinct and vanishing birds, etc. Verh. VI Int. Orn. Congr., Kopenhagen, pp. 503–534.
1928. Wild birds introduced or transplanted in North America. U.S. Dept. Agr. Techn. Bull. no. 61, 63 pp.

Pinkerton, J.
  1808–1814. A general collection of . . . voyages and travels, etc., 17 vols. London. [Great auk, **3**, p. 730.]
Pinto, O.
  1938–1944. Catalogo das Aves do Brasil. Pt. 1 (1938), 566 pp.; pt. 2 (1944), 700 pp. São Paulo.
Pitelka, F. A., and Bryant, M. D.
  1942. Available skeletons of the passenger pigeon. Condor, **44**, pp. 74–75.
Platt, M. I.
  1936. [Remains of great auk in Orkney.] Proc. Soc. Antiquaries Scotland, **70**, pp. 415–419.
Plessen, Baron V. von.
  1929. Meine Reise nach den Inseln zwischen Flores und Celebes. Journ. Orn. Berlin, **77**, p. 407.
Poole, E. L.
  1949. The extinct heath hen. Frontiers [Philadelphia], **13**, pp. 68–70.
Porter, S.
  1934. Notes on *Cyanoramphus parrakeets*. Avic. Mag. (4), **12**, pp. 273–285.
Potter, S. D.
  1949. The elusive fern bird [*Bowdleria*]. New Zealand Bird Notes, 3 (6), p. 162.
Potts, T. H.
  1870. On the birds of New Zealand. Trans. New Zealand Inst., **2**, pp. 40–78. [*Coturnix*, p. 66.]
  1871. On the birds of New Zealand. Trans. New Zealand Inst., **3**, pp. 59–109.
  1885. Oology of New Zealand. New Zealand Journ. Sci., **2**, pp. 475–484.
Pouget, M. de Compte.
  1875. Note sur la kagou [in captivity]. Bull. Mensuel Soc. Acclimat. Paris (3), **2**, pp. 162–171.
Prater, S. H.
  1921. The Arabian ostrich. Journ. Bombay Nat. Hist. Soc. **27**, pp. 602–605.
Preyer, W.
  1862a. Der Brillenalk *Plautus impennis* in europaeischen Sammlungen. Journ. für Orn., **10**, pp. 77–79.
  1862b. Über *Plautus impennis* Brunn. Journ. für Orn., **10**, pp. 110–124, 337–356.
Purchas, S.
  1905–1907. Hakluytus posthumus: or Purchas his Pilgrimes, etc., 1625. Hakluyt Society, extra ser. 2, 20 vols. Glasgow. [Great auk in New England, **18**, p. 305.]
Putnam, F. W.
  1868. Remains of great auk in Massachusetts. Proc. Essex Inst., **5**, p. 310 (note).
  1870. Remains of great auk in Massachusetts. Amer. Nat., **3** (10), p. 540 (note).
Pycraft, W. P.
  1909. [Skull of *Paleocorax moriorum*] Bull. Brit. Orn. Club, **23**, pp. 95–96.
Ramsay, E. P.
  1866. List of birds of Port Denison [Australia]. Ibis, 1866, pp. 325–336.
Rand, A. L.
  1936. The distribution and habits of Madagascar birds, etc. Bull. Amer. Mus. Nat. Hist., **72**, art. 5, pp. 143–499.
Reichenow, A.
  1891. A collection of birds from Fiji Islands (*Trichocichla rufa* described). Journ. für Orn., **34**, p. 130.
Reid, S. G.
  1884. Birds of Bermuda. U.S. Nat. Mus. Bull. 25, pp. 163–279.

REINHARDT, J.
1854. [*Alca impennis* listed from Greenland.] Journ. für Orn., **2**, p. 442.
RENSHAW, G.
1914. The pigeon Hollondais (*Alectroenas pulcherrima*). Avicult. Mag. (London), **6**, pp. 61–62.
1930. The black emu. Journ. Soc. Preservation Fauna Empire (n.s.), pt. 12, pp. 17–21.
1931. The Mauritius dodo. Journ. Soc. Preservation Fauna Empire (n.s.), pt. 15, pp. 14–21.
1933a. Poulet rouge and corbeau Indian [of Cauche and De Bry]. Journ. Soc. Preservation Fauna Empire (n.s.), pt. 19, pp. 16–20.
1933b. The dodo and aphanapteryx. Nature, **131**, p. 728.
1936. Poulet rouge or aphanapteryx. Journ. Soc. Preservation Fauna Empire (n.s.), pt. 29, pp. 60–61.
RETT, E. Z.
1938. Hailstones fatal to California condors. Condor, **40**, p. 225.
REVOIL, B. H.
1869. Chasse dans l'Amérique du Nord; Les pigeons. English translation in Bird Lore, **30**, pp. 317–320.
RICHARDSON, FRANK.
1949. Status of native land birds on Molokai. Pacific Sci., **3** (3), pp. 226–230.
1955. Reappearance of Newell's shearwater in Hawaii. Auk, **72**, p. 412.
RICHDALE, L. E.
1945(?). Vanishing birds of New Zealand, no. 5. Dunedin, N. Z., Otago Daily Times and Witness Newspapers Co.
RIDGWAY, R.
1873. [*Ectopistes* in Nevada.] Bull. Essex Inst., **5**, p. 178.
1876. Ornithology of Guadaloupe [*sic*] Island [Pacific]. Bull. U.S. Geol. Geogr. Surv. Terr., **2** (2), pp. 183–195.
1877. [Origin of avifauna of Guadalupe Island (Pacific Ocean).] Bull. Nuttall Orn. Club, **2**, pp. 58–66.
1896. On birds collected by Doctor W. L. Abbott in the Seychelles, Amirantes, Gloriosa, Assumption, Aldabra, and adjacent islands, etc. Proc. U.S. Nat. Mus., **18**, pp. 509–546
1916. The birds of North and Middle America. U.S. Nat. Mus. Bull. 50, pt. 7.
RIDGWAY, R., and FRIEDMANN, H.
1946. *Colinus virginianus insulanus,* Key West bobwhite. U.S. Nat. Mus. Bull. 50, pt. 10, p. 328.
RIPLEY, S. D.
1951. The thrushes. Postilla [Yale Peabody Mus. Nat. Hist.], **13**, pp. 1–48.
1952. Vanishing and extinct bird species of India. Journ. Bombay Nat. Hist. Soc., **50** (4), pp. 902–906.
RIPLEY, S. D., and BIRCKHEAD, H.
1942. On the fruit pigeons of *Ptilinopus purpuratus* group. Amer. Mus. Novit., no. 1192, pp. 1–14 (6).
RITCHIE, J.
1920. The influence of man on animal life in Scotland, 550 pp. Cambridge University Press.
ROBERTS, A.
1948. On a collection of birds and eggs from the Tristan da Cunha Islands. Ann. Transvaal Mus., **21** (1), pp. 55–63.
ROBERTS, T. S.
1918. Alleged re-discovery of the passenger pigeon. Science, **48**, p. 575.
1932. The birds of Minnesota, vol. 1, xxii+691 pp.; vol. 2, xx+821, pp. University of Minnesota, Minneapolis.

ROBERTSON, J. S.
1946. The eastern bristle bird in Queensland. Emu, **45**, pp. 265–270.
ROCK, J. F.
1913. The indigenous trees of the Hawaiian Islands, 518 pp. Honolulu.
ROGERS, R. A.
1926. The Lonely Island [Tristan da Cunha Island], 223 pp. London.
ROTHROCK, J. F.
1894. Forests of Pennsylvania. Proc. Amer. Philos. Soc., **33**, pp. 114–134.
1895. Forests of Pennsylvania. Journ. Franklin Inst., **140**, pp. 105–117. [Reprint.]
ROTHSCHILD, W.
1893. *Rallus muelleri,* sp. n. Ibis, 1893, p. 442.
1893–1900. The avifauna of Laysan and the neighboring islands, etc., xxxiv +
34 + 320 + 21 pp. London.
1903. [Original description of *Rallus wakensis.*] Bull. Brit. Orn. Club, **13**, p. 78.
1905. On extinct and living parrots of the West Indies, Bull. Brit. Orn. Club, **16**,
pp. 13, 14.
1907. Extinct birds, etc., 29 + 244 pp. London.
1919a. On one of the four original pictures from life of the Réunion or white
dodo. Ibis, 1919, pp. 78–79.
1919b. [No title.] Bull. Brit. Orn. Club, **39**, p. 83.
1928. Mounted example of great auk. Bull. Brit. Orn. Club, **49**, pp. 4–9.
ROTHSCHILD, W., and HARTERT, E.
1899. A review of the ornithology of the Galapagos. Novit. Zool., **6**, pp. 85–205.
ROWAN, W.
1926. Comments on two hybrid grouse and occurrence of *Tympanuchus americanus* in the Province of Alberta. Auk, **43**, pp. 333–336.
ROWLEY, G. D.
1877. *Somateria labradoria* (the pied duck). Rowley's Orn. Misc., **2**, pp. 205–223.
SALOMONSEN, F.
1933. *Neodrepanis hypoxantha* sp. nov. Bull. Brit. Orn. Club, **53**, pp. 182–183.
1934. Les Neodrepanis. Ois. Rev. Franç. Orn., **4** (1), pp. 1–9.
1935. 'Aves' *in* Zoology of the Faroes, edited by Ad. S. Jensen, et al. Copenhagen. [Reference to great auk, pp. 111.]
1945. Gejrfuglen et hundredaars minds [great auk]. Dyr I Natur og Museum,
Copenhagen, 1944–1945, pp. 101–110.
1950–1951. Grønlands Fugle [Birds of Greenland], 2 vols. Copenhagen.
SALVADORI, T.
1893. Catalogue of the Columbae or pigeons in the collection of the British
Museum (vol. 21 of series), p. 676.
1907. Notes on parrots. Ibis, 1907, pp. 311–322 (314).
SALVIN, O.
1861. [*Numenius borealis* in Guatemala.] Ibis, 1861, p. 356.
1873. Note on the *Fulica* alba of White. Ibis, 1873, p. 295.
1876. Avifauna of the Galapagos Archipelago. Trans. Zool. Soc. London, **9**, pp.
447–510.
SALVIN, O., and GODMAN, F. D.
1879–1904. Biologia Centrali-Americana, Aves, 4 vols. London.
SÁNCHEZ, J., and VILLADA, M. M.
1873. *Ectopistes migratorius* in Mexico. La Naturaleza (Soc. Mexicana de Historia Natural), **2** (Ann. 1871, 1872, 1873), pp. 250–253 (see also 254, 255).
SANFORD, L. C.
1903. The water fowl family, 598 pp. New York.
SARASIN, F.
1913. Nova Caledonia (Sarasin and Roux): Zoology, vol. 1, 75 pp. Weisbaden.

SASSI, M.
1939. Die wertvollsten Stücke der Wiener Vogelsammlung. Ann. Naturh. Mus. Wien, **50**, pp. 395–409.
SCHAANING, H. TH. L.
1927. Kjendte fugle-arter Norges Sten-og Jernalder, pp. 1–25 (=68–93). Stavajger Mus. Aarsh, **35**, 1924–1925. Stavanger Museum Femti Aar, 1877–1927, 200 pp.
SCHAUENSEE, R. DE.
1941. Rare and extinct birds in the collections of the Academy of Natural Sciences of Philadelphia. Proc. Acad. Nat. Sci. Philadelphia, **93**, pp. 281–324.
1948–1952. The birds of Colombia. Pt. 1, Caldasia [Bogota], **5**, no. 22, pp. 251–380; pt. 2, **5**, no. 23, pp. 381–644; pt. 3, **5**, no. 24, pp. 645–871; pt. 4, **5**, no. 25, pp. 873–1112; pt. 5, **5**, no. 26, pp. 1115–1214 (992, 294, 1004).
SCHAUINSLAND.
1899. Drei Monate auf einer Koralleninsel (Laysan). Bremen.
SCHERDLIN, P.
1926. Great auk in Strasbourg. Mus. Bull. Assoc. Philom. Alsace Lorraine, **7**.
SCHIØLER, E. L.
1925–1931. Danmarks Fugle, etc., 3 vols. Kjøbenhavn.
SCHLEGEL, H.
1858. Ueber einige ausgestorbene riesige Vögel, etc. Journ. für Orn., **6**, p. 367–383. (See Ibis, 1866, pp. 146–167, for account in English.)
1866. On some extinct gigantic birds of the Mascarene Islands. Ibis, 1866, pp. 146–167.
SCHNEE, P.
1912. [Notes on *Megapodius laperouse*.] Zeitschr. Naturw., **82**, pp. 467–468.
SCHØNWETTER, M.
1928. Eier von *Struthio camelus syriacus*. Orn. Monatsb., **36**, pp. 176–177.
SCHORGER, A. W.
1937. A great Wisconsin passenger pigeon nesting in 1871. Proc. Linn. Soc. New York, **48**, pp. 1–26.
1938a. The last passenger pigeon killed in Wisconsin. Auk, **55**, pp. 531–532.
1938b. Unpublished manuscript by Cotton Mather on the passenger pigeon. Auk, **55**, pp. 471–477.
1942a. Extinct and endangered mammals and birds of the upper Great Lakes region. Trans. Wisconsin Acad. Sci., Arts and Lett., **34**, pp. 23–44.
1942b. A clue to prairie chicken mortality. Passenger Pigeon, **4** (3), p. 83.
1943. The prairie chicken and sharp-tailed grouse in early Wisconsin. Trans. Wisconsin Acad. Sci., Arts and Lett., **35**, pp. 1–59.
1955. The passenger pigeon; its natural history and extinction. xiii+424 pp. Madison.
SCHWARTZ, C. W.
1945. Ecology of the prairie chicken in Missouri. Univ. Missouri Stud., 99 pp.
SCLATER, P. L.
1861. [No title.] Proc. Zool. Soc. London, 1861, p. 209.
1881. [Birds collected on Tristan da Cunha Island.] Report on the scientific results of the voyage of H. M. S. *Challenger* during the years 1873–76, Zoology, **2**, pp. 110–113.
SCLATER, W. L.
1915. The 'Mauritius hen' of Peter Mundy. Ibis, 1915, pp. 316–319.
SCLATER, W. L., and SALVIN, O.
1866–1869. Exotic ornithology, etc., 204 pp. London.
SCOTT, C. D.
1936. Who killed the condors? Nature Mag., **28**, pp. 368–437.

SCOTT, H. H.
  1924. Note on the King Island emu. Papers and Proc. Roy. Soc. Tasmania, 1923, pp. 103–107.
  1932. The extinct Tasmanian emu. Papers and Proc. Roy. Soc. Tasmania, 1931, pp. 108–110.
SCOTT, J. H.
  1883. Macquarie Island. Trans. New Zealand Inst., **15**, p. 484.
SCOTT, W. E.
  1940. Eskimo curlew record corrected [Wisconsin record for 1912 is *hudsonicus*]. Auk, **57**, pp. 566–567.
SCOTT, W. E. (EDIT.).
  1947. Silent wings, a memorial to passenger pigeon, 39 pp. Wisconsin Ornithological Society.
SCOTT, W. E. D.
  1888. Supplementary notes from the Gulf coast of Florida. Auk, **5**, pp. 183–188 (186).
  1889. [A summary of observations of the coast of Florida.] Auk, **6**, pp. 245–252.
SEALE, A.
  1901. Report of a mission to Guam. Occ. Papers B. P. Bishop Mus., **1** (3), pp. 17–62.
SEEBOHM, H.
  1890. Birds of the Bonin Islands. Ibis, 1890, pp. 95–108.
  1891. Birds of the Volcano Islands. Ibis, 1891, pp. 189–192.
SERVENTY, D. L., and WHITTELL, H. M.
  1951. A handbook of birds of Western Australia (with the exception of the Kimberley Division), 384 pp. Perth.
SETH-SMITH, D.
  1931. On the turquoise grass parrakeet. Avicult. Mag. (4), **9**, pp. 293–294.
  1932. Note on *Rhodonessa,* the pink headed duck. Avicult. Mag. (4), **10**, p. 118.
SETON, E. T. (See also under Thompson, E. E.)
  1886. Birds of western Manitoba (Canada). Auk, **3**, pp. 145–146.
SHARLAND, M.
  1947. Notes on two *Neophema* parrots. Emu, **46**, pp. 258–264.
SHARPE, R. B.
  1890. Catalogue of Birds in the British Museum, **13**.
  1894. Catalogue of Birds in the British Museum, **23**.
  1896. Catalogue of Birds in the British Museum, **24**.
  1906. History of the collections contained in natural history departments, British Museum, London, 1904–1912.
SHAW, H. S.
  1940. An early figure of the great auk. Auk, **57**, p. 112.
SHELLEY, G. E.
  1883. On the Columbidae of the Ethiopian region. Ibis, 1883, pp. 258–330 (258, 259).
SHEPPARD, T.
  1922. Unrecorded egg of great auk. Naturalist (London), 1922, p. 254.
SHUFELDT, R. W.
  1914. Osteology of the passenger pigeon, Auk, **31**, pp. 358–362.
  1915. Anatomical and other notes on the passenger pigeon (*Ectopistes*) lately living in the Cincinnati Zoological Garden. Auk, **32**, pp. 29–41.
  1916. The bird caves of Bermuda and their former inhabitants. Ibis (10), **4**, pp. 623–635.
  1921. Published figures and plates of the extinct passenger pigeon. Sci. Monthly, **12** (5), pp. 458–481.

476   EXTINCT AND VANISHING BIRDS OF THE WORLD

1922. [Subfossil remains of birds from Bermuda.] Ann. Carnegie Mus., **13**, pp. 333–418.
SIBBALD, R.
1684. Scotia illustrata, etc. 3 vols. Edinburgh.
SIMPSON, G. G.
1946. [Fossil penguins.] Bull. Amer. Mus. Hist., **87** (1), pp. 7–99.
SIMSON, F. B.
1884. [Notes on *Rhodonessa caryophyllacea*]. Ibis, 1884, pp. 271–275.
SLEVIN, J. R.
1936. Outposts of Baja California. Nat. Hist., **38**, pp. 65–73 (66).
SMITH, A. P.
1907. Summer notes from an Arizona camp. Condor, **9**, pp. 196–197.
SMITH, F. J.
1916. [Condor specimens.] Condor, **18**, p. 205.
SMITH, W. W.
1884. *Sceloglaux albifacies* (laughing owl). New Zealand Journ. Sci., **2**, pp. 86–87.
SNOW, D. W.
1950. The birds of Sao Tomé and Principé in the Gulf of Guinea. Ibis, 1950, pp. 579–595.
SNOW, H. J.
1910. In forbidden seas, etc., 303 pp. London.
SONNERAT, P.
1806. Voyages aux indes orientales et à la Chine, fait par ordue de Louis XVI depuis 1774 jusqu'en 1781, 4 vols. Paris. (First edition 2 vols., Paris, 1782.)
SOWERBY, A. DE C.
1922–1930. The naturalist in Manchuria, 3 vols. Tiensin. [Manchurian crane, **3**, p. 222.]
SPENCER, A, and KERSHAW, J. A.
1910. A collection of subfossil bird and marsupial remains from King Island, Bass Strait. Mem. Nat. Mus. Melbourne, **3**, pp. 5–35.
SPRUNT, A.
1954. Florida bird life, 531 pp. New York.
SPRUNT, A., and CHAMBERLAIN, E. B.
1949. South Carolina bird life. Contr. Charleston Mus., no. 11, 585 pp. Columbia, S. C.
STAHL, A.
1887. Beitrag zur Vogelfauna Portorico [notes on Puerto Rican Birds]. Ornis, **3**, pp. 448–453.
STEAD, E. F.
1938. The supposed flightless duck from Campbell Island. Trans. and Proc. Roy. Soc. New Zealand, **68**, pp. 100–101.
1948. Bird life on the Snares. New Zealand Bird Notes, **3** (3), pp. 69–80.
STEENSTRUP, J. J. S.
1855. Et Bidrag til Gierfuglens, *Alca impennis,* etc. Vid. Medd. Naturh. For. Kobenhavn, no. 3–7, pp. 33–115.
STEGMAN, B.
1930–1931. [Die Vögel des Amurlandes.] Journ. für Orn., **78** (4), 1930, pp. 389–471; **79** (2), 1931, pp. 137–236.
1936. Ueber das Flugvermögen der ausgestorbenen Scharbe *Phalacrocorax perspicillatus* Pall. Orn. Monatsb., **44** (5), pp. 140–153.
STEINER, H.
1936. [Embryo of the emu.] Rev. Suisse Zool. Genève, **43**, pp. 543–550.

STEJNEGER, L.
1883. [Status of Pallas's cormorant.] Proc. U.S. Nat. Mus., **6**, p. 65.
1885. Results of ornithological explorations in the Commander Islands and in Kamtschatka. U.S. Nat. Mus. Bull. **29**, 382 pp.
1890. Contributions to the history of Pallas' cormorant. Proc. U.S. Nat. Mus., **12**, pp. 83–88.
1936. Georg Wilhelm Steller, 623 pp. Harvard University Press.

STELFOX, A. W.
1938. Birds of Lagore about one thousand years ago. Irish Nat. Journ., **7**, pp. 37–43.

STOCK, C.
1929. Census of Pleistocene mammals of Rancho La Brea, etc. Journ. Mamm., **10**, pp. 281–289.

STODDARD, H. L.
1931. The bobwhite quail: Its habits, preservation, and increase, 559 pp. New York.

STOLPE, H.
1879. Fynd från Greby i Tanums socken och härad. In Montelius Göteborgs och Bohusläns fornmin. och hist., **2**, pp. 27–28 (note).

STOPHLET, J. J.
1946. Birds of Guam. Auk, **63**, pp. 534–540.

STOTT, K., JR.
1945. Breeding condors. Zoonooz (San Diego), **18** (2), pp. 3, 4.
1947. Notes on Saipan birds. Auk, **64**, pp. 523–527.

STREETS, T. H.
1876. [Description of *Chaulelasmus couesi*.] Bull. Nuttall Orn. Club, **1**, p. 46.
1877. Natural history of the Fanning Islands. Amer. Nat., **11** (2), pp. 65–72.

STRESEMANN, E.
1923a. Neu aufgefundene Abbildung der ausgestorbenen "Riesenralle" *Leguatia gigantea* Schlegel. Journ. für Orn., **71**, pp. 158–160.
1923b. *Leguatia gigantea* nach der Zeichnung von Francis Barlow. Journ. für Orn., **71**, pl. 6, pp. 451–456.
1923c. Nochmals *Leguatia gigantea* Schlegel. Journ. für Orn., **71**, pp. 511–512.
1932. La structure des rémiges chez quelques rales physiologiquement aptéres. Alauda (2), **4**, pp. 1–5.
1937. Ein neuer Fund von *Neodrepanis hypoxantha* Salom. Orn. Monatsb., **45**, pp. 135–136.
1939–1941. Die Vögel von Celebes. Journ. für Orn., **87** (3), pp. 299–425; **88** (1), pp. 1–135 (42–43); **88** (3), pp. 389–487; **89** (1), pp. 1–102 (31).
1948. The status of *Synallaxis sclateri* Cabanis. Auk, **65**, pp. 445–446.
1949. Birds collected in the North Pacific area during Capt. Cook's last voyage (1778 and 1779). Ibis, 1949, pp. 244–255.
1950. Birds collected during Capt. James Cook's last expedition (1776–1780) Auk, **67**, pp. 66–88.
1953. Birds collected by Capt. Dugald Carmichael on Tristan da Cunha 1816–1817. Ibis, 1953, pp. 146–147.
1954. Ausgestorbene und aussterbende Vogelarten, vertreten im Zoologischen Museum Berlin. Mitt. Zool Mus. Berlin, **30** (1), pp. 38–53.

STRICKLAND, H. E., and MELVILLE, A. G.
1948. The dodo and its kindred: or the history, affinities of the dodo and other extinct birds of the Islands Mauritius, Rodriguez, and Bourbon, 141 pp. London.

STROMER, E. VON.
1902. [Fossil ostrich from Egypt.] Zeitschr. deutsch. geol. Ges. Sitzb., **54**, p. 108.

SUTTON, G. M.
1928. Birds of Pymatuning Swamp and Conneaut Lake, Crawford Co., Pennsylvania. Ann. Carnegie Mus., **18**, pp. 19–239.
SWAEN, A. E. H.
1940. Voyage de François Leguat et de ses compagnons. Ardea, **29**, pp. 19–44.
SWANN, H. K.
1925. [Account of nest of *Polyborus lutosus*.] Monograph of the Birds of Prey, pt. 2, p. 73.
SWARTH, H. S.
1914. A distributional list of the birds of Arizona. Cooper Orn. Club, Pacific Coast Avif., no. 10, 133 pp.
1931. Avifauna of the Galapagos Islands. Occ. Pap. California Acad. Sci., **18**, pp. 1–299.
1933. Off shore migrants over the Pacific (notes on Ornis of Galapagos and islands off Baja California). Condor, **35**, pp. 39–40.
SWENK, M. H.
1916. The Eskimo curlew and its disappearance. Ann. Rep. Smithsonian Inst. for 1915, pp. 325–340.
1926. Eskimo curlew in Nebraska. Wilson Bull., **33**, pp. 117–118.
1934. The interior Carolina paroquet as a Nebraska bird. Nebraska Bird Rev., **2**, p. 55.
TAKA-TSUKASA, N.
1932. [Notes on *Megapodius laperouse*.] The Birds of Nippon, **1**, pt. 1, p. 8.
TANNEHILL, I. R.
1938. Hurricanes, their nature and history, x + 251 pp. Princeton, N. J.
TANNER, J. T.
1942. The ivory-billed woodpecker. Nat. Audubon Soc. Res. Rep. no. 1. New York. Reprinted by Dover Publications in 1966.
TAVERNER, P. A., and SUTTON, G. M.
1934. Birds of Churchill, Manitoba. Ann. Carnegie Mus., **23**, pp. 1–83.
TAYLOR, H. R.
1898. [Notes on nesting of *Gymnogyps californianus* made by his collectors.] Osprey, **3**, p. 39.
TETRE, J. B. DU.
1654. Histoire generale, des Isles de S. Christophe, de la Guadaloupe, de la Martinique, etc., 481 pp. Paris.
1667–1671. Histoire generale des Antilles, etc., 4 vols. Paris. (Birds, vol. 2.)
TEVIOTDALE, D.
1924. Excavations near the mouth of the Shag River, Otago. Journ. Polynesian Soc., **33**, pp. 3–10.
TEXAS GAME, FISH AND OYSTER COMMISSION.
1945. Principal game birds and mammals of Texas, their distribution and management, 149 pp. Austin.
THAYER, J. E., and BANGS, O.
1908. The present state of the Ornis of Guadaloupe Island [Guadalupe Island, NW. coast Mexico, Pacific]. Condor, **10**, pp. 101–106.
THOMPSON, E. E. [Seton, E. T.]
1891. The birds of Manitoba. Proc. U.S. Nat. Mus., **13**, pp. 457–643. [*Ectopistes* in Manitoba, p. 522; *Grus americana*, p. 491.]
THOMPSON, W.
1849–1856. Natural history of Ireland, 4 vols. (Great auk, **3**, p. 238.)
THOMSEN, T.
1917. Implements and artefacts of the North-east Greenlanders. Medd. Grønland, **44** (5), pp. 359–496.

THOMSON, C. W.
  1878. Voyage of the *Challenger,* 2 vols. London.
THOMSON, G. M.
  1922. The naturalization of animals and plants in New Zealand, 607 pp. Cambridge University.
THORBURN, W. A.
  1899. Fur seals and fur-seal islands of the North Pacific [see Jordan, D. S., et al., Birds of Guadalupe Island, **3,** p. 278].
TICEHURST, N. F.
  1908. [Great auk bones found near the Bay of Ayr, Orkney Islands]. Brit. Birds, **1,** p. 310.
TINBERGEN, N.
  1951. The study of instinct, 210 pp. Oxford.
TODD, W. E. C.
  1940. Birds of western Pennsylvania, 710 pp. Pittsburgh.
TOWNSEND, C. W.
  1905. The birds of Essex County, Massachusetts. Mem. Nuttall Orn. Club, no. 5, 195 pp.
  1907. Bones of great auk at Mt. Desert Island [Maine]. Journ. Maine Orn. Soc., **9,** p. 82.
  1923. [Last record for *Oceanodroma macrodactyla.*] Bull. Amer. Mus. Nat. Hist., **48,** p. 6.
TOWNSEND, C. W., and ALLEN, G. M.
  1907. Birds of Labrador. Proc. Boston Soc. Nat. Hist., **33** (7), pp. 277–428.
TOWNSEND, C. W., and WETMORE, A.
  1919. Reports on the scientific results of the expedition to the tropical Pacific in charge of Alexander Agassiz on the U.S. Fish Commission steamer *Albatross.* Bull. Mus. Comp. Zool., **63** (4), pp. 151–225. (Birds.)
TOWNSEND, J. K.
  1839. Narrative of a journey across the Rocky Mountains, etc. 352 pp.
TRAUTMAN, M. B.
  1935. Additional notes on Ohio birds. Auk, **52,** p. 321.
  1940. The birds of Buckeye Lake, Ohio. Misc. Publ. Mus. Zool. Univ. Michigan., no. 44, 466 pp.
TRAYLOR, M. A.
  1950. Sight record for *Diomedea albatrus* off San Francisco, February 1946. Condor, **52,** p. 90.
TRISTRAM, H. B.
  1880. Description of a new genus and species of owl from the Seychelles Islands. Ibis, 1880, pp. 456–459.
TUGARINOV, A. J.
  1929. Breeding of the least whimbrel *Mesocolopax minutus* Gould in Yakut Land [N-E. Siberia]. Journ. für Orn., Festschr. E. Hartert, **2,** pp. 136–142.
TURBOTT, E. G.
  1951. Winter observations on *Notornis.* Notornis [continuation of New Zealand Bird Notes], **4** (5), pp. 107–113.
TURBOTT, E. G., and BUDDLE, G. A.
  1948. Birds of Three Kings Islands. Rec. Auckland Mus., **3,** pp. 319–336 (327).
TURNBULL, W. P.
  1869. The birds of East Pennsylvania and New Jersey, pp. 10–62. Glasgow.
TURNER, L. M.
  1886. Contributions to the natural history of Alaska, **2** (5), Birds, pp. 115–196.
TYZZER, E. E.
  1929. Coccidiosis in gallinaceous birds. Amer. Journ. Hygiene, **10** (2).

UNDÅS, ISAK.
    1914. Bergens Tidende of 22 Nov. 1941, p. 12. See also Wiman and Hessland,
        1942, from which this is quoted.
USSHER, R. J.
    1897. Discovery of bones of the great auk in Co. Waterford, Ireland. Irish Nat.,
        6, p. 208.
    1899. Remains of *Plautus impennis* from Irish kitchen middens. Bull. Brit. Orn.
        Club, 8, p. 50.
    1902. Great auk bones in Co. Clare, Ireland. Irish Nat., 11, p. 188.
USSHER, R. J., and WARREN, R.
    1900. The birds of Ireland, etc., 410 pp. London.
VAHL, M. (ED.).
    1928. Greenland, 3 vols. Published by the commission for the direction of geo-
        logical and geographical investigations in Greenland. Copenhagen and
        London.
VAN CLEEF, J. S.
    1899. Flights of passenger pigeons. Forest and Stream, 52, p. 385.
VAN ROSSEM, A. J.
    1945. A distributional survey of the birds of Sonora, Mexico. Occ. Pap. Mus.
        Zool. Louisiana State Univ., no. 21, 379 pp. (105).
VAN TYNE, J.
    1936. The discovery of the nest of the Colima warbler (*Vermivora crissalis*).
        Misc. Publ. Mus. Zool. Univ. Michigan, no. 33, pp. 5–11.
    1948. Eskimo curlew and whimbrel collected in Newfoundland Labrador. Wil-
        son Bull., 60, p. 241.
    1951. Distribution of the Kirtland warbler. Proc. 10th Int. Orn. Congr., pp. 537–
        544.
VERRILL, A. E.
    1895. [Birds collected on Gough Island, S. Atlantic, with notes of G. Comer,
        collector.] Trans. Connecticut Acad. Arts and Sci., 9, pp. 430–478.
    1902. The Bermuda Islands. Trans. Connecticut Acad. Arts Sci., 11 (2), pp.
        413–849.
VESEY-FITZGERALD, D.
    1936. Birds of the Seychelles and other islands included in that colony. Victoria,
        Seychelles.
    1940. Birds of the Seychelles. Ibis, 1940, pp. 480–504. (Pt. II, the Sea Birds, by
        F. N. Betts.)
    1941. Further contributions to the ornithology of the Seychelles Islands. Ibis,
        1941, pp. 518–531.
VIEHMEYER, G.
    1938. Is the prairie chicken passing? Nebraska Bird Rev., 6, pp. 25–28.
VIGORS, N. A.
    1839. On zoology of the *Blossom* (see Beechey).
VIVIELLE, J.
    1926. L'enigme du voyage de François Leguat à l'île Rodriguez. Ministère de
        l'instruction publique et des beaux arts. Comité des travaux historiques et
        scientifiques. Bull. Section Geographie, 41. (Not sighted, see Swaen, 1940.)
WAITE, E. R.
    1909. Vertebrata of the Subantarctic islands of New Zealand, 2 vols. Wellington.
        (Vol. 2, pp. 551–584.)
WALKINSHAW, L. H.
    1946. Some prairie chicken observations in southern Michigan. Jack-Pine Warbler
        (Battle Creek, Mich.), 24, pp. 87–90.

WARNER, D. W.
   1948. The present status of the kagu (*Rhynochetos jubatus*) on New Caledonia. Auk, **65**, pp. 287–288.
WARREN, B. H.
   1890. Report on the birds of Pennsylvania, etc., 424 pp. Harrisburg.
WATSON, S.
   1876. Flora of Guadalupe Island (Pacific). Proc. Amer. Acad. Arts and Sci., **11**, pp. 105–121.
WEBB, J. J.
   1939. Inquisitiveness of California condor, etc. The Gull (San Francisco), **21**, p. 58.
WETMORE, A.
   1925a. Bird life among lava rock and coral sand [notes on islands NW. of the Hawaiian Islands]. Nat. Geogr. Mag., **48**, pp. 77–108.
   1925b. The Coues gadwall extinct. Condor, **27**, p. 36.
   1927. The birds of Porto Rico and the Virgin Islands. Scientific Survey of Porto Rico and the Virgin Islands, New York Acad. Sci., **9**, pts. 3, 4, pp. 245–598.
   1931a. Fossil birds (pp. 402–472), *in* A.O.U. Check-list of North American Birds, 4th ed.
   1931b. The avifauna of the Pleistocene in Florida. Smithsonian Misc. Coll., **85** (2), 41 pp.
   1932. Additional records of birds from cavern deposits in New Mexico. Condor, **34**, pp. 141–142.
   1933a. Status of *Minerva antiqua* Shuf., *Aquila ferox, A. lydekkeri* (=*Protostrix*). Amer. Mus. Novit., no. 680, pp. 1–4.
   1933b. Skeleton of Guadaloupe wren in U.S. National Museum, Condor, **35**, p. 206.
   1937. Bird remains from cave deposits on Great Exuma Island, Bahamas. Bull. Mus. Comp. Zool., **70** (12), pp. 427–441.
   1938. Bird remains from the West Indies. Auk, **55**, pp. 51–55.
   1939a. A record for the black capped petrel from Haiti. Auk, **56**, p. 73.
   1939b. Recent observations on the Eskimo curlew in Argentina. Auk, **56**, p. 475.
   1943. Evidence for the former occurrence of the ivory-billed woodpecker in Ohio. Wilson Bull., **55**, p. 127.
   1953. A record for *Neodrepanis hypoxantha* of Madagascar. Auk, **70**, p. 91.
   1956. A checklist of fossil and prehistoric birds of North America and the West Indies. Smithsonian Misc. Coll., **131** (5), 105 pp.
WEYGANDT, C.
   1906. Summer birds of Broadhead's Creek, Monroe County, Pennsylvania. Cassinia, **9**, pp. 6–23.
WHEATON, J. M.
   1882. Report on the birds of Ohio. Ohio Geol. Survey, **4**, pp. 187–628.
WHITE, JOHN.
   1790. Journal of a voyage to New South Wales. London.
WHITLEY, G.
   1934. The doom of the bird of Providence, Australian Zool., **8**, pp. 42–49.
WHITLOCK, F. L.
   1924. Journey to central Australia in search of the night parrot [*Geopsittacus*]. Emu, **23**, pp. 248–281.
WHITMAN, C. O.
   1919. The behavior of pigeons. Posthumous works of Charles Otis Whitman, ed. by H. A. Carr, vol. 3, xi+161 pp. Carnegie Institution of Washington.
WHITMEE, S. J.
   1874. Letter to P. L. Sclater on birds of Samoa [only record of habits of *Pareudiastes*]. Proc. Zool. Soc. London, 1874, pp. 183–186.

WHITTELL, H. M.
  1939. Recent records of the whipbird. Emu, **39**, pp. 129–131.
  1942. Elegant parrot [*Neophema elegans*] in Bridgetown District (West Australia). Emu, **42**, pp. 50–51.
  1943. The noisy scrub bird (*Atrichornis clamosus*). Emu, **42**, pp. 217–234.
WIDMANN, O.
  1907. A preliminary catalogue of the birds of Missouri [*Ectopistes,* pp. 84–85]. Trans. Acad. Sci. St. Louis, **17**, pp. 1–288.
WIGLESWORTH, L. W.
  1891. Aves Polynesiae. A catalogue of the birds of the Polynesian subregion. Abh. Zool., Anthrop., Ethnogr. Mus. Dresden, 1890–91, **6**, 92 pp.
WILKINS, G. H.
  1923. Birds collected by the *Quest* Expedition, etc. Ibis, 1923, pp. 474–513.
WILLETT, G.
  1933. Revised list of the birds of southwestern California. Pacific Coast Avif., no. 21, p. 39.
WILLIAMS, G. R.
  1953. Dispersal from New Zealand and Australia of some introduced European passerines. Ibis, 1953, pp. 676–692 (681).
WILLIAMSON, K.
  1941. Early drawings of the great auk and gannet made on the Isle of Man in 1652. Ibis, 1941, pp. 301–310. (Also Journ. Manx. Mus., **4**, 1939, pp. 168–172.)
WILLUGHBY, F.
  1678. The ornithology of Francis Willughby, etc. London.
WILSON, A.
  1808–1814. American ornithology; or the natural history of the birds of the United States, 9 vols. Philadelphia.
WILSON, A., and BONAPARTE, L.
  1832. American ornithology, etc., 3 vols. London.
WILSON, E. S.
  1934. Personal recollections of the passenger pigeon. Auk, **51**, pp. 157–168.
  1935. Additional notes on the passenger pigeon. Auk, **52**, pp. 412–413.
WILSON, H.
  1937. Notes on the night parrot [*Geopsittacus*] with reference to recent occurrences. Emu, **37**, pp. 79–87.
WILSON, R. S.
  1948. The summer bird life of Attu. Condor, **50**, pp. 124–129.
WILSON, S. B.
  1907. Notes on birds from Tahiti and the Society group. Ibis, 1907, pp. 373–379.
WILSON, S. B., and EVANS, A. H.
  1890–1899. Aves Hawiiensis. The birds of the Sandwich Islands, published in 7 parts with colored plates, text figs,. photos. London.
WILTON, D. W., PIRIE, J. H., and BROWN, R. N. R.
  1908. Report on the scientific results of the voyage of S. Y. *Scotia,* **4**, Zool., pt. 1. Edinburgh. [Zoological log, pp. 1–101; Gough Island, p. 75.]
WIMAN, C.
  1942. Über den Tarsometatarsus der *Alca impennis* L. Medd. Goteborg Mus. Zool. Avd. 94.
WIMAN, C., and HESSLAND, I.
  1942. On the garefowl, *Alca impennis* L., and the sternum of birds. Nov. Acta Reg. Soc. Sci. Uppsaliensis (4), **13** (2), 28 pp.

WINGE, H.
  1903. [*Alca impennis* in Danish kitchen middens]. Vid. Medd. Naturh. For.
        Kobenhavn, 1903, pp. 61–109 (89).
  1910. Om Plesiocetus og Sgvalodon fra Danmark. Vid. Medd. Naturh. For.
        Kobenhavn, 1909, p. 113.
  1912. [Remains of great auk in Scandinavia.] Vid. Medd. Naturh. For. Koben-
        havn., 63, pp. 184–193.
WITHERBY, H. F., ET AL.
  1938–1941. The handbook of British birds, 5 vols. London.
WOLFE, L. R.
  1950. Notes on the birds of Korea. Auk, 67, pp. 433–455.
WOLLEY, J. See Newton, A., 1905.
WOOD, C. A., and WETMORE, A.
  1926. A collection of birds from the Fiji Islands, pt. 3. Ibis, 1926, pp. 91–136
        (92).
WOOD, J. C.
  1910. The last passenger pigeons in Wayne Co., Michigan. Auk, 27, p. 208.
WOOD, N. A.
  1951. The birds of Michigan. Misc. Publ. Mus. Zool. Univ. Michigan, no. 75, 522
        pp.
WOOD, W.
  1635. New England's prospect, etc., 83 pp. London.
WORTH, C. B.
  1940. Egg volumes and incubation periods. Auk, 57, pp. 44–60.
WRIGHT, A. H.
  1910. Some early records of the passenger pigeon. Auk, 27, pp. 428–443.
  1911. Other early records of the passenger pigeon. Auk, 28, pp. 346–365, 427–
        449.
  1913. The passenger pigeon: Early historical records, 1534–1860. Bird Lore, 15,
        pp. 85–93.
WYMAN, J.
  1868. An account of some Kjoekkenmoeddings or shell-heaps in Maine and
        Massachusetts. Amer. Nat., 1 (11), pp. 561–584. (Remarks on great auk,
        p. 578.)
  1869. Second annual report, Peabody Museum of Archeology and Ethnology,
        Harvard University, p. 17. [Great auk on Cape Cod.]
YAMASHINA, Y.
  1936. Habits of *Janthoenas janthina* on the Seven Islands of Izu. Tori (Tokyo),
        9, pp. 222–231.
  1948. Notes on the Marianas mallard. Pacific Sci., 2, pp. 121–124.
YEATTER, R. E.
  1943. The prairie chicken in Illinois. Illinois Nat. Hist. Div. Bull. 22, art. 4, pp.
        377–415.
YOUNGWORTH, W.
  1933. Field notes from Sioux City, Iowa. Auk, 50, p. 124.
ZEUNER, F. E.
  1946. Dating the past; an introduction to geochronology, xviii + 444 pp. London.
ZIMMERMAN, E.
  1948. Insects of Hawaii, 5 vols. Honolulu. (Introduction to vol. 1.)

Additional Bibliography for the 1967 Edition

BENSON, C. W.
    1960. Birds of the Comoro Islands. Ibis, **103 B**, no. 1.

BLACKBURN, A.
    1965. [Status of Creadion on the South Cape Islands, New Zealand.] Notornis, **12**, no. 4.

BOND, J.
    1963. Eighth supplement to the check list of the birds of the West Indies of 1956.
    1964. Ninth supplement to the check list of the birds of the West Indies of 1956.
    1965. Tenth supplement to the check list of the birds of the West Indies of 1956.

BOWMAN, R. I.
    1961. Morphological differentiation and adaptation in Galápagos finches. University of California Publications in Zoology, **58**, pp. 25–271.

BUREAU OF SPORT FISHERIES AND WILDLIFE
    1966. Rare and endangered fish and wildlife of the United States. Resource Publ. no. 34. Washington, D. C.

CONDON, H. T.
    1962. [Re-discovery of the Eyrean grass-wren, *Amytornis goyderi*.] Emu, **62**, p. 216.

FLEMING, C. A.
    1964. in Landsborough—Thompson (ed.) A new dictionary of birds. London.

HAHN, P.
    1963. Where is that vanished bird? Royal Ontario Museum, Univ. of Toronto [Whereabouts of specimens of rare and extinct North American birds].

HUMPHREY, P. S., and RIPLEY, S. D.
    1962. [Notes on the pink-headed duck (*Rhodonessa*).] Postilla, Yale Univ. Peabody, no. 61, pp. **1–21**.

INTERNATIONAL UNION FOR CONSERVATION OF NATURE AND NATURAL RESOURCES
    1966. Red data book. Morges, Switzerland.

LEOPOLD, N.
    1963. Check list of the birds of Puerto Rico, Univ. of Puerto Rico Agricultural Experiment Station, Bull. no. 168.

MERTON, D. V.
    1965. [Status of *Creadion* on the South Cape Islands of New Zealand.] Notornis, **12**, no. 4.

MILLER, A. H. and McMILLAN, I. I. and E.
    1965. Current status and welfare of the California condor. National Audubon Society Research Report, no. 6.

MORGAN, ROBINSON and ASHTON
    1961. [Discovery of *Amytornis goyderi*.] Australian bird watcher, **1**, no. 6 [not seen].

OLIVER, W. R. B.
    1955. New Zealand birds, 661 p./Wellington. 2nd ed.

REYNARD, G. B.
    1962. [Re-discovery of *Caprimulgus noctitherus*.] The living bird. Cornell Laboratory of Ornithology, pp. 51–60.

RICHARDSON, F.
    1961. [*Phaeornis palmeri* seen.] Condor, **63**, p. 179.

RICHARDSON, F., and BOWLES, J.
 1964. A survey of the birds of Kauai, Hawaii. Bernice P. Bishop Museum, Honolulu, Bull. no. 227.

SCHWARTZ, A., and KLINIKOWSKI
 1963. [Discovery of *Leptotila wellsi.*] Proceedings Academy of Sciences of Philadelphia, **115**, no. 3.

SERVENTY, D. L.
 1962. [Re-discovery of *Atrichornis.*] Journal für Ornithologie, **103**, p. 213.

WEBSTER, H. O.
 1962. Discovery of the scrub-bird. Western Australian Naturalist, **8**, no. 3, p. 57; no. 4, p. 81.

WILLIAMS, G. R.
 1956. The kakapo (*Strigops*), a review of a nearly extinct species. Notornis, **7**, pp. 29–56.
 1962. Extinction and the land birds of New Zealand, Notornis, **10**, no. 1, pp. 15–32.

WINGATE, D.
 1964. Discovery of breeding black-capped petrels. Auk, **81**, no. 2.

APPENDIX

Museums in Which Extinct Birds Are to Be Found

AFRICA

Durban, South Africa
    Durban Museum
Tananarive, Madagascar
    Parc Botanique et Zoologique de Tananarive

AUSTRALIA

Adelaide, South Australia
    Public Library, Museum and Art Gallery of South Australia
Brisbane, Queensland
    Queensland Museum
Hobart, Tasmania
    University of Tasmania Museum
Launceston, Tasmania
    Queen Victoria Museum and Art Gallery
Melbourne
    National Museum of Natural History, Geology and Ethnology
Perth, Western Australia
    Public Library, Museum and Art Gallery of Western Australia
Sydney
    Australian Museum
    MacLeay Museum of Natural History

BELGIUM

Antwerp
    Natuurwetenschappelijk Museum der Stad Antwerpen
Brussels
    Musée Royal d'Histoire Naturelle de Belgique
Mons
    Musée d'Histoire Naturelle de la Ville
Tournai
    Musée d'Histoire Naturelle

BRITISH ISLES

Bath, England
    Bath Royal Literary and Scientific Institution

486

Cambridge, England
    University Museum of Zoology
Cardiff, Wales
    National Museum of Wales
Edinburgh, Scotland
    Royal Scottish Museum
Exeter, England
    Royal Albert Memorial Museum and Art Galleries
Hornsey, England
    Collection of O. J. Janson, 13 Fairfax Road
Huddersfield, England
    Tolsom Memorial Museum
Liverpool, England
    Derby Museum
London, England
    British Museum (Natural History)
Newcastle-upon-Tyne, England
    Hancock Museum
Norwich, England
    Norwich Castle Museum and Art Gallery
Nottingham, England
    Nottingham Public Natural History Museum
Oxford, England
Sheffield, England
    Weston Park Museum
Tring, England
    Zoological Museum
Wakefield, England
    Museum

CANADA

Kingston, Ontario
    Queen's University Museum
Ottawa, Ontario
    National Museum of Canada
Quebec
    Musée de la Province de Québec
St. John, New Brunswick
    The New Brunswick Museum
Toronto, Ontario
    Royal Ontario Museum
Vancouver, British Columbia
    Vancouver City Museum and Art Gallery
Winnipeg, Manitoba
    Manitoba Museum
    Museums of the University of Manitoba

CZECHOSLOVAKIA
Prague
   National Museum

CUBA
Guantánamo
   Collection of C. T. Ramsden
Havana

DENMARK
Copenhagen
   Zoologisches Museum der Universität
Uppsala
   Uppsala Universitets Zoologiska Institution

FINLAND
Helsinki

FRANCE
Autun
   Société d'Histoire Naturelle
Besançon
   Musée d'Histoire Naturelle
Bordeaux
   Musée d'Histoire Naturelle
Caen
   Musée d'Histoire Naturelle à Caen
Cassel
   Muséum
Lyons
   Muséum des Sciences Naturelles
Metz
   Musée Municipal
Montauban
   Musée d'Histoire Naturelle
Nancy
   Musée d'Histoire Naturelle
Nantes
   Muséum National d'Histoire Naturelle
Paris
   Musée Nationale d'Histoire Naturelle
Poitiers
   Musée d'Histoire Naturelle
Rouen
   Musée d'Histoire Naturelle

Strasbourg
    Musée Zoologique de l'Université et de la Ville
Toulouse
    Muséum d'Histoire Naturelle et Jardin Zoologique
Troyes
    Musée des Sciences Naturelles
Vernon
    Musée

GERMANY

Augsburg
    Naturwissenschaftliches Museum
Berlin
    Museum für Naturkunde der Universität Berlin
Bremen
    Deutsches Kolonial- und Uebersee-Museum
Delmenhorst
    Collection of M. Finsch
Dresden
    Staatliche Museen für Tierkunde und Völkerkunde
Frankfurt a. Main
    Natur-Museum "Senckenberg"
Halberstadt
    Museum Heineanum
Hamburg
    Hamburgisches Zoologisches Museum und Institut
Kiel
    Zoologisches Museum der Universität
Leipzig
    Zoologisches Museum der Universität Leipzig
Munich
    Zoologische Staatssammlung des Bayerischen Staates
Stettin
    Museum des Stadt Stettin
Stuttgart
    Württembergische Naturalien-Sammlung
Wiesbaden

GUADALOUPE

Point a Pitre
    Musée l'Herminier

HAWAIIAN ISLANDS

Honolulu
    Bernice P. Bishop Museum
    St. Louis College

Molokai
> Collection of Theodore Meyer

HOLLAND

Amsterdam
> Zoologisches Museum der Universiteit u. Zoologischer Garten

Leyden
> Rijksmuseum van Natuurlijke Historie

Nijmegen
> Museum van het St. Canisius-College

ITALY

Bologna
> Instituto di Zoologia della R. Università

Florence
> Il Palmerino

Genoa
> Museo Civico di Storia Naturale "Giacomo Doria"

Milan
> Museo Civico di Storia Naturale

Naples
> Museo Zoologico della R. Università

Pisa
> Museo Civico

Rome
> Museo Civico di Zoologia

Turin
> Museo di Zoologia

JAMAICA

Kingston
> The Institute of Jamaica

JAPAN

Tokyo
> Imperial University
> Ministry of Agriculture and Forestry
> Private Museum including collections of:
>> The Marquess Hachisuka
>> Prince Taka Tsukasa
>> The Marquess Yamashina
> Collection of Dr. Nagamichi Kuroda

MAURITIUS

Port Louis
> Museum Desjardins

NEW ZEALAND
Auckland
  Auckland Institute and Museum
Christchurch
  Canterbury Museum
Dunedin
  Otago University Museum
Wellington
  Dominion Museum

NORWAY
Bergen
  Bergens Museum
Oslo
  Zoologisches Museum der Universität

PUERTO RICO
Mayagüez
  Collection of Mr. Danforth

PORTUGAL
Lisbon
  Muséu Bocage

RUSSIA
Leningrad
  Muzej Zoologičeskogo Instĭtuta Akademii

SWEDEN
Gotenburg
  Natural History Museum
Lund
  Zoological Institute and Museum of the State University
Malmo
  Museum
Stockholm
  Royal Natural History Museum

SWITZERLAND
Basel
  Museum f. Völkerkunde
Geneva
  Muséum d'Histoire Naturelle
Lausanne
  Musée Zoologique de l'Université

Neuchâtel
    Musée d'Histoire Naturelle

UNITED STATES

Albany, New York
    New York State Museum
Alfred, New York
    Allen Steinheim Museum of Natural History
Andover, Massachusetts
    Phillips Andover Academy
Ann Arbor, Michigan
    Michigan Academy of Science, Arts and Letters
    Museum of Zoology, University of Michigan
Berkeley, California
    University of California Museum of Vertebrate Zoology
Bismarck, North Dakota
    Museum of Natural History of the State University and School of Mines
        of North Dakota
Bloomfield Hills, Michigan
    Cranbrook Institute of Science
Buffalo, New York
    Buffalo Museum of Science
Burlington, Vermont
    Robert Hull Fleming Museum (University of Vermont)
Cambridge, Massachusetts
    Museum of Comparative Zoology
Charleston, South Carolina
    Charleston Museum
Chicago, Illinois
    Chicago Academy of Sciences
    Chicago Natural History Museum
Cleveland, Ohio
    Cleveland Museum of Natural History
Davenport, Iowa
    Davenport Public Museum
Dayton, Ohio
    Dayton Public Library and Museum
Delaware, Ohio
    Ohio Wesleyan University Museums
Denver, Colorado
    Colorado Museum of Natural History
Fond du Lac, Wisconsin
    Fond du Lac Public Museum
Galesburg, Illinois
    Albert Hurd Museum of Natural History of Knox College

Granville, New York
    Pember Library and Museum
Green Bay, Wisconsin
    Neville Public Museum
Grinnell, Iowa
    Parker Museum of Natural History
Hastings, Nebraska
    Hastings Museum
Jupiter, Florida
    Collection of Mrs. Carlin
Lawrence, Kansas
    Dyche Museum of Natural History (University of Kansas)
Los Angeles, California
    Los Angeles Museum of History, Science and Art
Madison, Wisconsin
    Zoological Museum (University of Wisconsin)
Milwaukee, Wisconsin
    Milwaukee Public Museum
Minneapolis, Minnesota
    Museum of Natural History, University of Minnesota
Newark, New Jersey
    Newark Museum
New Haven, Connecticut
    Peabody Museum of Natural History (Yale University)
New London, Wisconsin
    New London Public Museum
New York, New York
    American Museum of Natural History
Northfield, Minnesota
    Carleton College Museum
Oakland, California
    Oakland Public Museum
Palo Alto, California
    Zoological Museum of Leland Stanford Junior University
Pemberton, New Jersey
    Collection of T. C. Shreve
Philadelphia, Pennsylvania
    Academy of Natural Sciences of Philadelphia
Pittsburgh, Pennsylvania
    The Carnegie Museum
Pittsfield, Massachusetts
    Berkshire Athenaeum and Museum
Princeton, New Jersey
    Natural Science Museum of Princeton University

Providence, Rhode Island
   (Roger Williams) Park Museum
Raleigh, North Carolina
   North Carolina State Museum
St. Johnsbury, Vermont
   The Fairbanks Museum of Natural Science
St. Paul, Minnesota
   Hamline University Museum
San Diego, California
   Natural History Museum
San Francisco, California
   California Academy of Sciences
   Pacific Museum of Ornithology
Santa Barbara, California
   Santa Barbara Museum of Natural History
Seattle, Washington
   Museum of the University of Washington
Springfield, Illinois
   Illinois State Museum
Springfield, Massachusetts
   Museum of Natural History
Syracuse, New York
   Natural Science Museum of Syracuse University
Trenton, New Jersey
   New Jersey State Museum
University, Alabama
   Alabama Museum of Natural History
Vermillion, South Dakota
   University of South Dakota Museum
Washington, D.C.
   United States National Museum
Worcester, Massachusetts
   Worcester National History Museum

Index

512   EXTINCT AND VANISHING BIRDS OF THE WORLD

Petrel, Guadalupe Island, 157
    West Indian, 150
Petrels, 4, 12
Petroica, 90
    macrocephala marrineri, 91
    traversi, 88
petrophila, Neophema, 337, 338, 341
    Neophema elegans, 337
Pezocrex herberti, 117
Pezophaps, 122, 125
    solitaria, 106, 120, 124
Pezoporus formosus, 343
    terrestris, 343
    wallicus, 343, 345
    wallicus flaviventris, 343
    wallicus leachi, 343
Phaebetria fusca, 136
Phaenicopterus sp., 106
phaeopygia, Pterodroma, 149
Phaeornis, 384
    obscurus, 386
    obscurus lanaiensis, 7, 50, 385, 388
    obscurus myadestina, 11, 50, 384, 385,
        386
    obscurus oahensis, 6, 50, 55, 384, 387
    obscurus obscurus, 385
    obscurus rutha, 7, 50, 385, 387
    palmeri, 11, 50, 384, 385, 387
Phaethon rubricauda, 82
    rubricauda melanorhynchus, 93
Phalacrocoracidae, 4
Phalacrocorax aristotelis, 288
    carbo, 160, 274, 288
    harrisi, 10, 160
    perspicillatus, 4, 99, 159
    punctatus featherstoni, 160
phasianellus, Pedioecetes, 189
Phasianidae, 4, 14
Phasianus colchicus, 200
Pheasant, Atjeh, 14
    western koklass, 14
Pheasants, 14
Philesturnus carunculatus rufusater, 102
philippensis, Rallus, 218, 221
phillipii, Pterodroma, 70
Phlegoenas canifrons, 18
Phoenix Islands, 86
Phygis solitarius, 84
Phylloscopus amoenus, 67
Picidae, 5, 20
picturata, Columba, 303
    Streptopelia, 106
    Streptopelia picturata, 303

Picus noguchii, 362
    principalis, 357
Pigeon, blue, 297
    Bonin, 74
    Bonin wood, 301, 302
    crested, of the Solomons, 311, 312
    Japanese wood, 301
    Lord Howe, 297
    Mauritius blue, 292, 293
    New Zealand, 295
    plain, 297
    passenger, 36, 39, 42, 43, 304, 306
    prairie, 264
    Puerto Rican blue, 297
    Riu Kiu wood, 302
    wild, 304
Pigeon hollandais, 292
Pigeons, 5, 7, 16, 70, 84, 107, 126
pileatus, Anous stolidus, 93
Pinguinus, 5
pinnata, Cupidonia, 190
pinnatus, Tympanuchus cupido, 190, 191,
    193, 195
Piopio, 388
Pipilo erythrophthalmus consobrinus, 6, 48,
    432
    maculatus consobrinus, 432
Pi-pi-pi-uk, 264
Plautus alle, 274
Plasmodium vaughani, 54
plateni, Aramidopsis, 15
    Rallus, 15
Platycercus pulcherrimus, 341
    rayneri, 334
platyrhynchos, Anas, 169, 170, 274, 287, 288
Plotus nanus, 115
plotus, Sula leucogaster, 93
Plover, double-banded, 270
    New Zealand sand, 247
    New Zealand shore, 257, 258
plumbea, Zenaida, 9
Pluvialis dominicus fulvus, 93
Plyctolophus productus, 312
podarginus, Otus, 80
Podiceps auritus, 109
    gadowi, 8, 109
    sp., 109
poecilopterus, Botaurus stellaris, 88
    Nesoclopeus, 7, 84
poecilorhyncha, Anas, 169
poliocephalus, Turdus, 373, 374
Poliolimnas cinereus brevipes, 5, 74, 240
pollens, Tyto, 34

A CATALOGUE OF SELECTED DOVER BOOKS
IN ALL FIELDS OF INTEREST

A CATALOGUE OF SELECTED DOVER
BOOKS IN ALL FIELDS OF INTEREST

RACKHAM'S COLOR ILLUSTRATIONS FOR WAGNER'S RING. Rackham's finest mature work—all 64 full-color watercolors in a faithful and lush interpretation of the *Ring*. Full-sized plates on coated stock of the paintings used by opera companies for authentic staging of Wagner. Captions aid in following complete Ring cycle. Introduction. 64 illustrations plus vignettes. 72pp. 8⅝ x 11¼.   23779-6 Pa. $6.00

CONTEMPORARY POLISH POSTERS IN FULL COLOR, edited by Joseph Czestochowski. 46 full-color examples of brilliant school of Polish graphic design, selected from world's first museum (near Warsaw) dedicated to poster art. Posters on circuses, films, plays, concerts all show cosmopolitan influences, free imagination. Introduction. 48pp. 9⅜ x 12¼.
23780-X Pa. $6.00

GRAPHIC WORKS OF EDVARD MUNCH, Edvard Munch. 90 haunting, evocative prints by first major Expressionist artist and one of the greatest graphic artists of his time: *The Scream, Anxiety, Death Chamber, The Kiss, Madonna,* etc. Introduction by Alfred Werner. 90pp. 9 x 12.
23765-6 Pa. $5.00

THE GOLDEN AGE OF THE POSTER, Hayward and Blanche Cirker. 70 extraordinary posters in full colors, from Maitres de l'Affiche, Mucha, Lautrec, Bradley, Cheret, Beardsley, many others. Total of 78pp. 9⅜ x 12¼.   22753-7 Pa. $5.95

THE NOTEBOOKS OF LEONARDO DA VINCI, edited by J. P. Richter. Extracts from manuscripts reveal great genius; on painting, sculpture, anatomy, sciences, geography, etc. Both Italian and English. 186 ms. pages reproduced, plus 500 additional drawings, including studies for *Last Supper,* Sforza monument, etc. 860pp. 7⅞ x 10¾. (Available in U.S. only)
22572-0, 22573-9 Pa., Two-vol. set $15.90

THE CODEX NUTTALL, as first edited by Zelia Nuttall. Only inexpensive edition, in full color, of a pre-Columbian Mexican (Mixtec) book. 88 color plates show kings, gods, heroes, temples, sacrifices. New explanatory, historical introduction by Arthur G. Miller. 96pp. 11⅜ x 8½. (Available in U.S. only)   23168-2 Pa. $7.50

UNE SEMAINE DE BONTÉ, A SURREALISTIC NOVEL IN COLLAGE, Max Ernst. Masterpiece created out of 19th-century periodical illustrations, explores worlds of terror and surprise. Some consider this Ernst's greatest work. 208pp. 8⅛ x 11.   23252-2 Pa. $5.00

CATALOGUE OF DOVER BOOKS

THE SENSE OF BEAUTY, George Santayana. Masterfully written discussion of nature of beauty, materials of beauty, form, expression; art, literature, social sciences all involved. 168pp. 5⅜ x 8½. 20238-0 Pa. $2.50

ON THE IMPROVEMENT OF THE UNDERSTANDING, Benedict Spinoza. Also contains *Ethics, Correspondence*, all in excellent R. Elwes translation. Basic works on entry to philosophy, pantheism, exchange of ideas with great contemporaries. 402pp. 5⅜ x 8½. 20250-X Pa. $3.75

THE TRAGIC SENSE OF LIFE, Miguel de Unamuno. Acknowledged masterpiece of existential literature, one of most important books of 20th century. Introduction by Madariaga. 367pp. 5⅜ x 8½.
20257-7 Pa. $3.50

THE GUIDE FOR THE PERPLEXED, Moses Maimonides. Great classic of medieval Judaism attempts to reconcile revealed religion (Pentateuch, commentaries) with Aristotelian philosophy. Important historically, still relevant in problems. Unabridged Friedlander translation. Total of 473pp. 5⅜ x 8½. 20351-4 Pa. $5.00

THE I CHING (THE BOOK OF CHANGES), translated by James Legge. Complete translation of basic text plus appendices by Confucius, and Chinese commentary of most penetrating divination manual ever prepared. Indispensable to study of early Oriental civilizations, to modern inquiring reader. 448pp. 5⅜ x 8½. 21062-6 Pa. $4.00

THE EGYPTIAN BOOK OF THE DEAD, E. A. Wallis Budge. Complete reproduction of Ani's papyrus, finest ever found. Full hieroglyphic text, interlinear transliteration, word for word translation, smooth translation. Basic work, for Egyptology, for modern study of psychic matters. Total of 533pp. 6½ x 9¼. (Available in U.S. only) 21866-X Pa. $4.95

THE GODS OF THE EGYPTIANS, E. A. Wallis Budge. Never excelled for richness, fullness: all gods, goddesses, demons, mythical figures of Ancient Egypt; their legends, rites, incarnations, variations, powers, etc. Many hieroglyphic texts cited. Over 225 illustrations, plus 6 color plates. Total of 988pp. 6⅛ x 9¼. (Available in U.S. only)
22055-9, 22056-7 Pa., Two-vol. set $12.00

THE ENGLISH AND SCOTTISH POPULAR BALLADS, Francis J. Child. Monumental, still unsuperseded; all known variants of Child ballads, commentary on origins, literary references, Continental parallels, other features. Added: papers by G. L. Kittredge, W. M. Hart. Total of 2761pp. 6½ x 9¼.
21409-5, 21410-9, 21411-7, 21412-5, 21413-3 Pa., Five-vol. set $37.50

CORAL GARDENS AND THEIR MAGIC, Bronsilaw Malinowski. Classic study of the methods of tilling the soil and of agricultural rites in the Trobriand Islands of Melanesia. Author is one of the most important figures in the field of modern social anthropology. 143 illustrations. Indexes. Total of 911pp. of text. 5⅝ x 8¼. (Available in U.S. only)
23597-1 Pa. $12.95

CATALOGUE OF DOVER BOOKS

AMERICAN BIRD ENGRAVINGS, Alexander Wilson et al. All 76 plates. from Wilson's *American Ornithology* (1808-14), most important ornithological work before Audubon, plus 27 plates from the supplement (1825-33) by Charles Bonaparte. Over 250 birds portrayed. 8 plates also reproduced in full color. 111pp. 9⅜ x 12½. 23195-X Pa. $6.00

CRUICKSHANK'S PHOTOGRAPHS OF BIRDS OF AMERICA, Allan D. Cruickshank. Great ornithologist, photographer presents 177 closeups, groupings, panoramas, flightings, etc., of about 150 different birds. Expanded *Wings in the Wilderness.* Introduction by Helen G. Cruickshank. 191pp. 8¼ x 11. 23497-5 Pa. $6.00

AMERICAN WILDLIFE AND PLANTS, A. C. Martin, et al. Describes food habits of more than 1000 species of mammals, birds, fish. Special treatment of important food plants. Over 300 illustrations. 500pp. 5⅜ x 8½. 20793-5 Pa. $4.95

THE PEOPLE CALLED SHAKERS, Edward D. Andrews. Lifetime of research, definitive study of Shakers: origins, beliefs, practices, dances, social organization, furniture and crafts, impact on 19th-century USA, present heritage. Indispensable to student of American history, collector. 33 illustrations. 351pp. 5⅜ x 8½. 21081-2 Pa. $4.00

OLD NEW YORK IN EARLY PHOTOGRAPHS, Mary Black. New York City as it was in 1853-1901, through 196 wonderful photographs from N.-Y. Historical Society. Great Blizzard, Lincoln's funeral procession, great buildings. 228pp. 9 x 12. 22907-6 Pa. $7.95

MR. LINCOLN'S CAMERA MAN: MATHEW BRADY, Roy Meredith. Over 300 Brady photos reproduced directly from original negatives, photos. Jackson, Webster, Grant, Lee, Carnegie, Barnum; Lincoln; Battle Smoke, Death of Rebel Sniper, Atlanta Just After Capture. Lively commentary. 368pp. 8⅜ x 11¼. 23021-X Pa. $6.95

TRAVELS OF WILLIAM BARTRAM, William Bartram. From 1773-8, Bartram explored Northern Florida, Georgia, Carolinas, and reported on wild life, plants, Indians, early settlers. Basic account for period, entertaining reading. Edited by Mark Van Doren. 13 illustrations. 141pp. 5⅜ x 8½. 20013-2 Pa. $4.50

THE GENTLEMAN AND CABINET MAKER'S DIRECTOR, Thomas Chippendale. Full reprint, 1762 style book, most influential of all time; chairs, tables, sofas, mirrors, cabinets, etc. 200 plates, plus 24 photographs of surviving pieces. 249pp. 9⅞ x 12¾. 21601-2 Pa. $6.50

AMERICAN CARRIAGES, SLEIGHS, SULKIES AND CARTS, edited by Don H. Berkebile. 168 Victorian illustrations from catalogues, trade journals, fully captioned. Useful for artists. Author is Assoc. Curator, Div. of Transportation of Smithsonian Institution. 168pp. 8½ x 9½.
23328-6 Pa. $5.00

ART FORMS IN NATURE, Ernst Haeckel. Multitude of strangely beautiful natural forms: Radiolaria, Foraminifera, jellyfishes, fungi, turtles, bats, etc. All 100 plates of the 19th-century evolutionist's *Kunstformen der Natur* (1904). 100pp. 9⅜ x 12¼. 22987-4 Pa. $4.50

CHILDREN: A PICTORIAL ARCHIVE FROM NINETEENTH-CENTURY SOURCES, edited by Carol Belanger Grafton. 242 rare, copyright-free wood engravings for artists and designers. Widest such selection available. All illustrations in line. 119pp. 8⅜ x 11¼.
23694-3 Pa. $3.50

WOMEN: A PICTORIAL ARCHIVE FROM NINETEENTH-CENTURY SOURCES, edited by Jim Harter. 391 copyright-free wood engravings for artists and designers selected from rare periodicals. Most extensive such collection available. All illustrations in line. 128pp. 9 x 12.
23703-6 Pa. $4.00

ARABIC ART IN COLOR, Prisse d'Avennes. From the greatest ornamentalists of all time—50 plates in color, rarely seen outside the Near East, rich in suggestion and stimulus. Includes 4 plates on covers. 46pp. 9⅜ x 12¼. 23658-7 Pa. $6.00

AUTHENTIC ALGERIAN CARPET DESIGNS AND MOTIFS, edited by June Beveridge. Algerian carpets are world famous. Dozens of geometrical motifs are charted on grids, color-coded, for weavers, needleworkers, craftsmen, designers. 53 illustrations plus 4 in color. 48pp. 8¼ x 11. (Available in U.S. only) 23650-1 Pa. $1.75

DICTIONARY OF AMERICAN PORTRAITS, edited by Hayward and Blanche Cirker. 4000 important Americans, earliest times to 1905, mostly in clear line. Politicians, writers, soldiers, scientists, inventors, industrialists, Indians, Blacks, women, outlaws, etc. Identificatory information. 756pp. 9¼ x 12¾. 21823-6 Clothbd. $40.00

HOW THE OTHER HALF LIVES, Jacob A. Riis. Journalistic record of filth, degradation, upward drive in New York immigrant slums, shops, around 1900. New edition includes 100 original Riis photos, monuments of early photography. 233pp. 10 x 7⅞. 22012-5 Pa. $6.00

NEW YORK IN THE THIRTIES, Berenice Abbott. Noted photographer's fascinating study of city shows new buildings that have become famous and old sights that have disappeared forever. Insightful commentary. 97 photographs. 97pp. 11⅜ x 10. 22967-X Pa. $4.50

MEN AT WORK, Lewis W. Hine. Famous photographic studies of construction workers, railroad men, factory workers and coal miners. New supplement of 18 photos on Empire State building construction. New introduction by Jonathan L. Doherty. Total of 69 photos. 63pp. 8 x 10¾.
23475-4 Pa. $3.00

THE ANATOMY OF THE HORSE, George Stubbs. Often considered the great masterpiece of animal anatomy. Full reproduction of 1766 edition, plus prospectus; original text and modernized text. 36 plates. Introduction by Eleanor Garvey. 121pp. 11 x 14¾. 23402-9 Pa. $6.00

BRIDGMAN'S LIFE DRAWING, George B. Bridgman. More than 500 illustrative drawings and text teach you to abstract the body into its major masses, use light and shade, proportion; as well as specific areas of anatomy, of which Bridgman is master. 192pp. 6½ x 9¼. (Available in U.S. only) 22710-3 Pa. $2.50

ART NOUVEAU DESIGNS IN COLOR, Alphonse Mucha, Maurice Verneuil, Georges Auriol. Full-color reproduction of *Combinaisons ornementales* (c. 1900) by Art Nouveau masters. Floral, animal, geometric, interlacings, swashes—borders, frames, spots—all incredibly beautiful. 60 plates, hundreds of designs. 9⅜ x 8-1/16. 22885-1 Pa. $4.00

FULL-COLOR FLORAL DESIGNS IN THE ART NOUVEAU STYLE, E. A. Seguy. 166 motifs, on 40 plates, from *Les fleurs et leurs applications decoratives* (1902): borders, circular designs, repeats, allovers, "spots." All in authentic Art Nouveau colors. 48pp. 9⅜ x 12¼.
23439-8 Pa. $5.00

A DIDEROT PICTORIAL ENCYCLOPEDIA OF TRADES AND INDUSTRY, edited by Charles C. Gillispie. 485 most interesting plates from the great French Encyclopedia of the 18th century show hundreds of working figures, artifacts, process, land and cityscapes; glassmaking, papermaking, metal extraction, construction, weaving, making furniture, clothing, wigs, dozens of other activities. Plates fully explained. 920pp. 9 x 12.
22284-5, 22285-3 Clothbd., Two-vol. set $40.00

HANDBOOK OF EARLY ADVERTISING ART, Clarence P. Hornung. Largest collection of copyright-free early and antique advertising art ever compiled. Over 6,000 illustrations, from Franklin's time to the 1890's for special effects, novelty. Valuable source, almost inexhaustible.
Pictorial Volume. Agriculture, the zodiac, animals, autos, birds, Christmas, fire engines, flowers, trees, musical instruments, ships, games and sports, much more. Arranged by subject matter and use. 237 plates. 288pp. 9 x 12.
20122-8 Clothbd. $13.50

Typographical Volume. Roman and Gothic faces ranging from 10 point to 300 point, "Barnum," German and Old English faces, script, logotypes, scrolls and flourishes, 1115 ornamental initials, 67 complete alphabets, more. 310 plates. 320pp. 9 x 12. 20123-6 Clothbd. $13.50

CALLIGRAPHY (CALLIGRAPHIA LATINA), J. G. Schwandner. High point of 18th-century ornamental calligraphy. Very ornate initials, scrolls, borders, cherubs, birds, lettered examples. 172pp. 9 x 13.
20475-8 Pa. $6.00

THE STANDARD BOOK OF QUILT MAKING AND COLLECTING, Marguerite Ickis. Full information, full-sized patterns for making 46 traditional quilts, also 150 other patterns. Quilted cloths, lame, satin quilts, etc. 483 illustrations. 273pp. 6⅞ x 9⅝. 20582-7 Pa. $3.95

ENCYCLOPEDIA OF VICTORIAN NEEDLEWORK, S. Caulfield, Blanche Saward. Simply inexhaustible gigantic alphabetical coverage of every traditional needlecraft—stitches, materials, methods, tools, types of work; definitions, many projects to be made. 1200 illustrations; double-columned text. 697pp. 8⅛ x 11. 22800-2, 22801-0 Pa., Two-vol. set $12.00

MECHANICK EXERCISES ON THE WHOLE ART OF PRINTING, Joseph Moxon. First complete book (1683-4) ever written about typography, a compendium of everything known about printing at the latter part of 17th century. Reprint of 2nd (1962) Oxford Univ. Press edition. 74 illustrations. Total of 550pp. 6⅛ x 9¼. 23617-X Pa. $7.95

PAPERMAKING, Dard Hunter. Definitive book on the subject by the foremost authority in the field. Chapters dealing with every aspect of history of craft in every part of the world. Over 320 illustrations. 2nd, revised and enlarged (1947) edition. 672pp. 5⅜ x 8½. 23619-6 Pa. $7.95

THE ART DECO STYLE, edited by Theodore Menten. Furniture, jewelry, metalwork, ceramics, fabrics, lighting fixtures, interior decors, exteriors, graphics from pure French sources. Best sampling around. Over 400 photographs. 183pp. 8⅜ x 11¼. 22824-X Pa. $5.00